JOURNAL DE PHYSIQUE IV
Volume 139
Proceedings
[...]ember 2006

CW00921746

From Regional Climate Modelling to the Exploration of Venus

ERCA – Volume 7

UNIVERSITE JOSEPH FOURIER GRENOBLE 1
SCIENCES. TECHNOLOGIE. SANTE

CNRS CENTRE NATIONAL DE LA RECHERCHE SCIENTIFIQUE

LGGE
Laboratoire de Glaciologie et Géophysique de l'Environnement

European Research Course on Atmospheres
Grenoble, France

Edited by: Claude Boutron

EDP SCIENCES

17 avenue du Hoggar, PA de Courtabœuf, B.P. 112, 91944 Les Ulis Cedex A, France

First pages of all issues in the series and full-text articles
in PDF format (from January 2002 onwards) are available to registered users at:
http://www.edpsciences.org

A first volume of ERCA was published in 1994 with the title:

Topics in Atmospheric and Interstellar Physics and Chemistry

A second volume of ERCA was published in 1996 with the title:

Physics and Chemistry of the Atmospheres of the Earth and other Objects of the Solar System

A third volume of ERCA was published in 1998 with the title:

From Urban Air Pollution to Extra-Solar Planets

A fourth volume of ERCA was published in 2000 with the title:

From Weather Forecasting to Exploring the Solar System

A fifth volume of ERCA was published in 2002 with the title:

From the Impact of Human Activities on our Climate and Environment to the Mysteries of Titan

A sixth volume of ERCA was published in 2004 with the title:

From Indoor Air Pollution to the Search for Earth-like Planets in the Cosmos

(see their abridged contents at the end of this volume)

Supported by

- Université Joseph Fourier de Grenoble
- Centre National de la Recherche Scientifique
- Ministère de l'Éducation Nationale, de l'Enseignement Supérieur et de la Recherche
- Abdus Salam International Centre for Theoretical Physics (ICTP)
- European Commission (Integrated Project ENSEMBLES, Marie Curie Research Training Network GREENCYCLES, European Network of Excellence ACCENT)
- Conseil Général de l'Isère
- Grenoble Alpes Métropole
- Ville de Grenoble

ERCA

Director: Claude BOUTRON, University of Grenoble/Institut Universitaire de France/CNRS

Deputy-Director: Christophe FERRARI, University of Grenoble/Institut Universitaire de France/Polytech Grenoble/CNRS

Secretary: Michèle POINSOT, CNRS

Scientific Advisory Committee (ERCA 2007)

F. ADAMS, Antwerpen
C. BARBANTE, Venice
M. BENISTON, Geneva
A. BERGER, Louvain
T. COX, Cambridge
P. CRUTZEN, Mainz
R. EBINGHAUS, Geesthacht
P. EBNER, Paris
C. ELICHEGARAY, Paris
A. FRIEND, Saclay
P. LAJ, Clermont-Ferrand
J. LUNINE, Tucson
J. PLANE, Leeds
M. QUANTE, Geesthacht
G. SCARPONI, Ancona
H. VAN DOP, Utrecht
E. WOLFF, Cambridge

Organising Committee (ERCA 2007)

S. ANQUETIN, CNRS Grenoble
L. CHARLET, University of Grenoble
R.J. DELMAS, CNRS Grenoble
A. DOMMERGUE, University of Grenoble
J.L. JAFFREZO, CNRS Grenoble
J. JOUZEL, CEA Saclay
G. KRINNER, CNRS Grenoble
J. LILENSTEN, CNRS Grenoble
A. SARKISSIAN, CNRS Verrières

ERCA 2006 at Observatoire de Haute Provence (photo by C. Boutron)

CONTRIBUTORS

ADAMS Freddy
Department of Chemistry, University of Antwerpen (UIA), Universiteitsplein 1, 2610 Wilrijk, Belgium
e-mail: Freddy.Adams@ua.ac.be

BARBANTE Carlo
Dipartimento di Scienze Ambientali, Universita Ca' Foscari di Venezia, Dorsoduro 2137, 30123 Venezia, Italy
e-mail: barbante@unive.it

BETTS Richard
Ecosystems and Climate Impacts, Hadley Centre for Climate Prediction and Research, Fitzroy Road, Exeter, Devon EX5 2SN, UK
e-mail: richard.betts@metoffice.gov.uk

BRENNINKMEIJER Carl
Max-Planck-Institute for Chemistry, Joh. Becherweg 27, 55128 Mainz, Germany
e-mail: carlb@mpch-mainz.mpg.de

BRIMBLECOMBE Peter
School of Environmental Sciences, University of East Anglia, Norwich NR4 7TJ, UK
e-mail: P.Brimblecombe@uea.ac.uk

COE Hugh
School of Earth, Atmospheric and Environmental Sciences, The University of Manchester, Sackville Street Building, Sackville Street, PO Box 88, Manchester M60 1QD, UK
e-mail: hugh.coe@manchester.ac.uk

CORADINI Marcello
Solar System Missions, European Space Agency, Science Directorate, 8-10 rue Mario Nikis, 75738 Paris Cedex 15, France
e-mail: Marcello.Coradini@esa.int

EBINGHAUS Ralf
Department for Environmental Chemistry, Institute for Coastal Research, GKSS Research Center, Max-Planck-Strasse 1, 21502 Geesthacht, Germany
e-mail: ralf.ebinghaus@gkss.de

FULLEKRUG Martin
Telecommunications, Space and Radio Group, Department of Electronic and Electrical Engineering, University of Bath, Bath BA2 7AY, UK
e-mail: eesmf@bath.ac.uk

GIORGI Filippo
Physics of Weather and Climate Section, The Abdus Salam International Centre for Theoretical Physics, PO Box 586, Strada Costiera 11, 34100 Trieste, Italy
e-mail: giorgi@ictp.trieste.it

HALL Nicholas
Laboratoire d'Étude des Transferts en Hydrologie et Environnement, École Nationale Supérieure d'Hydraulique et de Mécanique de Grenoble, 1025 rue de la Piscine, Domaine Universitaire, BP. 53, 38041 Grenoble Cedex 9, France
e-mail: Nick.Hall@hmg.inpg.fr

JITARU Petru
University Al I. Cuza of Iasi, Faculty of Chemistry, Department of Inorganic and Analytical Chemistry, 11 Carol I Blvd., 700506 Iasi, Romania/Institute for the Dynamics of Environmental Processes (CNR), Dorsoduro 2137, 30123 Venice, Italy
e-mail: petru.jitaru@uaic.ro

KECKHUT Philippe
Service d'Aéronomie, UMR 7620 CNRS/Université Pierre et Marie Curie/Université de Versailles-St. Quentin/Institut Pierre Simon Laplace, Réduit de Verrières, Route des Gatines, BP. 3, 91371 Verrières-le-Buisson Cedex, France
e-mail: philippe.keckhut@aerov.jussieu.fr

LIU Xiande
Chinese Research Academy of Environmental Sciences (CRAES), Anwai, Beiyuan fang 8, 100012 Beijing, China
e-mail: xiande.liu@gmail.com

LUCHT Wolfgang
Potsdam Institute for Climate Impact Research, PO Box 60 12 03, 14412 Potsdam, Germany
e-mail: wolfgang.lucht@pik-potsdam.de

LUNDSTEDT Henrik
Swedish Institute of Space Physics, ISES, Scheeev. 17, 22370 Lund, Sweden
e-mail: henrik@lund.irf.se

MARTENS Pim
Maastricht University/Open University Netherlands/Zuyd University, PO Box 616, 6200 MD Maastricht, The Netherlands
e-mail: p.martens@icis.unimaas.nl

MATTHIAS Volker
Department of Environmental Chemistry, Institute for Coastal Research, GKSS Research Center, 21502 Geesthacht, Germany
e-mail: volker.matthias@gkss.de

NOONE Kevin
International Geosphere-Biosphere Program, Royal Swedish Academy of Sciences, Box 50005, Lilla Frescativagen 4A, 10405 Stockholm, Sweden
e-mail: zippy@igbp.kva.se

OLESEN Jorgen
Danish Institute of Agricultural Sciences, Research Centre Foulum, Department of Agroecology, Blichers Allé 20, PO Box 50, 8830 Tjele, Denmark
e-mail: jorgene.olesen@agrsci.dk

PATRIS Julie
Université Paul Cezanne, FST Centre Universitaire de Montperrin, 6 avenue du Pigonnet, 13090 Aix-en-Provence, France
e-mail: julie.patris@univ-cezanne.fr

PAZMINO Andrea
Service d'Aéronomie, UMR 7620 CNRS/Université Pierre et Marie Curie/Université de Versailles-St. Quentin/Institut Pierre Simon Laplace, Réduit de Verrières, Route des Gatines, BP. 3, 91371 Verrières-le-Buisson Cedex, France
e-mail: andrea.pazmino@aero.jussieu.fr

PEYRILLÉ Philippe
Centre National de Recherches Météorologiques, Metéo-France, 42 avenue Gaspard Coriolis, 31057 Toulouse Cedex 01, France
e-mail: philippe.peyrille@cnrm.meteo.fr

PLANE John
School of Chemistry, University of Leeds, Woodhouse Lane, Leeds LS2 9JT, UK
e-mail: j.m.c.plane@leeds.ac.uk

QUANTE Markus
Department of Environmental Chemistry, Institute for Coastal Research, GKSS Research Center, 21502 Geesthacht, Germany
e-mail: markus.quante@gkss.de

RAES Frank
Climate Change Unit, Institute for Environment and Sustainability, EC Joint Research Center, TP 440, Via Enrico Fermi, 21020 Ispra-Varese, Italy
e-mail: frank.raes@jrc.it

SALIOT Alain
Laboratoire d'Océanographie et du Climat : Expérimentations et Approches Numériques, UMR 7159 CNRS/IRD/Université Pierre et Marie Curie/MNHN/Institut Pierre Simon Laplace, Case Courrier 100, 4 place Jussieu, 75252 Paris Cedex 05, France
e-mail: saliot@ccr.jussieu.fr

SARKISSIAN Alain
Service d'Aéronomie, UMR 7620 CNRS/Université Pierre et Marie Curie/Université de Versailles-St. Quentin/Institut Pierre Simon Laplace, Réduit de Verrières, Route des Gatines, BP. 3, 91371 Verrières-le-Buisson Cedex, France
e-mail: alain.sarkissian@aerov.jussieu.fr

SAUNDERS Russell W.
School of Chemistry, University of Leeds, Woodhouse Lane, Leeds LS2 9JT, UK
e-mail: r.w.saunders@leeds.ac.uk

SCHUEPBACH Evi
cabo3/Physical Geography, University of Berne, Hallerstrasse 12, 3012 Berne, Switzerland
e-mail: cabo@giub.unibe.ch

STOFAN Ellen
Proxemy Research, PO Box 338, Rectortown, VA 20140, USA
e-mail: ellen@proxemy.com

VAN DEN BROEKE Michiel
Institute for Marine and Atmospheric Research, Utrecht University, PO Box 80005, 3508 TA Utrecht, The Netherlands
e-mail: m.r.vandenbroeke@phys.uu.nl

WOLFF Eric
British Antarctic Survey, High Cross, Madingley Road, Cambridge CB3 OET, UK
e-mail: ewwo@bas.ac.uk

XIE Zhiyong
Department of Environmental Chemistry, Institute for Coastal Research, GKSS Research Center, 21502 Geesthacht, Germany
e-mail: zhiyong.xie@gkss.de

Foreword

This book is the seventh volume in the series of books published within the framework of the European Research Course on Atmospheres ("ERCA"), the advanced international research course organized every year in Grenoble, France. This course was initiated in 1993 by the University Joseph Fourier of Grenoble, in order to provide PhD students and more senior scientists from Europe and the rest of the world with a multidisciplinary course which covers: the physics and chemistry of the Earth's atmosphere; the climate system and climate change; human dimensions of environmental change; and the physics and chemistry of other planets and satellites in the solar system and beyond. Since 1993, fourteen sessions have been attended by more than 700 participants from 50 countries, selected from a very large number of applications. The fifteenth session will take place from 8 January to 10 February 2007.

Each session lasts five weeks, which is considerably longer than most other research courses at this level. The first four weeks are devoted to a comprehensive programme of lectures (about 120 hours), seminars, panel discussions, poster sessions and visits to research institutes. The fifth week takes place at Observatoire de Haute Provence, South of Grenoble, where the participants learn about various instruments used for atmospheric measurements and Astronomy, as well as visiting the Cadarache Research Center of the French Atomic Energy Commission (International Themonuclear Experimental Reactor programme). There are fifty lecturers per session, amongst them many renowned specialists from Europe and North America, including Nobel Laureate Paul Crutzen.

This new volume contains twenty-six chapters dealing with a truly wide range of topics. After an introductory chapter on Earth System Science written by the Director of the International Geosphere-Biosphere Programme, Kevin Noone, the following subjects are covered: the exploration of Venus and the other planets of the solar system; water in the Earth's atmosphere; the West African monsoon; regional climate modelling; forcings and feedbacks by land ecosystem changes on climate change; atmospheric electricity and climate change; solar magnetic activity; the contribution of the Antarctic ice sheet to global sea level change; climate and atmospheric records from ice cores; biogeochemical processes in the ocean and at the ocean-atmosphere interface; novel organic pollutants in the marine environment; inorganic aerosol formation in the Earth's lower and upper atmosphere; Asian dust events; elemental speciation analysis; the use of aircraft for atmospheric measurements; Rayleigh temperature lidars; DIAL lidars for ozone measurements; optical telescopes; sustainable health in a globalised world; adaptation and mitigation to climate change in agriculture; and the communication of air pollution science to the public and politicians.

I wish to thank again the authors for kindly agreeing to write the chapters of this new volume. I am also very grateful to Michele Poinsot for her major contribution to the success of ERCA, and Isabelle Houlbert for editing the ERCA book series (with the help of "éclairs au café, à la vanille et au chocolat").

Claude Boutron
2 November 2006
Professor at the University Joseph Fourier of Grenoble
Director of ERCA

DOI: 10.1051/jp4:2006139001

Contents

J. Phys. IV France **139** (2006) 1–8
© EDP Sciences, Les Ulis
DOI: 10.1051/jp4:2006139002

Earth system science: Putting together the "big picture" puzzle

K.J. Noone[1]

[1] *International Geosphere-Biosphere Programme, Royal Swedish Academy of Sciences, 10405 Stockholm, Sweden*

Abstract. The Earth behaves as a highly coupled, interdependent system of components and processes – all of which operate on a multitude of time and spatial scales. Humans – rather than simply affecting or being affected by the natural environment – are a central component in the Earth system. Within the Earth system, there are feedbacks and teleconnections that operate on a planetary scale, and in which humans are directly involved. This article presents examples of such feedback systems and of human interactions in the Earth system, and describes some of the research infrastructure necessary to pursue Earth System Science.

1. INTRODUCTION

One of the enduring legacies of the Apollo space program of the 1960s was the images of our planet Earth taken from the perspective of a tiny spacecraft in orbit around the moon. We saw a single orb floating in space. Continents, oceans, clouds, and the fantastically thin shell that is our atmosphere all appear together as a beautiful whole. It was the first time that millions of humans had the opportunity to see our home planet as a single entity; to get a glimpse of a connected system that transcended the political, social, economic and even scientific boundaries we had projected onto it. From a scientific perspective, these images helped to prod us to rediscover the science of the Earth System.

Back in the time of Sir Isaac Newton, science was called *natural philosophy*. In the three centuries since, we have divided science up into a multitude of focused disciplines. Look in any University catalog and you will find departments of chemistry, physics, economics, meteorology, law, and many, many more. We have sliced the "Big Picture" up into a large number of smaller, more manageable bits. There are some good reasons for taking this reductionist approach. Taking this focused approach is a necessary step in developing an understanding of the processes that determine how complex systems behave. If we view the complex earth system as a large puzzle, the reductionist approach helps us to bring out the details of each one of the pieces of the puzzle. However, regardless of how much detail we achieve for each of the individual pieces, seeing the "Big Picture" is not possible unless all the puzzle bits are assembled. Earth System Science can be seen as the science of putting together the big picture (Figure 1).

In its full sense, Earth System Science includes aspects of many disciplines; from the natural sciences through economics and social sciences. One way of thinking about this coupled system is to use the *Drivers-Pressure-State-Impact-Response* (DPSIR) framework, illustrated in Figure 2.

A *driver* (such as automobile use) creates a *pressure* on the system (e.g., CO_2 emissions) that changes the *state* of the system (e.g., increasing atmospheric CO_2 concentrations), causing an *impact* (e.g., a surface temperature increase) leading to a *response* (e.g., increasing intensity of Atlantic hurricanes), which may feed back on the original driver. This particular set of relationships is an example of a feedback loop involving natural systems (which natural scientists may view as their "territory"), but the model could just as easily be applied to societal responses (which social scientists and economists may view at their "territory"). The true response to global environmental change is a combination of both natural and societal responses. Earth System Science must couple the natural and social sciences to be successful.

Figure 1. Putting together the Earth System puzzle.

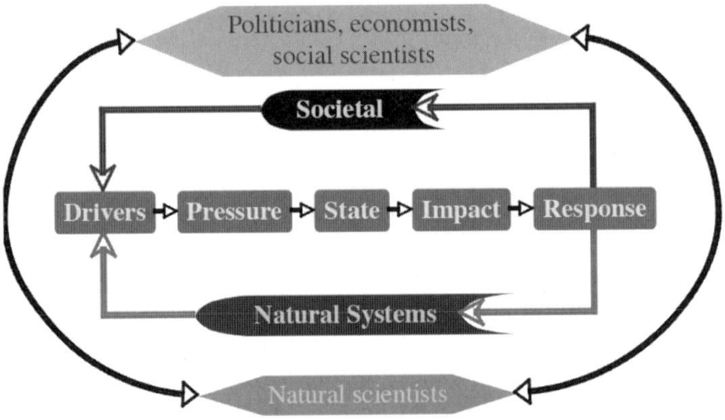

Figure 2. The DPSIR framework.

The approach of this article will be to present a few examples of planetary-scale feedbacks and teleconnections that illustrate the connectedness of the Earth system, and to describe the kind of research infrastructure needed to make Earth System Science possible.

2. FEEDBACKS AND TELECONNECTIONS IN THE EARTH SYSTEM

2.1 The "CLAW hypothesis"

One of the best examples of a potential planetary feedback loop involves plankton in the ocean, chemical reactions in the atmosphere, cloud physics, and radiative transfer (Figure 3). The loop starts in the ocean, where some marine phytoplankton (the lower right box in the figure) produce *dimethyl sulfide* (DMS), a reduced sulfur compound. When these plankton are eaten (or die and dissolve), this compound is released into the surface seawater (the lower left box in Fig. 3). A plus sign between boxes in the figure means that increasing concentrations of these phytoplankton would lead to increased DMS concentrations in the seawater.

Through the action of wind and waves, this DMS is transferred into the atmosphere, where it reacts to produce compounds that can eventually form (or modify existing) aerosol particles. More DMS in the air would presumably lead to more (or larger) aerosol particles. In marine clouds, each droplet started its life as an aerosol particle. Increasing concentrations of particles would presumably lead to increasing cloud droplet concentrations – which for a given amount of condensed water, would lead to

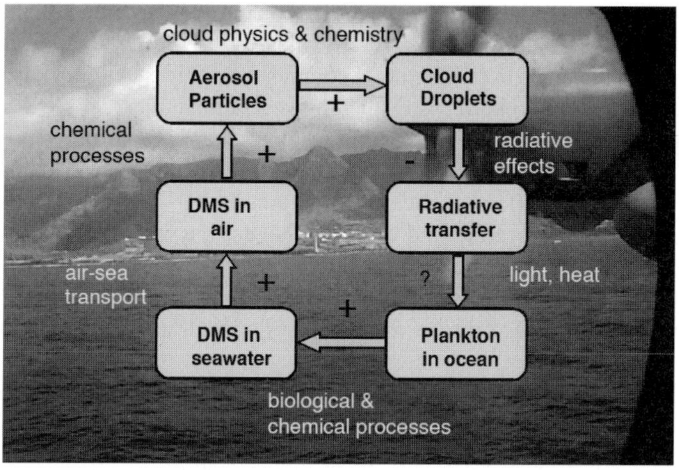

Figure 3. The CLAW hypothesis.

smaller droplets. Smaller droplets would cause a cloud to become more reflective, and less incoming solar radiation would be transmitted through the cloud (the minus sign in the figure).

How plankton would react to this decrease in solar radiation is still unknown. If plankton were to react by producing even more DMS, we would have a positive feedback, since more DMS would lead to even brighter clouds, and less sunlight reaching the surface. On the other hand, if plankton reacted by producing less DMS, we would have a stabilizing feedback loop – a kind of planetary thermostat. Less DMS would lead to fewer particles, fewer cloud droplets, and finally less reflective clouds. The result would be more solar radiation reaching the surface, which in this case would stimulate plankton to produce more DMS (and subsequently brighter clouds), thus creating a kind of thermostat. This theory has been called the "CLAW hypothesis" after the authors who first proposed it [1]. Researching the processes behind this potential planetary feedback loop requires many disciplines; marine biologists, meteorologists, atmospheric chemists, cloud physicists, radiative transfer specialists. It is a good example of the necessity for a multidisciplinary approach to understanding the complex earth system.

2.2 Who stole the snow in Tibet?

Another potential feedback loop with planetary consequences links deforestation in the Amazon with snow cover in Tibet. One part of the loop is similar to the one described above, but involves vegetation in the Amazon basin, rather than marine phytoplankton. Vegetation produces particles directly, as well as emitting volatile organic compounds (VOCs). Like DMS in the marine atmosphere, these VOCs can oxidize to form compounds that can form or modify existing particles. These particles then form cloud droplets. In the case of the Amazon, changing particle concentration (and even chemistry) can not only influence the brightness of the clouds (which is relatively less important over the Amazon compared with marine clouds, since clouds over the Amazon are already very bright), but more importantly, how likely they are to produce rain, and the strength of the convective circulation.

There are a few very important differences between the marine case and the Amazon. Vegetation in the Amazon basin emits many different kinds of VOCs, which could lead to many different organic compounds in the aerosol particles. The chemistry of these particles is much richer than compared with the marine case [2]. Much of the water in the atmosphere over the central Amazon basin is recycled through the vegetation. Thus, the vegetation not only determines the concentration of the aerosol particles in the atmosphere, but also the water vapor concentration – everything needed to make a cloud.

Recent research has shown that deforestation and biomass burning (in this case burning vegetation to prepare for the next season's planting) have caused changes in convection and precipitation over the Amazon basin [3]. These changes in precipitation complete the feedback loop, since the availability of water influences the amount and kind of VOCs that the vegetation emits [2].

While at first glance it would appear that this feedback loop would have only a regional effect, the influence of changes in Amazonia in fact can be felt much farther away. Calculations show that deforesting the Amazon could affect surface temperatures as far away as Tibet (J. Foley, personal communication). Large-scale deforestation drastically changes the surface energy balance, leading to a weakening of deep convection. This weakening of convection in the tropics has a number of subsequent effects such as a weakening and northward shift in the Inter-Tropical Convergence Zone (ITCZ), which causes changes in the jet stream that directs the trajectory of mid-latitude weather systems, ultimately influencing weather in Tibet.

The Amazon provides another example of a feedback loop that requires biology, hydrology, atmospheric chemistry, and many other disciplines to begin to understand all the processes involved in the feedback sysem. It also illustrates the fact that there are processes that connect different parts of the globe, and that changes in one region of the globe may have effects on the other side of the world.

2.3 Acidifying the Ocean

The final example of planetary-scale feedbacks and teleconnections is one of ocean acidification. Not all of the CO_2 that is emitted into the atmosphere by the combustion of fossil fuels and land use change remains there. Some is taken up by the terrestrial biosphere, and some is taken up by the oceans. A recent estimate for the 1990s [4] shows that of the 6.3 PgC/yr emitted into the atmosphere by cement production and the combustion of fossil fuels, 3.2 PgC/yr accumulates in the atmosphere, 1.4 PgC/yr is transferred to the terrestrial biosphere, and 1.7 PgC/yr is transferred into the oceans.

The uptake of CO_2 into the oceans is the result of two processes: a "solubility pump" and a "biological pump" [5]. Since CO_2 is more soluble in colder, more saline water, the "solubility pumping" of carbon dioxide into the oceans is controlled by the formation of cold, dense waters in the North Atlantic and Southern Oceans. The results of this process can be seen in maps of anthropogenically-produced CO_2 in the oceans [6], which show maxima in concentrations in these locations. The "biological pumping" of CO_2 into the oceans is due both to phytoplankton photosynthesis and the production of organic matter, and to the action of calcifying organisms that convert dissolved carbonate into $CaCO_3$. Three important mineral phases of calcium carbonate are calcite, aragonite, and high-magnesium calcite.

As CO_2 increases in the atmosphere, more will be dissolved in the oceans. The result will be a decrease in ocean pH, or an increase in ocean acidity. This increasing ocean acidity can affect the extent to which calcifying organisms are able to sequester carbon. It has been shown that corals (as well as other organisms that form aragonite) are threatened by increasing ocean acidity [7]. Experiments have also shown that two organisms that produce calcite (a more stable form of calcium carbonate compared with aragonite) also are adversely affected by increasing acidity levels [8].

These effects (a kind of destabilizing feedback loop) could have potentially very serious consequences for the carbon cycle. Any reduction in the ability of marine organisms to form carbonate shells would mean that less CO_2 would be transferred from the atmosphere to the ocean – the efficiency of the marine "biological pump" would be reduced. In fact, a very substantial step change in the behavior of the biological pump could occur if ocean acidity reached the point at which aragonite becomes soluble. If this were to occur, then the organisms that produce this form of calcium carbonate would no longer be able to do so at all, causing a drastic (and very long-lasting) reduction in the rate at which CO_2 is removed from the atmosphere – making the anthropogenic greenhouse effect even worse. Recent model calculations indicate that under the IPCC IS92a "business as usual" emissions scenario, surface waters in the Southern Ocean will become undersaturated with respect to aragonite by 2050 [9] – less than a human generation from now. By 2100, this undersaturation could extend throughout the Southern Ocean, and even into the subarctic Pacific. In addition to its effect on atmospheric CO_2, such a step change in the life cycle of these calcifying organisms could have extremely serious consequences for marine ecosystems and for marine biogeochemical cycles and biodiversity.

2.4 Summary of feedbacks and teleconnections

The three examples of planetary feedback loops presented here are serve to illustrate the fact that the Earth functions as a connected system. No part of the planet is entirely disconnected from the others, and often (as in the case of Amazonian deforestation and surface temperature changes in Tibet) processes at work in one region can have consequences far away.

These examples come from the realm of biogeochemistry and physics, but in each case also involve human actions as part of the feedback loop. The drivers of Amazonian deforestation are economic and social, and the consequences of environmental changes in Tibet and the Himalayas will also have social and economic consequences for people in those regions. Doing Earth System Science research requires coupling the natural and social science aspects and approaches.

3. STRUCTURES SUPPORTING EARTH SYSTEM SCIENCE RESEARCH

As mentioned in the introduction, our current educational and research structure is built up around a multitude of different disciplines. Pursuing Earth System Science research requires a support structure that allows and facilitates international, interdisciplinary research – a support structure not generally found in most universities and research organizations.

The International Council for Science (ICSU) recognized the need for international organizations to support multidisciplinary research, and has established four global programmes to address different aspects of Earth system science.

The kind of research described in this article falls naturally into the profile of one of these programmes – the International Geosphere-Biosphere Programme (IGBP). The vision of IGBP is to provide scientific knowledge to improve the sustainability of the living Earth. IGBP aims to describe and understand the interactive physical, chemical and biological processes that regulate the total Earth System, the unique environment that it provides for life, the changes that are occurring in this system, and the manner in which they are influenced by human actions. By encouraging scientists to cross their disciplinary, institutional and political boundaries, IGBP promotes scientific consensus and a more holistic understanding of the changing global environment. The IGBP research structure (Figure 4) comprises a suite of nine research projects focused on the major Earth System components (land, ocean and atmosphere), the interface between them (land-ocean, land-atmosphere and ocean-atmosphere) and system-wide integration (Earth System modelling and palaeo-environmental studies). Each of these components has at least one international research project aimed at investigating issues related to the component area, and to collaborating with the other projects to enable a "big picture" analysis to be undertaken. Table 1 provides contact information for the projects.

To help support a structure that facilitates truly integrated Earth System Science research, the four global environmental change Programmes of ICSU formed the Earth System Science Partnership (ESSP) in 2001. Besides IGBP, the partnership consists of DIVERSITAS, an international programme of biodiversity science, the International Human Dimension Programme on Global Environmental Change (IHDP), which takes a social science perspective on global change and works at the interface between science and practice, and the World Climate Research Programme (WCRP), which identifies knowledge gaps, prioritizes needs and leads world-wide research into climate variability and climate change to meet end-user requirements and policy needs.

Together, the ESSP has four research projects and projects aimed at capacity building and regional studies (Table 2). The aim of ESSP is to address the biogeophysical and socioeconomic factors that

Figure 4. The IGBP scientific structure.

Table 1. The IGBP Programme and projects.

Programme or Project	**Component**	**Website**
IGBP	System-level perspective	www.igbp.kva.se
IGAC	Atmosphere	www.igac.noaa.gov
GLP	Land	www.globallandproject.org
GLOBEC	Ocean	www.globec.org
IMBER	Ocean	www.imber.info
SOLAS	Ocean-Atmosphere	www.solas-int.org
LOICZ	Land-Ocean	www.loicz.org
iLEAPS	Land-Atmosphere	www.atm.helsinki.fi/ILEAPS
PAGES	Integration	www.pages-igbp.org
AIMES	Integration/ Prediction	www.aimes.ucar.edu

influence the Earth System and, in turn, the consequences for human societies, wellbeing and health. Its mission is to integrate the intellectual capital of disciplines required to enhance the understanding of the complexity of the Earth system and to explore the policy relevance of its main findings for mutually inclusive ecological, social and economic sustainability.

The four global environmental change research programmes, the Earth System Science Partnership, and all the component projects provide the scientific infrastructure that allows thousands of researchers from more than 70 countries around the world to engage in international, multidisciplinary research into Earth system issues. They provide the structure for the endeavor of scientific integration, Earth systems-level analysis, and interaction with stakeholders and society.

Table 2. The Earth System Science Partnership and projects.

Programme or Project	Component	Website
ESSP	Partnership	www.essp.org
DIVERSITAS	Biodiversity	www.diversitas-international.org
IGBP	Biogeochemistry	www.igbp.kva.se
IHDP	Human dimensions	www.ihdp.uni-bonn.de
WCRP	Climate	wcrp.wmo.int
GCP	Carbon cycle	www.globalcarbonproject.org
GECAFS	Food systems	www.gecafs.org
GWSP	Water systems	www.gwsp.org
GECHH	Health	(no web site yet)
START	Capacity building	www.start.org
MAIRS	Integrated regional study	www.mairs-essp.org

4. SUMMARY REMARKS

Increasingly, the scientific community is viewing the Earth as a complex, interdependent, coupled human-environmental system. This article attempts to give some examples of the kinds of Earth systems-level issues that require a multidisciplinary, international scientific approach and infrastructure to understand. This approach requires scientists to transcend the traditional structures upon which most of our educational and research institutions are built. The scientific infrastructure of the global environmental change programmes described in the article provides the framework for Earth System Science research – for putting together the "big picture" puzzle.

References

[1] R. J. Charlson, J. E. Lovelock, M. O. Andreae, and S. G. Warren, Nature, 655 (1987).

[2] J. Kesselmeier, U. Kuhn, A. Wolf, M. O. Andreae, P. Ciccioli, E. Brancaleoni, M. Frattoni, A. Guenther, J. Greenberg, P. D. Vasconcellos, T. de Oliva, T. Tavares, and P. Artaxo, Atmospheric Environment, 4063 (2000).

[3] M. O. Andreae, D. Rosenfeld, P. Artaxo, A. A. Costa, G. P. Frank, K. M. Longo, and M. A. F. Silva Dias, Science, 1337 (2004).

[4] I. C. Prentice, G. D. Farquhar, M. J. R. Fasham, M. L. Goulden, M. Heimann, V. J. Jaramillo, H. S. Kheshgi, C. Le Quere, R. J. Scholes, D. W. R. Wallace, D. Archer, M. R. Ashmore, O. Aumont, D. Baker, M. Battle, M. Bender, L. P. Bopp, P. Bosquet, K. Caldeira, P. Ciais, P. M. Cox, W. Cramer, F. Dentener, I. G. Enting, C. B. Field, P. Friedlingstein, E. A. Holland, R. A. Houghton, J. I. House, A. Ishida, A. K. Jain, I. A. Janssens, F. Joos, T. Kaminski, C. D. Keeling, R. F. Keeling, D. W. Kicklighter, K. E. Khohfeld, W. Knorr, R. Law, T. Lenton, K. Lindsay, E. Maier-Reimer, A. C. Manning, R. J. Matear, A. D. McGuire, J. M. Melillo, R. Meyer, M. Mund, J. C. Orr, S. Piper, K. Plattner, P. J. Rayner, S. Stich, R. Slater, S. Taguchi, P. P. Tans, H. Q. Tian, M. F. Weirig, T. Whorf, and A. Yool, in IPCC 2001: Climate Change 2001: The Scientific Basis. Contribution of Working Group I to the Third Assessment Report of the Intergovernmental Panel on Climate Change, edited by J. T. Houghton, et al., (Cambridge University Press, Cambridge, United Kingdom and New York, NY, USA, 2001).

[5] P. Falkowski, R. J. Scholes, E. Boyle, J. Canadell, D. Canfield, J. Elser, N. Gruber, K. Hibbard, P. Högberg, S. Linder, F. T. MacKenzie, B. Moore III, T. Pedersen, Y. Rosenthal, S. Seitzinger, V. Smetacek, and W. Steffen, Science, 291 (2000).

[6] C. L. Sabine, R. A. Feely, N. Gruber, R. M. Key, K. H. Lee, J. L. Bullister, R. Wanninkhof, C. S. Wong, D. W. R. Wallace, B. Tilbrook, F. J. Millero, T.-H. Peng, A. Kozyr, T. Ono, and A. F. Rios, Science, 367 (2004).

[7] J. A. Kleypas, R. W. Buddemeier, D. Archer, J.-P. Gattuso, C. Langdon, and B. N. Opdyke, Science, 118 (1999).

[8] U. Riebesell, I. Zondervan, B. Rost, P. D. Tortell, R. E. Zeebe, and F. M. M. Morel, Nature, 364 (2000).

[9] J. C. Orr, V. J. Fabry, O. Aumont, L. Bopp, S. C. Doney, R. A. Feely, A. Gnanadesikan, N. Gruber, A. Ishida, F. Joos, R. M. Key, K. Lindsay, E. Maier-Reimer, R. Matear, P. Monfray, A. Mouchet, R. G. Najjar, G.-K. Plattner, K. B. Rodgers, C. L. Sabine, J. L. Sarmiento, R. Schlitzer, R. D. Slater, I. J. Totterdell, M.-F. Weirig, Y. Yamanaka, and A. Yool, Nature, 681 (2005).

J. Phys. IV France **139** (2006) 9–19
© EDP Sciences, Les Ulis
DOI: 10.1051/jp4:2006139003

Venus: Divergent outcomes of terrestrial planet formation

E.R. Stofan[1]

[1] *Proxemy Research, Rectortown VA 20140, USA & Department of Earth Sciences, UCL, Gower St., London WC1E 6BT, UK*
e-mail: ellen@proxemy.com

Abstract. Although Venus is often referred to as the most Earth-like of the terrestrial planets, its runaway greenhouse has resulted in a dry, hot, uninhabitable surface. Its surface geology is complex, with volcanoes ranging from <5 to >500 km across, lava flows fields >800 km across, mountain belts, rift zones, and terrains unique to Venus such as tesserae and coronae. It surface has an average crater retention age similar to Earth's continents, but the random nature of the impact crater population renders it useless in providing time constraints for the geologic history of Venus. At some point in the past, Venus lost an ocean's worth of water. If this water persisted on the surface for long periods of time in Venus's early history, life may have evolved. Untangling the complex history of Venus, and what it reveals for the evolution of habitable planets, will require future missions to the surface.

1. INTRODUCTION

Venus has long been a subject of interest, due to its prominent position in the morning or evening sky. Despite the extreme difference in surface temperature and pressure, Venus and Earth are often referred to as twins, due to their similar size, mass and likely composition. The CO_2 dominated greenhouse atmosphere of Venus drove surface temperatures up to a relatively stable 482°C and a surface pressure of 90 bars at some unknown time in the past, making the Venus of today an unlikely site for life. Measurements of atmospheric hydrogen and deuterium by Pioneer Venus indicated that Venus lost an ocean of water at some point in the past; though whether water was ever stable on the surface and for how long remains a subject of debate. The history of volatiles on Venus will ultimately be the key to understanding its evolution, and what the implications are of these very different twins for understanding how habitable worlds form.

Exploration of Venus began with the study of transits of the planet, which enabled determination of the orbital period, diameter and presence of an atmosphere [1]. Ground-based radio telescopes were used in the 1960's to refine the rotation period and diameter of Venus, and produce the first images of the surface. In 1960's, flybys of the Mariner 2 and 5 missions determined the surface temperature, data on atmospheric composition, and the lack of a magnetic field. At this same time, the Soviet Union began its intense exploration of Venus, sending multiple probes to the atmosphere and then to the surface. In 1972, the Venera 8 lander measured a granitic-type surface composition, and returned a panoramic image of the surface. The Veneras 9, 10, 11, 13 and 14 landers also returned surface images, and found more basaltic-type surface compositions (Figure 1).

The U.S. Pioneer Venus mission, with an orbiter and four entry probes, measured atmospheric circulation, composition, pressure and temperature. The orbiter also returned radar images of the surface and a global topographic map (resolution ~150 km). This was followed by the Soviet Veneras 15 and 16 orbiters, which obtained 5-10 km resolution radar images of the northern hemisphere of Venus, and discovered two new types of surface features: tessera terrain and coronae. A global radar map of Venus was provided by NASA's Magellan mission, which mapped Venus from 1990-1994 (Figure 2). Magellan returned radar images at 120 m resolution and topographic data with 1-10 km resolution, as well as gravity data. In 2006, the European Space Agency's Venus Express orbiter began analyzing the atmosphere of Venus, with a focus on composition and the circulation.

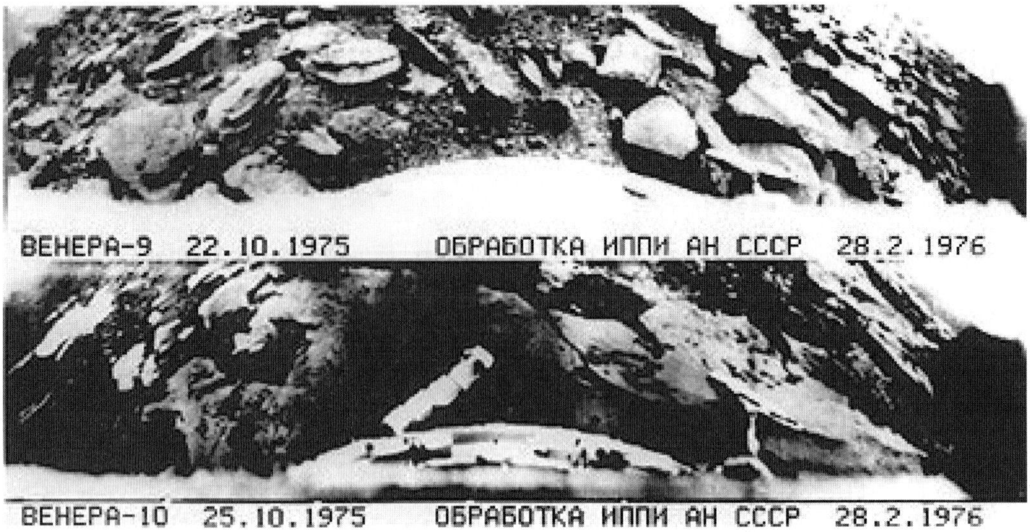

Figure 1. Venera 9 and 10 image panoramas of the venusian surface. Abundant platy rocks and some soil can be seen at both sites. The landers measured surface compositions similar to terrestrial tholeiitic basalts.

Here, we review the major discoveries of these missions, and the current theories of how Venus has reached its current state. Despite the large amount of available data, future missions to the surface are necessary to fully understand the implications of Venus for the evolution of habitable worlds.

2. THE ATMOSPHERE OF VENUS

The atmosphere of Venus is composed of about 97% carbon dioxide, with smaller amounts of nitrogen, sulfur dioxide, water vapor, carbon monoxide, and noble gases [1]. This runaway greenhouse results in temperature variations on the surface of less than 10°K due to altitude and a yearly variation of less than 1°K. The thick clouds enveloping Venus extend upward from about 48 km above the surface. They are composed primarily of sulfur dioxide and sulfuric acid. The clouds super-rotate: they revolve around the surface in 4 days, and in a direction opposite to the very slow (243 day) retrograde rotation of the surface. Below about 48 km, the atmosphere is hazy; descent probes should be able to view the surface below about 15 km altitude. Winds near the surface are on the order of 1 m/s, which in the dense atmosphere is sufficient to move particles to form dunes and produce some limited amount of erosion. The detailed composition of the lower atmosphere, and how it interacts with the surface, is still largely unknown. A radar-bright coating of the surface over about 3.5 km altitude is thought to be metallic in composition, formed by the interaction of the atmosphere with the surface.

3. SURFACE GEOLOGY

Though Earth-like volcanoes and mountain belts can be found on Venus, the Magellan data conclusively demonstrated that Venus does not have any trace of an earth-like system of plate tectonics [2]. It lacks the organized arrangement of tectonic and volcanic features that on Earth clearly delineate plate boundaries. The lack of plate tectonics is probably due to the lack of water on Venus; water on Earth facilitates subduction that drives plate tectonics. Internal heat drives geologic processes: Earth loses the majority of its internal heat via plate tectonics. Venus is nearly the same size as Earth, and therefore must be contain a relatively similar inventory of radioactive elements that produce a similar amount of internal heat. In the absence of plate tectonics, how does Venus lose its heat?

Figure 2. Magellan SAR (synthetic aperture radar) image data of a hemisphere of Venus. Radar-bright areas are rough or topographic faces oriented toward the radar; smooth areas appear radar-dark. The Dali-Diana rift system cutting across the equatorial region of Venus is seen.

3.1 Volcanism

From Magellan data, we know that the surface of Venus is dominated by volcanism [3]. Constructional volcanic features include shields, domes, cones, some with large summit calderas. Large flow fields, some over >1000 km^2, have also been mapped. Most of these features are very similar in morphology to volcanic features on Earth and Mars, though a few classes of features have been found with very unusual morphologies. Volcanism is produced by heating within or below the lithosphere of the planet producing magma, which then rises and erupts onto the surface. Therefore, understanding the styles, history and distribution of volcanism provides clues to the thermal evolution of the planet.

Shield- shaped volcanoes from <5 km to over 800 km in diameter are scattered across the surface, with volcanoes <10 km in diameter numbering in the 10,000s. About 180 large volcanoes (diameter >150 km) have been mapped, many with flows and summit features similar to terrestrial volcanoes [4] (Figure 3). Most of these volcanoes seem to grow from a combination of short flows originating at the summit and longer flows originating from vents on the flans of the volcanoes, similar to Mt. Etna on Earth [5]. Unlike on Earth, where volcanoes delineate plate boundaries, volcanoes on Venus are generally scattered across the surface. Some large volcanoes are located on dome-shaped topographic rises, interpreted to be hotspots underlain by mantle plumes, similar to Hawaii on Earth. Volcanoes on Venus tend to be larger than those on Earth, with generally very low slopes. Like on Mars, the lack of plate tectonics allows volcanoes to grow to larger sizes, as they remain stationary over mantle plumes.

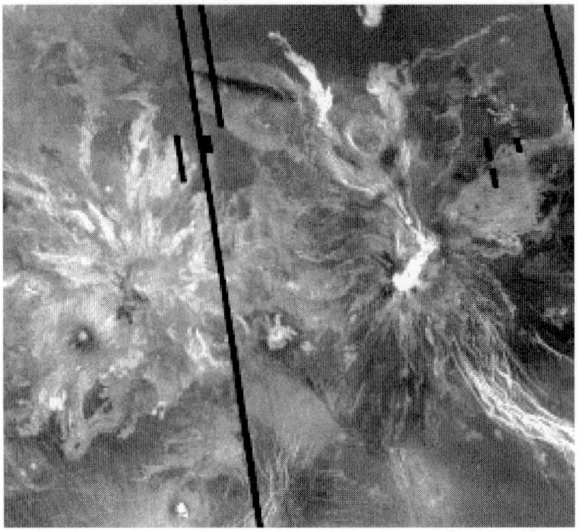

Figure 3. The volcanoes Sif (left) and Gula (right) Montes. Both volcanoes are over 2 km high and over 350 km across. Sif Mons has a summit caldera, while Gula's summit is cut by a radar-bright rift.

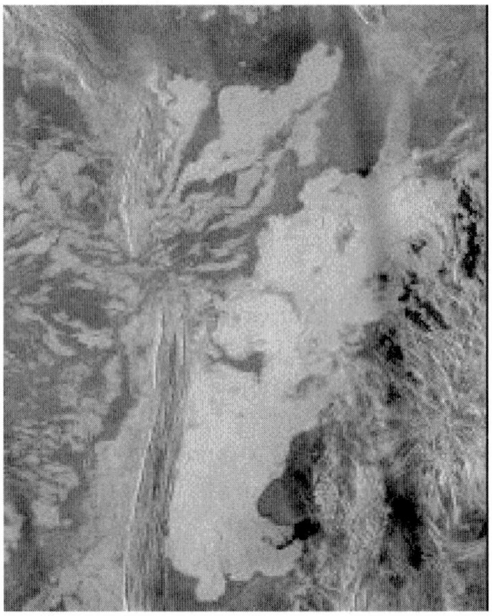

Figure 4. Magellan SAR image of Kaiwan Fluctus, a large lava flow field. The flow field extends for over 500 km, and appears to have been partially dammed by the approximately N-S belt of ridges.

Over 100 large flow fields have been identified on Venus [6] (Figure 4). These flows cover areas similar to terrestrial flood basalts such as the Deccan Traps or the Columbia River Basalts. Many of these large flow fields are located near extension zones, while others are at coronae or large volcanoes. Terrestrial geologists argue about the emplacement rates for large flow fields; studies of these features on Venus where erosion rates are low may help constrain their formation.

Several unusual types of volcanic features were identified in Magellan data, including 'pancakes', 'ticks' and very long channels. Pancake domes or steep-sided domes are flat-topped, ~20–80 km across

Figure 5. Two steep-sided or pancake domes in the northern hemisphere of Venus, adjacent to a smaller 'tick' (to the right). The central dome is 50 km across, while the smaller tick with scalloped edge sis 25 km across.

volcanoes (Figure 5). They are similar in shape to silicic does on Earth, but on Venus their smooth upper surfaces seem to indicate emplacement by more basaltic lavas. Ticks are steep-sided domes whose sides have undergone collapse [7], producing a scalloped edge (Figure 5). Very long, radar dark channels were identified in the plains of Venus, resembling meandering terrestrial rivers. One of the channels, Baltis Vallis, is over 6000 km long [8]. It is thought that these channels are formed by very fluid lavas of unusual composition, perhaps high in sulfur or carbonates.

The volcanic features that dominate the surface of Venus clearly contribute to its overall heat loss, through both convective and conductive processes. Their link to underlying mantle circulation is still not clearly understood, although hotspots clearly play a role in the formation of some volcanoes. Magellan gravity data indicates that some hotspots on Venus may still be active [9]. No changes were seen in SAR images of volcanoes over the 4-year length of the Magellan mission that would indicate ongoing eruptions, but declining SO_2 values detected by Pioneer Venus were interpreted as possible evidence of a recent volcanic eruption [10].

3.2 Tectonism

Tectonics have also played a role in shaping the surface, with complexly deformed tessera plateaus, compressional mountain belts in Ishtar Terra, and rift zones in the equatorial region. Most of these features, however, are produced by relatively limited horizontal motion, likely related to underlying mantle circulation.

Two types of features produced by crustal shortening have been observed: mountain belts and ridge belts. Mountain belts are elevated >1.5 km above the surrounding surface, and have sinuous ridges and folds spaced 5-10 km apart. Mountain belts (Akna, Freyja, Danu and Maxwell Montes) are found surrounding the highland plateau Ishtar Terra in the northern hemisphere of Venus (Figure 6). Ishtar

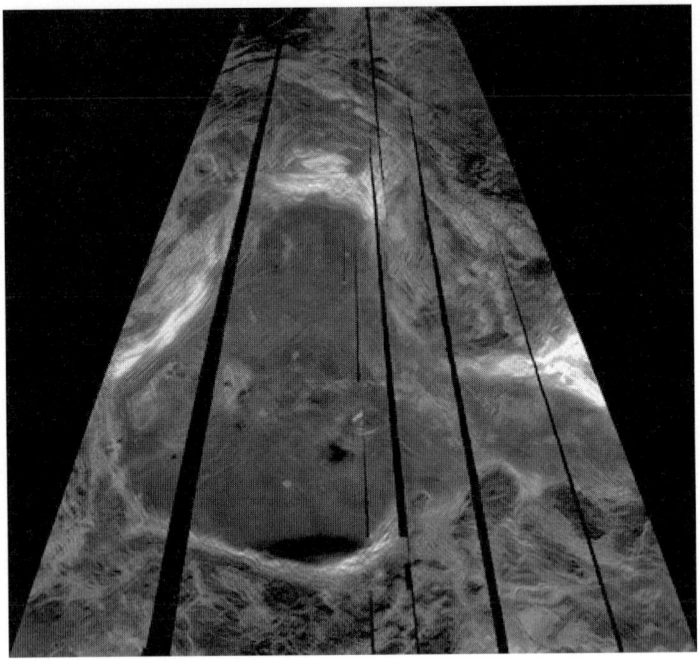

Figure 6. Magellan SAR mosaic of Ishtar Terra in the northern hemisphere of Venus. Ishtar, about the size of Australia, is composed of a high plateau (Lakshmi Planum) surrounded by the Freyja Montes to the north, Danu Montes to the south, and the Akna Montes to the west. To the east (out of the image), lie the Maxwell Montes. The vertical black stripes on the image are data gaps.

has been suggested to form over a mantle downwelling or a mantle upwelling. Ridge belts are narrow (<200 km across), low (<~1 km elevation) of ridges located in the plains. Two sets of ridge belts are found: one in the northern hemisphere and one in the southern hemisphere. These sets of ridges are thought to be linked to downwelling mantle causing crustal shortening.

Belts of extensional features are also located in the plains of Venus, along with more developed rift systems. Rifts, with troughs >1 km deep, are primarily located in the equatorial region. Rifts extend for 1000's of kilometers, and are about 200 km wide (Figure 2). The venusian rifts are thought to be more analogous to continental rifts on Earth rather than oceanic rift zones, in that relatively small amounts of crustal extension have occurred. It is not known whether any of the rift systems are still active.

Tessera terrain was first identified in Venera 15/16 radar images of Venus, characterized by complex intersecting sets of ridges and fractures [11] (Figure 7). Tessera occurs as isolated patches in the plains, as well as highland plateaus, >1000 km across and over 1.5 km high. The highly deformed surface of tessera appears to have undergone both extensional and compressional deformation. Although the margins of many highland plateaus are embayed by surrounding plains, suggesting that they are old, the tessera surface does not contain a statistically significant increase in impact craters. This suggests that while the plateaus may be relatively old, they continued to internally deform over a long period of time. The origin of tessera plateaus is controversial; they have been suggested to form over mantle downwellings or upwellings.

Coronae are volcano-tectonic circular features (diameters 90-2600 km) that were first seen in Soviet Venera radar images of Venus [11, 12]. They have a ring of concentric ridges and fractures, surrounding an interior region with volcanic flows, radial fractures and small volcanoes (Figure 8). They are often surrounded by extensive flow deposits, although some coronae have very little associated volcanism. Most coronae are topographic highs, although some are depressions. These complicated characteristics

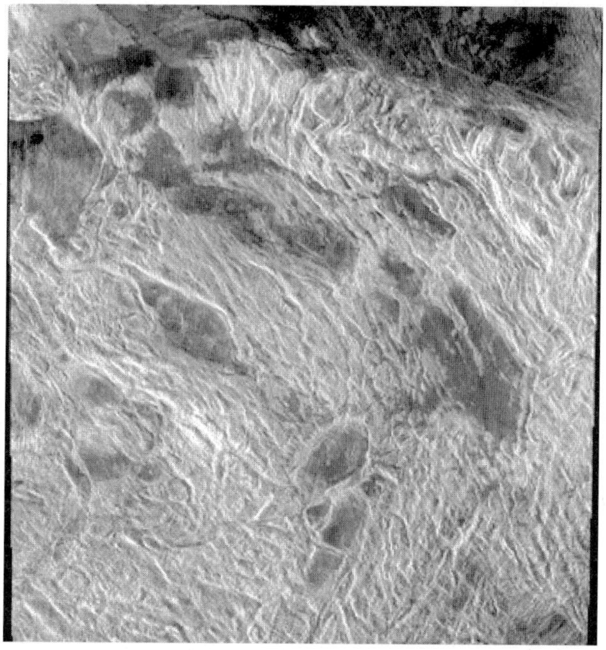

Figure 7. Magellan SAR image of northern Alpha Regio, a tessera highland plateau. Alpha rises about 2 km above the surrounding plains, and has complex, crosscutting sets of ridges and fractures. This image is about 500 km across.

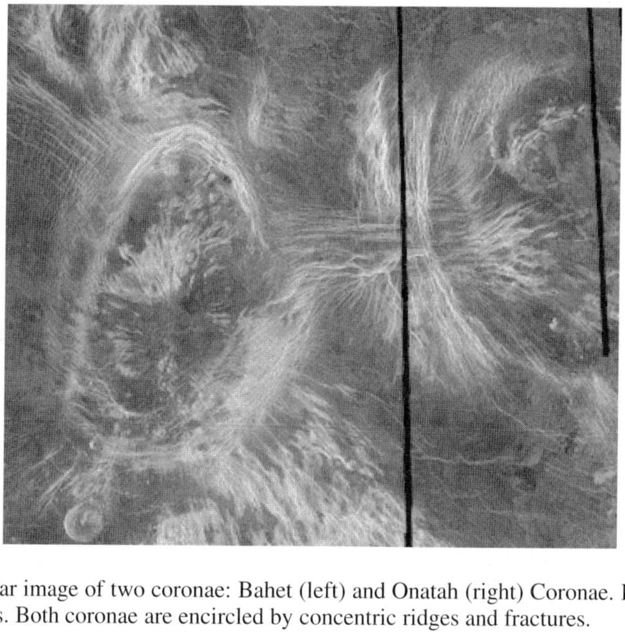

Figure 8. Magellan radar image of two coronae: Bahet (left) and Onatah (right) Coronae. Bahet is 230 km across; Onatah is 350 km across. Both coronae are encircled by concentric ridges and fractures.

are most consistent with formation over small-scale, shallow mantle plumes, some in the presence of depleted mantle [13]. Coronae can account for approximately 15% of heat loss on Venus [13]. It is not known whether coronae are still active.

3.3 Impact cratering

940 impact craters have been found in Magellan data, which covers about 98% of the venusian surface. The impact craters have diameters from $\sim$ 3 km - 268.7 km; craters $<$ 3 km are not expected and the population of craters less than about 30 km is reduced due screening of impactors by the dense atmosphere of Venus. When impactors $<$ 1 km break up in the atmosphere, they produce a shock wave that pulverizes the upper few m of the surface [14], producing a radar-dark or radar-bright splotch on the venusian surface. Approximately 400 such features have been mapped, along with a number of craters surrounded by a parabolic halo of material. The haloes are believed to form when fine-grained material associated with the impact settled out of the atmosphere [15]. Both the splotches and haloed craters are considered by some to be relatively young, on the idea that all craters initially have haloes that are subsequently eroded or buried. Another unique feature of venusian craters are outflows extending from the ejecta blankets of some craters (Figure 9). These are believed to form due to fluidization of the ejecta in the dense atmosphere of Venus [15].

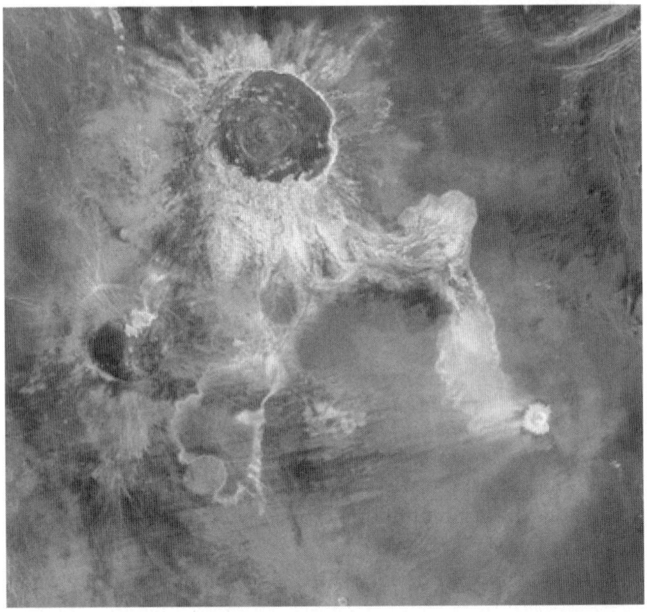

Figure 9. Magellan SAR image of the 175 km diameter crater Isabella. The crater is surrounded by bright ejecta, some of which has formed an extensive flow to the southeast. Just to the southeast of Isabella lies a 20 km diameter impact crater, surrounded by a faint, radar-bright parabolic halo.

The most intriguing finding of the Magellan mission was the seemingly random, pristine nature of the impact crater population (e.g., [15]). The population gives an average surface age of about 750 my [16], however, the randomness of the population prevents relative or absolute age dating of different regions of Venus [17]. The crater population led some workers to suggest that the planet was catastrophically resurfaced about 750 my ago, possibly due to mantle overturn, with little current geologic activity (e.g., [18]). This process could repeat episodically, due to gradual thickening of a stagnant lid [19]. However, other workers have interpreted the crater population to indicate a more gradual resurfacing process, occurring over 10's-100's mys (e.g., [20]). Rapid resurfacing of the planet has implications for the atmosphere: greatly increased amounts of volcanism over relatively short time periods could cause the surface temperatures to temporarily rise by hundreds of degrees [21]. Data from future missions will be required to solve this controversy.

3.4 Stratigraphic history

Without an impact crater population to use for relative or absolute dating, the stratigraphic history of Venus is difficult to constrain. Basilevsky and others [22] have suggested that the recent (<750 my) history of Venus is one in which specific geologic processes dominated at given times. For example, they propose that all plains of a specific morphology (i.e., lobate) formed at a specific time, all wrinkle ridges formed within a short time period globally, etc. Guest and Stofan [23] suggested a more non-directional history, in which similar geologic processes (e.g., wrinkle ridge formation, small volcano formation) overlap in time, more similar to Earth's geologic history. This non-directional theory does not exclude the possibility of some geologic activity being confined to relatively short periods of time, for example, the formation of the tessera plateaus. Geologic evidence can be found to support both of these end-member theories; additional data is needed to fully understand where the surface history of Venus falls between them.

4. POTENTIAL HABITABILITY?

The dry, extremely hot surface of Venus is considered to be uninhabitable at present, although it has been suggested that the clouds of Venus could harbor life given their relatively long lifetimes [24]. However, Venus may have harbored life in the past, prior to the formation of the atmospheric greenhouse. Early Venus is interpreted to have abundant water that might have been stable on the surface for long periods of time, before being lost through some runaway process. These long-lived oceans may have harbored life, traces of which may still remain. Additional evidence for an early wet Venus may come from rock compositions. For example, the high-standing tessera terrain on Venus may be granitic in composition, produced when water was more abundant.

Determining when and how much water was lost from the surface and interior of Venus is critical, as it will help to define the width and stability of the habitable zone around solar-type stars. If Venus had oceans that harbored life for the first few billions of years of its history, the continuously habitable zone would have been much wider than if Venus lost its water within a few hundred million years after formation.

5. UNANSWERED QUESTIONS AND FUTURE EXPLORATION OF VENUS

Despite the volumes of data returned from Magellan and the promise of new insights into the chemistry and circulation of the atmosphere from Venus Express, more detailed measurements in the atmosphere and on the surface of Venus are required to address critical questions on the evolution of Venus and its implication for the formation of Earth-like planets. Particularly critical are constraining the atmospheric evolution through in situ isotopic measurements of atmospheric gases. Differentiating between models of surface evolution will require more detailed measurements of the geochemistry and mineralogy of surface rocks. Landed missions on the surface, hopefully with the capability of visiting multiple sites, will most importantly use both atmospheric and surface geochemical instruments to determine the history of water on Venus.

Another critical measurement if the detailed structure of the interior of Venus, which can only be gained through detailed seismic data. While Venus is undoubtedly still geologically active, it would take at least 6 months to one year of seismic monitoring to be certain of gaining adequate seismic data. Long-term survivability on the surface of Venus is currently a technological challenge, which will hopefully be overcome in the cooing decades. Our return to the surface of Venus is critical to understanding the past, and possibly the future, of Earth.

References

[1] Cattermole, P., Venus: Geological Story. Baltimore: John Hopkins University Press, 250 pp. (1998).

[2] Solomon, S. C., Smrekar, S. E., Bindschadler, D. L., Grimm, R. E., Kaula, W. M., McGill, G. E., Phillips, R. J., Saunders, R. S., Schubert, G., Squyres, S. W. and Stofan, E. R., Venus tectonics: An overview of Magellan observations, *J. Geophys. Res.*, **97**, 13, 199-13, 256 (1992).

[3] Head, J. W., Crumpler, L. S., Aubele, J. C., Guest, J. E. and Saunders, R. S., Venus Volcanism: Classification of Volcanic Features and Structures, Associations, and Global Distribution from Magellan Data, *J. Geophys. Res.* **97**, 13, 153-13, 197 (1992).

[4] Crumpler, L. S., Aubele, J. C., Senske, D. A., Keddie, S. T., Magee, K. P. and Head, J. W., Volcanoes and centers of volcanism on Venus, In *Venus II*, Edited By S. W. Bougher, D. M. Hunten, R. J. Phillips, Univ. of Ariz. Press, Tucson (1997).

[5] Stofan, E. R., Guest, J. E. and Copp, D. L., Development of large volcanoes on Venus: Constraints from Sif, Gula and Kunapipi Montes, *Icarus*, **152**, 75-95 (2001).

[6] Magee, K. P. and Head, J. W., Large flow fields on Venus: Implications for plumes, rift associations and resurfacing, *GSA Spec. Paper* **352**, 81-101 (2001).

[7] Guest, J. E., Bulmer, M. H., Aubele, J., Beratan, K., Greeley, R., Head, J. W., Michaels, G., Weitz, C. and Wiles, C., Small volcanic edifices and volcanism in the plains of Venus. *J. Geophys. Res.* **97**, 15, 949-15, 966 (1992).

[8] Baker, V. R., Komatsu, G., Parker, T. J., Gulick, V. C., Kargel, J. S. and Lewis, J. S., Channels and valleys on Venus - Preliminary analysis of Magellan data, *J. Geophys. Res.* **97**, 13, 421-13, 444 (1992).

[9] Smrekar, S. E., Evidence for active hotspots on Venus from analysis of Magellan gravity data, *Icarus*, **112**, 2-26 (1994).

[10] Esposito, L. W., Sulfur dioxide: Episodic injection shows evidence for active Venus volcanism, *Science* **223**, 1072-1074 (1984).

[11] Basilevsky, A. T., Pronin, A. A., Ronca, L. B., Kryuchkov, V. P., Sukhanov, A. L. and Markov, M. S., Styles of tectonic deformation on Venus: Analysis of Veneras 15 and 16 data, *J. Geophys. Res.* **91**, 399-411 (1986).

[12] Stofan, E. R., Bindschadler, D. L. Head, J. W. and Parmentier, E. M., Corona structures on Venus: Models of origin, *J. Geophys. Res.*, **96**, 20933-20, 946 (1991).

[13] Smrekar, S. E. and Stofan, E. R., Coupled upwelling and delamination: A new mechanism for coronae formation and heat loss on Venus, *Science*, **277**, 1289-1294 (1997).

[14] Zahnle, K. J., Airburst origin of dark shadows on Venus, *J. Geophys. Res.*, **97**, 10, 243-10, 255 (1992).

[15] Schaber, G. G., Strom, R. G., Moore, H. J., Soderblom, L. A., Kirk, R. L. Chadwick, D. J. Dawson, D. D. Gaddis, L. R. Boyce, J. M. and Russell, J., 1992, Geology and distribution of impact craters on Venus: What are they telling us?, *J. Geophys. Res.* **97**, 13, 257-13, 302 (1992).

[16] McKinnon, W. B., Zahnle, K. J., Ivanov, B. A. and Melosh, H. J., Cratering on Venus: Models and observations, in *Venus II*, eds. Brougher, S. W., Hunten D. M. and Phillips, R. J. University of Arizona Press, Tucson, 969-1014 (1997).

[17] Campbell, B. A., Surface formation rates and impact crater densities on Venus, *J. Geophys. Res.*, **104**, 21, 951-21, 955 (1999).

[18] Strom, R. G., Schaber, G. G. and Dawson, D. D., The global resurfacing of Venus, *J. Geophys. Res.* **99**, 10, 899-10, 926 (1994).

[19] Solomatov, V. S. and Moresi, L. N., Scaling of time-dependent stagnant lid convection: Application to small-scale convection on Earth and other terrestrial planets, *J. Geophys. Res.*, **105**, 21, 795-21, 817, (2000).

[20] Phillips, R. J., Raubertas, R. F., Arvidson, R. E., Sarkar, I. C., Herrick, R. R., Izenberg, N. and Grimm, R. E., Impact craters and Venus resurfacing history, *J. Geophys. Res.* **97**, 15923 (1992).

[21] Bullock, M. A. and Grinspoon, D. H., The recent evolution of climate on venus. *Icarus* **150**, 19, 037-19, 048 (2001).

[22] Basilevsky, A. T., Head, J. W., Schaber, G. G. and Strom, R. G., The resurfacing history of Venus, in *Venus II*, eds. Brougher, S. W., Hunten, D. M. and Phillips, R. J., University of Arizona Press, Tucson, 1047-1085 (1997).

[23] Guest, J. E. and Stofan, E. R., A new view of the stratigraphic history of Venus: Icarus 139, 55-66 (1999).

[24] Grinspoon, D., Lonely Planets. New York: Harper Collins, 440 pp., (2003).

J. Phys. IV France **139** (2006) 21–35
© EDP Sciences, Les Ulis
DOI: 10.1051/jp4:2006139004

Future steps in the exploration of the Solar System

M. Coradini[1]

[1] *European Space Agency, Science Directorate, 8–10 rue Mario Nikis, 75738 Paris Cedex 15, France*

Abstract. The last ten years have been characterized by a renewed worldwide effort of exploration of the Solar System, which has produced spectacular new data and concluded the "discovery" phase. The next decades will focus on refinement of knowledge and will tackle specific problems such as the boundary conditions for the onset and maintenance of life on planetary bodies. Final understanding of the formation and evolution of planetary systems is expected, also based on the observation of extra solar systems and possible detection of "habitable" extra solar planets.

1. INTRODUCTION

Spacecraft are currently present on many planets in the Solar system, including the Moon, Mars, the Saturn/Titan system, Venus and, tomorrow, Mercury. We are, thus, acquiring data on all the major solid bodies and their atmospheres in the solar system: Venus, Mars and Titan. The potential benefits for the understanding of the evolution and fate of the fourth solid body atmosphere in the Solar System, that of our Earth, are apparent.

Planetology helps us to put in context the particular planet on which we happen to live. On the other hand, participating in missions closing in on the Sun has given us a new view of our own star, which ultimately controls our lives. Also the study of our magnetosphere, the magnetic bubble which travels with our Earth and protects it from the outbursts of our star and from the steady flux of cosmic rays, is another area where a large number of international missions made important contributions, following up a series of earlier small missions. Among these, a mission flying four identical spacecraft in formation, the CLUSTER mission, is allowing for the first synchronous study in three dimensions of particles and fields in our magnetosphere. Here, ESA had to fight also against bad luck, because the first Cluster mission was lost in the failure of the first Ariane 5 launch in 1996. However, the decision was quickly taken and acted upon to fly a replica of the mission. This took place in 1999 and since then Cluster is flying with success. In 2003-2004 the mission was enriched by two Chinese spacecraft (the Double Star mission), which carry many European experiments.

Ulysses (1990), still operating, has been exploring the heliosphere, the bubble of particle, gas, radiation and magnetic field travelling with our Sun through interstellar space.

But perhaps the best example ever of a successful cooperative mission is given by SOHO, still operational after a decade in orbit. Thanks to SOHO a number of mysteries and questions about the inner and outer structures of our Sun have been answered.

In planetary research, competition with NASA has mostly given way to cooperation, for example through the imaginative Cassini-Huygens mission. However, Mars Express, a European mission, and certainly the cheapest mission ever sent to Mars, has been producing first class scientific data in spite of the loss of Beagle 2, with breathtaking three-dimensional high-resolution images, and the discovery of water and methane, the chemical prerequisite/markers of possible biotic activity. Mars Express is in good company of a large number of US spacecraft and it can even dialogue with two NASA landers, Spirit and Opportunity, which are revealing unthinkable details of the ancient Martian chemistry, mineralogy and climate.

In recent years NASA has found worldwide challenges from Russia, Japan, China and even India. However, in the last two decades Europe has made gigantic steps to become the second largest space fairing organization in the world. In particular, the scientific missions of ESA have become a constant presence in the Solar System and in the vicinity of our planet.

Globally, there are now in orbit fifteen ESA scientific spacecraft, of which nine are directly operated by ESA. At the time of writing, ESA is flying a total of 17 satellites. Attentive observers may have noticed that the Science Programme, since the inception of Horizons 2000 (1985), has been doing more and more missions with less and less funds. In spite of the decrease in resources, the Science Programme of ESA has recently had three "*anni mirabiles*" in a row: 2003, 2004 and 2005. Mars Express, SMART-1 and the first Double Star satellite in 2003, Rosetta and the second Double Star satellite in 2004. There were also events not related to launches: these include the arrival of Cassini-Huygens near Saturn, and the insertion of SMART-1 in lunar orbit (15 November 2004). The year 2005 starts with the Huygens mission, which has crowned a long cruise and fifteen years of expectations by European scientists. In mid 2005 Venus Express was launched successfully and it now orbits the planet carrying out novel scientific observations of the atmospheres.

In this context, we assume that by 2015 the Aurora [exploration] programme will have reached its cruise-speed of about 200 MEuro/year and will be focused on Mars. A simple division of responsibilities between the Science Programme and Aurora will be achieved: access to Mars and its surface and technologies aiming at the development of human exploration of Mars should be the responsibility of Aurora; scientific, robotic exploration should be the responsibility of the Science Programme. Of course, Aurora missions will continue to offer opportunities for scientists. A similarly symmetric situation is also present at NASA where the exploration initiative will probably characterize the next decades.

2. PLANETARY FORMATION AND THE DEVELOPMENT OF LIFE

From the Big Bang, to the formation of stars and galaxies, the question of why, after which succession of events, and under which conditions, life on Earth originated is a question that fascinates mankind. Equally captivating is the question of whether life exists elsewhere in the Universe, in which form, on which kind of planets and linked to which type of stars. As we are working on theories to explain the physical processes by which life might appear and evolve on a planet, we are in the somewhat peculiar situation in which only one planet hosting life is presently known. No other sign of life has ever been detected neither on the other planets or satellites in the solar system nor elsewhere in the Universe. Hence, for the present time life on Earth provides a single case example to guide physical, chemical and biological investigations.

A decade ago, as the solar system was the only planetary system known, theories were developed to account for the formation and evolution of such systems. Since then, the discovery of over 130 planets orbiting stars other than our Sun, has taught us the limit of such an approach! The formation of many of these systems was simply not possible within the framework of the theories accepted as little as ten years ago. Based on this recent example, we can only wonder to which scientific and philosophical revolution the discovery of life on another planet will lead.

We are now at a unique moment in mankind's history. For the first time since the dawn of philosophical and scientific thought, it is within our grasp to answer, rigorously and quantitatively, two fundamental questions:

- are there other forms of life in the Solar System and have they an independent origin from the ones which developed on Earth?
- are there other planets orbiting other stars similar to our own Earth, and could they harbour life?

Related questions crucial to the understanding of the whole process, are the following:

- What are the conditions for stars to form and where do they form?
- How do they evolve as a function of environment?
- Are there any specific characteristics in stars that host planets?
- What are the conditions for planets to form around stars?
- What are the different kinds of planets orbiting stars? What is their mass range? Are there planets similar to those of the Solar System?
- Which planets are surrounded by an atmosphere? What are the characteristics of these atmospheres?
- What are the conditions for life (of any form) to appear on these planets?

For the first time we are going to be the able to build instruments that allow us to directly investigate how unique the Earth is and whether or not we are alone in the Universe. Discovering Earth's sisters and possibly life is the first step in the fundamental quest of understanding which succession of events lead to the emergence and survival of life on Earth. For this, we need to understand how, where and when stars form from gas and dust and how, where and when planets emerge from this process. This is certainly one of the most important scientific goals that Europe and the US could set themselves.

3. FROM GAS AND DUST TO STARS AND PLANETS

The atoms from which stars and planets are actually formed have gone through a succession of violent processes from the very early times, when the Universe followed by the first generation of stars formed. Most of the objects we are observing today are made from the ashes of stars that no longer exist. This of course also applies to mankind, as we are literally stardust. The stars that produced the carbon we have in our bodies, and the oxygen we breathe, have formed, evolved and died long ago. The way stars and planetary systems form remains almost unknown: while our understanding of stellar evolution is making giant leaps forward, we are still lacking a comprehensive theory explaining why and how stars and, apparently quite often, planetary systems form from interstellar matter. Both are of course closely linked and the formation of planets has to be considered in the wider context of star formation and circumstellar disk evolution. Magnetic fields and turbulence are often invoked as playing a key role in the birth process. The large diversity among the orbital characteristics of the exo-planets is pointing towards the importance of planet-disk and planet-planet interactions. These interactions can lead to such surprising results as large scale inward migration of giant planets and/or pumping of the orbital eccentricity which then rises the question of the long term stability of these systems. The problem is therefore essentially to establish which basic characteristics of the star formation process determine the bulk properties of the planetary system eventually emerging several tens of millions of years later.

The star and planet formation processes requires a multi-wavelength approach, mostly from near-infrared to millimetre wavelengths. A large part of this wavelength range is absorbed by the atmosphere, and only observable from space. With ESA's ISO mission, the upcoming Herschel observatory, the NASA/ESA James Webb Space Telescope, and with ESO's ground-based facilities, including the joint Europe-U.S. ALMA project, the star formation community is in a very strong position. However, to resolve the protostars and their associated disk in the nearest star forming regions, a spatial resolution of the order of 0.01 arcsec will be needed together with high- and low-resolution spectroscopy capabilities in order to characterize line emission and dust mineralogy.

The first detection of a planet orbiting a solar-type star, performed by a European team, occurred only 10 years ago. Since then, more than 130 planets have been discovered, many of them having unexpected orbital characteristics. These discoveries have sparked a large number of observational efforts, all over the world, to find more of these objects as well as theoretical studies aimed at explaining their characteristics.

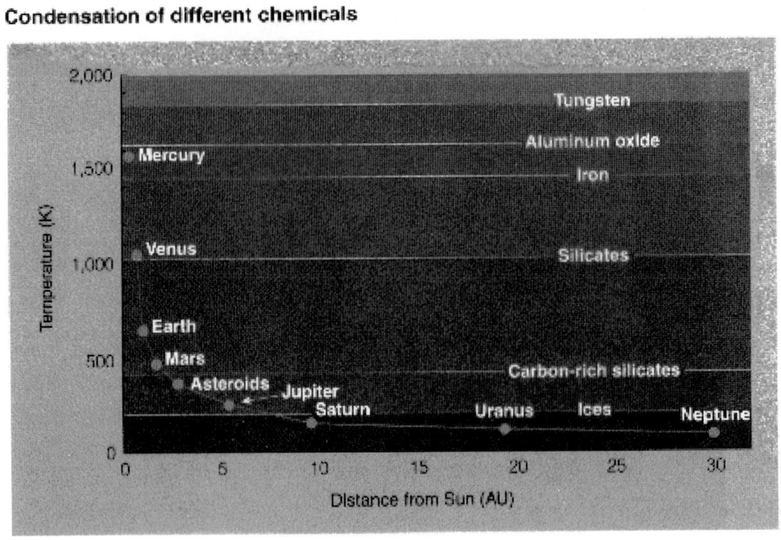

Figure 1. How temperature gradients within the Solar System determine condensation of different chemicals.

To understand the origin of the solar system in general and of the Earth in particular, it is essential to place our planetary system into the overall context of planetary system formation. To guide the theory, a complete census of all the planets from the largest to the smallest out to distances as large as possible is required. This can be achieved by making use of a variety of detection techniques ranging from the high precision measurement of radial velocities, high accuracy astrometry to detect the tiny reflex motion of the star in the plane of the sky, photometry to measure the changes of brightness during a transit or during a gravitational lensing event. A large and complete sample will tell us which stars are most likely to host which kinds of planets. It will for example allow to quantify the influence of the chemical characteristics of host stars (is metallicity a key factor for planetary formation?), and of their position and motion with respect to the galactic plane and the global rotation of the Galaxy. The statistical analysis of planets orbital parameters and mass will unravel correlations which might point towards the key physical mechanisms involved in the formation and evolution of these systems. Undoubtedly, we will have to discover planets with masses and temperatures compatible with the formation of an atmosphere and the presence of liquid water, i.e. planets in the "habitable zone".

After the coming decade, mainly devoted to the statistical exploration of planetary populations and the understanding of the best conditions for planetary system formation, the development of new observational techniques will allow us, within the time frame 2015-2025, to isolate the photons coming from the planet from those stemming from the star. This will represent a major step forward in our capabilities to study exo-planets and will open an entirely new era: the age of planetary spectroscopy during which the physical characteristics of the atmosphere of these bodies, such as temperature or chemical composition, can be measured. With these spectroscopic capabilities, we will also have the means to search in the spectrum for possible markers of biological activities.

On a longer timescale, a complete census of all terrestrial planets within 100 pc of the sun would be highly desirable. This could be achieved, for example, using high precision astrometry. Further, the direct detection of these planets followed by high-resolution spectroscopy with large telescope at IR, optical and UV wavelengths and ultimately spatially resolved imaging will mark the rising of yet another entirely new field: Comparative exo-planetology.

4. LIFE AND HABITABILITY IN THE SOLAR SYSTEM

The quest for evidence of a second, independent genesis of life in the solar system must begin with an understanding of what makes a planet habitable and how the habitable conditions change (either improving or degrading) with time. For instance, the environmental conditions on the Earth today are not the same as when life first arose on this planet. The early Earth, with its oxygen-less atmosphere, high UV radiation, high temperatures and slightly acidic waters, could not support evolved life forms. However, life could not have arisen on a planet with the environmental conditions that exist on Earth today.

We can define the basic habitable conditions for life, as we know it, to appear on a planet. For life to *appear*, a planet needs: liquid water, a source of carbon, a source of energy and a source of nutrients (the nutrients include N, P, S, Mg, K, Ca, Na and Fe). For life to *continue to exist* the nutrients need to be renewed and this can only be done by active geological processes, such as recycling of the crust by some form of tectonic activity. For life to *evolve*, however, the environmental conditions on a planet need to evolve as well. On Earth the phenomenon of habitat evolution is related to the parallel processes of geological evolution and the interaction of life processes with the planet, leading to oxygen and a protective ozone layer in the atmosphere. However, a major problem on Earth is that plate tectonics has eliminated all of the first 500 million years (My) of rock history and severely altered the subsequent 500 My of history, *i.e.* the crucial first billion years when life arose and took a foothold.

This gap in our knowledge can be filled by studying other planets that did not develop plate tectonics and, therefore, still have a record of the early environmental conditions. Mars is an ideal goal. Although the present conditions at the surface of the planet are not conducive to the long-term sustenance of life, Mars had an early history that was similar to that of the early Earth and conditions that were suitable for the appearance of life. A major question is: how did continued evolution of the planet affect the habitable environment and what did happen to the planet to make its surface *uninhabitable* today?

Mars missions can address the key scientific questions regarding habitability in the Solar System, such as: what were the conditions during the earliest period in the history of the terrestrial planets when the planets became habitable and when life appeared, at least on Earth? What were the major causes in the degradation of habitability on Mars? Was there or is there still life on the planet?

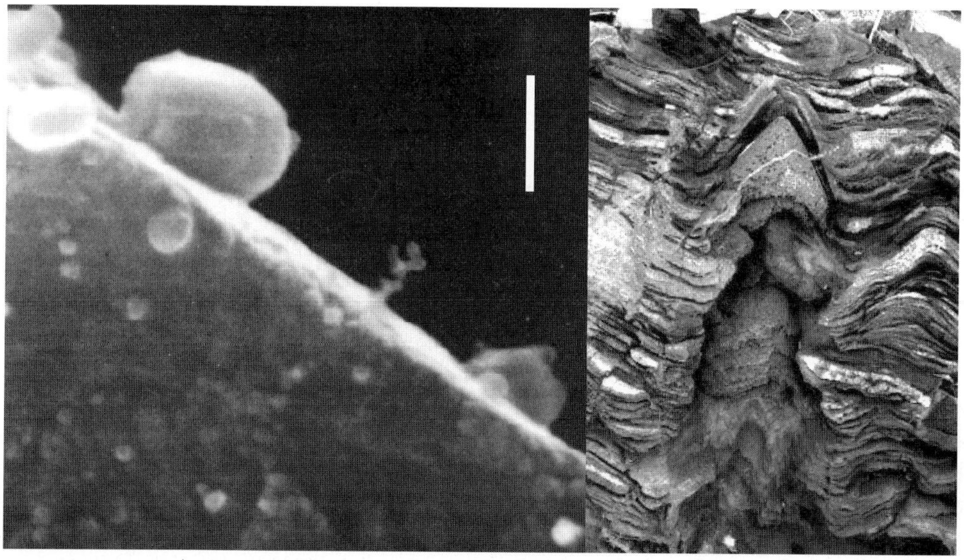

Figure 2. Example of what to search on extraterrestrial surfaces: Nanobacteria (left) and/or primordial fossils.

The future missions will have to investigate the structure, geochemistry and mineralogy of rocks in various geological locations on Mars in order to identify their origin and geological history. More generally, they need to gather information about the mechanisms that controlled the evolution of the Martian environment and the history of water on Mars. A key point is the need to place any *in situ* measurements in context, e.g. did the rocks form in a liquid water environment or not? Such investigations should also include science packages to search for evidence of extinct or extant life. Additional geophysical investigations of the deep and crustal structure of the planet are also needed to understand its actual activity. Measurements about present climatic conditions are also required to trace evolution and habitat conditions back in time. Access to specific, selected locations on Mars, including rough and high terrain or the subsurface, will be essential to investigate different geological and environmental settings and, thus to favour the identification of traces of life. This goal may require the development of new technologies, such as capable rovers, precision landing or deep drilling. Orbiting spacecraft could be used to carry out remote sensing of the planet, its atmosphere and climate, or its plasma magnetic environment whilst also acting as a relay satellite. Monitoring of the present environment will also be needed to understand the present condition of the habitat and also in preparation for future manned missions.

Ultimately a high priority goal that is achievable in the 2015-2025 time frame is a sample return mission from Mars, returning samples from selected locations that have been studied by *in situ* missions. While *in situ* measurements at multiple locations provide invaluable information, there are some investigations that require terrestrial laboratory analyses (i.e., isotopic measurements, microfossil identification, and age dating, etc.).

The search for habitability in the solar system through a study of Europa is also a high priority. In particular, the determination of its internal structure and especially, its potential, active, internal heat sources. Analysis of the composition of the ocean and icy crust is of paramount importance for determining the availability of nutrients. The plasma and radiation environment around Jupiter and its interaction with Europa would also provide important information regarding the survivability of any life throughout the satellite's history. These science goals could be achieved by a dedicated Europa orbiter. A Europa lander is highly desirable, but may not be technologically feasible within the 2015-2025 timeframe.

5. HOW DOES THE SOLAR SYSTEM WORK?

The search for the origins of life discussed above must begin in our own solar system. Understanding how the Sun behaves over a range of timescales, how the planets can be shielded from its radiative and plasma output, why the nine solar system planets are so different from each other, and what the small bodies such as comets and asteroids can tell us about our origin are only a few of the aspects that are involved in this question. The generic circumstances under which planets are habitable are unknown, but must depend on the radiative output and magnetic activity of the neighbouring star, the behaviour of the space environment surrounding the planet, the material from which the planet originally accreted etc.

Yet the exploration of the solar system addresses scientific questions of fundamental importance beyond the origins of life. Why do the Sun (and by inference many stars) generate magnetic fields? Why do these fields result in high temperature corona and winds? How do planetary atmosphere and magnetospheres respond to the interaction with the solar wind? Why do planets and moons have such different atmospheres and surfaces from each other? What determines the (past or present) presence of water on planets? What are comets and asteroids made from and what does this tell us about the origin of the solar system?

The Sun dominates the solar system. Its radiation provides the means to sustain life, but its continual and occasionally violent activity provides the means to destroy it. Both are critically important areas to be studied. Only in the solar system can we establish the "zero order truths" concerning the Sun, it's all

Figure 3. A solar eruption photographed by the ESA-NASA SOHO spacecraft and the schematic Earth's magnetic protection (not in scale).

important magnetic field, and the interaction of the solar wind with the planetary environments that can then be extended to planetary systems elsewhere in the Universe.

The US, Europe and Japan are international leaders in this field. The SOHO and Ulysses missions have led the way in the past decade in the exploration of the Sun and the distant heliosphere, the Cluster mission has provided the first multi-point measurements of the Earth's space environment, and the study of the plasma environments of other planets is being undertaken by the Cassini-Huygens mission. In the future, Europe will play a major role with the Solar Orbiter mission designed to explore the Sun from the unprecedented close distance of 30 million km (20% of the average Sun-Earth separation), while Japan with the Solar-B and the US with the STEREO mission and the other missions of the Living with a Star programme will contribute to consolidate our understanding of the Sun.

The varying magnetic field of the Sun is directly responsible for changes in the solar ultraviolet and X-ray emission, and is also closely related to the physics of long-term solar cycles and their possible forcing role in climatic evolutions. It is responsible for the solar activity that leads to the solar wind plasma interacting with the planetary environments. The solar magnetic field is continuously generated and destroyed on timescales ranging from fractions of a second to decades, and fills the heliosphere, a volume of space that extends to at least 10 billion km away from the Sun. These topics will remain major scientific challenges in the 2015-25 timeframe.

The expansion of the Sun's atmosphere fills the heliosphere with plasma and magnetic field that are collectively known as the solar wind. In this medium, processes that are generic to all of astrophysics (heating, the acceleration of particles and turbulence) can be studied with ease. These processes also dictate how the Sun's magnetic field interacts with planetary environments. Some planets have magnetic fields that provide partial shielding (Earth, the gaseous giants and Mercury). Others have atmospheres with (Mars) and without (Venus) weak remnant magnetic fields. All are different, so the solar system provides a vast range of laboratories for studying the potential interaction of planets with the winds from stars.

While the scales of planetary magnetosphere are vast (up to 10 million km at Jupiter), the inescapable fact is that the interaction between the magnetic fields of the planet and the Sun occurs over a range of scales between a few km and a few planetary radii! Similar hierarchies of scales are likely to arise in other fundamental processes such as turbulence, magnetic field annihilation and particle acceleration, leading to the astonishing diversity of structures and dynamical behaviours that characterize most astrophysical media. Measurements have never been made on the smallest required scales and the fundamental aspect of the plasma universe electrodynamics – the cross-scale coupling - has thus been inaccessible. A comprehensive set of magnetospheric measurements, involving a hierarchy of 3D satellite configurations, thus a 4D matrix of s/c, is vital for understanding of these generic physical processes and provides an exciting prospect in the future years.

The Jovian magnetosphere is another wonderful laboratory for studying how plasmas behave in space. With its rapid rotation, strong magnetic field and internal sources of plasma, it has been compared to binary stellar systems and even pulsars. It is the most accessible environment for studying the fundamental processes not addressed above such as plasma/neutral and plasma/satellite interactions, magnetodisk stability, the relaxation of rotational energy and associated energetic processes, and loss of angular momentum by magnetoplasma interactions, the last two being important in understanding accretion mechanisms. A Jovian mission involving at least three spacecraft with an optimized plasma payload will permit the first fundamental advances in understanding the structure and dynamics of this fascinating environment.

The boundary with interstellar space, called the heliopause, is the final frontier of the solar system. Were a spacecraft to pass through it, we would enter the interstellar medium, a completely distinct environment from the solar system that has never been sampled in situ. Going there would provide the first ground truth measurements of what the interstellar medium really looks like, and the interplay between the various components of the interstellar medium (plasma, dust, magnetic fields and neutral atoms) and the solar system.

6. THE GIANT PLANETS AND THEIR ENVIRONMENTS

Giant planets, together with their rings, their geophysically diverse satellites, and complex environments of dust, gas and plasma, can be considered as "mini-systems" inside the solar system. Their study is a key step in building up a firm understanding of the formation of planetary systems. While at present it is only by in-situ exploration in the solar system that we can provide strong constraints on the formation scenarios of the giant planets and study them in detail as physical systems, in the next 20 years such investigations will also benefit from complementary studies of extra-solar planetary systems. Indeed, while giant planet systems play a key role in the evolution of planetary systems in general, their central bodies, the giant planets themselves, are connected to the "hot Jupiter" family of giant exo-planets.

The study of giant planet systems (i.e. the planets and their satellites) addresses many key scientific questions:

- How were they formed from the Solar Nebula? One wishes to test relevant formation scenarios such as disc instability versus core accretion.
- What is their internal structure, and, in particular, do they have a solid core, and of what size? This first pair of questions can be answered by carrying out deep atmospheric soundings through remote sensing and in-situ investigations, coupled with accurate measurements of the planetary gravitational and magnetic fields.
- What are the processes involved in the formation and evolution of the atmosphere of these planets and their moons such as Titan? Although we still await the results of the Huygens probe at this time, it is clear that these results will raise many new scientific questions of high priority. Again, one needs a combination of remote sensing and atmospheric probe measurements of the deep atmospheres.
- What is the internal and sub-surface structure of their satellites, especially the icy ones, what is geological history, and how does this reflect their formation? Here one needs to study the gravitational and magnetic fields, as well as the surface morphology, topology, mineralogy and composition.

- How are their complex plasma, gas and dust environments coupled to the central object, its satellites and rings, and the interplanetary medium? In addition to in-situ measurements of these quantities, one needs to relate these measurements to plasma injection from the solar wind, moons (e.g. Io) and the planet itself, to the role of planetary rotation, and of the consequences of any magnetospheric activity such as aurora.

The vastness of the range of topics needing to be studied requires a staggered approach with a series of missions a planet such as Jupiter. Measurements of many different physical quantities: atmosphere, gravitational and magnetic fields, plasmas, planetary surfaces are needed.

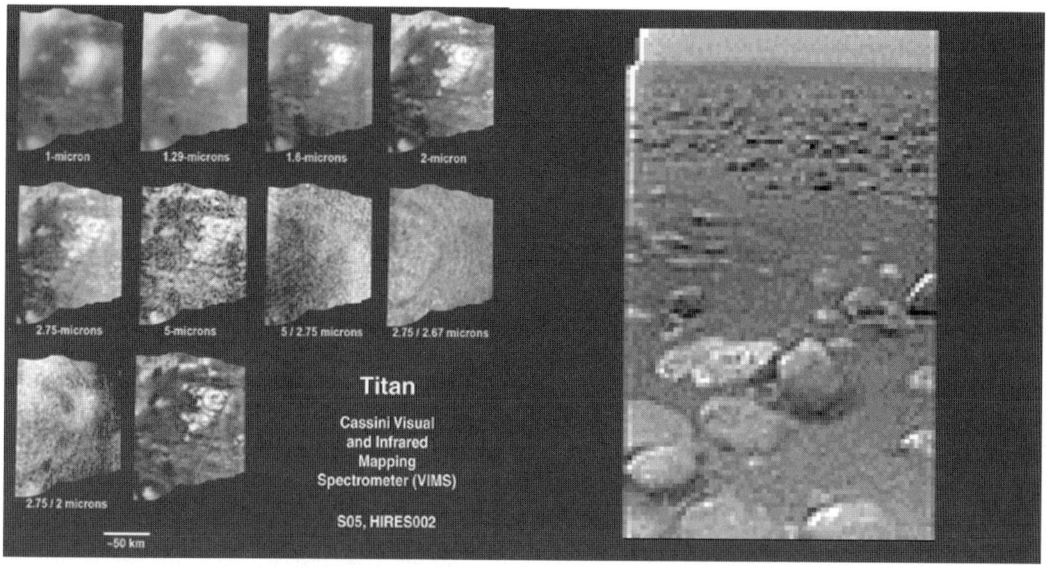

Figure 4. "Volcanic" plume observed on Titan by the Cassini VIMS spectrometer (left). High resolution image of the Titan's surface taken during Huygens probe touch down (right).

7. ASTEROIDS AND SMALL BODIES

As the primitive, leftover building blocks of planet formation, small bodies of the solar system offer clues to the chemical mixture from which the planets formed. Holding unique information on the initial conditions and early history of the solar nebula, their study is essential to understand the processes by which interstellar material becomes new planetary life-bearing systems.

The natural next step of exploration of small solar system bodies is a sample return mission of material from Near-Earth Asteroids (NEAs). NEAs are dynamically connected to the family of main belt asteroids, of which they can be considered to be extensions.

By choosing an object belonging to one of the most primitive classes of this family and by analysing samples taken in various, well determined, geological contexts, many key long-standing questions can be answered.

- What was the composition and the physical properties of the building blocks of the terrestrial planets?
- What are the processes occurring in the Solar nebula accompanying planetary formation?
- What is the nature and origin of the organic materials in primitive asteroids?
- Are there lessons for our understanding of the origin of life in the solar system?

- Do asteroids of primitive classes contain pre-solar material yet unknown in meteoritic samples?
- Do they contain chondrules, the main component of carbonaceous chondrite meteorite class, which formation process is strongly debated?
- How do the elemental, mineralogical and isotopic properties of the asteroids samples vary with geological context on the surface?
- How space weathering and impacts do affect asteroid surface composition?
- What is the timeline and duration of major events, such as agglomeration, heating and degassing, aqueous alteration?
- How did the various asteroids and meteorite classes form and acquire their present properties and how are asteroids and meteoritic classes related each other?

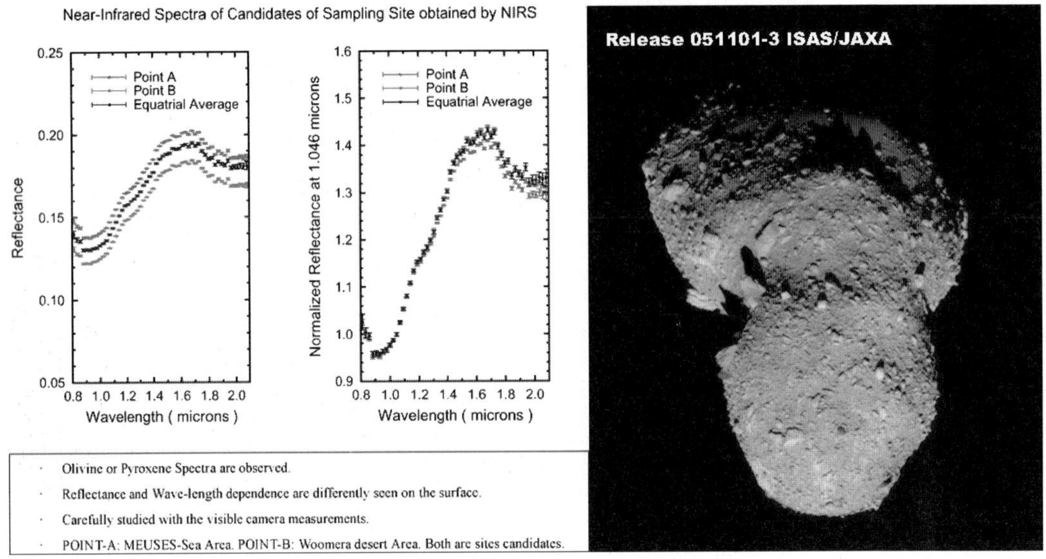

Figure 5. Image of asteroid Itokawa taken be the JAXA spacecraft Muses. On the left mineralogical composition of potential landing site, obtained with the on-board infrared spectrometer, is shown.

The return to Earth of an asteroid sample is by far the most efficient approach to address these questions as a full set of extensive and unique diagnostics can be applied. Combined with detailed imaging and spectroscopic investigations of the parent body, laboratory analysis of asteroid samples will improve the interpretation of all meteoritic data and provide a new understanding of all astronomical spectra of asteroids acquired so far.

Clearly, a full understanding of the asteroid and meteorite populations, histories and relationships requires sample return missions to asteroids belonging to various spectral classes. The JAXA MUSES-C/HAYABUSA mission towards asteroid 25143 Itokawa will be the first asteroid sample return mission which will unravel in 2007 the nature of differentiated S-type asteroid material. But only a sample return mission towards the most primitive objects will address the key questions regarding the origin of the solar system. Ultimately, exploration of bodies such as Kuiper belt objects, the likely building blocks of giant planets cores, is desirable, but is unlikely to be feasible in the next two decades.

8. THE INTERNATIONAL EXPLORATION INITIATIVE

The exploration of space has been one of humankind's greatest achievements in the past century. Only sixty-six years after the invention of flight, a man walked on the moon - a stunning feat of human ingenuity and endeavour. But in the last thirty-four years, this journey of exploration has lost its momentum. The human race remains stuck in near-earth orbit. The human exploration of remote worlds and the settlement of space remain distant dreams.

Today, there are a number of competing visions for the future of space exploration. Some of these are bold and expansive, envisioning a deep and long-term commitment to the discovery and settlement of space. Other visions are more constrained, focused on near-term and incremental progress. The decisions that the world's space-faring nations make about which type of vision to pursue will determine what the next era of space exploration looks like.

After the Columbia accident of February 2003, the White House National Security Council initiated an interagency review to examine priorities for the overall US civil space sector. The outcome of this review confirmed that the US should pursue human space flight in the long run, redirected NASA's priorities on exploration and away from the Shuttle (by 2010) and ISS (by 2016) and announced that the Crew Exploration Vehicle will succeed the Space Shuttle as the US human space vehicle. As such, NASA would not anymore be developing a crew rescue vehicle for the ISS and would thus be departing from the Orbital Space Plane (OSP).

Both the Shuttle and the ISS' standing in the US changed after this announcement, as they became "liabilities" for NASA instead of "assets" as they were until now.

Knowing that there would be no short-term possibilities to increase substantially the NASA funding for exploration due to an increasing public deficit as well as the fact that both the Shuttle and ISS would probably eat up at least 50% to 60% of the NASA annual budget for the next 6 to 7 years, the interagency review called for a "sustained and affordable" space exploration programme. Both words were used to call for a series of technology demonstrations, in particular for the Crew Exploration Vehicle, while NASA would continue to assemble and operate the ISS. Once freed of the ISS commitments, in the middle of the next decade, then NASA would eventually step up dramatically its funding allocations to exploration to perform the first human flight of the CEV as well as to engage into a new generation of space exploration technology development.

At the moment, NASA is refocusing all of its scientific priorities around the space exploration vision. NASA is becoming an "exploration agency", which conducts science to sustain a human space flight effort to the Moon and Mars. Until now, NASA was conducting human space flight to enlarge its scientific agenda. This difference establishes the parameters within which NASA is now revamping its scientific approach.

The recently released policy directive, "A Renewed Spirit of Discovery: The President's Vision for U. S. Space Exploration," seeks to advance the U. S. scientific, security and economic interest through a program of space exploration which will robotically explore the solar system and extend human presence to the Moon, Mars and beyond. NASA's implementation of this vision will be guided by compelling questions of scientific and societal importance, including the origin of our Solar System and the search for life beyond Earth. The Exploration Roadmap identifies four key targets: the Moon, Mars, the outer Solar System, and extra-solar planets. First, a lunar investigation will set up exploration test beds, search for resources, and study the geological record of the early Solar System. Human missions to the Moon will serve as precursors for human missions to Mars and other destinations, but will also be driven by their support for furthering science. The second key target is the search for past and present water and life on Mars. Following on from discoveries by Spirit and Opportunity, by the end of the decade there will have been an additional rover, lander and orbiter studying Mars. These will set the stage for a sample return mission and increasingly complex robotic investigations in the next decade, and an eventual human landing. The third key target is the study of underground oceans, biological chemistry, and their potential for life in the outer Solar System. The National Aeronautics and Space Administration (NASA) has begun to develop a hierarchy of performance requirements (essentially, objectives) for its proposed exploration mission to define it more fully than the current budget projection does. Nevertheless, that level of technical detail is not sufficient to allow the Congressional Budget

Office to perform an independent cost estimate. NASA's initial, or Level 0, exploration requirements include the following:

- Implement a safe, sustained, and affordable robotic and human program to explore the solar system and beyond and to extend the human presence across space;
- Acquire a transportation system for space exploration to convey crews and cargo from the Earth's surface to exploration destinations and return them safely;
- Finish assembling the International Space Station–by the end of the decade, according to NASA's plans–including the U.S. components that support the President's space exploration goals and the components that are being provided by foreign partners;
- Pursue opportunities for international participation to support U.S. space exploration goals; and
- Seek commercial arrangements for providing transportation and other services to support the International Space Station and exploration missions beyond low-Earth orbit.

In a position paper issued by the National Space Society (NSS), a return to the Moon should be considered a high space program priority, in order to begin development of the knowledge and identification of the industries unique to the Moon. The Moon is a repository of the history and possible future of our planet, and the six Apollo landings only scratched the surface of that treasure. According to NSS, the Moon's far side, permanently shielded from the noisy Earth, is an ideal site for future radio astronomy. Unique products may be producible in the nearly limitless extreme vacuum of the lunar surface, and the Moon's remoteness is the ultimate isolation for biologically hazardous experiments. Initial return missions as recently proposed by the President and NASA can be done through space operations using the existing launch infrastructure and assets developed by the shuttle and International Space Station programs, plus existing expendable launch vehicles, with a minimum of new research and development programs. The lessons learned from international cooperation during ISS construction and operations can be improved upon and extended to human missions to the Moon, Mars and elsewhere. Initial missions could place scientific equipment on the Moon and return samples from areas never explored, such as the polar regions. Extent of water and other volatiles important to lunar industrialization could be determined. As future reusable launch systems begin operations, reducing cost and enabling higher flight rates, Earth-Moon traffic can become routine. With humans on the Moon again, NASA's space activities would take on new vigor and public interest.

The European response to the US exploration initiative is the Aurora programme.

The objective of the Aurora Programme is first to formulate and then to implement a European long-term plan for the robotic and human exploration of solar system bodies holding promise for traces of life. The Programme will also provide for the missions and technology necessary to complement those planned in the existing ESA and national programmes, in order to bring about a coherent European framework for exploration and to progressively develop a unified European approach. From the dawn of humankind the need to explore has driven expansion across our planet. Today this expansion continues towards other planets in the solar system by means of robotic spacecraft - virtual explorers. But will human expansion continue? In the public consciousness this is only a matter of time. By 2025 an international human mission to Mars may be a reality. It may use the Moon as a way station and to prepare for the great leap. The feasibility of such a mission is being assessed, however, the necessary technology and capability still need to be developed. Having reached maturity in human spaceflight, thanks to its activities in the ISS framework, Europe will have to decide whether to play a key role in the next step or join later as a junior partner. Given the timespan of such a human mission, Europe also faces the issue of how to exploit the industrial know-how developed in the ISS framework and to orient it toward the new mission. Which areas of expertise Europe wants to lead in the future has to be decided soon; this cannot be left to our future partners. Over the next 20 years robotic missions will prepare for human missions, by collecting as much scientific and engineering data as possible, without human scientists *in situ*. These robotic missions will contribute and demonstrate the technologies needed to put humans on Mars and return them safely to our planet. Some of the key technologies for a human mission

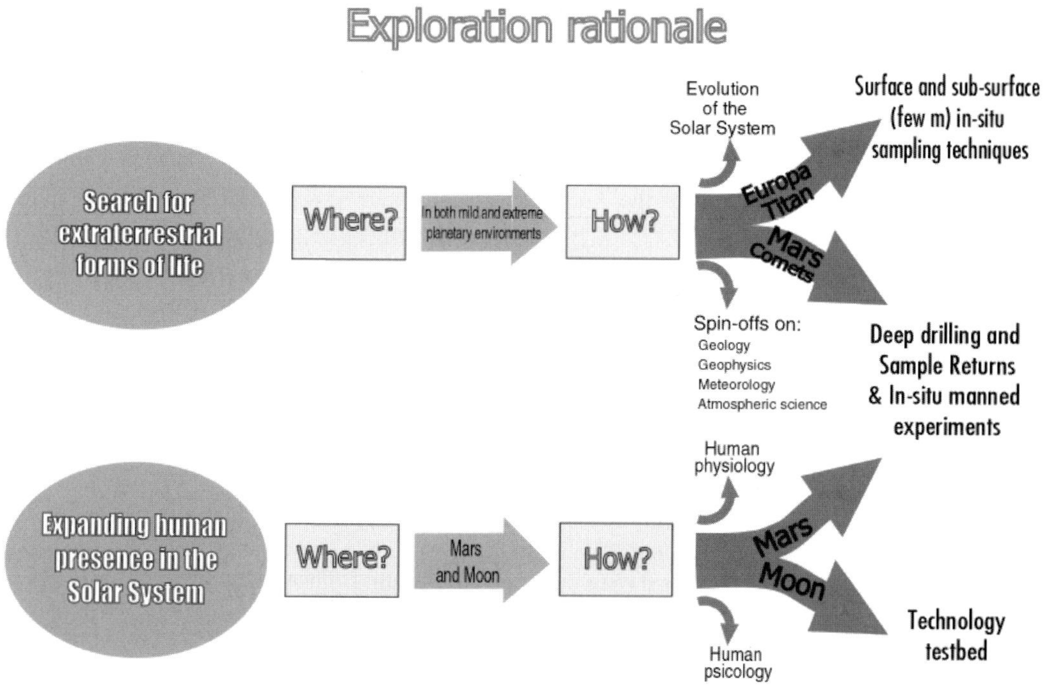

Figure 6. The European exploration rationale.

Figure 7. The ExoMars rover will be ESA's field biologist on Mars.Its aim is to further characterise the biological environment on Mars in preparation for robotic missions and then human exploration.

are also very important to the search for life *in situ* on the red planet and on other solar systems, planets and moons.

These missions will carry sophisticated exobiology payloads and provide answers to some key questions on the origin of life in the solar system and possible causes for its extinction. These precursor missions will also greatly advance our technology capacity making Aurora a genuine programme for innovation. Spin-offs are expected in sensor technology; information technology, in particular spacecraft autonomy (signal return times from Mars); biochemical technology (searching for life means

understanding what life is on our planet and what different forms it may take, how it can be identified, not contaminated and viceversa); navigation and communication technology (precison landing and large volume of data transmission); propulsion; power generation, conversion, transmission, conditioning and storage; thermal control; extreme temperature and radiation hardened electronics; *in situ* resource utilisation, aerothermodynamics; etc.
By its very nature the Programme is, therefore, multidisciplinary across many sectors of science, technology and space activities. Thus the Aurora Programme can be seen as a road map for human exploration, from which a large number of scientific as well technology spin-offs will emerge, driven by the goal of exploration.

ExoMars is the first Aurora Flagship mission to be assessed. Its aim is to further characterise the biological environment on Mars in preparation for robotic missions and then human exploration. Data from the mission will also provide invaluable input for broader studies of exobiology - the search for life on other planets. This mission calls for the development of a Mars orbiter, a descent module and a Mars rover. The Mars orbiter will have to be capable of reaching Mars and putting itself into orbit around the planet. On board will be a Mars rover within a descent module. After their release and landing on the surface of Mars, the orbiter will transfer itself into a more suitable orbit where it will be able to operate as a data relay satellite. Initially it will act as a data relay for the ExoMars rover but its life may be extended to serve future missions. The Mars descent module will deliver the rover to a specific location by using an inflatable braking device or parachute system. Both systems are sufficiently robust to survive the stresses of atmospheric entry and their landing accuracy will be sufficient for this mission. Using conventional solar arrays to generate electricity, the Rover will be able to travel a few kilometres over the rocky orange-red surface of Mars. The vehicle will be capable of operating autonomously by using onboard software and will navigate by using optical sensors. Included in its approximately 40 kg exobiology payload will be a lightweight drilling system, a sampling and handling device, and a set of scientific instruments to search for signs of past or present life.

In order to be successful ExoMars will require advanced technology in the following areas:

- rover systems
- landing systems
- an inflatable braking device
- power supply
- autonomy and navigation

Although this presents a considerable technological challenge for European and Canadian industry, it will bring to fruition many years of technological development both at ESA and national level.

Acknowledgements

I am grateful to Giovanni Bignami, who, as chairman of the ESA space science advisory committee, was the real driving force of the Cosmic Vision planning exercise. I also want to express my gratitude to Frederic Nordlund, head of the ESA Washington office, for his stimulating analysis of the US exploration initiative.

References

A Renewed Spirit of Discovery: The President's Vision for U. S. Space Exploration. US Congress Library.
Cosmic Vision: Space Science for Europe 2015-2025. ESA BR-247, 2006.
A collection of papers on **Titan and the Huygens probe** can be found in Nature 438, 8 December 2005.
A collection of very recent papers on **minor bodies** can be found in Asteroids, Comets and Meteors, proceeding of the IAU Symposium 229, Cmbridge University Press, 2006.

References on the origin of life

Bada, J. L. 1995. Origins of homochirality. *Nature* 374 (6523):5945.
Baltscheffsky, H., C. Blomberg, H. Liljenstrom, B. I. Lindahl, and P. Arhem. 1997. On the origin and evolution of life: an introduction. *Journal of Theoretical Biology* 187 (4):4539.
Bernal, John Desmond. 1967. *The origin of life*. London: Weidenfeld & Nicolson.
Conrad, M. 1997. Origin of life and the underlying physics of the universe. *Biosystems* 42 (2-3):17790.
De Duve, Christian. 1995. *Vital dust: life as a cosmic imperative*. New York: Basic Books.
Florkin, Marcel, ed. 1960. *Aspects of the origin of life*. Oxford, New York, Pergamon Press.
Fox, Sidney W. 1972. *Molecular evolution and the origin of life*. San Francisco: Freeman.
Horgan, J. 1996. The world according to RNA. Experiments lend support to the leading theory of life's origin. *Scientific American* 274 (1):2730.
Keefe, A. D., S. L. Miller, G. McDonald, and J. Bada. 1995. Investigation of the prebiotic synthesis of amino acids and RNA bases from CO2 using FeS/H2S as a reducing agent. *Proceedings of the National Academy of Sciences of the United States of America* 92 (25):119046.
Lahav, Noam. 1999. *Biogenesis: theories of life's origin*. New York: Oxford University Press.
Westall F.: *Life on the Early Earth: A Sedimentary View,* Science 15 April 2005: Vol. 308. no. 5720, pp. 366–367.

References on the origin of Solar System

Shirley J.H. & Rhodes W. Fairbridge Editors: *Encyclopedia of Planetary Sciences,*Chapman and Hall, 1998.
Johnson T.V., McFadden L.A. & Weissman P: The Solar System, Acdemic press, 2000.
C.M.O'D. Alexander, A.P. Boss and R.W. Carlson: *The early evolution of the inner solar system: A meteoritic perspective*, Science 293, 64-68, 2001.
Carlson R.W. and E.H. Hauri: *Extending the 107 Pd- 107 Ag Chronometer to Low Pd/Ag Meteorites with the MC-ICPMS* , Geochim. Cosmochim. Acta 65, 1839-1848, 2001.
Carlson R.W. and G.W. Lugmair: *Timescales for planetesimal formation and differentiation based on extinct and extant radioisotopes,* in Origin of the Earth and Moon, ed. K. Righter and R. Canup, University of Arizona Press, Tucson, pp. 25-44, 2000.
Wadhwa M., and G. Srinivasan: *Time scales of planetesimal differentiation in the early solar system* Meteorites and the Early Solar System II, Sept., 2004.

J. Phys. IV France **139** (2006) 37–61
© EDP Sciences, Les Ulis
DOI: 10.1051/jp4:2006139005

Water in the Earth's atmosphere

M. Quante[1] and V. Matthias[1]

[1] *GKSS Research Center, Institute for Coastal Research, 21502 Geesthacht, Germany*
e-mail: markus.quante@gkss.de

Abstract. Water is the key to our existence on this planet and it is involved in nearly all biological, geological, and chemical processes. Life on Earth depends very much on the remarkable properties of water. The availability of freshwater is for many regions one of the key concerns in connection with global climate change. The atmosphere contains only about 0.001% of the water available on our planet. Despite this small amount its horizontal and vertical distribution plays a key role in the global water cycle and the Earth's climate. The atmosphere has direct connections to most of the other reservoirs and steers the redistribution of water between them with an average turnover time of about 10 days. Evaporation over the oceans exceeds precipitation and over land evapotranspiration amounts only to 2/3 of the precipitation reaching the ground. Consequently, there is a net flux of water from the oceans towards the continents, of course via the atmosphere, which has the largest overall volume of fluxes. Water is present in the atmosphere as solid, liquid, or gas. Water vapour is the most important greenhouse gas in the atmosphere and, in addition, changes of water phase and cloud-radiation interaction contribute strongly to the global energy cycle. Water is also a physically and chemically integral part of other biogeochemical cycles. Although there have been large efforts and improvements in recent years, uncertainties in quantifying the components of the atmospheric water cycle still exist. Observational capabilities on the global scale are not satisfactory at present, but the advent of new satellites devoted to the global observation of precipitation and cloud systems along with dedicated modelling projects certainly will improve the situation. Progress is urgently needed to adequately contribute to the answer of one of the central questions in the context of global warming: Is the hydrological cycle accelerating?

1. INTRODUCTION

Water is an extraordinary substance, due to its in many respects unusual behaviour it is irreplaceable in a multitude of natural and engineering processes. Water enters into the operation of the climate system in a remarkable variety of ways [1]. Although the atmosphere contains only 0.001% of the available water on our planet, this compartment of the global water cycle has a particular significance. The atmosphere is in direct contact to all other reservoirs of the water cycle, with the exception of the ground water. Therefore, the atmosphere plays a central role in the redistribution of water.

Driven by solar radiation, that for a large extent is used for evaporation, there is a perpetual exchange of water between the oceans, the atmosphere and the land surface. About 90% of the water in the atmosphere emanates from the oceans, lakes and other open waterbodies. Via atmospheric transport and relevant transfer processes a part of the water which evaporated over the oceans reaches the land areas, where it may precipitate and support the existence of life. In addition, weather and climate of a region is significantly influenced by water vapour, clouds and precipitation. Water vapour absorbs and emits strongly in the infrared part of the electromagnetic spectrum. It is the most important greenhouse gas in the atmosphere [2]. Clouds, which backscatter radiation in the visible spectral range and absorb in the infrared, cover large areas of the sky and thereby modify, in a dominant way, the radiation balance and thus climate [3-7]. Water is also important because it occurs in its three phases in the atmosphere, ice, liquid and gas. During phase changes considerable amounts of energy are bounded or released, and this plays an important role in the local energy budget, thus influencing atmospheric dynamics. The related latent heat flux amounts to about 80 W/m^2 on global average for a mean insulation of 342 W/m^2 [8].

Water in the atmosphere is also in physical (clouds, precipitation) as in chemical respect integral part of biogeochemical cycles, which themselves are important components of the climate system [9-11]. Tropospheric chemistry is largely influenced by clouds [12-16]. By the processes of rainout and washout of pollutants (gases and aerosols) the water cycle contributes crucially to the self-cleansing of the atmosphere. Back on earth precipitated water is a powerful geologic force shaping and altering the landscape by weathering [17] and erosion [18,19], partly because of its acidity and partly because water expansion during freezing.

In the stratosphere water in its solid phase as constituent of polar stratospheric clouds (PSCs) plays a critical role in heterogeneous chlorine chemistry and, thus, in Antarctic/Arctic ozone depletion [20,21].

Knowledge of the atmospheric aspects of the water cycle is also a fundamental basis for hydrological research in support of the operational water management community and policymakers, which deal with the availability of drinking water on one side and flooding risks on the other [22-25].

To provide profound knowledge on processes controlling the atmospheric water cycle is a great challenge for atmospheric sciences. As major components of the atmospheric branch of the hydrological cycle, in addition to atmospheric storage, the water fluxes connected to precipitation and evaporation are to be determined to an acceptable accuracy. The change of global or regional climate, as anticipated nowadays, is certainly accompanied by a change in components of the water cycle. Due to the lack of reliable data for oceanic precipitation and surface evaporation our quantitative knowledge of atmospheric water storage and water fluxes and their temporal behaviour is still fairly limited. Progress is definitely needed, since one of the central questions in the context of future global warming is: to what extent will the hydrological cycle accelerate? In this overview the main topics around water in the Earth's atmosphere are covered, several references are included, which could serve as a starting point for more detailed assessments.

2. WATER

Water is the only chemical compound on this planet that occurs naturally in all three physical states: solid, liquid and vapour. It existed long before any form of life evolved, but, since life developed, its properties came to exert a controlling influence over many biochemical and physiological processes that make our life possible [26]. It is because of this close connection to life that many missions to other planets concentrate on the detection of water. Here, we start with a few general remarks on water, since in atmospheric sciences, it is not widely perceived, that there are still basic questions to be answered concerning the physical details of one of its most important substances. The water molecule as well as local molecular structures have been, and still are, the subject of intensive theoretical and experimental study.

Water exhibits an impressive array of anomalies in its physical properties, among these are the well-known expansion when it freezes at normal pressure and the presence of a liquid-phase density maximum at +4°C. Compared to "similar" substances water has a high melting point, high boiling point and a high critical point. Other features are the high specific heat capacity and surface tension of liquid water. Additionally, the temperature dependence of its compressibility, heat capacity and thermal expansion coefficient show an unusual progression. All these anomalies have important consequences which make up for the special role water plays in the atmosphere. E.g. evaporation, latent heat fluxes, saturation pressures, cloud droplet and precipitation formation, as well as supercooling would all look much different without these peculiar properties.

Water seems, at first sight, to be a very simple V-shaped molecule, consisting of just two (light) hydrogen atoms attached to an (relatively heavy) oxygen atom by strong covalent bonds. There are very few molecules that are smaller or lighter. The approximately 16-fold difference in mass gives rise to its ease of rotation and the significant relative movements of the hydrogen nuclei. The principal vibrations and rotations are at frequencies in the terrestrial radiation spectrum, they make water vapour to be the most important greenhouse gas on Earth. These frequencies are modified in the condensed states, where intermolecular effects become important. Ultimately, the detailed interaction of a water molecule with its

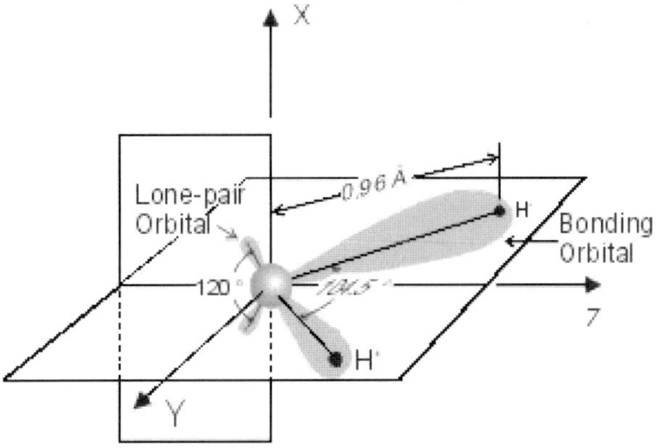

Figure 1. Geometry of a free H_2O molecule in equilibrium.

neighbours, also called "water structure", is of key importance for understanding the unique properties of aqueous systems on the macroscopic scale.

An isolated H_2O molecule is sketched in figure 1, the equilibrium geometry shows an O-H bond length of 0.957 Å, and the H-O-H angle is 104.52°. It is of crucial importance for the properties of water and ice that the molecule is bent. The bent form results in a dipole moment and determines the ways the molecules fit together. The water molecule contains ten electrons, two from the hydrogen atom. Because of the presence of the hydrogen nuclei the electronic charge is not distributed symmetrically around the oxygen nucleus, it is drawn towards the hydrogen nuclei with the consequence of a dipole moment.

The polarity of each water molecule results in a weak attraction between it and other water molecules. Linus Pauling [27] called the favourable attraction between two water molecules a hydrogen bond. Each water molecule can simultaneously participate in four such bonds, sharing its two hydrogen atoms with two neighbouring water molecules and sharing two other hydrogen atoms associated with two other neighbours. Only 5% of the total energy is in the hydrogen bond, which still is about 17 times stronger than typical bonds between molecules in liquid phase. Ice is a tetrahedral ordered array of such hydrogen bonded water molecules, resulting in a network of tetrahedrons. Liquid water is a disordered network of such bonded waters. Considerable revision has occurred to details of how the liquid is geometrically organized by hydrogen bonding, and the earlier views that water should be seen as modified versions of ice had to be modified. Although the hydrogen atoms are often shown along lines connecting the oxygen atoms, this is now thought unlikely even in ice, with non-linearity, distances and variance; all increasing with temperature [28]. In liquid water, the lifetime of a hydrogen bond is in the femtosecond to picosecond order, and it is broken in the course of translational and rotational motions [29]. It has proven difficult to transform this insight into a quantitatively accurate molecular theory of liquid water. A detailed description of the hydrogen bond network in liquid water is the key to understanding its unusual properties.

The present consensus seems to be that liquid water is a macroscopic network of molecules connected by frequent but transient hydrogen bonds, which allow unbonded neighbours to occur in numbers that vary with temperature and pressure. Anomalous properties of water arise from the competition between relatively bulky ways of connecting molecules into local patterns characterized by strong bonds and nearly tetrahedral angles and more compact arrangements characterized by more strain and bond breakage.

The most recent discussion on the molecular structure of liquid water is about, whether it consists mainly of structures with two strong hydrogen bonds (with the possibility that water molecules form chains or closed rings), in contrast to the commonly accepted semi-tetrahedral structure close to the four bonds found in the tetrahedral structure of ice [30-33]. These details are important for a better understanding of the exceptional properties of liquid water, and at the end for a clearer answer to why water is essential for life including the many processes determining climate.

As already mentioned, in ice H_2O molecules arrange in a tetrahedral hydrogen bonding with little distortion. Currently there are 15 different solid phases of water known, the most common one under atmospheric pressure and temperature conditions is the Ih phase, the normal hexagonal crystalline ice. So, virtually all ice in the Earth's atmosphere is ice Ih with the exception of only a small amount of ice Ic (cubic ice, formed at low temperatures, 130 –150 K, and stable to 200 K). Under high pressure other phases of ice exist, the tetrahedral hydrogen-bonding is still the preference, but some bending and deviations from linearity occur, since a higher packing density is sought for [34]. Ice in the atmosphere under real world conditions appears in many different crystal shapes [35], many of them with different types of impurities [36].

There are several general references on liquid water [26, 34, 37-43], while only a few more general texts on ice can be recommended [44-46]. An excellent and well kept source of information on liquid water is the internet site "www.lsbu.ac.uk/water/index.html" compiled by M. F. Chaplin from London South Bank University. The books by Ball [47] and Kandel [48] should also be mentioned, which set water into a broader perspective, they tell the story of water from the big bang through the rise of civilization to present.

The phase water occurs in depends on the temperature and pressure it is exposed to. At the normal range of atmospheric pressures and temperatures on Earth water can exist in all its three basic states, as is evident from its phase diagram in Figure 2, which shows the phase transition curves as function of temperature and partial pressure (see [49]). The Earth trajectory in figure 2 driven by an increasing water vapour greenhouse effect intercepts the water phase curves in the vicinity of the triple point of water (273.16 K), allowing the formation of a complex hydrological cycle. The curve for Venus, in contrast, because of a start with a considerably warmer primitive surface temperature does not intercept any of the water phase transitions at all, water stays as a gas. The state curve for Mars starts at a relatively low temperature (~240 K; not shown here) and rapidly intercepts the vapour-ice phase transition. Martian water occurs in relatively small concentrations but can exist either as gas or as ice. In the upcoming sections the water on Earth will be the subject.

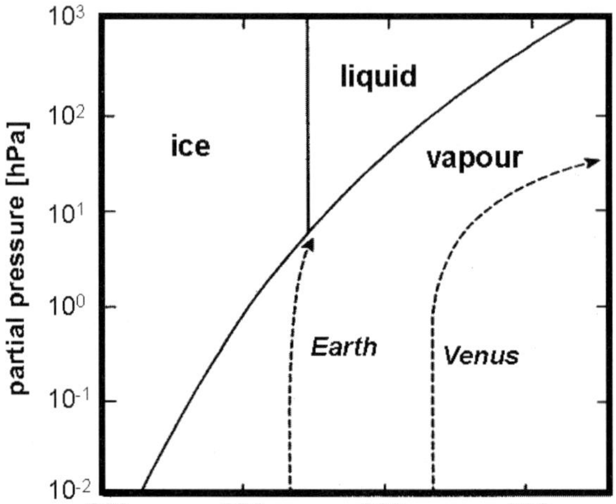

Figure 2. Phase diagram of water illustrating the possible occurrence of the three states of water for the range of temperatures observed at the Earth's surface and in the lower atmosphere (~190K to 325 K and 0 to 50 hPa partial pressure range at the surface).

3. THE GLOBAL HYDROLOGICAL CYCLE

Water is present in all compartments of the Earth's system which determine the climate of our planet. Appoximately 71% of the Earth's surface is covered with salty water in the oceans. Some parts of the continents are covered by fresh water (in lakes and rivers) and solid water (as ice or snow). Some water is existent in vegetation and in the atmosphere. Several studies deal with a quantitative description of the available water and its movement (e.g. [50-52]), a comprehensive review of freshwater resources can be found in [53-55]. Table 1 provides estimates of the water distribution, beside the volume and its share of the total water also the share of fresh water is given, which for obvious reasons is of special interest for mankind. It can be seen that only less than 3% of the Earth's available fresh water occurs in lakes and streams. So the exchange between the reservoirs is of vital importance for life.

Table 1. Distribution of Water on Earth (from [50]).

Form of water	*Area covered* ($1000\ km^2$)	*Volume* ($1000\ km^3$)	Share of world reserves (%) *of total water reserves*	Share of world reserves (%) *of reserves of fresh water*
World ocean	361,300	1,338,000	96.5	-
Total groundwater[(1)]	134,800	23,400	1.7	-
Fresh ground water	134,800	10,530	0.76	30.1
Soil moisture	82,000	16.5	0.001	0.05
Glaciers and permanent snow cover	16,000	24,000	1.74	68.7
Antarctica	14,000	22,000	1.56	61.7
Greenland	1,800	2,300	0.17	6.68
Arctic islands	230	83.5	0.006	0.24
Mountainous areas	220	40.6	0.003	0.12
Ground ice in zones of permafrost strata	21,000	300	0.022	0.86
Water reserves in				
Lakes	2,000	180	0.013	-
Fresh water	1,240	91	0.007	0.26
Salt water	820	85.4	0.006	-
Marsh water	2,700	11.47	0.0008	0.03
Water in rivers	148,800	2.12	0.0002	0.006
Biological water	510,000	1.12	0.0001	0.003
Atmospheric water	510,000	12.9	0.001	0.04
Total water reserves[(2)]	**510,000**	**1,390,000**	**100**	-
Fresh water[(2)]	148,800	35,000	2.35	100

(1) Not including ground-water reserves in Antarctica.
(2) Deviations are due to rounding.

The hydrological cycle is the perpetual movement of water throughout the various components of the Earth's climate system. Although it is a continuum, its description usually begins with the evaporation of water from the ocean driven mainly by solar radiation. Under the influence of certain changes in temperature and/or pressure in the atmosphere, the moisture might condense and eventually return to the Earth in form of rain, hail, sleet or snow. Some of the water that precipitates over land does not evaporate again and enters lakes or rivers or infiltrates into the soil and contributes to the ground water. The water is brought back to the oceans mainly by rivers and ground water flow. It are these streams in addition to the precipitation over the oceans, which close the water cycle. A sketch of the various fluxes within the water cycle is presented in figure 3. It has to be mentioned that the hydrological cycle is characterized by its variability in space and time, the time scales among the different reservoirs vary considerably, the atmosphere having the fastest overturning times (see below). Quite a large amount of water is stored for long times in the antartic and arctic ice sheets.

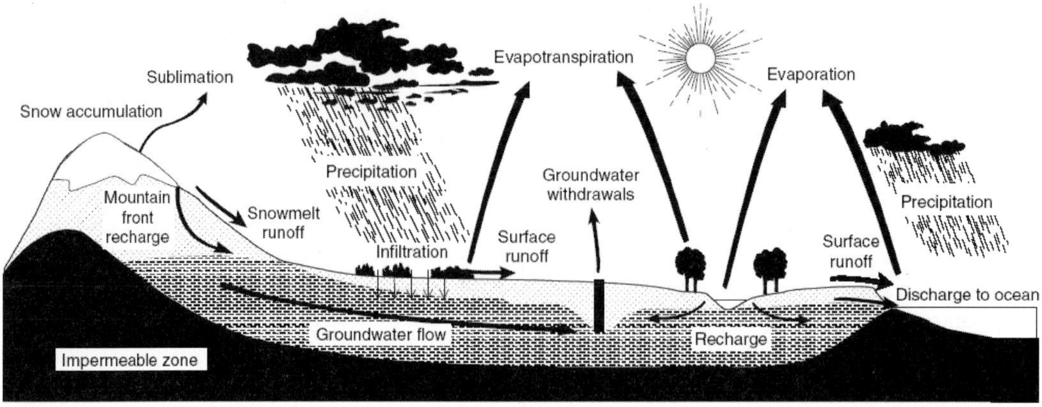

Figure 3. Schematic diagram of the various fluxes within the hydrological cycle (figure provided by T.C. Pagano).

The water cycle is intricately intertwined with many other global and regional environmental cycles [10], most prominently the global energy cycle ([56, 57]. Additional information on the overall global water cycle can be found in e.g. [1, 49, 52, 58, 59].

4. THE WATER BUDGET OF THE ATMOSPHERE

The atmospheric water budget can be described by the equation (e.g. [51])

$$\partial W/\partial t + \partial W_c/\partial t = -\mathrm{div}_h \mathbf{Q} - \mathrm{div}_h \mathbf{Q_c} + (E - P) \quad (1)$$

here:

- W is the water vapour content of a vertical column, which extends from the ground to the top of the atmosphere (this quantity is known as precipitable water).
- W_c respectively denotes the column storage of liquid water and ice.
- $\mathbf{Q}$ is the vertically integrated two-dimensional water vapour flux.
- $\mathbf{Q}_c$ is the vertically integrated two-dimensional water flux in the liquid and solid phases.
- E denotes evaporation and
- P is the precipitation.
- h subscript stands for horizontal

Generally, the water content in the liquid and solid phases and the related fluxes in the atmosphere are small ($\partial W_c/\partial t << \partial W/\partial t$ and $\mathbf{Q_c}, << \mathbf{Q}$), often the water budget is approximated by the following simplified equation

$$\partial W/\partial t = -\mathrm{div}_h \mathbf{Q} + (E - P), \quad (2)$$

which is schematically illustrated in figure 4. The excess of evaporated water compared to precipitation is balanced by the local rate of water vapour storage and by the horizontal transport of water vapour into and out of the considered column. By spatial averaging the atmospheric water budget can be derived for selected regions (e.g. for the purpose of water budget studies for river catchments).

The first overview requires the consideration of global averages of the water budget components which, for this purpose, should be divided into those representing the oceans and those for the continental areas. The scheme in figure 5 provides the parts of the global water fluxes through the marine and terrestrial atmosphere as well as the respective water storage. The numbers given in figure 5

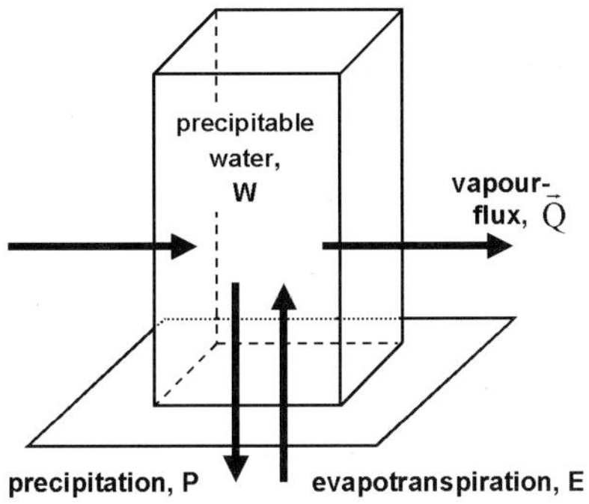

Figure 4. Schematic of the terms of a simplified water balance in an atmospheric column.

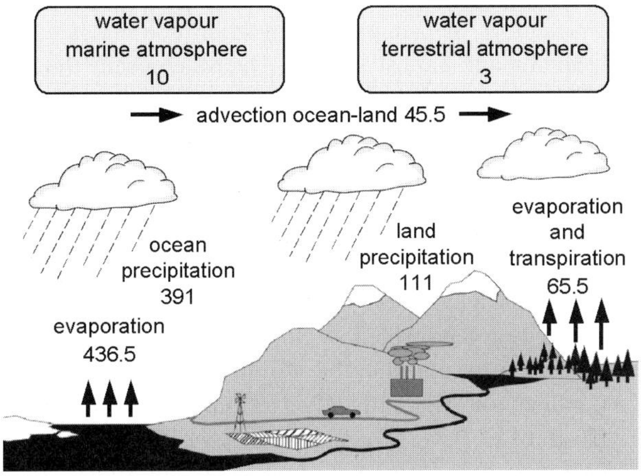

Figure 5. Schematic of the atmospheric branch of the water cycle. Storage (in boxes) is given in 10^{15} kg and fluxes in 10^{15} kg/a. Considering a liquid water density of 10^3 kg m^{-3}: 10^{15} kg correspond to 10^3 km^3 (data from [55]).

Table 2. Global hydrological fluxes in the atmosphere.

	Chahine [58] [10^{15} kg/a]	Shiklomanov [61] [10^{15} kg/a]	Oki [52] [10^{15} kg/a]	Oki and Kanae [55] [10^{15} kg/a]	Trenberth et al. [62] [10^{15} kg/a]
Evaporation over oceans	434	503	431	436.5	433
Precipitation over oceans	398	458	391	391	399
Evapotranspiration over land	71	74	75	65.5	69
Precipitation over land	107	119	115	111	103
Net flux ocean → land	36	45	40	45.5	34

are those published by Oki and Kanae [55], within the expected error margins they are roughly comparable to those in [52, 58, 60-62]; Table 2 compares hydrological fluxes in the atmosphere as published by different authors.

The given equation (2) can only lead to reliable results if input data is of adequate quality [52]. Large errors are to be expected in the determination of $\mathrm{div_h}\, \mathbf{Q}$. By combining all data sources, it is now possible to approximate the global water cycle on an annual mean. Quite large errors are still to be expected if water budgets for smaller spatial scales (regional and smaller) are to be obtained or if the averaging time is less than a year [63].

Although the given values are still not totally certain, it can be deduced from figure 5, that, on average, over the oceans evaporation considerably exceeds precipitation. The opposite is true over the land areas, where precipitation amount is much larger than evaporation. Therefore, on average, there is a net transport of water from the oceans towards the land. Based on Oki's data [52], this amounts to about 9.3% of the water which evaporates over the oceans. In other words, almost 35% of the precipitation reaching the land surfaces was originally evaporated over the oceans and subsequently transported by the large scale wind systems towards the continents.

The mean residence time of water in a reservoir can be estimated from the ratio of the mass in the reservoir to the flux of water out of the reservoir. For the atmosphere this estimation leads to a value of about 10 days as a mean residence time. In other words, the atmosphere exchanges its water about 36 times per year and therefore it is the reservoir of the water cycle with the largest rates of exchange.

The fluxes in figure 5 are given in units of km^3/year. Weighting those values with the appropriate areas allows the presentation of mean annual column heights of precipitation and evaporation, respectively. The global average of the annual precipitation column height as well as that of the evaporation column height amounts to almost 1 meter. Over the oceans the mean annual precipitation according to Oki's data [52] sums up to about 1157 mm and the mean annual evaporation to about 1275 mm. The respective heights for the land areas are 669 mm for precipitation and 436 mm for evaporation. Thus the mean evaporation height is roughly three times as high over the oceans compared to the land areas. The evaporation ratio E/P for land amounts to 0.65, which means that on average only about 2/3 of the precipitation reaching the land surfaces evaporates there again. Roads [63] presents precipitation and evaporation data including some error estimates for nine representative climate regions distributed over the globe. Estimates of average annual precipitation and evaporation for the different continents based on data tabled in Peixoto and Oort [51] are provided by Pagano and Sorooshian [59]. As can be expected, considerable differences exist on the continental scale, which range from extremely low values for Antarctica (P: 169 mm, E: 28 mm) to the highest values found for South America (P: 1564 mm, E: 946). The land-atmosphere interaction leads to high variability and seasonality in the governing processes. As an example for the mid-latitudes of the northern hemisphere the German Weather Service reports for Germany an average annual precipitation column height of 779 mm for the time period from 1961 to 1990 and a respective evaporation column height of 481 mm.

5. WATER VAPOUR

Water vapour, which accounts only for roughly 0.25% of the mass of the atmosphere, is a highly variable constituent in space and time. The inhomogeneous water vapour distribution is pronounced along the vertical coordinate, its concentration decreases drastically with the height above the surface. But also near the ground the concentrations vary by more than three orders of magnitude from 10 parts per million by volume in the coldest regions of the Earth's atmosphere up to as much as 5% by volume in the warmest regions. The latter value is only reached in very hot and humid air masses in the tropics. The tropical atmosphere contains more than three times as much water in comparison to the extratropical atmosphere. Expressed as specific humidity (mass of water vapour in g per 1 kg of humid air), the values near the ground vary between 18 to 19 g/kg in the tropics and 1 g/kg in the polar regions. The large scale distribution pattern of water vapour principally follows that of the temperature. Since the equilibrium vapour pressure strongly increases with temperature (Clausius-Clapeyron-equation), warm air masses can contain many more water molecules compared to colder ones before saturation (equilibrium vapour pressure) is reached. The region with highest humidity on Earth is therefore located over the Western Equatorial Pacific, the area with the highest observed sea surface temperatures. But

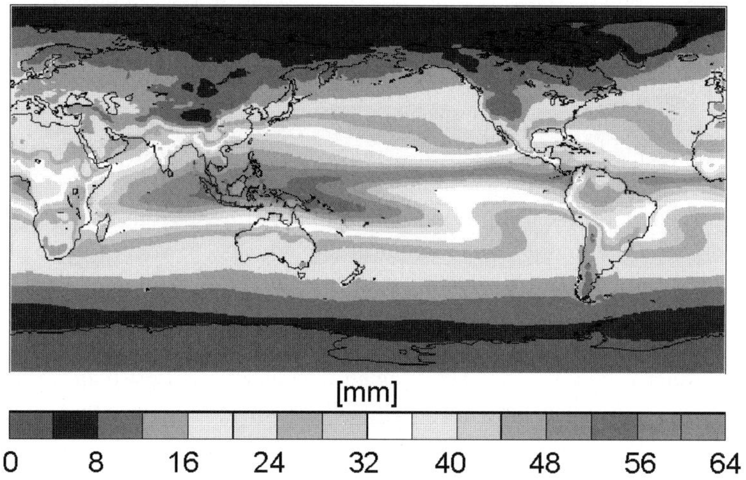

Figure 6. The global distribution of total atmospheric water vapour above the Earth's surface (precipitable water). The given values are averages for the period 1988-1997 and include data from both satellite and radiosonde observations (after [64]).

there are also exceptions to the temperature related distribution of water vapour. Over the larger scale deserts the water vapour concentration in air is extremely low despite high temperatures, mainly due to large scale sinking motions over these parts of the continents.

If the total water vapour content of the atmosphere would condense, precipitate and stay homogeneously distributed at the surface, a column with a height of about 25 mm would result. Figure 6 shows the global distribution of precipitable water based on a multi-year averaging period. The continuous decrease, with only a few exceptions, of the atmospheric water content from equatorial latitudes with about 50 mm towards the poles with typical values around 5 mm is obvious. The exceptions from zonal symmetry are associated with the geographical location of the large mountain ranges along the coasts of the continents. In general, the precipitable water is higher over the oceans than over the continents. For several scientific assessments the fields of relative humidity (ratio of the actual to the equilibrium water vapour pressure) are of interest. A related climatology can be found in [65]. A fairly short satellite based climatology of relative humidity in the upper troposphere is provided by Gettelman et al. [66]. Global surface data of relative humidity for the period from 1974 to 2004 is evaluated by Dai [67].

The mean values for a period of ten years as shown in figure 6, may lead to the impression that the global humidity field behaves smoothly. This is only true if a multi-year averaging is used. The inspection of humidity distributions on a daily basis reveals a significantly more complex pattern, which is related to the position of cylones and the actual wind fields. The analysis of global weather data on winds (storms) and their water content revealed that there are major conduits in the atmospheric circulation system, which transport large amounts of water in narrow streams from the tropics through the midlatitudes toward higher latitudes [68, 69]. For these plumes the term "atmospheric rivers" has been coined. There are about three to five of these rivers in the sky in each hemisphere at any time, they contain 95% of meridional water vapor flux at 35° latitude, but in less than 10% of the zonal circumference. Not only do atmospheric rivers play a crucial role in the global water budget, under certain circumstances these events can also lead to heavy coastal rainfall and flooding [70]. In a temporal sense, in general, the water vapour distribution changes with seasonal changes in temperature, which are more pronounced in the Northern hemisphere than in the Southern hemisphere. Of course, the temporal variability is closely related to special events in the atmospheric circulation, such as monsoons and ENSO (El Nino Southern Oscillation). Variability within a year is primarily due to monsoon events, while year-to-year variability is attributable to ENSO. In general, interannual variability is less pronounced compared to interseasonal variability [71]. A characteristic of the ongoing global climate change is the tendency toward an increase in tropospheric water vapour [2], which accompanies an increase in global mean temperature and an increase in sea surface temperature. However, there is a lack of reliable measurements for larger time periods on global scale to support this statement in a quantitative manner. Data for the last three to four decades of the twentieth century indicate an increase of the water vapour content for the lower troposphere of the northern hemisphere [72].

The non-uniform distribution of water vapour in the atmosphere is even more pronounced in the vertical direction. Here, the generally decreasing temperature with altitude is the crucial factor. The water vapour concentration varies over four orders of magnitude, ranging from one to a few percent by volume near the ground, to a few parts per million (ppm) by volume in the stratosphere. Almost half of the atmospheric water vapour is found below an altitude of 1.5 km. Less than 5% occurs above 5 km and less than 1% in the stratosphere above approximately 12 km [64]. Table 3 provides the water vapour content for different height bands of the troposphere.

Table 3. Water vapour column content in mm for different altitude bands in the troposphere. The values are hemispheric resp. global averages for time period 1988 – 1992 (after [73]).

Pressure range [hPa]	Altitude band [km]	Northern hemisphere [mm]	Southern hemisphere [mm]	Global [mm]
500 -300	5.5 – 9	1.5	1.4	1.5
700 -500	3 – 5.5	5.0	4.2	4.6
ground -700	0 – 3	19.4	18.4	18.9

Although of minor importance on an amount basis, the water vapour in the upper troposphere/lower stratosphere region (UTLS) needs to be considered (reported values of the water vapour content of the lower stratosphere are typically in the range of 3–7 ppm). Beside its importance for the radiation budget, water vapour plays a key role in the chemistry of this sensitive part of the atmosphere. Changes in stratospheric water vapour could considerably modify the circulation of the extratropical troposphere [74]. There is strong evidence for a considerable increase of the water vapour concentration in the lower stratosphere. Continuous observations suggest that the stratospheric water vapour has increased by about 1% per year for a period of 20 years until the mid-1990s [75-77]. The trend may even have lasted longer, probably up to half a century [78].

The distribution and transport of water vapour above the planetary boundary layer is closely linked to the general circulation of the atmosphere. A thorough discussion of the zonal and meridional transports can be found in [51]. The upward vertical transport in the equatorial regions is coupled to the ascending branches of the Hadley cells. In the mid-latitudes and higher latitudes this transport is connected to extratropical cyclones. On the regional and local scale the most effective transport of water vapour takes place via convection.

Water vapour enters the atmosphere by evaporation. During this process liquid water or ice at the surface is transferred to the gaseous phase. As can be seen in figure 5, evaporation over the oceans is the dominant source for the atmospheric water budget. The rate of evaporation depends on several factor, for example on the local energy budget, the availability of water, the actual vapour pressure, the turbulent exchange of air near the surface, the surface structure, as well as the natural cover. Strictly, the term evaporation is used only for the phase change over open water surfaces. This includes the water on the surface of vegetation (intercepted water). Plants also give off water vapour to the atmosphere through their leaves or needles (90% through their stomata), this process is called transpiration. The rate of transpiration depends beside on meteorological parameters (solar radiation, humidity, temperature, wind) strongly on the type of plant, the habitat, the season and on soil parameters. Evaporation from the surface and transpiration are not easy to distinguish above vegetated surfaces, therefore, often the expression evapotranspiration is used for the sum of land surface evaporation, interception, evaporation and transpiration. Furthermore, the term potential evaporation is used in contrast to the actual evopotranspiration to denote the amount of water that could be evaporated and transpired if there was sufficient water available. For many land areas, because of an insufficient water supply, the actual evapotranspiration is far below the potential one. Of the 30-years average of 481 mm for evaporation over Germany (previously mentioned), according to the German Weather Service, 328 mm are due to transpiration, 72 mm from evaporation of intercepted water and 42 mm evaporated at the surface. But on global scale 90% of atmospheric water comes from evaporation, while the remaining 10% is from transpiration.

6. CLOUDS

Clouds can be seen as the connecting link between water vapour on one side and precipitation on the other. Precipitation is exclusively produced by clouds, but it also has to be mentioned, that not all clouds lead to precipitation. Clouds are the visible evidence for the existence of the liquid or solid phase of water in the atmosphere. Although clouds on average cover more than 60% of the Earth's surface, the amount of water they contain is comparatively small. It accounts for only 0.25–0.3% of the total water in the atmosphere. Despite this relatively small amount of water, clouds play a crucial role in the global water cycle. The microphysical processes in clouds eventually form large cloud particles, which may start falling as rain, snow or graupel. Precipitation is an effective path to bring water from the atmosphere back to the oceans or land surfaces. Beside this vital role, clouds contribute to the vertical and horizontal redistribution of water vapour in the atmosphere. As a result of their significance in the radiation and energy budget of the Earth [6, 7, 79], in many regions of the globe clouds determine the rates of evaporation and influence regional and local circulation systems through the release of latent heat or heating- and cooling rates associated with radiative processes.

Substantial requirement for the effective formation of clouds are the water vapour saturation of the environment and the existence of suited cloud condensation nuclei (CCN) and ice nuclei (IN), respectively [80-83]. Water vapour saturation can be reached in several ways. In the majority of cases of cloud formation, saturation is the result of lifting of air masses with subsequent (adiabatic) cooling. Corresponding vertical motions are mainly due to thermal convection (cumulus, cumulonimbus), active and passive lifting in connection with movements of frontal systems (cirrostratus, altostratus, nimbostratus), and forced lifting by mountain ranges (orographic lifting). Microphysical processes during cloud formation and evolution are numerous and complex (see e.g. [84-86]). Here, in addition, aerosol physics and chemical aspects play an important role [87-89]. Thorough knowledge of the microphysics of clouds is crucial for understanding the formation of precipitation [81].

Many different types of clouds exist and detailed classification schemes have been developed, but will not be presented here. In general, clouds are classified as high-level, mid-level, and low-level clouds (stratiform clouds) and as clouds with large vertical extent (convective clouds). According to the phase of the cloud particles liquid water-, ice water-, and mixed phase clouds can be distinguished. Often a distinction between precipitating and non-precipitating clouds is also made. Global average amounts for different cloud types according to surface observation climatologies [90] are shown in table 4. The most common types are stratocumulus, altocumulus and cirriform clouds, the dominance of low-level stratus and stratocumulus over large areas of the oceans is obvious in the data. The annual average total cloud cover from surface observations (1982-1991) is 64% (54% over land and 68% over the oceans) [90]. The annual total cloud amount from the International Satellite Cloud Climatology Project (ISCCP) considering data from 1986 to 1993 is 68% (58% over land and 72% over oceans) [91].

The liquid water content or ice water content of clouds is highly variable. Typical values are (range in brackets): marine stratocumulus 0.4 gm^{-3} (0.1 – 0.6 gm^{-3}), continental stratocumulus 0.3 gm^{-3} (0.03

Table 4. Cloud type amounts from surface observations; cloud overlap is possible (from [90]).

Cloud type	Annual average amount [%]	
	Land	Ocean
Stratus	5	11
Stratocumulus	12	22
Cumulus	5	12
Cumulonimbus	4	4
Nimbostratus	5	6
Altostratus	4	
Altocumulus	17	22
Cirriform	22	13

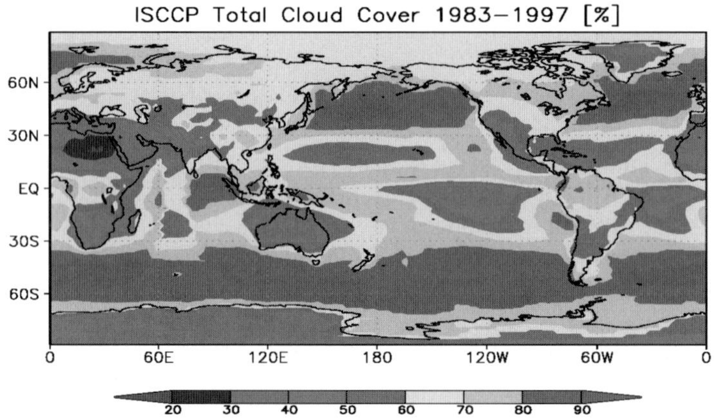

Figure 7. Annual average cloud amount (1983-1997) in % from the International Satellite Cloud Climatology Project (ISCCP), which uses data from geostationary and polar-orbiting satellites [91].

– 0.45 gm^{-3}); cumulus 1 gm^{-3} (0.5 – 2.5 gm^{-3}), cumulus congestus und cumulonimbus up to 4 gm^{-3}; cirrus 0.02 gm^{-3} (0.0001-0.3 gm^{-3}).

The global distribution of cloud amount as annual average for the time period 1983 – 1997 is shown in figure 7. As can be expected, the cloud cover is continuously high in the equatorial belt due to strong convection along the Inter Tropical Convergence Zone (ITCZ). High cloud amounts also occur in the regions of the extratropical storm tracks along the polar fronts in mid-latitudes (50-60°). Minima of cloudiness are observed in the zones of downward motion in the subtropics associated with the Hadley cells. Lowest values of cloud amount are found over the desert areas. A further examination of satellite cloud climatologies (figures not shown here) reveals in the tropics and subtropics the existence of low level, often quite homogeneous, stratocumulus fields at the western rims of the large continents over ocean areas, which are typically relatively cold. Largest coverage with high clouds is found in the tropics, many of these are sheared off the tops of deep cumulonimbus towers. Consistent global climatologies for cloud water content over land and ocean areas are currently not available and only rough estimates are possible. The preliminary evaluation of the ISCCP data [91] with respect to cloud water content come to a global average cloud water path falling in the range of 60 to 80 g/m^2 (or 0.006 – 0.008 cm) (pers. communication W. Rossow).

Due to their tremendous influence on the solar and terrestrial radiation and the formation of precipitation, clouds are an important factor determining climate. The different types of clouds are embedded in the climate system by a multitude of dynamical, thermodynamical and related feedback processes [92-93]. Substantial knowledge on changes in cloud properties over time periods of decades would lead to an improved understanding of their role in current and future climate change.

Unfortunately the data currently available is not sufficient to allow for reliable statements on changes in global cloud cover for longer periods back in time. I.e. large inadequacies exist in monitoring long-term changes in global cloudiness with surface and satellite observations [94]. Changes in other cloud parameters are even more difficult to assess on a global scale. Nevertheless, for some larger regions there is evidence of an increase of cloud amount. The IPCC [95] reports an estimated increase of about 2% in cloud cover over land for the last one hundred years. This increase is in many regions significantly correlated with a change in the daily temperature range (maximum minus minimum temperature). For ocean areas only a few ship-based observations can be used for robust estimates on regional changes in cloudiness. Long-term upward trends in altostratus and nimbostratus clouds are found for the mid-latitude North Pacific and North Atlantic Oceans [96,97] found an increase in total sky cover of approximately 2%, and an increase of approximately 4% in low cloud cover over the oceans in his analyses of ship reports between 1952 and 1995. Surface observations have been analysed to document changes in cirrus clouds [98], and low-, mid-, and upper-level cloud cover [99], and it is found that upper-level cloud cover may have declined by 1.5% (of sky cover) over global land from 1971 to 1996. High ice clouds in the tropics, which play a special role in controlling climate of that region, show an increase in cover of about 2% since 1978 [100]. To assess the impact of these changes on climate is not an easy task. Relevant studies need to consider possible changes in cloud height and thickness, in cloud overlap, and in microphysical and radiative properties, which are, if at all, only rudimentarily known on global scale. Changes in radiative properties and the life time of clouds are closely coupled to the distribution of aerosols, which is also likely to be altered in a changing climate. All of these topics are presently subjects of intensive research.

7. PRECIPITATION

Via precipitation, the water, which originally evaporated at ground level, is brought back from the atmosphere to the Earth's surface. Precipitation includes rain, snow, sleet, grauple, and hail. Although precipitation and its distribution in space and time is essential for life on Earth, the cloud processes leading to precipitation size particles are not known in full detail. The description of the relevant microphysics and related modelling activities are one of the major tasks of cloud physics. In the centre of interest are the growth processes, which eventually lead to particle sizes allowing for terminal velocities sufficient for the particles to reach the ground before they evaporate. In the case of water droplets, particles with radii larger than 0.1 mm are formally called rain drops. During the development of precipitation various macro- and microphysical processes are involved, which can not be treated here in detail (see e.g. [81,84]). The size distribution and number concentration of cloud particles play an essential role during the formation process, as does the vertical wind component (updrafts). Also, the temperature at cloud level plays an important role, as it essentially determines the phase of cloud particles. In pure water clouds (warm rain process, Bowen-Ludlam-process) precipitation formation results from coalescence (merging of water droplets of typically different sizes after collision, which is favoured by differing relative fall velocities [101]. In mixed phase clouds, consisting of supercooled liquid droplets and ice crystals, the Bergeron-Findeisen-process is the dominant way of precipitation formation. Ice crystals acquire water molecules from nearby supercooled water droplets. As these ice crystals gain mass they may begin to fall, acquiring more mass as coalescence occurs between the crystal and neighbouring water droplets. The resulting precipitation can reach the ground either in liquid or solid phase depending on the local atmospheric conditions. Precipitation from pure ice clouds is the result of ice crystal growth by sublimation of water vapour and by aggregation. The precipitation efficiency of clouds, on average, is in the order of 30%, thus only the minor part of the cloud water is transferred to precipitation. It should also be mentioned that a non negligible fraction of particles falling from clouds evaporate before they reach the surface.

In general, related to the external forces supporting the formation, precipitation is distinguished according to convective, stratiform (occurs as a consequence of slow ascent of air in synoptic systems,

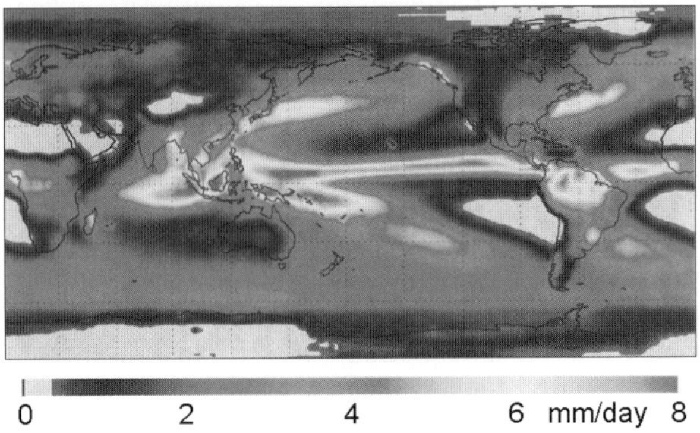

Figure 8. Annual mean precipitation in mm/day. The data is representative for the 23 year period from 1979 to 2001 and is based on a merged analysis that incorporates precipitation estimates from low-orbit satellite microwave data, geosynchronous-orbit satellite infrared data, and surface rain gauge observations [103].

i.e. warm fronts), and orographic precipitation. Stratiform compared to convective precipitation typically covers larger areas and has a much longer duration, it occurs typically with frontal systems [86,102]. Convective precipitation falls as showers, with rapidly changing intensity, which can be very high. It occurs briefly and only over smaller areas, as convective clouds have limited horizontal extent. Convective precipitation is most important in the tropics. Graupel and hail always indicate convection. In midlatitudes, convective precipitation is associated with cold fronts (often behind the front) and squall lines.

Although most precipitation falls over the oceans, precipitation reaching the land surfaces is of crucial importance for life on Earth. On long-term average 2/3 of the water precipitated over land returns back to the atmosphere via evapotranspiration, the rest contributes to the surface runoff or eventually reaches the groundwater. The distribution pattern of precipitation shows tremendous spatial variation, that is caused by or largely attributable to the general circulation, the temperature distribution, the non-uniform land-ocean distribution, and orographic conditions. The high spatial and temporal variability of precipitation has a large impact on vegetation, droughts, and flooding. Figure 8 shows the global distribution of annual means of precipitation expressed in mm per day. The global and also the regional distribution of precipitation occurrence has a similarity with the distribution of cloudiness; but this is not necessarily true for the amount of precipitation. Generally, the annual mean of precipitation amount decreases from the equator towards the poles. But larger inhomogeneities exist in this averaged distribution, explained by the aspects mentioned above. About 2/3 of global precipitation occurs in the latitude band between 30°N and 30°S. Most intensive belts of precipitation are connected to the pronounced convection in the ITCZ and the Southern Pacific Convergence Zone (SPCZ). Here a persistent precipitation amount of more than 2000 mm per year can be found, in some regions the value of 3000 mm per year is exceeded (i.e. the equatorial regions of South America, Africa, and Indonesia). A secondary maximum in precipitation amount in both hemispheres occurs over the midlatitudes along the tracks of the extratropical cyclones, where considerable precipitation is produced by their frontal systems.

Extremely dry areas can be found in the subtropical regions that are under the influence of large, almost permanent, anticyclones. Huge parts of the subtropical continents, such as Africa and Australia, are covered by deserts, where precipitation is very low. Over the polar regions the water vapour content of the atmosphere is extremely low and accordingly the amounts of precipitation are typically very low with annual amounts less than 200 mm per year.

The global average annual precipitation amount, as already mentioned further above, is about 990 mm per year. The world record in annual precipitation amount is reported for Cherrapunji (India,

Khasia mountains), where 26,461 mm/year were recorded from August 1860 to July 1861. On the other hand, there are areas that do not receive any precipitation, sometimes for many consecutive years, e.g. the region Assuan in Egypt received almost no rain over the course of 20 years between 1901 to 1920.

In many regions around the Earth the temporal variations in precipitation activity are noticeable. Beside pronounced daily cycles there are strong seasonal cycles, as well as non-periodical variability. A prominent seasonal phenomenon is the tropical rain fall which is associated with the monsoon circulation. Remarkable non-periodical deviations from longer term means of precipitation amounts can be observed in regions that are influenced by El-Nino-Southern Oscillation (ENSO).

The spatial variability of precipitation is most notably pronounced on small scales (i.e. strong shower activity). These extreme events are of foremost interest for water- and traffic authorities as well as for the agricultural sector.

A part of the precipitation on regional scales comes from precipitation recycling [104]. Precipitation recycling denotes that part of the precipitation in a region, which originates from water evaporated in that region (the contribution of local evaporation to local precipitation); the other part is formed from water vapour advected into the area. Any study on precipitation recycling concerns how the atmospheric branch of the water cycle works, namely, what happens to water vapour molecules after they evaporate from the surface, and where will they precipitate? In general it can be stated that the part of the precipitation from regionally evaporated water gets lower when the considered region is smaller. The recycling ratio varies strongly between winter and summer. In summer the importance of horizontal transport of water vapour declines. According to Trenberth [105] the contribution by precipitation recycling on the 500 km scale accounts for about 10% on global average; on the 1000 km scale this value is about 20%. The latter value means that on average 80% of the humidity, which contributes to the precipitation, comes from distant regions. The associated atmospheric transport covers distances of more than 1000 km (see figure 9). Quantitative descriptions of regional water cycles need, of course, to consider the local conditions. However, the results of Trenberth [105] underline the importance of long-distance transport of water vapour for the global distribution of precipitation.

In connection with the increase of global temperature over the last decades and the increase in water vapour concentration in the lower troposphere, as observed for the northern hemisphere, the question arises, whether an associated change in precipitation occurs. The IPCC report [95] states with high confidence an increase of precipitation on the order of 5 to 10% since 1900 for the mid-latitudes and

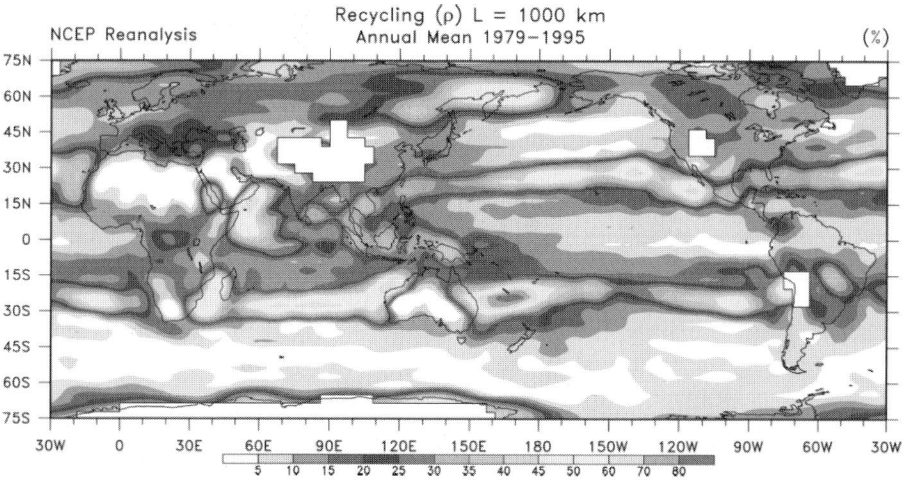

Figure 9. Estimate of the annual mean recycling ratio of the percentage precipitation coming from evaporation within a length scale of 1000 km (from [106]).

higher latitudes of the northern hemisphere. This increase is most likely attributable to strong or even extreme events [106].

8. ATMOSPHERIC WATER AND GLOBAL CHANGE

Amongst the highest priorities in Earth science and environmental policy issues confronting society are the potential changes in the Earth's water cycle due to climate change. Key questions are aiming on the availability of water on one hand and possible flood events on the other. Both topics a closely related to the development of the atmospheric water cycle in a warmer world. By now it is generally agreed upon that the Earth's climate will undergo changes in response to natural variability, including solar variability, and to increasing concentrations of greenhouse gases and aerosols. Furthermore, agreement is widespread that these changes may profoundly affect atmospheric water vapour concentrations, clouds, and precipitation patterns. For example, a warmer climate, directly leading to increased evaporation, may well accelerate the hydrologic cycle, resulting in an increase in the amount of moisture circulating through the atmosphere. Detecting changes is not an easy task given the natural variability of the climate system. Using estimates of natural variability in precipitation (P), evaporation (E), and discharge (R) in combination with model results for a global warming scenario Ziegler et al. [107] determined that data records having lengths on the order of 35 to 70 years are needed to detect significant changes in global terrestrial P, E, and R that might be caused by a warming-induced intensification in the global hydrological cycle. Longer records are generally needed to detect predicted changes in P, E, and R at the continental scale. Huntington [108] has compiled historical trends in hydrologic variables and reviews the current state of science, the paper concludes that although data are often incomplete in spatial and temporal sense the weight of evidence indicates an ongoing intensification of the water cycle.

Discussions of global climate change tend to focus on increasing surface temperature, but quite a few studies are addressing also the water cycle-climate change relation, although their number may not be adequate regarding its potential importance. Several of these research projects are based on predictions using global coupled climate models [109,110], and for a number of those extreme precipitation events [111,112] and their spatial distribution pattern [113] is of main interest. Many uncertainties remain, however concerning regional aspects, as illustrated by the inconsistent results given by current climate models regarding the 20th century [114] and the future distribution of precipitation. It should be mentioned that processes related to the water cycle, like evapotranspiration, cloud- and precipitation formation, are amongst the more uncertain ones represented in the models. These are especially important processes, since they are involved in feedback loops. Water vapour is found to provide the largest positive feedback in models (e.g. [115]), and the feedback from clouds is generally not consistent in between models [116]. Nevertheless, a combination of modelling and data analysis [117] as well as model sensitivity studies and ensemble evaluation might give first indications, whether the water cycle "shifts gear" [118].

Will global warming lead to more evaporation and hence to an increased water vapour content of the atmosphere? What do recent observations suggest? Studies based on radiosonde data from the end of the past century, which saw an increase of the mean temperature over land and over the oceans [95], revealed the presence of regional moistening trends for the lower troposphere since the mid-1970s [72,119]. Based on satellite measurements Trenberth et al. [120] observed increases in precipitable water over the global oceans since the mid-1980s. But caution is advised by Trenberth et al [120] for extracting information on trends in tropospheric water vapour from global reanalysis products, since they suffer from spurious variability and trends related to changing data quality and coverage. Satellite data presented in [121] supports column-integrated moistening trends for the years from mid-1980s. A most recently published evaluation of *in situ* surface air and dewpoint temperature data found very significantly increasing trends in global and Northern Hemispheric specific humidity [67]. For the period from 1974 to 2004 the global annual specific humidity increased by 0.06 $g\,kg^{-1}$ $decade^{-1}$. Global changes in relative humidities are reported to be small for this time segment, mainly due to concurrently increasing temperatures.

While it seems to be established that evaporation over the ocean increases with increasing SST, as it can be inferred from HOAPS data [122] for 1988 to 2002 (Bakan personal communication, 2006), the situation over land is far from being settled [123]. Here, the interpretation of reported recent decreasing trends in pan-evaporation is a reason for controversy [123]. The expectation is that a warming would increase evaporation. Two proposals exist for an explanation of the so-called pan-evaporation paradox. Pan-evaporation (potential evaporation) may be indicative of increasing actual evapotranspiration because of cooler and more humid air is surrounding the pan [124,125]. A second explanation relates the decline in pan evaporation to those in diurnal temperature range and global solar radiation (possibly due to increased cloudiness and aerosol concentrations, e.g. [126], implying that actual evapotranspiration is also declining due to a reduced availability of energy at the surface [127,128]. Reduced wind speed near the surface due to changes of vegetation surrounding the observational sites may also to be considered. A more recent analysis by Wild et al. [129] of surface temperature and solar radiation records from the late 20th century based on 30 years of energy balance data including records from more than 2000 sites indicates that the observed intensification of the hydrological cycle outside the tropics was likely caused by the transfer of moist air from the oceans rather than from evaporation over land. These authors found no indications of increased radiative heating between 1960 and 1990, thus ruling out increased atmospheric moisture from land surface evaporation. A mechanism which might reduce evapotranspiration is a direct effect of an increase of atmospheric carbon dioxide (CO_2). Notably, elevated CO_2 concentrations induce stomata closure and therefore reduce transpiration [130]. The analysis by Gedney et al. [131] suggests that stomatal-closure effects are already having a direct influence on the water balance over land.

What do models forecast for the future water vapour content? All climate models predict that the concentration of water vapour in the troposphere will increase markedly in the future. In particular, they produce increases in water vapour concentrations that are comparable to those predicted by fixing relative humidity [2]. An evaluation of most recent runs of 20 coupled climate models performed for the Fourth Assessment Report of the Intergovernmental Panel on Climate Change (AR4) by Held and Soden [132] supports the finding of an increase of global mean column integrated water vapour with increasing global mean surface temperature which follows a Clausius-Clapeyron scaling (column integrated water vapour is dominated by the lower troposphere). One important aspect of the climate debate over the last 15 years centers around the humidity of the upper troposphere. Actually a negative feedback mechanism has been postulated [133]. As climate models predict that the concentration of water vapour in the UT might double by the end of the century, given the proposed increases of greenhouse gases, a clarification of UT-water vapour trends based on observations accompanied by modelling is desirable. The moistening is not easy to detect using conventional observing systems. The study by Soden et al. [121] uses multi-channel satellite measurements from 1988 to 2004 to detect distinct radiative signatures of UT-moistening in order to test one aspect of water vapour feedback. They could show that the observations were accurately captured by climate model simulations driven by observed SST, a clear evidence that GCMs are properly representing climate feedback from upper tropospheric water vapour. Their observed radiance record requires a global moistening of the UT in response to global warming that is equivalent to the assumption of constant relative humidity. The results of Soden et al. [121] eliminate one potential uncertainty within climate models, and lend further credence to model projections of future water vapour concentrations as those reported by Held and Soden [132].

What about precipitation in a warmer climate? The increased atmospheric moisture content associated with a warming might be expected to lead to increased global mean precipitation. However, precipitation is also strongly influenced by changes in the tropospheric energy budget and the atmospheric circulation, so spatio-temporal patterns of precipitation change are likely to be complex. The Third Assessment Report of Intergovernmental Panel on Climate Change [95] has reported various changes in regional precipitation, there was no simple overall pattern beside that the increase occurs predominately at higher latitudes. Global terrestrial annual mean precipitation showed a small upward trend over the 20th century of approximately 2.1 mm/decade (based on the Global Historic Climatology Network, GHCN, data; to be reported in the IPCC Fourth Assessment Report, AR4, in 2007). However, the record is characterised by large interdecadal variability, and global terrestrial annual mean precipitation based on the New et al. [134] record shows almost no trend since 1940. By use of an ensemble of global models simulating the 20th century, Bosilovich et al. [135] find an increasing trend in precipitation over ocean areas and a decreasing trend over land. The decreases over land are not uniform, however, with most decreases being over the tropics. The assessment of a modification of precipitation intensity is not so conclusive. One of the major problems in examining the climate record for changes in extremes is a lack of high-quality, long-term data [136].

What are the expected changes in precipitation in a future climate? As mentioned above, modelling precipitation without supporting data assimilation is not an easy task. In addition precipition is not characterized by an averaged amount, but frequency, intensity and spatial distribution are also of interest. In the sense that all models in use for AR4 as well as earlier studies show considerable changes for precipitation quantities related to an increase in global temperature [95,109,110] a robust result is available. The models all show an increase in global mean precipitation, but this forecast is far less consistent as the one for temperature projections when the percentage change of amount is considered. This is not surprising since a complex energy balance relationship is expected to be responsible for changes in the mean hydrologic cycle [117].

As an example for projected changes in mean precipitation, in figure 10 results from the ECHAM 5 model for the northern hemisphere winter months are displayed. The simulations followed the climate change scenario A1B of IPCC (a mid-range positive radiative forcing scenario), here the difference of 30 year averages at the end of the 21st century and the end of the 20th century are shown. It can be seen that according to this simulation changes are to be expected everywhere on the globe, although with

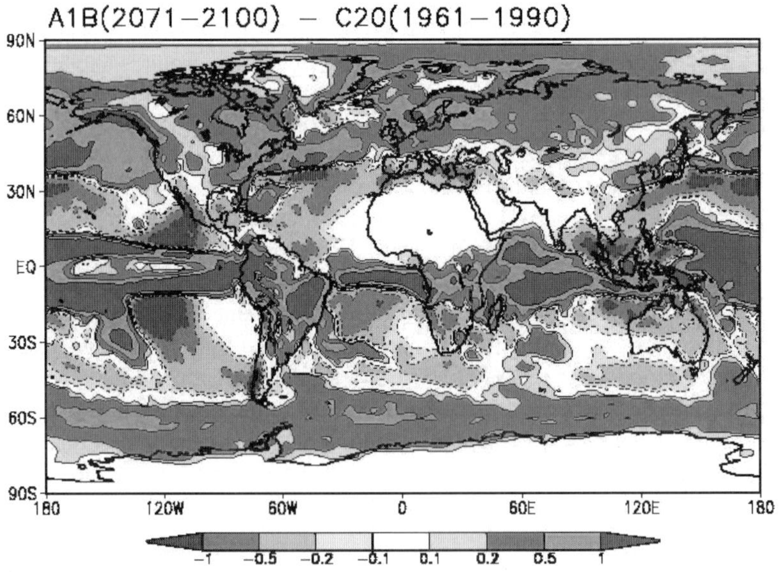

Figure 10. Geographical distribution of the simulated change in mean rates of precipitation for winter months (in mm/day) from the last 30 years of the 20th century to the last 30 years of the current century. Model calculations were made using the ECHAM 5 model under the IPCC A1B scenario (graph from E. Röckner).

variable amount. The most pronounced increases in precipitation are found along the equatorial belt. Increases in precipitation are also forecasted for higher latitudes in both hemispheres, while less rain is expected to occur in the subtropical areas around 30° latitude. For the summer months (not shown here) the potentially drier regions cover large parts of Europe (E. Röckner, personal communication). Experiments conducted using a suite of regional climate models for Europe underline that a warming may regionally lead to strong changes in the hydrological cycle [137]. But the changes occur regionally as well as seasonally with varying intensity and sign and are strongly dependent on the model in use.

It has been widely projected that extremes in precipitation could increase more than the annual or seasonal mean. Since changes in precipitation extremes are postulated to be related to changes in atmospheric temperature [117], under warmer atmospheric conditions, increases in extreme precipitation events could be expected. It has been argued that the enhancement of extremes is principally caused by enhancement of atmospheric moisture content [138]. A general increase in the intensity of precipitation in a future warmer climate has been reported in e.g. [95,111,112,139,140]. Several extreme indices which relate to precipitation have been defined [141,142], the discussion around them can not be repeated here. But it is clear that increases of precipitation intensity do not have a uniform spatial distribution, the actual pattern is related to the relative importance of dynamical to thermodynamical processes, which itself is varying between different models [113,143]. Thermodynamical changes are due to changes in atmospheric water content, while dynamic changes are due to changes in atmospheric motion (circulation, advection, updrafts). In a multi-model evaluation Meehl et al. [113] found that in the tropics increases in water vapour associated with positive SST anomalies produce increased precipitation intensity over most land areas. In midlatitudes the pattern of precipitation increase is related only in part to increased water vapour content but also to changes in atmospheric circulation. Advective effects contribute to greatest precipitation intensity increases over large parts of North America, Northern Europe and other high-latitude regions. In a study using 6 climate models Emori and Brown [143] looked into the relative increase in precipitation mean and extremes. In contrast to earlier reported assessments (e.g. [95]), they found over most parts of mid- to high latitudes, that mean and extreme precipitation increase in comparable magnitude mainly due to a comparable thermodynamic increase. And the dynamic influence on the difference between mean and extreme precipitation plays a secondary role [143]. Barnett et al. [112] reported that the frequency of extremely wet days is likely to increase. They deduced from an ensemble of 53 model versions of the HadSM3 coupled general circulation model, that under $2 \times CO_2$ conditions the global- and ensemble average extremely wet days (according to their statistical definition) become twice as common.

But model results still should be considered only as indicative. Dai [114], examined the newest generation of 18 coupled climate models using available observations, one concluding statements is: "The results show that considerable improvements in precipitation simulations are still desirable for the latest generation of world's coupled climate models". In a similar assessment Sun et al. [144] specify: "Although the models examined here are able to simulate the land precipitation amount well, most of them are unable to reproduce the spatial patterns of the precipitation frequency and intensity". So, caution is advised in too strictly interpreting the currently predicted patterns of precipitation intensity. But it is highly probable that regionally and locally severe changes of the atmospheric part of the hydrological cycle are to be expected, if global warming can not be mitigated by political and technological measures.

9. SOME FINAL REMARKS

The tremendous importance of water in the atmosphere for the global water cycle and the climate of the Earth is still confronted with considerable uncertainties in the quantification of the various storage and flux components and their variability and changes in the branches of the atmospheric water cycle. Our quantitative knowledge of these components is still fairly limited because of a lack of reliable data, which is due to deficiencies in global coverage of high-quality measuring systems for the acquisition of precipitation, water vapour and cloud parameter fields [62]. These fields would be needed to form a basis for the construction of consistent climatological data series and for modelling studies concerning the water exchange between the reservoirs. For an improvement of weather forecast and climate simulation studies in connection with global change an enhanced knowledge of the frequency, duration, and intensity of precipitation on global and regional scale is needed.

A few programmes do exist which are devoted to an improvement of the understanding of the global water cycle. The programmes are also intensely concerned with its atmospheric branch. One of the more comprehensive activities is the Global Energy and Water Cycle Experiment (GEWEX), a core project of the World Climate Research Programme (WCRP). GEWEX, an integrated program of research, observations and science activities, began in 1988 and is currently in its second phase (see www.gewex.org). Its strategy is to combine results of observations and modelling of the water and energy cycle for the atmosphere, the land surfaces and the upper ocean layers. The ultimate goal of GEWEX is the improved prediction of global and regional climate change. As central activities, the continental scale experiments should be mentioned. These concentrate on the development of the best available water and energy budgets for selected regions of the globe. Quite a few of the sub projects within GEWEX are ultimately devoted to the atmospheric branch of the water cycle.

Based on all currently ongoing research activities, an improvement of the quantitative determination of the components of the atmospheric water cycle might be expected in the near future. An improvement of the parameterised physics and the horizontal as well as the vertical resolution of the models, which are used for four dimensional data assimilation projects (4D-VAR; re-analysis; e.g. [145]) will have a substantial impact on further development in this field. Extended studies on theses aspects are currently under way at several operational or research centres such as the European Centre for Medium-Range Weather Forecasts (ECMWF) or the National Center for Atmospheric Research (NCAR).

The extension of the global observing system within the upcoming years will specifically enhance the progress in the quantitative determination of precipitation over the ocean areas and knowledge of vertical profiles of cloud water and cloud ice content. Here a combination of satellite based active remote sensing like the recently launched CLOUDSAT [146] and CALIPSO Missions with passive sensors is a key to advancement. Extended ground based networks of Doppler radar systems, wind profilers, water vapour lidars in conjunction with radiosondes (improved with respect to water vapour measurements) will permit further comprehensive and advanced studies of the water budget for subsystems on continental scale, creating a sound basis for the evaluation of models used for regional and global predictions of the atmospheric water cycle in a changing climate. Reliable regional predictions of future availability of fresh water are needed for the development of adaptation strategies in response to global warming. Authoritative advice at an early stage is highly desirable for the prevention of potential regional and international conflicts about water.

Acknowledgements

MQ thanks Prof. Claude Boutron for have been given the opportunity to contribute to ERCA 2005 and 2006. The help of Dr. Thomas C. Pagano, NRCS Portland, Oregon, and Drs. Stephan Bakan and Erich Roeckner, Max-Planck-Institute for Meteorology, Hamburg, by providing unpublished material is gratefully acknowledged. We also thank Prof. Peter Hupfer, Humboldt University, Berlin, for many useful comments on parts of the manuscript.

References

[1] Pierrehumbert R. T., *Nature* **419** (2002) 191-198.
[2] Held I. M. and Soden B. J., *Annual Rev. Energy Environm.* **25** (2000) 441-475.
[3] Fouquart Y., Buriez J. C., Herman M. and Kandel R. S., *Rev. Geophys.* **28** (1990) 145-166.
[4] Arking, A., *Bull. Amer. Meteoro. Soc.* **72** (1991) 795-813.

[5] Hartmann D. L., "Radiative effects of clouds on earth's climate" Aerosol-Cloud-Climate Interactions, P. V. Hobbs Ed., (Academic Press, San Diego, 1993) pp. 151-173.
[6] Quante M., *Journal de Physique IV* **121** (2004) 61-86.
[7] Raschke E., Ohmura A., Rossow W. B., Carlson B. E., Zhang Y.-C., Stubenrauch C., Kottek M. and Wild M., *Int. J. Climatol.* **25** (2005) 1103-1125.
[8] Kiehl, J. T. and Trenberth K. E., *Bull. Am. Met. Soc.* **78** (1997) 197-208.
[9] Charlson, R. J., Global Biogeochemical Cycles (International Geophysi cs Series, Vol 50, Academic Press Inc., 1992) 392pp.
[10] Jacobson M., Charlson R. J., Rodhe H. and Orians G. H., Earth System Science: From Biogeochemical Cycles to Global Changes. (Academic Press, San Diego, 2000) 527pp.
[11] Kabat P., Claussen M., Dirmeyer P. A., Gash J. H. C., Bravo de Guenni L., Meybeck M., Pielke Sr R. A., Vörösmarty C. J., Hutjes R. W. A. and Lütkemeier S. Eds., Vegetation, Water, Humans and the Climate. (Springer Verlag, Berlin, 2004) 600pp.
[12] Madronich S., *J. Geophys. Res.* **92** (1987) 9740-9752.
[13] Lelieveld J. and Crutzen P., *J. Atmos. Chem.* **12** (1991) 229-267.
[14] Prather M. and Jacob D., Geophys. Res. Lett. **24** (1997) 3189-3192.
[15] Barth M. C., Hess P. G. and Madronich S., *J. Geophys. Res.* **107** (2002) 4126 doi:10.1029/2001JD000468.
[16] Tie X., Madronich S., Walters S., Zhang R., Rasch P. and Collins W., *J. Geophys. Res.* **108** (2003) 4642 doi:10.1029/2003JD003659.
[17] Bland W. and Rolls D., Weathering: An Introduction to Scientific Principles. (Hodder Arnold, London, 1998) 288pp.
[18] Morgan R. P. C., Soil erosion and conservation. (3rd ed., Blackwell Publishing, Malden, MA, 2005) 320pp.
[19] Kinnell P. I. A., *Hydrological Processes* **19** (2005) 2815-2844.
[20] Peter T., *Annu. Rev. Phys. Chem.* **48** (1997) 785-822.
[21] Solomon S., *Rev. Geophys.* **37** (1999) 275-316.
[22] Arnell N., Global Warming, River Flows and Water Resources. (John Wiley and Sons, Chichester, UK, 1996).
[23] Lettenmaier D. P., The role of climate in water resources planning and management, R. Lawford et al. Eds. Water: Science, Policy, and Management. (Water Resources Monograph No. 16, AGU Press, 2004) pp.247-266.
[24] Garbrecht J. D. and Piechota T. C. Eds., Climate Variations, Climate Change, and Water Resources Engineering. (American Society of Civil Engineers, Reston, Virginia, 2006) 198pp.
[25] Lozán J. L., Graßl H., Hupfer P., Menzel L. and Schönwiese Ch.-D. Eds., GLOBAL CHANGE: Enough Water for all? (Wissenschaftliche Auswertungen, Hamburg, 2007) 400pp.
[26] Franks F., Water – a matrix of life. (2nd edition, RSC Paperbacks, Royal Society of Chemistry, Cambridge, UK, 2000) 225pp.
[27] Pauling L., The Nature of the Chemical Bond. (2nd edition, Cornell University Press, New York., 1948).
[28] Modig K., Pfrommer B. G. and Halle B., *Phys. Rev. Lett.* **90** (2003) 075502.
[29] Sutmann G. and Vallauri R., J. Mol. Liq. 98-99 (2002) 215-226.
[30] Wernet Ph., Nordlund D., Bergmann U., Cavalleri M., Odelius M., Ogasawara H., Näslund L. Å., Hirsch T. K., Ojamäe L., Glatzel P., Pettersson L. G. M. and Nilsson A., *Science* **304** (2004) 995-999.
[31] Smith J. D., Cappa C. D., Wilson K. R., Messer B. M., Cohen R. C. and Saykally R. J., *Science* **306** (2004) 851-853.
[32] Smith, J. D., Cappa C. D., Wilson K. R., Cohen R. C., Geissler P. L. and Saykally R. J., *Proc. Natl. Acad. Sci. USA* **102** (2005) 14171-14174.

[33] Head-Gordon T. and Johnson M. E., *Proc. Natl. Acad. Sci. USA* **103** (2006) 7973-7977.
[34] Stillinger F. H., Science 209 (1980) 451-457.
[35] Hallet J., Arnott W. P., Bailey M. P. and Hallet J. T., Ice crystals in cirrus. D. Lynch, K. Sassen, D.O'C. Starr and G. Stephens Eds., Cirrus. (Oxford University Press, New York, 2002) pp.41-77.
[36] Baker M. B., Ice in the troposphere, J. S. Wettlaufer, et al. Eds., Ice Physics and the Natural Environment (NATO ASI Series Vol. I 56, Springer, 1999) pp.121-142.
[37] Eisenberg D. and Kauzmann W., The Structure and Properties of Water. (Clarendon Press, Oxford, 1969).
[38] Eisenberg D. and Kauzmann W., The Structure and Properties of Water. (New edition, Oxford University Press, New York, 2005) 314pp.
[39] Franks F. Ed., Water Science Reviews. (Vol. 1-4, Cambridge University Press, Cambridge, 1985-1990).
[40] Debenedetti P. G., Metastable Liquids. (Princeton Univ. Press, Princeton, USA, 1996) 400pp.
[41] Robinson G. W., Zhu S.-B., Singh, S. and Evans M. W., Water in Biology, Chemistry, and Physics. (World Scientific, Singapore, 1996) 509pp.
[42] Mishima O. and Stanley H. E., *Nature* **396** (1998) 329-335.
[43] Stanley H. E., *Pramana – J. Phys.* **53** (1999) 53-83.
[44] Hobbs P. V., Ice Physics. (Clarendon Press, Oxford, 1974) 837pp.
[45] Wettlaufer J. S., Dash J. G. and Untersteiner N. Eds., Ice Physics and the Natural Environment. (NATO ASI Series Vol. I 56, Springer, Berlin, 1999) 355pp.
[46] Petrenko V. and Whitworth R., The Physics of Ice. (Oxford UniversityPress, New York, 1999) 390pp.
[47] Ball P., Life's Matrix: A Biography of Water. (University of California Press, Berkeley, 2001) 417pp.
[48] Kandel R., Water from heaven – The story of water from the big bang to the rise of civilization and beyond. (Columbia University Press, New York, 2003) 312pp.
[49] Webster P. J., *Rev. Geophys.* **32** (1994) 427-476.
[50] Korzun V. I., Sokolov A. A., Budyko M. I., Voskresensky K. P., Kalinin G. P., Konoplyansev A. A., Korotkevich E. S., Kuzin P. S. and L'vovich M. I. Eds., World Water Balance and Water Resources of the Earth. UNESCO, Paris, 1978).
[51] Peixoto J. P. and Oort A. H., Physics of Climate. (American Institute of Physics, New York, NY, 1992) 520pp.
[52] Oki T., The Global Water Cycle. K. A. Browning and R. J. Gurney Eds., Global Energy and Water Cycles. (Cambridge University Press, Cambridge, 1999) pp.10-29.
[53] Gleick P. H. Ed., Water in Crisis: A Guide to the World's Fresh Water Resources. (Oxford University Press, New York, 1993) 504pp.
[54] Shiklomanov I. A. and Rodda J. C. Eds. World water resources at the beginning of the 21st century. (Cambridge University Press, New York, 2003) 435pp.
[55] Oki T. and Kanae S., *Science* **313** (2006) 1068-1072.
[56] Rosen R. D., The global energy cycle. K. A. Browning and R. J. Gurney Eds., Global Energy and Water Cycles. (Cambridge University Press, Cambridge, 1999) pp.1-9.
[57] Trenberth K. E. and Stepaniak D. P., *Quart. J. Roy. Meteor. Soc.* **130** (2004) 2677-2701.
[58] Chahine M. T., *Nature* **359** (1992) 373-380.
[59] Pagano T. and Sorooshian S., Global Water Cycle (Fundamental, Theory, Mechanisms). M. G. Anderson Ed., Encyclopedia of Hydrological Sciences. (Vol 5, John Wiley & Sons, Ltd., Chichester, England, 2006) 2697-2711pp.
[60] Trenberth K. E. and Guillemot C. J., *Climate Dyn.* **14** (1998) 213-231.

[61] Shiklomanov I. A., World Water Resources, a new appraisal and assessment for the 21st century – summary. (United Nations Educational, Scientific and Cultural Organization,UNESCO, 7 Place de Fontenoy, 75352 Paris 07 SP, 1998) 37pp.
[62] Trenberth K. E., Smith L., Qian T., Dai A. and Fasullo J., *J. Hydrometeor.* (2006) submitted.
[63] Roads J., *GEWEX Newsletter* **12-1** (2002).
[64] Seidel D. J., Water Vapor: Distribution and Trends. M. C. MacCracken and J. S. Perry Eds., Encyclopedia of Global Environmental Change. (John Wiley & Sons, Ltd, Chichester, 2002).
[65] Peixoto J. P. and Oort A. H. *J. Climate* **9** (1996) 3443-3463.
[66] Gettelman A., Collins W. D., Fetzer E. J., Eldering A. and Irion F. W., *J. Climate* (2006) in press.
[67] Dai A., *J. Climate* **19** (2006) 3589–3606.
[68] Zhu Y. and Newell R. E., *Mon. Wea. Rev.* **126** (1998) 725-735.
[69] Ralph F. M., Neiman P. J. and Wick G. A., *Mon. Wea. Rev.* **132** (2004) 1721-1745.
[70] Ralph F. M., Neiman P. J., Wick P. J., Gutman S. I., Dettinger S. I., Cayan S. I. and White A. B., *Geophys. Res. Lett.* **33** (2006) L13801, doi:10.1029/2006GL026689.
[71] Amenu G. G. and Kumar A. B., *Bull. Amer. Meteor. Soc.* **86** (2005) 245-256.
[72] Ross R. J. and Elliott W. P., *J. Climate* **14** (2001) 1602-1612.
[73] Randel D. L, Vonder Haar T. H., Ringerud M. A., Stephens G. L., Greenwald T. J. and Combs C. L., *Bull. Amer. Meteorol. Soc.* **77** (1996) 1233-1246.
[74] Joshi M. M., Charlton A. J. and Scaife A. A., *Geophys. Res. Lett.* **33** (2006) L09806, doi:10.1029/2006GL025983.
[75] Oltmans S. J. and Hofmann D. J., *Nature* **374** (1995) 146.
[76] Kley D., Russell III J. M. and Phillips C. Eds., SPARC Assessment of Upper Tropospheric and Stratospheric Water Vapour. (WCRP – 113, WMO/TD - No. 1043, SPARC Report No.2, 2000) 312pp.
[77] Oltmans S. J., Volmel H., Hofmann D. J., Rosenlof K. and Kley D., *Geophys. Res. Lett.* **27** (2000) 3453-3456.
[78] Rosenlof K. H., Oltmans S. J., Kley D. Russell III J. M., Chiou E.-W., Chu W. P. Johnson D. G., Kelly K. K., Michelsen H. A., Nedoluha G. E., Remsberg E. E., Toon G. C., McCormick M. P.,*Geophys. Res. Lett.* **28** (2001) 1195-1198.
[79] Kiehl J. T., *Physics Today* **47** (1994) 36-42.
[80] Hudson J. G., *J. Appl. Meteor.* **332** (1993) 596-607.
[81] Young K. C., Microphysical processes in clouds (Oxford University Press, New York, 1993) 427pp.
[82] Vali G., Principles of ice nucleation. R. E. Lee, Jr., G. J. Warren and L. V. Gusta Eds., Biological Ice Nucleation and its Applications (APS Press, St. Paul, 1995) pp.1-28.
[83] Szyrmer W. and Zawadzki I., *Bull. Amer. Meteoro. Soc.* **78** (1997) 209-228.
[84] Pruppacher H. R. and Klett J. D., Microphysics of clouds and precipitation. (2nd edition, Kluwer Academic Publishers, Dordrecht, 1997) 954pp.
[85] Seinfeld J. H. and Pandis S. N., Atmospheric Chemistry and Physics: From Air Pollution to Climate Change. (John Wiley & Sons, Inc, New York, 1998) 1326pp.
[86] Wallace J. M. and Hobbs P. V., Atmospheric Sciences: An introductory survey. (2nd edition, Academic Press, 2006) 209-269.
[87] Flossmann A. I. and Laj P., Aerosols, gases and microphysics of clouds. C. Boutron Ed., From Urban Air Pollution to Extra-Solar Planets, (EDP Sciences, Les Ulis, 1998) pp89-119.
[88] Charlson R. J, Seinfeld J. H., Nenes A., Kulmala M., Laaksonen A. and Facchini M. C., *Science* **292** (2001) 2025-2026.
[89] Harrison R. G. and Carslaw K. S., *Rev. Geophys.* **41** (2003) 1012 doi:10.1029/2002RG000114.
[90] Warren S. G. and Hahn C. J., Clouds/Climatology. J. Holton, J. Pyle and J. Curry Eds., Encyclopedia of Atmospheric Sciences. (Academic Press, San Diego, 2002) 476-483pp.
[91] Rossow W. B. and R. A. Schiffer, *Bull. Amer. Meteor. Soc.* **80** (1999) 2261-2287.

[92] Zhang M., Cloud-climate feedback: how much do we know? Observation, Theory and Modeling of Atmospheric Variability. (World Scientific Co. Pte. Ltd., River Edge, NJ, 2004) pp.161-183.
[93] Stephens G. L. *J. Climate* **18** (2005) 237-273.
[94] Dai A., Karl T. R., Sun B. and Trenberth K. E., *Bull. Amer. Meteor. Soc.* **87** (2006) 597-606.
[95] Houghton T., Ding Y., Griggs D. J., Noguer M., van der Linden P. J. and Xiaosu D. Eds., Climate Change 2001: The Scientific Basis. (Cambridge University Press, Cambridge, 2001) 944pp.
[96] Parungo F., Boatman J. F., Sievering H., Wilkison S. W. and Hicks B. B., *J. Climate* **7** (1994) 434-440.
[97] Norris J. R., *J. Climate* **12** (1999) 1864-1870.
[98] Minnis P., Ayers J. K., Palikonda R. and Phan D., *J. Climate* **17** (2004) 1671-1685.
[99] Norris J. R., *J. Geophys. Res.* **110** (2005) D08206, doi:10.1029/2004JD005600.
[100] Wylie D. P., Menzel W. P., Jackson D. and Bates J. J., "Trends in High Clouds Over the Last 20 Years". Proceedings 12th Conference on Satellite Meteorology and Oceanography, 9-13 February 2003, Long Beach, California, (American Meteorological Society, Boston, 2003), P1.4.
[101] Beard K. V. and Ochs III H. T., *J. Appl. Meteor.* **32** (1993) 608-625.
[102] Gedzelman S. D. and Arnold R., *Mon. Wea. Rev.* **121** (1993) 1957-1978.
[103] Adler R. F., Huffman G. J., Chang A., Ferraro R., Xie P., Janowiak J., Rudolf B., Schneider U., Curtis S., Bolvin D., Gruber A., Susskind J., Arkin P. and Nelkin E., *J. Hydrometeor.* **4** (2003) 1147-1167.
[104] Eltahir E. A. B. and Bras R. L., *Rev. Geophys.* **34** (1996) 367-378.
[105] Trenberth K. E., *J. Climate* **12** (1999) 1368-1381.
[106] Trenberth K. E., Dai A., Rasmussen R. M. and Parsons D. B., *Bull. Amer. Meteor. Soc.* **84** (2003) 1205-1217.
[107] Ziegler A. D., Sheffield J., Maurer E. P., Nijssen B., Wood E. F. and Lettenmaier D. P., *J. Climate* **16** (2003) 535-547.
[108] Huntington T. G., *J. Hydrology* **319** (2006) 83-95.
[109] Douville H., Chauvin F., Planton S., Royer J. F., Salas-Melia D. and Tyteca S., *Clim. Dyn.* **20** (2002) 45-68.
[110] Wetherald R. T. and Manabe S., *J. Geophys. Res.* **107** (2002) 4379, doi:10.1029/2001JD001195.
[111] Räisänen J., *Clim. Dyn.* **24** (2005) 309-323.
[112] Barnett D. N., Brown S. J., Murphy J. M., Sexton D. M. H. and Webb M. J., *Clim. Dyn.*, **26**, (2006) 489-511.
[113] Meehl G., Arblaster, J. and Tebaldi, C., 2005: Understanding future patterns of increased precipitation intensity in climate model simulations. *Geophys. Res. Lett.* , **32**:L18719
[114] Dai A., *J. Climate* **19** (2006) 4605-4630.
[115] Soden B. J. and Held I. M., *J. Climate* **19** (2006) 3354-3360.
[116] Ringer, M. A., McAvaney B. J., Andronova N., Buja L. E., Esch M., Ingram W. J., Li B., Quaas J., Roeckner E., Senior C. A., Soden B. J., Volodin E. M., Webb M. J. and Williams K. D., *Geophys. Res. Lett.* **33** (2006) L07718, doi:10.1029/2005GL025370.
[117] Allan M. R. and Ingram W. J., *Nature* **419** (2002) 224-232.
[118] Stocker T. F. and Raible C. C., Nature 434 (2005) 830-833.
[119] Zhai P. and Eskridge R. E., *J. Climate* **10** (1997) 2643-2652.
[120] Trenberth K. E., Fasullo J. and Smith L. *Clim. Dyn.* **24** (2005) 741-758.
[121] Soden B. J., Jackson D. L., Ramaswamy V., Schwarzkopf M. D. and Huang X., *Science* **310** (2005) 841-844.
[122] Klepp C.-P., Fennig K., Bakan S. and Graßl H., HOAPS-II gobal ocean precipitation data base. Eumetsat Proceedings, P.44, Second International Precipitation Working Group Workshop (ISBN 92-9110-070-6, 2005) pp.169-176.
[123] Ohmura A. and Wild M., *Science* **298** (2002) 1345-1346.

[124] Brusaert W. and Parlange M. B., *Nature* **396** (1998) p.30.
[125] Hobbins M. T., Ramirez J. A. and Brown T. C., *Geophys. Res. Lett.* **31** (2004) L13503, doi:10.1029/2004GL019846.
[126] Liepert B., Feichter J., Lohmann U. and Roeckner E., *Geophys. Res. Lett.*, **31**, (2004) L06207, doi:10.1029/2003GL019060,
[127] Peterson T. C., Golubev V. S. and Groisman P. Y., *Nature* **377** (1995) 687-688.
[128] Roderick M. L. and Farquhar G. D., *Science* **298** (2002) 1410-1411.
[129] Wild M., Ohmura A., Gilgen H. and Rosenfeld D., *Geophys. Res. Lett.* **31** (2004) L11201, doi:10.1029/2003GL019188, 2004
[130] Field C., Jackson R. and Mooney H., *Plant Cell Environ.* **18** (1995) 1214-1255.
[131] Gedney N., Cox P. M., Betts R. A., Boucher O., Huntingford C. and Stott P. A., *Nature* **439** (2006) 835-838.
[132] Held I. M. and Soden B. J., *J. Climate* (2006) submitted
[133] Lindzen R. S., *Bull. Amer. Meteor. Soc.* **71** (1990) 288-299.
[134] New M. G., Hylme M. and Jones P. D., *J. Climate* **13** (2000) 2217-2238.
[135] Bosilovich M. G., Schubert S. D. and Walker G. K., *J. Climate* **18** (2005) 1591-1608.
[136] Easterling D. R., Evans J. L., Groisman P. Y., Karl T. R., Kunkel K. E. and Ambenje, P., *Bull. Amer. Meteor. Soc.* **81** (2000) 417-425.
[137] Jacob D., and Hagemann S., Intensification of the hydrological cycle – An important signal of climate change. J. L. Lozán, H. Graßl, P. Hupfer, L. Menzel and Ch.-D. Schönwiese Eds., GLOBAL CHANGE: Enough Water for all? (Wissenschaftliche Auswertungen, Hamburg, 2007) 143-146.
[138] Trenberth K. E., *Climate Change* **42** (1999) 327-339.
[139] Wehner M. F., *J. Climate* **17** (2004) 4281-4290.
[140] Watterson I. G. and Dix M. R., *J. Geophys. Res.,* **108** (2003) 4379, doi:10.1029/2002JD002928.
[141] Frich P., Alexander L. V., Della-Marta P., Gleason B., Haylock M., Klein Tank A. M. G. and Peterson T., *Clim. Res.* **19** (2002) 193-212.
[142] Tebaldi C., Hayhoe K., Arblaster J. M. and Meehl G. A., *Climatic Change* (2006) in press
[143] Emori S. and Brown S. J., *Geophys. Res. Lett.* **32** (2005) L17706, doi: 10.1029/2005GL023272.
[144] Sun Y., Solomon S., Dai A. and Portmann R. W., J. Climate 19 (2006) 916-934.
[145] Uppala S. M. and 44 authors, *Quart. J. Roy. Meteor. Soc.* **131** (2005) 2961-3012.
[146] Stephens G. L., Vane D. G., Boain R. J., Mace G. G., Sassen K., Wang Z., Illingworth A. J., O'Connor E. J., Rossow W. B., Durden S. L., Miller S. D., Austin R. T., Benedetti A., Mitrescu C. and CloudSat Science Team, *Bull. Amer. Meteor. Soc.* **83** (2002) 1771-1790.

J. Phys. IV France **139** (2006) 63–80
© EDP Sciences, Les Ulis
DOI: 10.1051/jp4:2006139006

Take a glass of water [1] . . . – concepts from physical chemistry used in describing the behaviour of aerosol and cloud droplets –

F. Raes[1]

[1] *European Commission, DG Joint Research Centre, Institute for Environment and Sustainability, 21020 Ispra (VA), Italy*

Abstract. This is the text of a lesson explaining the concepts of kinetic gas theory, diffusion theory and thermodynamics used in describing the formation and growth of aerosol and cloud droplets in the atmosphere. The classical Koehler theory and its extensions are derived in a step-wise approach, starting from clean laboratory systems and ending with the polluted atmosphere. The meaning of the Koehler theory is analysed. Some comments follow regarding the central role the theory still plays in atmospheric research today; it indeed forms a formal framework to link air pollution and climate change.

1. INTRODUCTION

Why would a student be interested in aerosols and clouds? Air pollution and climate change, of course, and we will say something about these issues at the end of this lesson. But surely the ever changing colours of the sky and the towering clouds, among the largest objects on our Planet, offer inspiration and joy when dealing with the subject. Furthermore it is stimulating to realize that aerosol and clouds sit at the interface between the liquid and the gas phase, and that, generally speaking, the most interesting things occur at interfaces. This is equally true for phases as for countries, cultures and people such as scientists and policy makers, for instance.

The theory of aerosol and cloud droplet formation and growth is based on the original work of Koehler, and is well established and treated in text books such as Pruppacher and Klett [2] and Seinfeld and Pandis [3]. Its theory is based on concepts from gas phase kinetics, diffusion theory and thermodynamics [4].

In this lesson we present the main concepts of these theories to the interested student. The approach is quantitative, but we aim at a presentation such that the concepts can be visualized and qualitatively understood. The scope is to create a "feel" for the subject, before the fundamentals in the text books are addressed.

Aerosol particles in the atmosphere are very often droplets of aqueous solutions. Clouds droplets always form on aerosol particles, so they are themselves aqueous solutions. To explain the formation and growth of aerosol and cloud droplets we therefore need to understand the behaviour of aqueous solutions and the transfer of water vapour and other vapours between the gas phase and the solution. We'll approach the subject step by step, starting with pure water, passing to aqueous solutions and droplets and finishing with aerosol and cloud droplets in the real atmosphere.

2. PURE WATER IN A CLOSED BEAKER

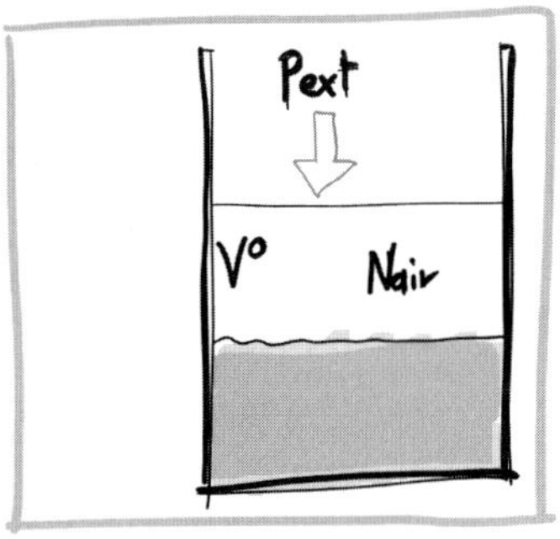

Take a glass of water, a beaker closed with an air-tight lid. The beaker is partly filled with pure water and partly with air. The lid is weightless and movable, so that the air pressure in the beaker is always equal to the external, say atmospheric pressure, (p_{ext}). The temperature (T) is the same in the air as in the liquid and constant.

Let's assume that we can arrange an experiment so that initially the air above the water is completely dry and occupies a volume V^o. When the system is left on its own, we observe that the volume will expand from V^o to a final volume V^e.

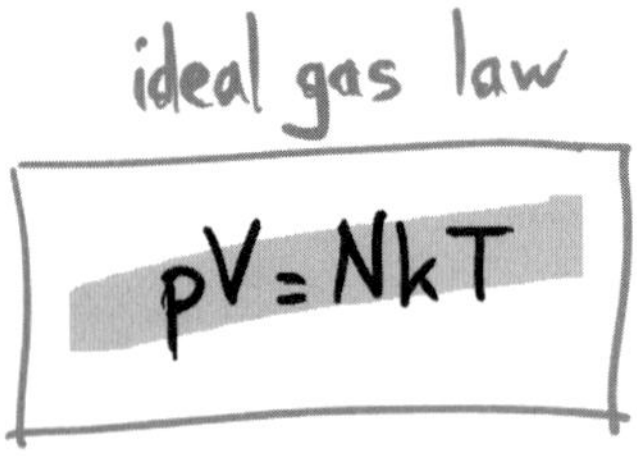

What happens is that water molecules evaporate from the liquid phase into the gas phase, increase the total number of molecules in the volume, which expands in order to remain in equilibrium with the external pressure. The fact that the expansion stops means that there must be an equilibrium number of vapour molecules N_v^e that can stay in the volume above the liquid. Using the ideal gas law, we can write

$$N_v^e = \frac{p_{ext}}{kT}(V^e - V^o) \tag{1}$$

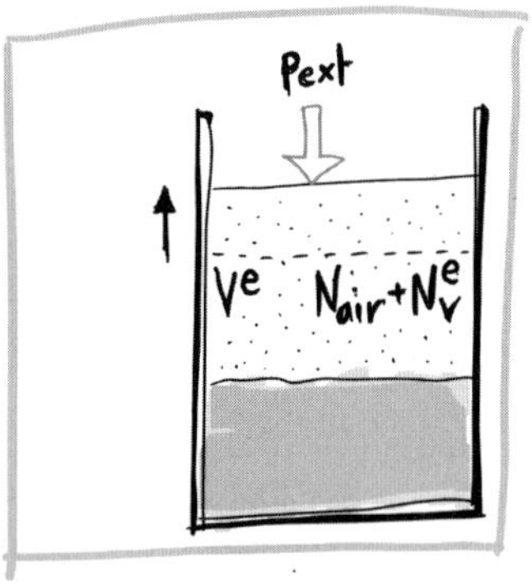

and the corresponding vapour pressure, the *equilibrium vapour pressure of water* at temperature T, equals:

$$p_v^e(\infty) = \frac{N_v^e}{V^e}kT = \frac{p_{ext}}{V^e}(V^e - V^o) \tag{2}$$

The infinity sign ∞ is used to indicate that we deal with a flat surface: a droplet with an infinite diameter.

If the experiment were feasible, it would be a way to determine the equilibrium vapour pressure of liquids.

We will now look in more detail how the equilibrium is achieved.

2.1 Gas Kinetics

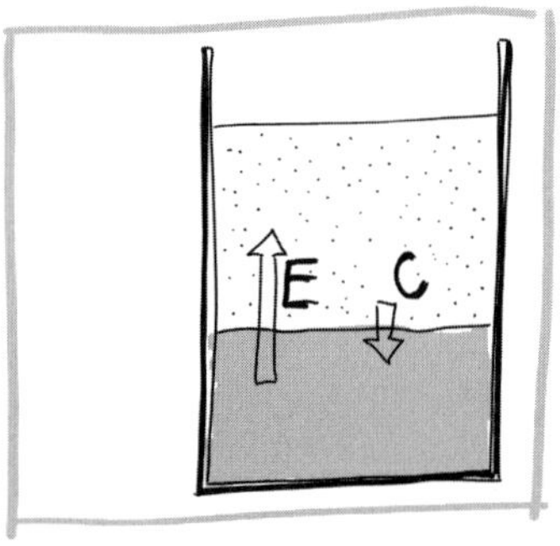

The process observed in the closed beaker can be described as follows. At each moment in time, molecules evaporate from the liquid at a rate E and others condense from the gas phase with a rate C.

In reality, since water vapour molecules originate at the surface of the liquid, their concentration n_v will always be higher near that surface than higher up, at least till equilibrium is reached. We'll come back to this point later, but let us assume for a moment that water molecules are immediately and homogeneously mixed throughout the volume above the liquid. The change in time of the number of vapour molecules $N_v(t)$ in the volume can then be written as:

$$\frac{dN_v(t)}{dt} = E - C(t) \qquad [\text{molecules.s}^{-1}]\ (3)$$

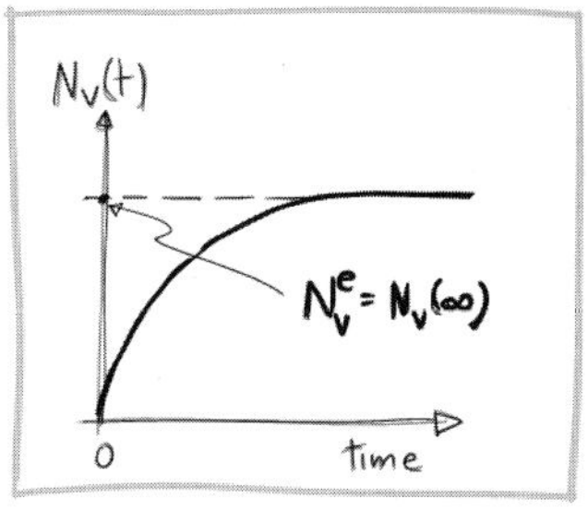

E, the evaporation rate, is solely determined by the properties of the liquid: i.e. the thermal motion of its molecules (its temperature T) and the strength of the attractive forces between the molecules. As T increases, the thermal motion increases and the chance that some of the molecules overcome the intermolecular forces and escape into the gas phase also increases.
C, the condensation rate, is determined by the *concentration* of vapour molecules close to the surface, that can be kicked into the liquid upon collision with another water or air molecule. Kinetic gas theory tells that:

$$C(t) = \frac{\bar{c} S n_v(t)}{4} \qquad (4)$$

where $\bar{c}$ is the mean velocity of vapour molecules (function of T and the mass of the molecules only), S the surface area of the liquid, and $n_v(t)$ the concentration of vapour molecules.
When there is equilibrium between evaporation and condensation $(dN_v(t)/dt = 0)$ we derive from Eq. 3 and 4 that the corresponding number of vapour molecules in the volume V^e, and the equilibrium vapour pressure equal:

$$N_v^e = \frac{4E}{\bar{c}\,S} V^e \qquad (5)$$

and

$$p_v^e(\infty) = \frac{4E}{\bar{c}\,S} kT \qquad (6)$$

Equation 6 shows that the $p_v^e(\infty)$, like E, is a property of the liquid only. (Note that when equilibrium is reached the concentration of vapour molecules is homogeneous throughout the volume $(n_v(t = \infty) = n_v^e$ everywhere), and hence Eqs. 5 and 6 are valid, independent of the assumption we made about the homogeneity of $n_v(t)$ while reaching the equilibrium.)

Equation 2 can also written as follows:

$$V^e = \frac{V^o p_{ext}}{p_{ext} - p_v^e(\infty)} \qquad (7)$$

which shows that:
if $V^o = 0$, hence if the lid rests on the liquid surface, there will be no expansion, and this is because liquid molecules, even if they are knocked around by their neighbours, are not given the space to escape.
- if $p_v^e\ \infty = p_{ext}$, V^e becomes infinite and the lid is literally blown off the beaker. Hence, whereas the external pressure p_{ext} is the pressure exerted *on* the liquid, the equilibrium vapour pressure $p_v^e(\infty)$ can be seen as a pressure *of* the liquid. At $p_v^e(\infty) = p_{ext}$ this pressure is able to push away p_{ext} not only at the surface but throughout the liquid, and bubbles will form. At $p_v^e(\infty) = p_{ext}$, the liquid boils...

2.2 Diffusion

As said before, while the system in our beaker is achieving equilibrium, a gradient in vapour molecules will exist, with the higher concentrations near the surface of the liquid. The air is filled with vapour molecules by a flux F_D of molecules that is driven by this gradient.

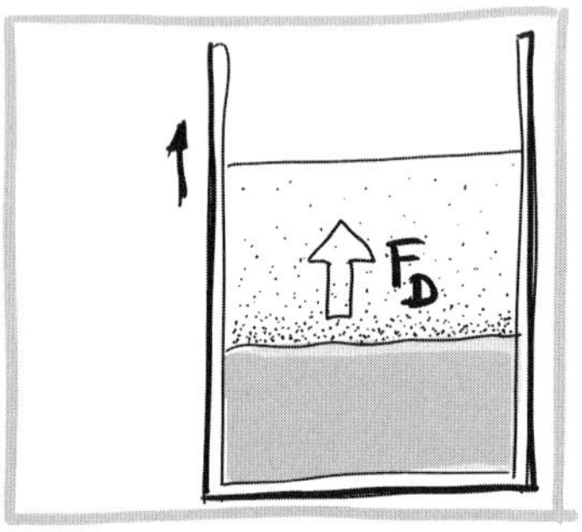

This process is called diffusion and is described by diffusion theory, which considers the gas as a continuum rather than individual molecules. The diffusion flux F_D is written as

$$F_D = -D_v \frac{\partial n_v(x,t)}{\partial x} \qquad [\text{molecules.s}^{-1}.\text{cm}^{-2}] \qquad (8)$$

At all times the vapor concentration $n_v(x,t)$ satisfies the diffusion equation:

$$\frac{\partial n_v(x,t)}{\partial t} = D \frac{\partial^2}{\partial x^2} n_v(x,t) \qquad (9)$$

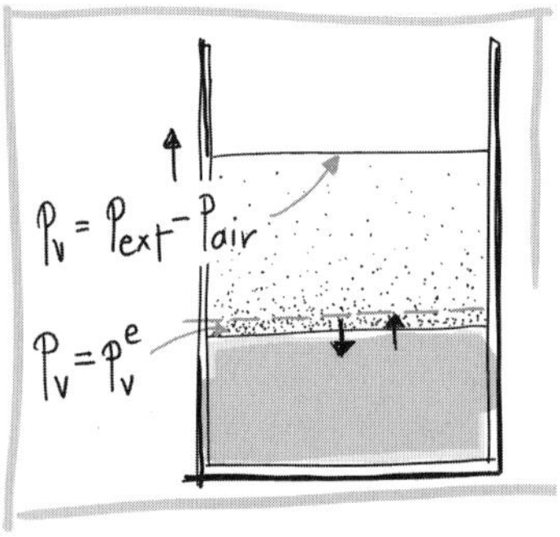

To solve this equation it is reasonable to assume, as one of the boundary conditions, that at each moment in time $n_v(0,t) = n_v^e$, hence that in a very shallow layer above the liquid, the vapour is in equilibrium with the liquid, as a result of the local evaporation and condensation. This assumption will be systematically used when we further develop the description of gas – liquid transfer in solutions and droplets.

2.3 Thermodynamics

The observation that water in a closed beaker spontaneously evaporates and that its molecules diffuse throughout the volume above the liquid points at the fundamental fact that systems, when left on their own, tend to increase their degree of disorder or entropy. This is simply so because there are always more possibilities for a system to be in a disordered state than in an ordered state. Hence, if a system can move with the same

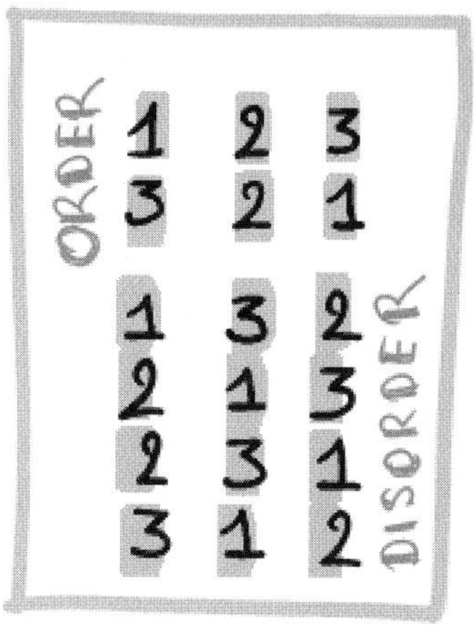

easiness from one state to the other, it is more likely to find it in a disordered state. If we start with the system in an ordered state, then it is natural to find it in a disordered state later on. The system will eventually stop changing and be in equilibrium with its surroundings when it reached its highest state of disorder, which is also the most probable one.
In thermodynamics the increase of disorder or entropy (dS) is related to the addition of heat to the system (dq); heat clearly tends to randomize molecular motion in systems. The relationship is given by the second law of thermodynamics:.

$$dS \geq \frac{dq}{T} \tag{10}$$

and the system will be in equilibrium with its environment when

$$dS = \frac{dq}{T} \tag{11}$$

In our beaker, we don't add heat, but evaporation happens anyway because we start with dry air above the liquid. The entropy increases, in accordance with $dS \geq 0$ because molecules in the gas phase are in a higher state of disorder than molecules in the liquid phase.
The theory of thermodynamics shows how the condition of equilibrium (Eq. 11) can be expressed in terms of the Gibbs free energy G of the system. The tendency of a system to increase its entropy is thus translated into a tendency to lower its free energy. Free energy is a more useful concept than entropy, because it is that part of the internal energy that is related to work, rather than to work and heat together.
It follows that for a system at constant temperature and pressure and on which no work is performed other that expansion work, the condition for equilibrium becomes:

$$dG = 0 \tag{12}$$

The change in free energy dG in our closed beaker experiment is the work that is performed by the system to cancel the difference in chemical potential between the liquid and gas phase. The chemical potential of a compound is nothing else than the free energy per mol of the compound. A difference in chemical potential drives chemical reactions or phase changes, very much like a difference in temperature or pressure drives exchange of heat or matter. During the process of evaporation:

$$A_V dG = \mu_v dn_v + \mu_l dn_l \tag{13}$$

where μ_v and μ_l are the chemical potentials of water in the gas and liquid phase respectively, n_v an n_l the respective number of molecules, and A_V Avogadro's number.

The Gibbs free energy G is defined in such a way that, when at constant temperature and pressure work is done on the system in optimally controlled (= reversible) conditions, we can say that this work will partly lead to an increase of the systems free energy (dG) and partly be used to expand its volume (pdV) so that the pressure in the system remains constant. So we have $dw_{max} = dG + pdV$ *or* $dG = dw_{max} - pdV = dw_{e,max}$. *Optimal control and reversibility mean that during the process the system is always in equilibrium with its surroundings. The latter equation thus shows that, at equilibrium, the change in free energy equals the maximised amount of non-expansion work that can be done on the system. When the same amount of work* $dw_{e,max}$ *is done on the system in uncontrolled (= irreversible) conditions, part of it will degrade in heat which will lead to a reduction in dG. In this case we have* $dG < dw_{e,max}$.

Since $dn_l = -dn_v$

$$A_V dG = (\mu_v - \mu_l) dn_v \tag{14}$$

where μ_v and μ_l are the chemical potentials of water in the gas and liquid phase respectively, n_v an n_l the respective number of molecules, and A_V Avogadro's number.

The evaluation of $\mu_v - \mu_l$ requires some elaboration which is in fact at the core of our understanding of the gas-liquid transition and so we will make it in the Appendix 1. The result is that

$$dG = kT \ln \frac{p_v}{p_v^e(\infty)} dn_l \tag{15}$$

where p_v and $p_v^e(\infty)$ are the vapour pressure and equilibrium vapour pressure over a flat surface respectively.

• So far we have discussed the observation that, when we have a liquid in a closed volume, there will be a spontaneous build up of a vapour pressure that after a while settles at the so-called equilibrium vapour pressure. We also saw that we can reasonably assume that *at each point in time* the vapour pressure right above the surface of the liquid equals the equilibrium vapour pressure. We related the equilibrium vapour pressure to the evaporation rate, which we know is determined by properties of the liquid only. We also showed how thermodynamics can be used to describe this spontaneous process and how equilibrium is reached.

We will now use this understanding to describe what happens when we, step by step, render the system more complex.

3. AN AQEOUS SOLUTIONS IN A CLOSED BEAKER

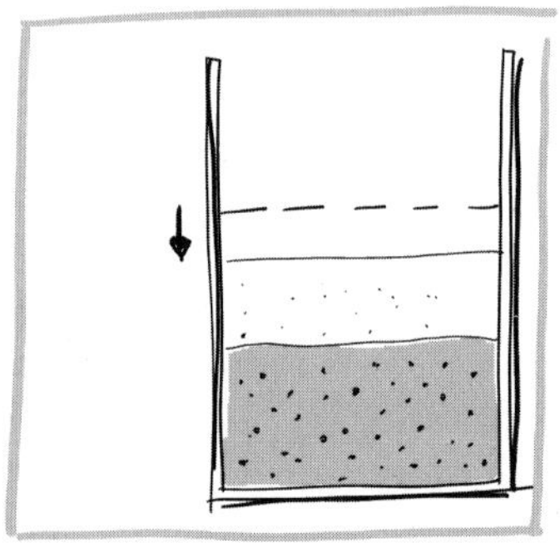

If we fill the beaker with an aqueous solution of a substance which itself is not volatile (i.e. solid salt) we notice that the lid on the beaker will not be pushed as high as in the case with pure water. This means that the equilibrium vapour pressure of a solution is smaller than that of pure water, and it is observed that:

$$p_{v,solution}^e(\infty) = x_l p_v^e(\infty) \qquad \text{(16, \textbf{Raoult})}$$

where x_l is the molecular fraction of water $\dfrac{n_l}{n_l + n_v}$.

This behaviour can easily be explained by the fact that when the aqueous molecular fraction is x_l, then also the fraction of molecules of water present at the liquid-gas interface is x_l. Hence the number of molecules that are able to escape into the gas phase and the equilibrium vapour pressure are also reduced by the factor x_l.

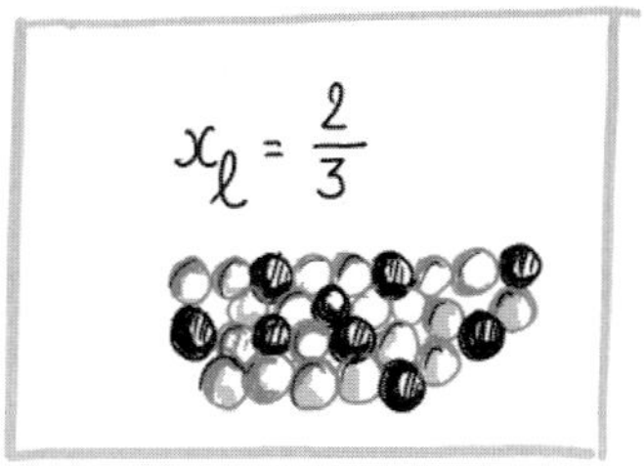

Often the molecular interactions between water and solute molecules are stronger than interactions between water molecules themselves, and this will prevent water molecules even more from escaping the liquid. This deviation from the ideal behaviour is felt especially in concentrated solutions. It is accounted for by introducing the so-called activity coefficient γ_l in Eq. 16:

$$p^e_{v,solution}(\infty) = \gamma_l x_l p^e_v(\infty) \tag{17}$$

γ_l depends on the concentration of all soluble substances that might be present in the solution. The more concentrated the solution is, the more $p^e_{v,solution}(\infty)$ will deviate from the ideal behaviour expressed in Eq. 16.

4. A CLOUD OF DROPLETS IN A CLOSED BEAKER

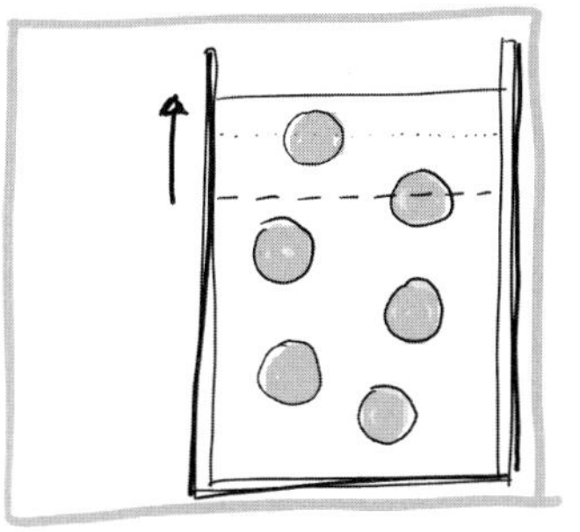

If we disperse the pure water in the beaker as a cloud of small droplets and let it attain equilibrium with its vapour, we see that the lid is lifted up more than when the same amount of water would be present as bulk liquid.

The reason is *not* that there is more surface area in the case of the mist, since Eq. 6 tells that $p^e_v(\infty)$ is dependent on the evaporation rate per unit surface (E/S) only. The reason is that in case of a small droplet a liquid molecule, sitting at the liquid-gas interface, interacts with less neighboring molecules than in the case of a flat surface. Hence it will more easily escape to the gas phase.

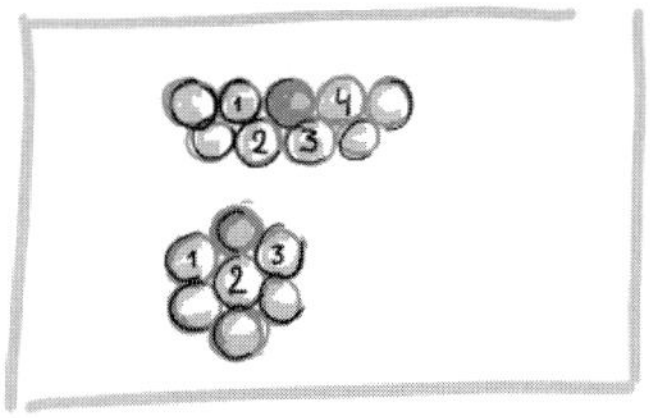

The enhancement of the equilibrium vapour pressure over a droplet with diameter D_p is described by the so-called Kelvin equation:

$$p^e_v(D_p) = p^e_v(\infty)\exp\left(\frac{4\sigma v_l}{kT\,D_p}\right) \qquad \text{(18, \textbf{Kelvin})}$$

where σ is the surface tension of water, v_l the volume of one molecule of liquid water. Since this paper is about the formation of droplets, we cannot *not* derive the Kelvin equation, and we will do it in Appendix 2.

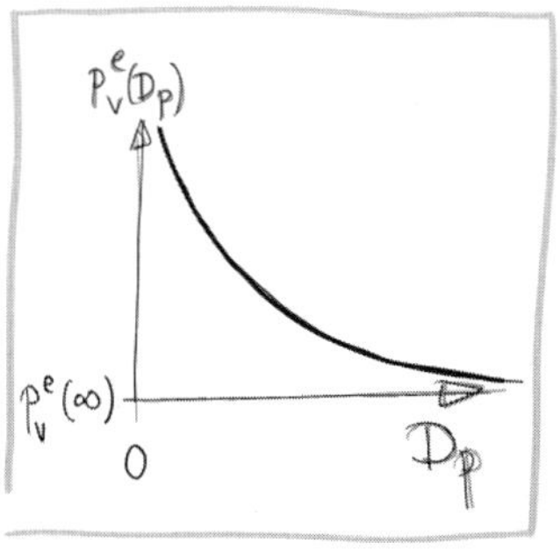

Applied to water droplets with $D_p > 0.05\ \mu$m and T = 298 K Eq. 18 can be approximated within 5% by:

$$p^e_v(D_p) \approx p^e_v(\infty)\left(1 + \frac{4\sigma v_l}{kT\,D_p}\right) \tag{19}$$

which expression we will use in the following derivations.

Cloud droplets in the atmosphere always form on aerosol particles which are often made of soluble salts or concentrated salt solutions. So it is clear that a combination of the Raoult and Kelvin equations will give us a tool to describe quantitatively the properties and behaviour of aerosol and cloud droplets. Doing so, i.e.. combining Eq. 16 and 18 yields

$$p^e_{v,solution}(D_p) \approx \gamma_l x_l p^e_v(\infty)\left(1 + \frac{4\sigma v_l}{kT D_p}\right) \quad (20)$$

Assuming that aerosol droplets with Dp $> 0,05$ m are dilute solutions, hence that the number of soluble molecules in the droplet (n_s) is much smaller than the number of water molecules (n_l), we can write:

$$x_l = \frac{n_l}{n_l + i_s n_s} \approx 1 - \frac{i_s n_s}{n_l}. \quad (21)$$

in which we also considered that the soluble molecule s will completely dissociate in a number of ions i_s.
Combining Eq. 20 and 21 and considering that $\gamma_l \to 1$since we deal with dilute solutions, gives:

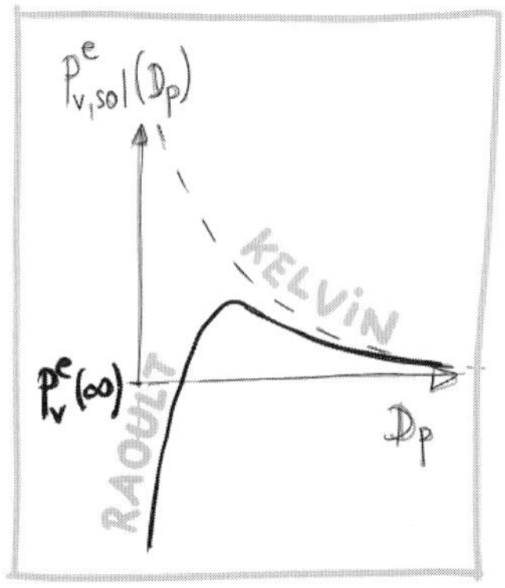

$$p^e_{v,solution}(D_p) \approx p^e_v(\infty)\left(1 + \frac{4\sigma v_l}{kT D_p} - \frac{i_s n_s}{n_l}\right) \quad (22)$$

Equation 22 is an approximate form of the so-called Koehler Equation. It requires only a small mathematical effort to show that the full Koehler equation, valid for a dilute solution of a single compound s, can be written in terms of the diameter of the droplet as follows:

$$p^e_{v,solution}(D_p) = p^e_v(\infty)\exp\left(\frac{4\sigma v_l}{kT D_p} - \frac{6 n_s v_l}{\pi D_p^3}\right) \quad (23, \textbf{Koehler})$$

This equation relates the equilibrium vapour pressure in a cloud of solution droplets to the diameter D_p of the droplets and the number of molecules of soluble material n_s that each of them contains.
In case of the presence of various compounds of soluble material $s, a, \ldots$ and insoluble material i, Eq. 22 can readily be generalized to

$$p^e_{v,solution}(D_p) \approx p^e_v(\infty)\left(1 + \frac{4\sigma v_l}{kTD_p} - \frac{i_s n_s}{n_l} - \frac{i_a n_a}{n_l} - \frac{n_i}{n_l}\ldots\right) \tag{24}$$

in which the total number of molecules satisfies the condition

$$n_l v_l + n_s v_s + n_a v_a + n_i v_i + \ldots = \frac{\pi D_p^3}{6} \tag{25}$$

The equations 23 and 24 are the basis for our understanding the bahaviour of aerosol and cloud droplets in the atmosphere, and we will analyse their meaning more thoroughly later.

5. WATER IN A BEAKER OPEN TO THE ATMOSPHERE

We now remove the lid from our beaker. In terms of temperature and pressure nothing changes compared to our previous experiment: we assume the liquid takes the temperature of the atmosphere and that the pressure on the liquid is the atmospheric pressure. The difference is that the vapour pressure in the surroundings of the liquid is no longer completely governed by evaporation of that liquid but by the presence of "wet stuff" elsewhere, by the presence of the ocean, or a cloud, or vegetation, etc. Hence the vapour pressure in the vicinity of the liquid is "fixed" at the ambient vapour pressure $p_v = p_{v,ambient}$, which is controlled by macroscopic features of the environment in which our beaker sits.

The ratio of $p_{v,ambient}$ to $p^e_v(\infty)$ is called the ambient saturation ratio $S_{ambient}$. It is important to notice that $S_{ambient}$ depends on the temperature, through the strong dependence of $p^e_v(\infty)$ on temperature.

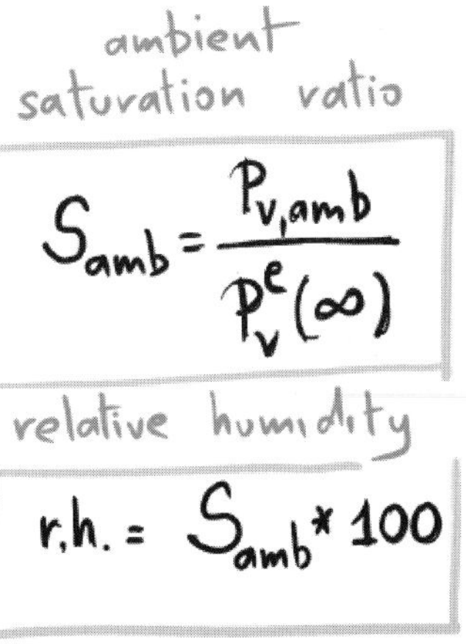

For the following discussions we will make use of the assumption that the vapour pressure in a shallow layer right above the surface equals the equilibrium vapour pressure (see our discussion following Eq. 9) and that the exchange of vapour molecules between the liquid and the open atmosphere is governed by diffusion from this layer to the environment.

The steady state diffusion flux F_D to the droplet can be calculated using Eqs. 9 and 10, considering spherical coordinates and integrating from the surface of the droplet, where $p_v = p^e_{v,solution}(D_p)$ to infinity, where $p_v = p_{v,ambient}$.

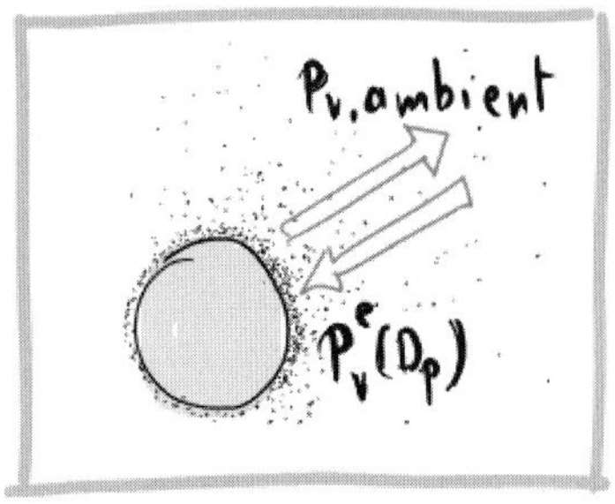

The result is:

$$F_D = \frac{2\pi D D_p}{kT}(p_{v,ambient} - p^e_{v,solution}(T, D_p)) \qquad (26)$$

The direction of the flux to the droplet is controlled on the one hand by macroscopic features in the environment ($p_{v,ambient}$) and on the other by microscopic properties of the droplet, notably its diameter and the presence of soluble material ($p^e_{v,solution}(T, D_p)$).

- So far we have focused on how the properties of a solution droplet determine its equilibrium vapour pressure. In applying our knowledge to the atmosphere, however, we will be more interested how changing ambient conditions will effect the properties of the droplets. Considering Eq. 26, we see that when $p_{v,ambient} = p^e_{v,solution}(T, D_p)$, no net exchange of water molecules will take place, and the droplet is in equilibrium its surroundings,

when $p_{v,ambient} < p^e_{v,solution}(T, D_p)$, water molecules will evaporate and the droplet will shrink, when $p_{v,ambient} > p^e_{v,solution}(T, D_p)$, water molecules will condense and the droplet will grow.

6. IDEAL AEROSOL AND CLOUD DROPLETS: THE CLASSICAL KOEHLER THEORY

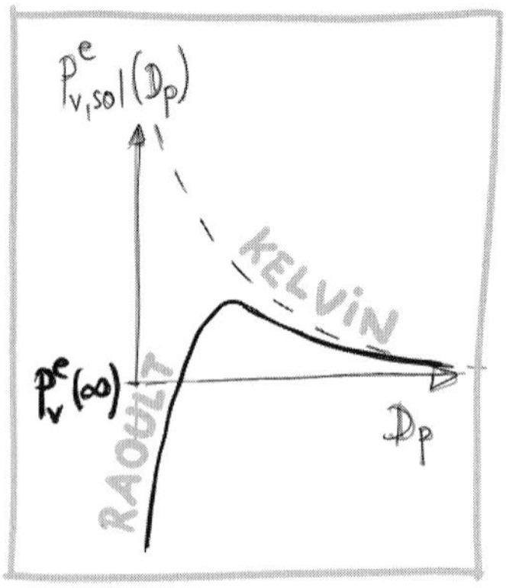

Considering just the Kelvin and the Raoult equations we can already see that we deal with complex behaviour when studying an aqueous solution droplet. For instance, when the droplet grows by condensation, according to Kelvin the vapour pressure right above its surface ($p^e_{v,solution}(D_p)$) will decrease. At the same time, however, the solution becomes more dilute, and according to Raoult the vapour pressure right above its surface will increase.

The Koehler equation (Eq. 23) allows to determine which of these two effects will prevail, and under what conditions. In the figure we have plotted the ratio $p^e_{v,solution}(D_p)/p^e_v(\infty)$ as a function of the droplet diameter D_p, for a given temperature and a given amount of soluble material in the droplet n_s.

The Figure illustrates how for very large droplets the first term in the exponential of Eq. 23 will dominate the second, and the Kelvin effect prevails. For very small droplets, the second term dominates and the Raoult effects prevails.

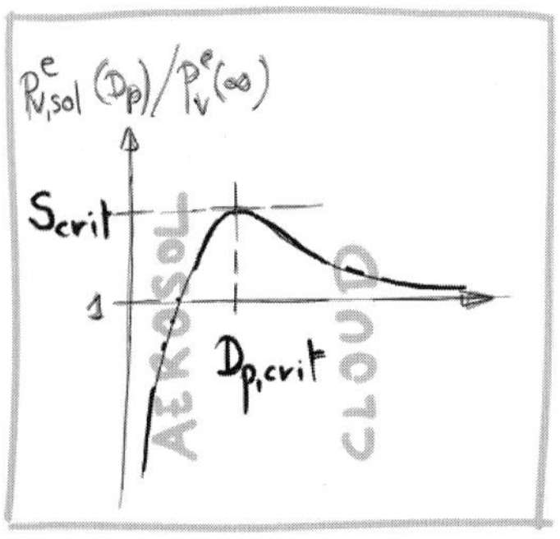

The separation between the two droplet size domains is at the maximum in the Koehler curve, where $p^e_{v,solution}(D_{p,crit})/p^e_v(\infty) = S_{crit}$. Assuming that all droplets contain the same amount of soluble material, $D_{p.crit}$ can be seen as the separation between aerosol droplets and cloud droplets based on their size. In the atmosphere different droplets will contain different amounts of soluble material. We will see the implication of that in a moment.

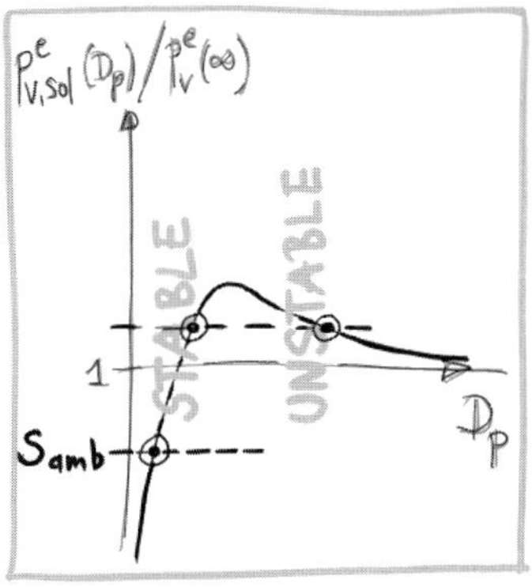

We now consider the ambient vapour pressure $p_{v,ambient}$, or rather $S_{ambient}$. We assume for the moment that $S_{ambient}$ is totally determined by macrospcopic features in the environment, so that it is independent of the droplet size. As such it can be represented by a horizontal line on top of the Koehler curve. In the points where this line crosses the Koehler curve $p_{v,ambient} = p^e_{v,solution}(D_p)$, which means that the droplet with the corresponding diameter D_p is in equilibrium with its environment.

We can see that when the ambient air is sub-saturated ($S_{ambient} < 1$) only aerosol droplets ($D_p < D_{p,crit}$) can be in equilibrium, and that this equilibrium is stable. Indeed; if there were a small increase in size $p^e_{v,solution}(D_p)$ would become larger than $p_{v,ambient}$ and the droplet would evaporate again.

When the ambient air is super-saturated ($S_{ambient} > 1$), but still lower than S_{crit}, the Kohler curve shows that aerosol droplets ($D_p < D_{p,crit}$) *and* cloud droplets ($D_p > D_{p,crit}$) can be in equilibrium with their environment. The aerosol droplets are in a stable equilibrium as before. The cloud droplets, however, are in an unstable equilibrium: once they grow they will keep growing because $p^e_{v,solution}(D_p)$ stays $< p_{v,ambient}$.

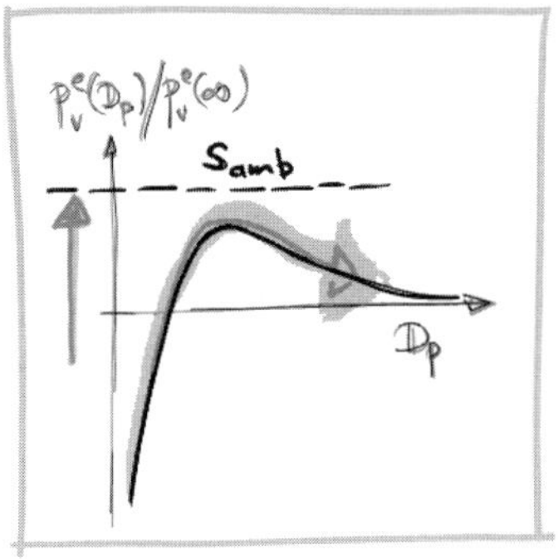

When $S_{ambient}$ in an air parcel gradually increases from sub-saturated to super-saturated, something interesting will happen the moment it becomes larger than S_{crit}. Initially, the aerosol droplets would be in equilibrium with the surroundings. As the $S_{ambient}$ increases they would collect vapour molecules and grow, but always remain in equilibrium (i.e the moment $S_{ambient}$ would decrease, they would shrink accordingly). The moment $S_{ambient}$ becomes larger than S_{crit}, the droplets will pass from a stable to an unstable equilibrium and they will remain there only shortly. Indeed, since they were growing $p^e_{v,solution}(D_p)$ will quickly become smaller than $p_{v,ambient}$, so that there will be condensation which will make the droplets grow even more, desperately in search for equilibrium.

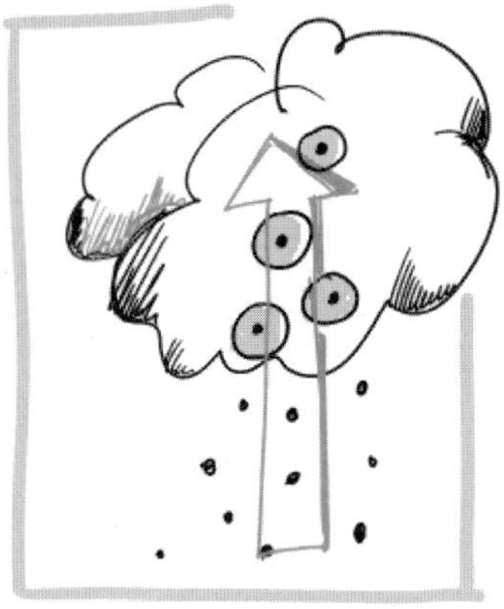

Classical Koehler theory hence predicts a critical behaviour the moment $S_{ambient} > S_{crit}$, which leads to the creation of a wide size gap between the stable aerosol droplets and the cloud droplets that have grown away.

What we have just described is the process of cloud droplet formation or *activation*. In the atmosphere an air parcel usually becomes supersaturated, not by increasing $p_{v,ambient}$ but by decreasing the temperature an hence by decreasing $p_v^e(\infty)$.

Supersaturated conditions can develop when an air parcel is moving upwards and cools, or when an air parcel moves horizontally from a warm to a colder surface, or when it cools by loosing energy by radiation. In all these cases cloud droplets will form if the air parcel contains aerosol particles.

Models describing the dynamics of cloud droplet formation and growth cannot assume that $S_{ambient}$ remains constant during the process of activation and subsequent growth. $S_{ambient}$ will eventually decrease since vapour molecules in the environment are consumed in the condensation process.

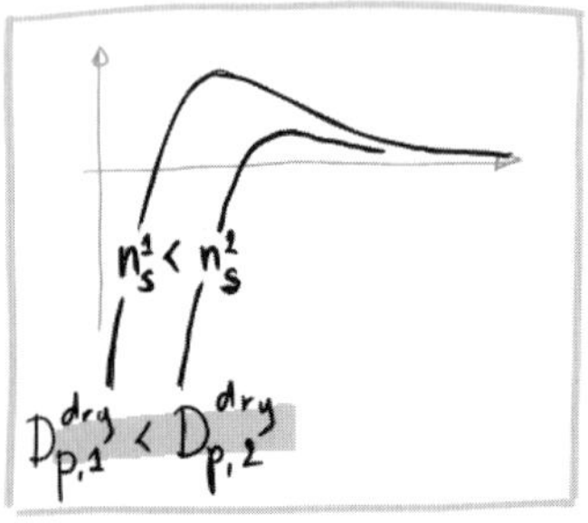

Before going further in the analysis, it is important to see how the classical Koehler curve depends on the number of solute molecules n_s in the droplet. n_s can be related to the so-called dry diameter of the particle D_p^{dry} which is the diameter of the particle when $S_{ambient} = 0$ and the particle consist only of the dry soluble material.

It follows from the Koehler equation that the larger n_s, the lower S_{crit} and the larger $D_{p,crit}$. Hence as $S_{ambient}$ rises, e.g. during an updraft of the air parcel, the particles with the largest dry diameter will be activated first and the smaller ones subsequently. $S_{ambient}$ will eventually stop rising, because of depletion of the water vapour. Its maximum value S^* will corresponds with a certain n_s^* or dry diameter D_p^{*dry}. All particles with $D_p^{dry} > D_p^{*dry}$ are called cloud condensation nuclei as they will become cloud droplets, whereas the other will remain the so-called interstitial aerosol droplets.

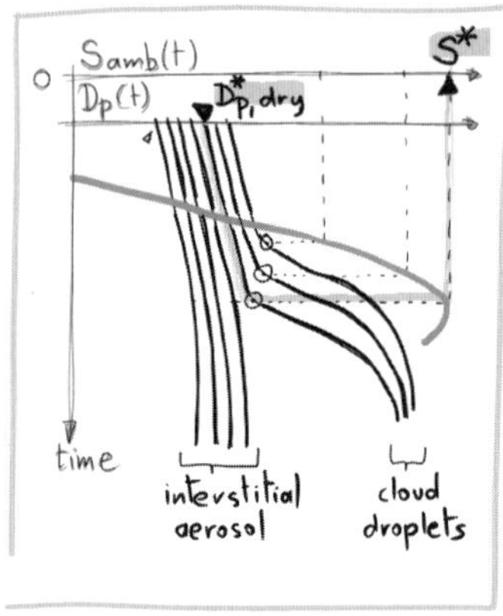

We have already mentioned that $S_{ambient}$ increases by macroscropic processes such as the lifting and cooling of the air parcel, or the rate of radiative cooling, etc. It is now also clear that the $S_{ambient}$ will be limited by the number and size and chemical composition of the particles present in the air parcel. Hence D_p^{*dry} and S^* and the number of cloud condensation nuclei depend both on macroscopic and microscopic features. This makes of the cloud condensation nuclei a rather poorly defined set of particles. In the real atmosphere the highest S^* is found in the strong updrafts within cumulus clouds, where it reaches values up to about 1.02. The corresponding D_p^{*dry} of soluble salt particles is around 10 nm. In the weaker updrafts of stratus clouds S^* reaches at most 1.005 and D_p^{*dry} of soluble particles around 70 nm.

- There is discussion between the "macroscopists", dealing with atmospheric dynamics, and the "microscopists", dealing with aerosol physics and chemistry, about which of the processes would determine cloud droplet properties such as their number and diameter. The answer can probably not be given, as both "-ists" still have problems to describe all details of their processes in a three dimensional model of the atmosphere and do the required sensitivity analyses.

7. REAL AEROSOL AND CLOUD DROPLETS: BEYOND THE CLASSICAL KOEHLER THEORY

We recall that the Koehler equation (Eq. 23) was first developed for describing the effect on droplet behaviour of soluble material that itself has no vapour pressure. In atmospheric applications it is applied to e.g. sodium chloride, a main component of sea salt particles, or ammonium sulfate, a main component of continental pollution aerosol. We have seen that the Koehler theory predicts a critical behaviour, whereby cloud droplets are suddenly formed once a certain level of supersaturation is arrived at. This critical behaviour is reflected by the fact that clouds can have a distinct base, for instance over some parts of the oceans where wind velocities are high and cloud droplets primarily form on sea salt particles. In polluted areas over the continents distinct cloud bases often do not exists. When landing with an airplane into a polluted boundary layer (like the one over the Milano airports, for instance) one can often notice clouds that develop out of the polluted layer without a clear transition between the pollution aerosol or haze and the cloud. This suggests that cloud activation might not always be as critical as the classical Koehler theory predicts.

Over the years, until very recently, extensions to the classical Koehler theory have been made in an attempt to better describe the observations. With hindsight these extensions can all be reduced to a use of Eq. 24.

It has been pointed out already by Junge [5] that in a polluted atmosphere, soluble material is often mixed with insoluble

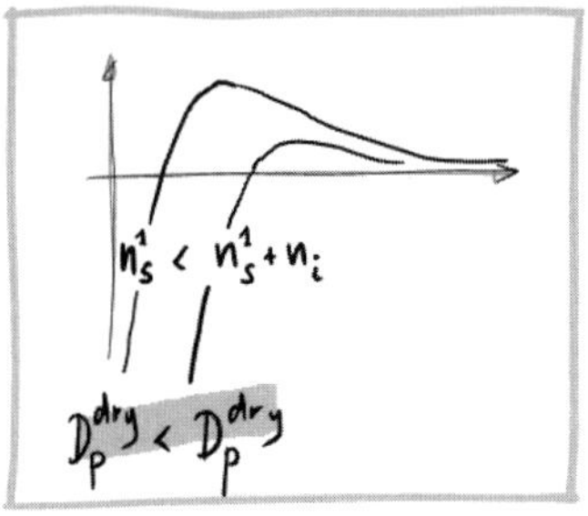

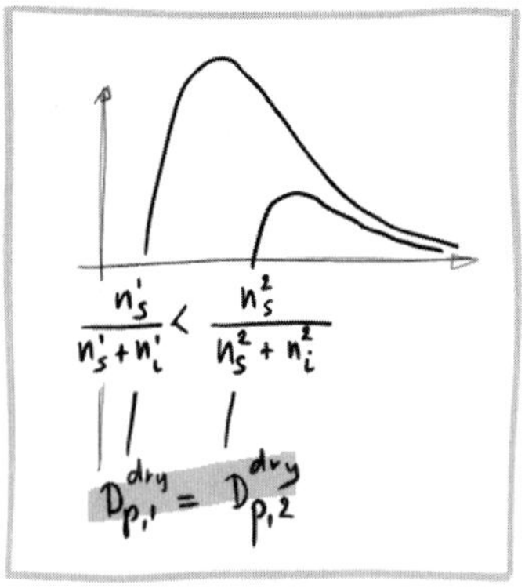

material within the same particle. Combustion. e.g., might easily lead to particles containing ammonium sulfate and elemental carbon. Equation 24 says that adding a number of molecules of insoluble material n_i to a solution droplet, while keeping its diameter D_p and number of molecules of soluble material n_s constant, will reduce the vapour pressure right above its surface. This is because in keeping D_p and n_s constant adding n_i must make n_l go down. Hence the solution remaining in the droplet becomes more concentrated and the Raoult effect will be stronger.

The way we described this is of course not the way it will happen in the real atmosphere. It is more likely that the insoluble and soluble material are already mixed in the dry state, before the particle takes up water and becomes activated. So it is more useful to start by considering particles with a given dry diameter and various fractions of soluble material $\frac{n_s}{n_s - n_i}$.

Doing this it follows from the same Eq. 24 that the lower the fraction of soluble material in the original particles the higher S_{crit}, and the higher the ambient super-saturation required to activate them.

This is a useful extension of the Koehler theory but does not predict a less critical behaviour, on the contrary.

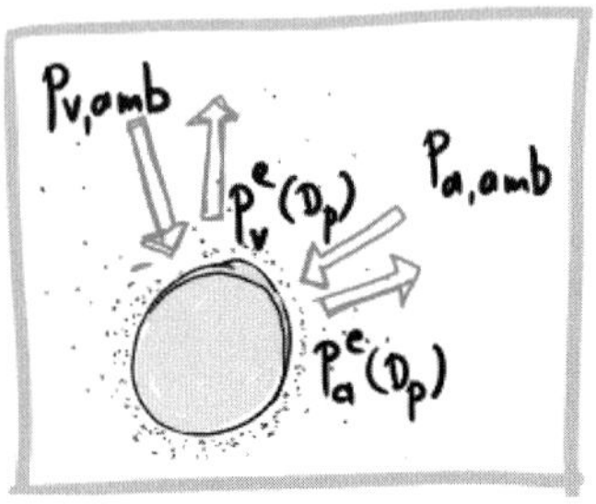

■ Kulmala et al [6] pointed out that a polluted atmosphere might also contain soluble gases, notably nitric acid derived from photo-oxidation of nitrogen oxides, also emitted by combustion sources, and ammonia emitted in agricultural practices. Being soluble, these gases will enter the growing aerosol droplets and partition themselves between the droplets and the gas phase according to Henry's law. As such they will add a term $-\frac{i_a n_a}{n_l}$ in the right hand side of Eq. 24 and help reduce the vapour pressure right above the droplets. Henry's law is written as

$$p^e_{a,solution}(\infty) = H_a x_a \qquad (27)$$

where $p^e_{a,solution}(\infty)$ is the equilibrium vapour pressure of the gas over the solution, H_a Henry's law constant, and

$$x_a = \frac{n_a}{n_l + n_s + n_a}$$

In the case of droplets, we need to consider the Kelvin effect and obtain

$$p^e_{a,solution}(D_p) = H_a x_a \exp\left(\frac{4\sigma v_a}{kT D_p}\right) \qquad (28)$$

When the droplet is at equilibrium with its environment Eq. 28 relates the ambient vapour pressure of the soluble gas ($p_{a,ambient} = p^e_{a,solution}(D_p)$) to the number of its molecules in the droplet n_a needed in Eq. 24.

The presence of soluble gases in the atmosphere and their (immediate) dissolution while the particles are taking up water will thus lower S_{crit} as well as the ambient super-saturation required to activate them. An important consequence of this analysis is that S_{crit} might disappear under realistic concentrations of HNO_3 or NH_3, and aerosol droplets might even grow without activation into cloud droplets even in sub-saturated conditions. That would lead to a less clear distinction between aerosol and cloud droplets in terms of their size, and might explain the gradual transition between a pollution haze and clouds observed in pollutated air masses.

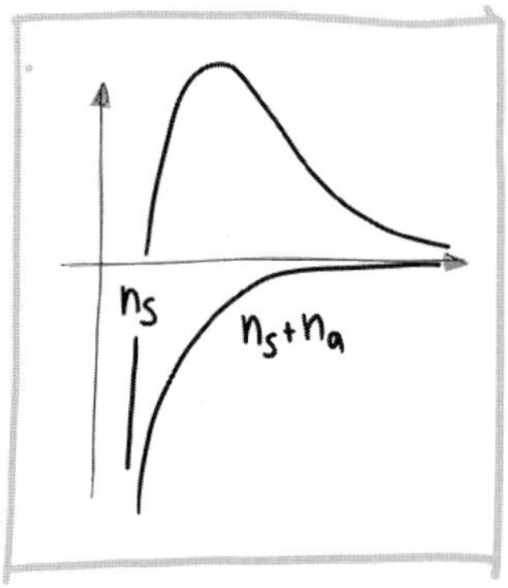

■ Finally a third effect. It is known that certain organic substances dissolved in water are able to reduce the surface tension σ of water. (See Eq. A8 in App. 2 for a discussion on the surface tension). Surface will therefore more easily form, which might lead to the creation of foams. These substances are called surfactants: *surf*ace *act*ive *a*gen*ts*. They are typically large organic molecules with a soluble head and an insoluble tail.

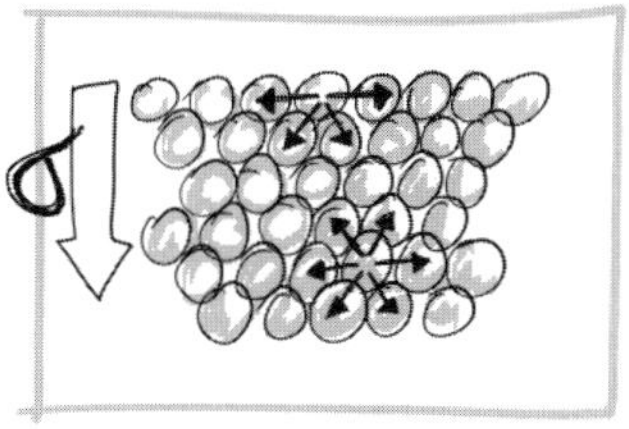

Facchini et al. [5] pointed out that surfactants are also present in the atmosphere, and that they do dissolve into cloud droplets where they might lower the surface tension. According to Kelvin, this will lower the equilibrium vapour pressure of water over the droplet. However such surfactants introduce also quite some mass of insoluble material into the droplet: their tails. The effect of adding this insoluble material will depend very much on whether surfactants are already mixed with the other aerosol material in the dry state, or whether they are dissolving in the droplet while it is taking up water. The overall effect on the required ambient super-saturation to activate droplets containing surfactants seems to be dominated however by the lowering of the surface tension.

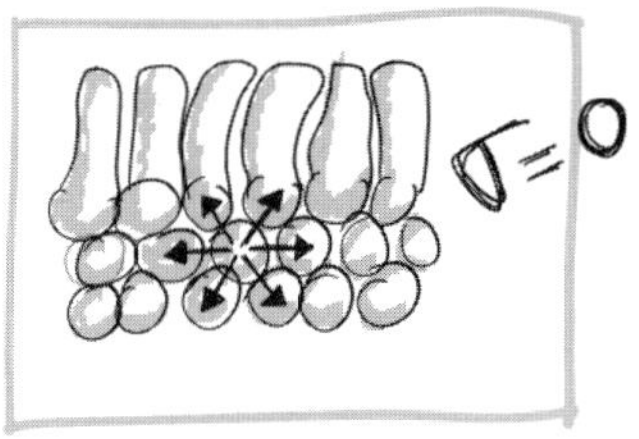

8. CONCLUSION

The expansion of the classical Koehler theory and the testing of its relevance against observations in the real atmosphere is front-line research in the field of atmospheric physics today. We talked about discussions between "macroscopists" and "microscopists" before. It seems that the "microscopists" are now further split between physicists and chemists. In a recent paper, for instance, Dusek et al. [6] have studied a wide variety of air masses (polluted, clean ...) and looked at the potential of aerosol particles to activate. They found that activation is primarily dependent on the dry diameter of the aerosol particles, and that the effects discussed above, which are related to the chemical composition of the aerosol, are of secondary importance, or cancel each other out.

The stakes of these scientific discussions are high. Aerosols and clouds are at the nexus between two important societal problems: (local) air pollution and (global) climate change [7]. If all the effects mentioned above indeed play a role, it might be very difficult to describe cloud formation and its links to atmospheric pollution in climate models. It would indeed require a tremendous amount of information about the physical and chemical nature of the aerosol particles and the gases surrounding them. On the other hand there is an urgent demand for assessments and model studies to develop policies and address air pollution and climate change in an integrated way. We need to proceed with simplified descriptions to put into climate models, but it is important to clarify which effects dominate before simplifications are made. Such clarifications can come from observations, from sensitivity analyses of models, from intuition, an most likely from a combination of all of this.

9. APPENDIX 1

One can derive within the theory of thermodynamics that

$$\left(\frac{\partial G}{\partial p}\right)_{T,n_i} = V \tag{A1}$$

or when expressed per mol of compound i in the system

$$\left(\frac{\partial \mu}{\partial p}\right)_{T,n_i} = v_i \tag{A2}$$

Where in this appendix v_i is the *molar* volume, and not the *molecular* volume as in the main text.
This equation gives the pressure dependence of the chemical potential at constant temperature and constant composition.

In an ideal gas $v_v = \frac{RT}{p_v}$, where $R =$ the gas constant $= A_v k$. Substituting this in Eq. A2 and integrating from a reference pressure p_{ref} to an arbitrary pressure p_v, yields the expression for the chemical potential of a vapour:

$$\mu_v = \mu_{ref} + RT \ln \frac{p_v}{p_{ref}} \tag{A3}$$

To derive the chemical potential of the liquid μ_l we start from the fact that at equilibrium ($dG = 0$) it must be equal to that of the vapour. Hence, and using Eq. A3

$$\mu_l^e = \mu_v^e = \mu_{ref} + RT \ln \frac{p_v^e}{p_{ref}} \tag{A4}$$

The fact to consider now is that properties of liquids (and solids) are only weakly dependent on the pressure (at least in the range of pressure changes observed in the atmosphere) and so Eq. A4 also holds for liquids not in equilibrium with their vapour:

$$\mu_l = \mu_l^e = \mu_{ref} + RT \ln \frac{p_v^e}{p_{ref}} \tag{A5}$$

Hence, in the same way the chemical potential of a gas is related to its vapour pressure (Eq. A3), so is the chemical potential of a liquid related to its own pressure, which is the equilibrium vapour (see our discussion in occasion of Eq. 8).

Subtracting Eq. A5 from Eq. A3 gives the expression we where searching for:

$$\mu_v - \mu_l = RT \ln \frac{p_v}{p_v^e} \tag{A6}$$

At equilibrium $p_v = p_v^e$, as it should since it was used as a condition in deriving the equation. The importance of Eq. A6 is that it gives the difference in the chemical potential of a liquid and that of its vapour in all conditions, and we can use it to describe how a system will move towards equilibrium.

10. APPENDIX 2: Derivation of the Kelvin equation

We can imagine that creating droplets out of just vapour, must involve some more "effort" or "work" than letting evaporate water. It is precisely the consideration and quanitification of that work that leads to the Kelvin equation.

So, rather than calculating the energetics of dispersing the bulk liquid into a cloud, we will calculate the energetics of forming liquid droplets out of the vapour phase, a process which is called *nucleation*. The end-result, i.e the conditions under which the droplets exists in equilibrium with their surroundings must be the same.

We still work in our closed beaker at constant temperature and pressure. What we need to do is to calculate the change in free energy dG during the nucleation process.

The first piece of work is that needed to cancel the difference in chemical potential during the phase transition. We can use Eq. 15, but considering that we now change from gas to liquid, rather than from liquid to gas. So:

$$dG = -kT \ln \frac{p_v}{p_v^e(\infty)} dn_l \tag{A7}$$

There is however a second piece of work that is done by the system, which is the work needed to create the surface of the liquid. When a droplet grows, surface is created and this implies indeed work. It can be understood as follows. In the bulk of the liquid, each molecule is surrounded by other molecules and the sum of the intermolecular force on that molecule is zero. A molecule at the surface of the liquid, however, is only surrounded by molecules at one side and hence it feels a net force towards the bulk of the liquid. (See figures near the end of chapter 6) So, each time the surface increases molecules must be brought from within the bulk against that force to the surface, which implies work which is described by:

$$dG = \sigma ds \tag{A8}$$

where σ is the surface tension of water and ds the increase in the surface area.

So the total change in free energy when moving n_l molecules from the gas phase to the droplet phase must be written as:

$$dG = -kT \ln \frac{p_v}{p_v^e(\infty)} dn_l + \sigma ds \quad \text{(A9)}$$

Both dn_l and ds can easily be expressed in terms of the increase of the diameter of the droplet dD_p, from which follows:

$$dG = -\frac{\pi k T D_p^2}{2 v_l} \ln \frac{p_v}{p_v^e(\infty)} dD_p + 2\pi\sigma D_p dD_p \quad \text{(A10)}$$

The diameter of a droplet in equilibrium with its vapour is calculated by setting Eq. A10 to zero. It is easily seen that there can be a physical solution (Dp > 0) to this equation only if $p_v > p_v^e(\infty)$, and this solution is

$$D_p = \frac{4\sigma v_l}{kT \ln p_v / p_v^e(\infty)} \quad \text{(A11)}$$

This turns into the Kelvin Equation (Eq. 23) when we consider that at equilibrium $p_v = p_v^e(Dp)$. Saying that a droplet can exist only when the ambient vapour pressure is larger than the vapour pressure above a flat surface, is the same as saying that the equilibrium vapour pressure over a droplet is larger that the equilibrium vapour pressure over a flat surface.

Acknowledgments

With thanks to Rita Van Dingenen and Francesca Barnaba for proof reading and commenting the manuscript.

References

[1] This title is vaguely referring to two books that I know do exist but never had the occasion to read: "Consider a spherical cow" by John Harte and "Clouds in a Glass of Beer" by Craig F, Boren.
[2] Pruppacher H. and Klett J., "Microphysics of Clouds and Precipitation" D. Reidel Publishing Company, (1980).
[3] Seinfeld J. and Pandis S. "Atmopsheric Chemistry and Physics" John Wiley and Sons, (1998).
[4] Atkins P.W., Physical Chemistry, 2nd Edition, Oxford University Press (1982).
[5] Junge C.E., Ann. Meteor. 3 (1950) 128, mentioned in [2].
[6] Kulmala M. et al., "The effect of atmospheric nitric acid vapor on cloud condensation nucleus activation.", J. Geophys. Res., 98, (1998), p. 22949.
[7] Facchini M.C. et al., "Cloud albedo enhancement by surface-active organic solutes in growing droplets" Nature 401 (1999), 257.
[8] Dusek et al., "Size matters more than chemistry for cloud-nucleating ability of aerosol particles", Science, 312 (2006), 1375.
[9] Raes F., "Climate change and air pollution – Research and Policy" International Geosphere Biosphere Programme Newsletter No 65, (2006).

J. Phys. IV France **139** (2006) 81–99
© EDP Sciences, Les Ulis
DOI: 10.1051/jp4:2006139007

Dynamics of the West African monsoon

N.M.J. Hall[1] and P. Peyrillé[2]

[1] *CNRS/LTHE, BP. 53, 38041 Grenoble Cedex 9, France*
[2] *CNRM, Météo France, 42 avenue Gaspard Coriolis, 31057 Toulouse Cedex 01, France*

Abstract. A review is given of the dynamical mechanisms responsible for the monsoon circulation over West Africa. Features of the circulation are first described, including the seasonal displacement of the rain bands, the structure of the heat low over the Sahara, the meridional circulation to the south and the associated zonal jets. Simple theories for the zonal-mean meridional circulation are then presented, using the principles of angular momentum conservation, thermal wind balance and moist convective equilibrium. The application of these theories to the West African monsoon reveals a sensitivity to the low-level meridional gradient of equivalent potential temperature, which helps explain observed variability in the monsoon onset. Processes leading to east-west asymmetries in the circulation are also described, and mechanisms linking West African rainfall anomalies with remote events in the tropics are discussed. These dynamical considerations are then placed in the broader context of the ongoing AMMA research program.

1. INTRODUCTION

The word 'monsoon' (meaning 'season') is generally associated with seasonal rain in the tropics. The arrival of monsoon rains in many tropical countries is the most important event in the seasonal cycle, and local agriculture and water resources can be greatly affected by the timing and intensity of this event. Associated with the seasonal cycle in rainfall are large-scale changes in atmospheric circulation both near the surface and in the upper level jet streams. These circulation changes occur in many locations, both within and outside the tropical belt. They represent the large-scale dynamical response to seasonal changes in heating patterns caused by the contrast in the thermal capacities of the land surface and the ocean. The precise nature of the effect varies from a relatively subtle change in the divergent component of the circulation as a continent heats up in the summer (e.g. the North American monsoon) to a total reversal in the direction of the prevailing wind at all levels (the Indian monsoon). The interplay between the divergent and rotational components of the heating response can lead to different types of monsoon for different continental geometries, but the basic mechanism is as follows.

A continental surface warms rapidly in the early summer relative to the adjacent ocean. Heat is transmitted to the atmospheric column above it, mainly by deep convection. Convective rainfall therefore plays an important part in the setup of a monsoon circulation and is not just a consequence of it. The warming of the atmospheric column over the land implies a raising of pressure surfaces at upper levels compared to the adjacent atmospheric column over the ocean. Upper level air is accelerated down the pressure gradient towards the ocean, evacuating the column over the land and creating low pressure at the surface. This is called a 'thermal low' and is a key feature of monsoon circulations. The low level pressure gradient then gives rise to a circulation in the opposite direction near the surface, creating a circulation cell, towards the continent at low levels and away from it at high levels. In this way a low-level supply of moist oceanic air is set up to maintain convection over the continent. This purely divergent pattern comes into large scale balance as positive vorticity develops at low levels and negative vorticity at upper levels, setting up cyclonic and anticyclonic circulations respectively.

Different parts of the tropics have their different monsoons, and each affects the global scale divergent (Hadley and Walker cells) and nondivergent (trade wind) circulations in such a way as to influence one another and to control the seasonal cycle of rainfall and convection. From here on we will concentrate on the so called 'West African Monsoon' (WAM), which is smoothly connected to the Indian monsoon upper level outflow, but also possesses some distinctive features of its own. The WAM is currently under intensive investigation by way of a number of national and international collaborative projects collectively known as AMMA [1] (Analyse Multidisciplinaire de la Mousson Africain or African Monsoon Multidisciplinary Analysis, to show but two of the many languages served by this acronym). At the time of writing an observational field program is underway to study the 2006 monsoon. The WAM is a complicated system which involves many interactions between the atmosphere, the land surface and the ocean and which is influenced by processes occurring over a range of spatial scales, down to individual rain events on the order of tens of kilometers. Rainfall is a highly intermittent variable, displaying a large degree of variability over short distances. It is also variable on intraseasonal timescales, with active-break behaviour in the displacement of the rain band. And it is famously variable on interannual timescales, with decadal signatures in drought in the Sahel region. The socioeconomic consequences, and the importance of understanding and predicting these variations cannot be overstated. Most of the effects mentioned above are, however, beyond the scope of this article, in which we try to provide a primer on the basic dynamical mechanisms underpinning the the atmospheric circulation of the monsoon system.

After giving a basic description of the the various elements that comprise the WAM in section 2, a review will be given of the dynamical ideas that have been exploited to explain various aspects of the WAM. First in section 3 the vertical-meridional structure is addressed and then in section 4 we discuss variations in longitude and in time. Section 5 contains a discussion and some details of the AMMA field program.

2. A BRIEF DESCRIPTION OF THE WEST AFRICAN MONSOON

From the Guinea coast to the Sahara desert, West Africa is characterised by east-west orientated climatic zones. With no large mountains to disturb this zonally banded structure, the associated monsoon circulation is also relatively uniform in the east-west direction. The WAM is characterised by the migration of zonally banded rainfall from the Guinea coast to the Sahel and back again, resulting in two rainy seasons per year in the south and one in the north. The associated circulation is, however, more complicated than the generic 'monsoon' described above. The meridional contrast between the hot, dry desert and the (relatively) cool, moist gulf region is responsible for most of the features particular to the west African region. In 1975, Charney [2] proposed a mechanism in which desert regions sustain themselves because a subsidiary circulation is naturally set up in which the essential thermal equilibrium is between the adiabatic warming of descending air and radiative cooling. The radiative deficit is aggravated by the increased albedo of the desert terrain, and the subsidence ensures dry conditions, which maintains the desert. In winter, the Sahara desert is the largest region of subsidence in the world and the thermal equilibrium described above is largely obtained. In spring and early summer, the tropical divergent circulation shifts. The Hadley cell reverses so that the main descending branch is to the south and the developing Walker circulation associated with the Indian monsoon outflow pushes the Saharan subsidence zone west and finally south into the Gulf of Guinea. Above about 5 km the Sahara is still characterised by subsidence, but at lower levels intense surface heating, redistributed vertically by dry convection, creates a 'heat low', and an ascending region with surface convergence and mid-tropospheric divergence. This ascending region is well mixed by the dry convection and constitutes an air mass known as the Saharan air layer (SAL). The SAL sits beneath the upper level descent as shown in fig. 1, and displays uniform potential temperature θ and equivalent potential temperature θ_e.[1] This can be seen in the synoptic observations based on aircraft dropsondes [4] illustrated in fig. 2. In the shallow

[1] Most of our discussion in this article will be in terms of θ_e. Like potential temperature θ, it is a point measure of the heat content of air with the dimensions of temperature, but it also includes the moisture content. It is materially conserved under reversible moist adiabatic processes. The vertical profile of θ_e gives information about upright convective instability.

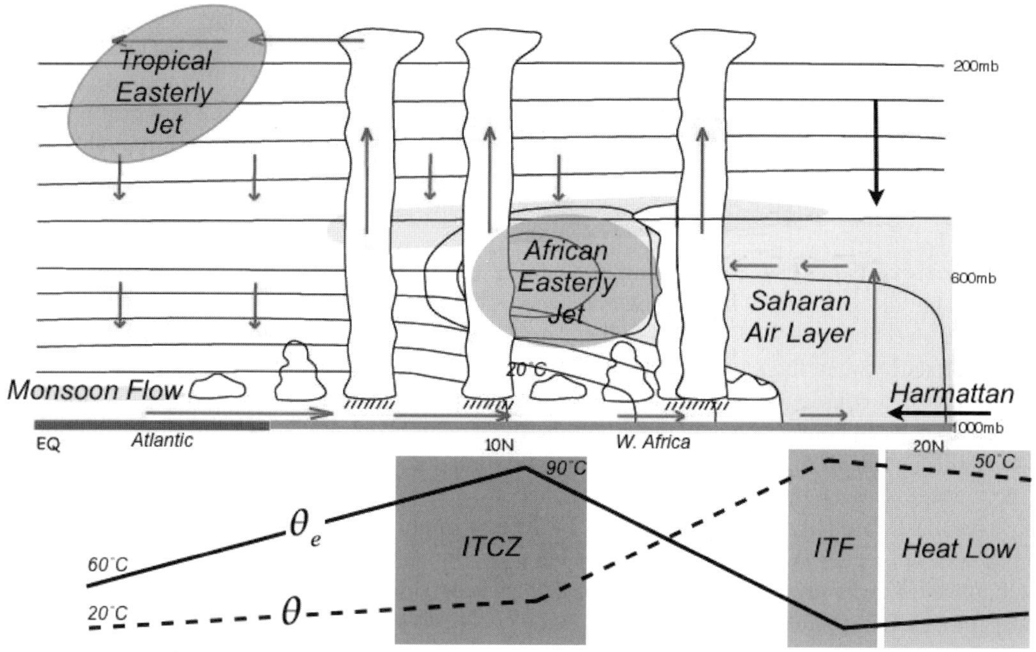

Figure 1. Schematic latitudinal section showing elements of a fully developed West African Monsoon, with typical meridional profiles of θ and θ_e. Adapted from Mohr and Thorncroft [3].

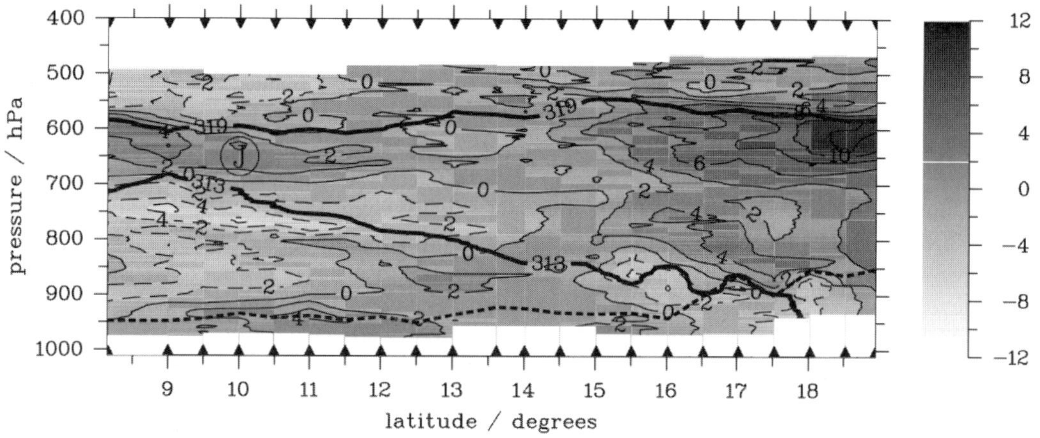

Figure 2. Dropsonde data from an aircraft transect at 2.2^{0}E, showing meridional wind by shading and light contours every 2 ms^{-1}, negative contours (southward flow) dashed. Also shown are the limits of the SAL (thick lines), the top of the mixed layer (thick dashed line) and the position of the AEJ (circle). From the Jet2000 experiment, Parker et al. [4].

SAL-formation region above the heat low, sensible heating exceeds radiative cooling requiring ascent for thermal equilibrium [5]. Aerosols mixed up from the surface by dry convection also play a role in the radiative heat balance [6]. Deep moist convection is inhibited by the dry conditions and the descent above.

As shown in fig. 1, the meridional circulation set up by the shallow heat low transports Saharan air south. As the SAL extends southwards it lifts above the cooler moister incoming southerly monsoon flux, which also forms part of the dynamical response to the heat low. The confluence of these two currents: the monsoon flux from the south and the dry 'Harmattan' winds from the north, is known as the inter-tropical front (ITF) and is also shown in fig. 1. These features are also seen in the observational section shown in fig. 2. The monsoon flux brings oceanic air onto the continent and penetrates further north as the heat low deepens and the monsoon onset progresses. The inter-tropical convergence zone (ITCZ) - the local expression of the ascending branch of the Hadley cell - also migrates north, behind the ITF.

In summary we thus have a heat low with low level convergence, shallow well mixed ascending air, forming a hot dry northerly wind that overlays moist southerlies south of the ITF, and to the south of this deep convection in the ITCZ which follows the ITF northwards as the monsoon develops.

The offset between the positions of the ITF and the ITCZ is an interesting feature of the African monsoon. It is a consequence of having a desert to the north of moist southerlies. Between the ITF and the ITCZ deep convection is inhibited by the overlaying warm dry SAL. Indeed, the inhibition is such that it is difficult for normal sized convective storms to break through, and in the Sahel region it is more usual for deep moist convection to occur in relatively large scale energetic organised systems known as mesoscale convective systems (MCS). These account for more than 90% of Sahel rainfall [7], and episodes in which Saharan air penetrates south may influence the generation of MCSs [8]. Further south in the ITCZ proper, convective available potential energy is at a maximum and tropical convection occurs more freely in smaller scale outbreaks and squall line structures.

The global zonal mean Hadley circulation has an ascending branch north of the equator. The air then flows south and is deflected westwards by the Coriolis force creating a belt of upper level equatorial easterlies consistent with the conservation of angular momentum. This tropical easterly jet (TEJ) is in evidence over the African continent and the Gulf of Guinea at about 10^0N where the meridional temperature gradient is positive below the tropopause and negative above it. Thermal wind balance thus constrains the core of the TEJ at about 16 km altitude [9] [10]. At lower levels over West Africa the temperature structure particular to the continent creates a distinct seasonal easterly jet at lower levels known as the African easterly jet (AEJ). Below about 650 mb the hot Sahara means that temperature increases northwards, implying easterly vertical shear. Surface westerlies are thus overlain by easterlies. In the dry air over the Sahara there is a steep vertical temperature lapse rate, whereas further south the moister air cools more slowly with height, so at about 650 mb the meridional temperature gradient becomes positive, implying westerly shear with height, or weakening easterlies. The AEJ thus develops at this transition point along with the monsoon flow in summer at about 15^0N. The jet core is situated at the level where a dry (Saharan) adiabat crosses a moist (Guinea coast) adiabat [11]. The AEJ is the site of development of African easterly waves (AEWs), which draw on the energy (both potential and kinetic) of the circulation and contribute to variability in monsoon flow and rainfall.

So far our description has been limited to an equilibrium view of the various features of the WAM and how they interact. The seasonal cycle of the WAM also deserves particular attention both because of its importance and its scientific interest. It appears that threshold effects are at work and we can talk of a monsoon 'onset' that is relatively rapid and variable from year to year. Sultan and Janicot [12] define two phases for the development of the WAM: the 'pre-onset' and the 'onset'. In the pre-onset the average (10^0W-10^0E) of the ITF crosses 15^0N on its northward migration while the ITCZ is still in the south. In the onset the ITCZ moves from 5^0N to 10^0N. The delay, and subsequent jump in the position of the ITCZ compared to the ITF corresponds to a jump in the mean latitude of the rains. The retreat back south at the end of the season is not so obviously discontinuous. According to Sultan and Janicot, the average onset date is 24 June, with a standard deviation of 8 days - large enough to be an important consideration for crop planting. Fig. 3 shows the seasonal development in the position of the rain band. The apparent 'jump' from one state to another also displays interannual variability, with the Sahel rains actually failing altogether some years. There are many possible factors to take into account

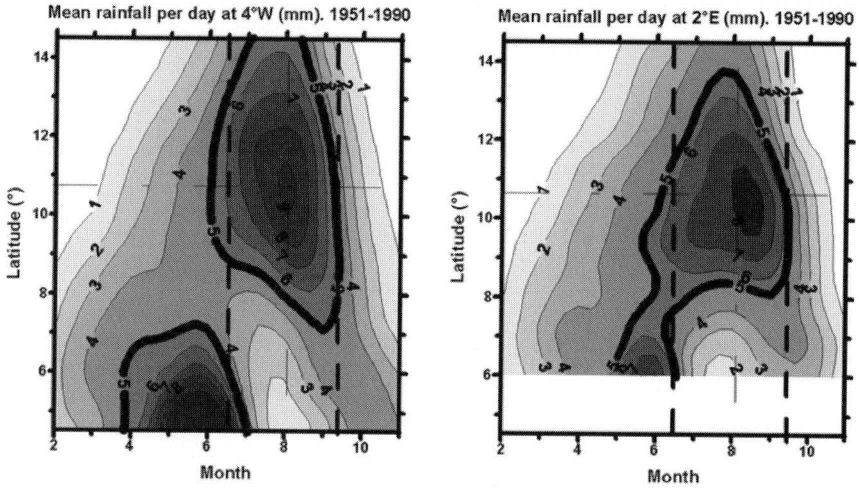

Figure 3. Latitude-time diagrams showing north-south migration of mean daily rainfall in mm (1951-1990) for 4^0W and 2^0E. Numbers on x-axis indicate start of month. From Lebel et al. [13].

in explaining this behaviour, including surface feedbacks, dynamical instability, sea surface temperature (SST) variability and the influence of transient tropical waves. In the following section we will consider the basic physics of the maintenance of a meridional circulation relevant to the WAM and start to isolate the influence of some of these factors.

3. THE ZONALLY UNIFORM MONSOON

The banded structure of west African surface conditions, from the Gulf to the Sahara, means that the WAM lends itself naturally to an understanding based on zonally uniform conditions. Progress in understanding the zonal mean circulation in the tropics, specifically the Hadley cell and associated jets, is therefore useful. Understanding the Hadley cell is one of the fundamental problems in dynamical meteorology, and discussions of this topic are often couched in terms of angular momentum. In the absence of viscous forces, angular momentum is conserved for a ring of air that circles the globe at a given latitude. We must consider a zonal average rather than a sector to ensure that there are no net pressure torques. In terms of the zonal average wind u, the angular momentum per unit mass is given by

$$M = \Omega a^2 \cos^2 \phi + ua \cos \phi, \tag{1}$$

where ϕ is the latitude, Ω is the earth's rotation rate and a the earth's radius. If air moves from one latitude to another it must change its zonal velocity u in order to conserve its angular momentum M. Displacement towards the pole induces acceleration towards the east. The air spins up as it approaches the axis of rotation, resulting in subtropical westerlies. The converse is true for air moving towards the equator resulting in equatorial easterlies. These induced accelerations are consistent with the action of the Coriolis force, which deviates the flow to the right in the northern hemisphere and to the left in the southern hemisphere (conservation of angular momentum is sometimes used to 'explain' the Coriolis force, see Persson [14] for an elegant discussion). Rising tropical air will also spin down relative to the earth and is concurrently accelerated to the west by the Coriolis force, but in our thin atmosphere this effect is minor compared to the effect of changing latitude. To get an idea of the wind speeds induced, consider a ring of air at rest at the equator. If it is displaced to 10^0N, according to (1) it must now have an eastward velocity of $14\,\mathrm{m\,s^{-1}}$. By 20^0N it is $43\,\mathrm{m\,s^{-1}}$ and by 30^0N it is $72\,\mathrm{m\,s^{-1}}$. Further north these speeds become unrealistic as the flow becomes dynamically unstable. We should note here that as well as conserving angular momentum, the 'spun up' subtropical westerlies have acquired kinetic energy. This is due to the work done by pressure forces acting on the fluid during its northward displacement (it is equivalent to the work done by the traditional ice skater as she pulls here arms in). Again, the converse applies for southward displacements: kinetic energy is lost for a displacement in the opposite direction to the pressure force. If there were no meridional pressure gradients, the air would not penetrate very far north before being diverted back to where it came from by the Coriolis force.

Two questions arise. First, what sets up the pressure gradients that maintain the meridional circulation and second, how far north can this circulation penetrate? The traditional ice skater is unable to help us with these fundamentals. We are effectively asking her why she is moving her arms in and out, how strong she is and how fast she can spin before she loses her balance and falls over. The answers to these questions lie in the thermal forcing of the tropical atmosphere and the parameters appropriate for the earth and more specifically West Africa. But before leaving our discussion of angular momentum we must first discuss a constraint on realisable circulations associated with its conservation. Following Hide [15], Shcneider [16] and Held and Hou [17], we note that for a conservative quantity subject to downgradient diffusive fluxes it is impossible to have an isolated extremum in the interior of the fluid. It would be eroded by any diffusion, however weak. This is the case for M. Any maximum in M must be situated on the lower boundary where surface stress can compensate for diffusion. We will call this the 'attainability condition', as it limits the types of flow that can be spun up from rest by thermal forcing (which exerts no torque). In particular, it is clear that the resting value of M at the equator cannot be exceeded. Thermal forcing is thus unable to produce equatorial westerlies in the free atmosphere.

Thermal forcing can, however, produce temperature gradients that lead to pressure gradients, forcing air across latitudes and creating a meridional circulation with associated balanced zonal jets. Energetically speaking, a thermally direct heat engine is created. Air is warmed by surface fluxes at low levels and high pressure. It rises and it cools radiatively at high levels and low pressure. It descends and repeats the cycle. This is a Carnot cycle that extracts heat from the thermal forcing and converts it to kinetic energy (through the creation of pressure gradients), which maintains the Hadley cell against friction. The question still remains, does this happen, and if so under what circumstances? One can equally well imagine a purely zonal flow with no meridional circulation at all, that is locally (at each latitude) in thermal equilibrium. Angular momentum conservation would not be an issue for such a flow as there would be no meridional displacements. It would, however, have to adhere to the attainability condition. For steady inviscid flow the zonal wind obeys the momentum equation for gradient wind balance,

$$2\Omega \sin \phi u + \frac{u^2 \tan \phi}{a} = -\frac{1}{\rho a}\frac{\partial p}{\partial \phi}, \tag{2}$$

where the term on the right expresses pressure gradients set up at thermal equilibrium (which can be expressed in terms of equilibrium temperature gradients by application of hydrostatic balance). One solution of (2) is simply no flow and no heating therefore no pressure gradients at or near the equator. Equation (2) also supports westerlies at the equator, which are forbidden by the attainability condition. At the equator we are thus faced with a choice of solutions that either have no flow, no meridional flow and a zonal flow that violates the attainability condition and thus cannot be obtained by spinning up from rest, or a flow with both meridional and zonal circulation that is not in thermal equilibrium and conserves angular momentum. Away from the equator the choice is not quite so stark as westerly zonal flow can be supported without creating extrema in angular momentum. It is common to refer to this zonal-flow-only' regime as the 'thermal equilibrium' regime, and to the regime with meridional flow that conserves angular momentum as the 'angular momentum conserving' regime. In truth, angular momentum conservation is always required, but the point is somewhat moot in the case with no meridional circulation. Held and Hou [17] set about finding symmetric solutions about the equator for the angular momentum conserving regime in the inviscid limit. Without going into mathematical details here, their basic procedure was as follows. First define a heating profile. Held and Hou chose restoration

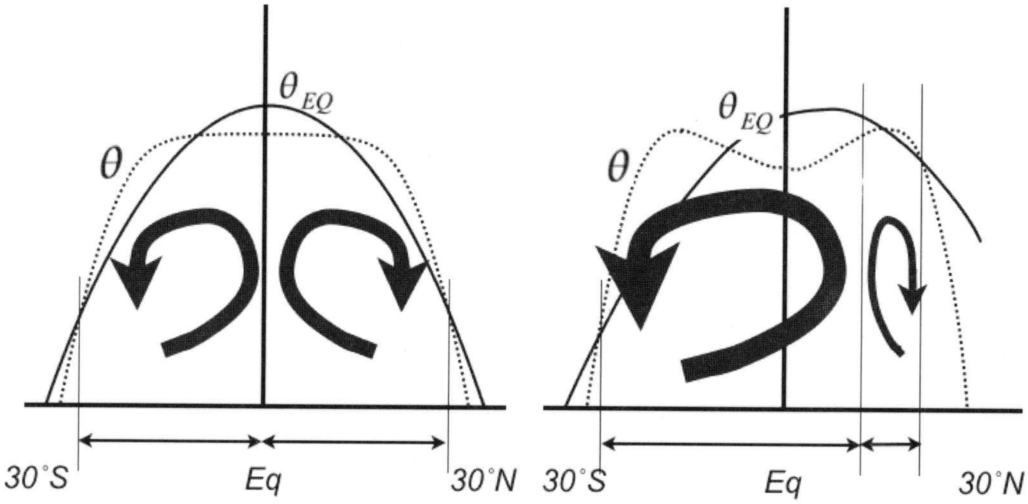

Figure 4. Schematic showing symmetric and asymmetric solutions for the Held-Hou model of the Hadley circulation [17] [18]. The widths of the meridional cells are indicated and the sense of the angular momentum-conserving circulation is shown. Solid curves show the mean tropospheric temperature for the solution and dotted curves show the equilibrium temperature used to force the circulation. Adapted from James [19].

to an idealised symmetric equilibrium temperature profile θ_{EQ}. The forcing in the thermodynamic equation is thus $(\theta_{EQ} - \theta)/\tau$ where τ is the timescale of the restoration. The thermal equilibrium solution corresponds to $\theta = \theta_{EQ}$. In this case there is no heating, no vertical motion to balance the heating and thus no meridional circulation. If $\theta \neq \theta_{EQ}$ there must be a meridional circulation and it must conserve angular momentum. Application of hydrostatic balance transforms (2) into a thermal wind balance equation for the vertical gradient of the zonal wind in terms of horizontal temperature gradients. This is integrated in the vertical to give an expression for the upper level zonal wind in terms of the average tropospheric temperature and then the angular momentum conserving latitudinal profile of zonal wind from (1) is substituted in. The resulting equation is integrated north from the equator to a limiting latitude. This limiting latitude is determined as part of the solution. It is the latitude at which the net heating matches the net cooling (the integral of $\theta - \theta_{EQ}$ vanishes). Continuity of θ is also required. The solution is shown schematically in fig. 4.

We have now found the extent of the Hadley cell. Held and Hou found that it is proportional to the square root of the height of the tropopause and the pole-equator temperature difference, and inversely proportional to the earth's rotation rate. A slower rotating planet with a deeper atmosphere or a stronger equatorial heating might have a Hadley cell that reaches almost to the pole, but in our case it reaches about 27^0N. This coincides with the latitude beyond which angular momentum conserving zonal winds start to become 'unrealistic'. Further diagnostics can be produced including surface winds and the strength of the meridional circulation. Held and Hou went on to evaluate the effects of viscosity, verifying that the thermal equilibrium solution is not valid at the equator as a limit for vanishing viscosity. They also showed that viscosity increases the wind at the surface and decreases it aloft, and strengthens the meridional circulation.

One difficulty with this theory is that it turns out that relatively large pole-equator temperature differences are needed to produce a realistically strong meridional circulation. This point was taken up by Lindzen and Hou [18] who also remarked that the Hadley cell is in fact almost never in a state of symmetry with two equal an opposite overturning circulations. The Hadley cell spends most of its time overturning either one way or the other, and the complementary cell on the other side of the ascending region is usually very weak. In boreal winter, the cell rises just north of the equator and descends in the northern subtropics. In the summer it rises further north and descends south of the equator. Transition between winter and summer states occurs relatively rapidly at the equinoxes. Lindzen and Hou analyse an asymmetric situation (also shown in fig. 4) with a smooth idealised heating centred on 10^0N. The procedure described above yields a very asymmetric response with the strongest ascent near the latitude of maximum heating but the division between the two cells considerably further north. The northern cell is very weak and the southern cell, which descends south of the equator, dominates. It is much stronger than a single cell in the symmetric setup of Held and Hou, and is consistent with a more realistic parameter value for the pole-equator temperature contrast. Another consequence of off-axis heating is the appearance of strong easterlies at the equator as angular momentum is transferred from the northern hemisphere. In fact this effect is exaggerated in the inviscid theory but the addition of diffusion makes it more realistic.

Another factor that can influence the strength of the meridional circulation is the meridional profile of the thermal forcing. Hou and Lindzen [20] redistributed the same total heating into a narrower latitude band and again produced much stronger Hadley cells. This is particularly effective if the heating is redistributed asymmetrically into the dominant 'winter' cell for off-axis forcing. So if the ITCZ followed a single latitude circle, rather than meandering as it does, it might create a stronger Hadley circulation. The observed Hadley circulation reflects the zonally averaged thermal forcing. On the other hand, it is clear from all these results that the time-average response to thermal forcing can be stronger than the response to the time-averaged thermal forcing. Due to the nonlinearity present in the dynamical equations a smoothly varying seasonal cycle in thermal forcing can give rise to an abruptly shifting response in terms of the meridional circulation. This is already reminiscent of the behaviour of monsoon circulations, but it can be taken further.

We can explain the relatively strong response to off-equator forcing in terms of conserved quantities. The equator is a special place. At the equator the gradient of resting angular momentum is zero, and the planetary vorticity $2\Omega \sin\phi$ is zero. An induced meridional flow at the equator advects no angular momentum and the angular momentum-conserving response just off the equator can easily generate a large enough gradient of relative vorticity to satisfy the conservation of absolute vorticity $\zeta = 2\Omega \sin\phi - \partial u/\partial y$. Such a flow is thus dynamically consistent with the conservation of M and ζ. A modest thermal forcing can produce a modest dynamical response. Conservation of ζ becomes increasingly difficult away from the equator. Stronger meridional circulations are required to satisfy the constraints imposed by the conservation laws. For a thermal forcing isolated at 10^0N, for example, the choice outlined above between a 'thermal equilibrium' solution and an 'angular momentum conserving' solution with a strong meridional circulation becomes more relevant and realistic than was the case for equatorially forced flows. The thermal forcing must cross a certain threshold before any meridional circulation can be sustained. Plumb and Hou [21] formalised this threshold behaviour and showed that the two regimes cannot coexist. Either we have thermal equilibrium with no meridional circulation - and the equilibrium temperature structure must be such that the attainability condition mentioned above is not violated, or we have an angular momentum conserving finite amplitude meridional circulation. The transition between the two regimes becomes more dramatic as we move further from the equator.

Conservation of angular momentum is only valid for a zonal ring of air in the free atmosphere. Application to the West African region depends on the assumption of no net zonal pressure gradient across the region. Alternatively it is possible to generalise the theory to three dimensions. Emmanuel [22] notes that the attainability condition discussed above can be expressed in terms of the absolute vorticity ζ at the tropopause where the vertical velocity vanishes. For it to be possible to attain a given flow in a time dependent problem, ζ must have the same sign as the planetary vorticity $2\Omega \sin\phi$ (so in the northern hemisphere it must be positive). This is because the divergent source of vorticity in the vorticity equation is proportional to ζ, so divergence can never force ζ to change sign. Armed with this condition, criteria can be deduced for the onset of monsoon circulations. As before, balance conditions are used to deduce the tropopause winds. The attainability condition is violated if ζ at the tropopause drops to zero. In such 'supercritical' conditions a thermally direct overturning circulation must result.

A further condition can be imposed concerning the thermal structure of the atmosphere. Up until now we have considered idealised heating distributions that mimic the effect of adjustment under deep convection and radiation. The heating structure is, nevertheless, imposed on the system. The lack of interaction between the heating and the response can be considered an unrealistic feature of these simple models. An alternative to this approach is to try to simulate a realistic temperature structure directly. Emmanuel [22] makes use of the observation that the tropical atmosphere is always close to neutrality for vertical moist convection [23]. On a large scale the vertical temperature structure approximately follows a moist adiabatic lapse rate from the cloud base to the tropopause. This is a powerful constraint. It means that once the near-surface equivalent potential temperature is specified, the entire thermal structure of the troposphere follows. The condition for the onset of a monsoon-type circulation, that $\zeta = 0$ at the tropopause, can thus be expressed in terms of sub-cloud layer θ_e.

All the elements are now in place for a theory of the African monsoon onset. Eltahir and Gong [24] took advantage of all these theoretical developments to provide insight into what makes the WAM special, and why it might vary considerably from year to year. Their explanation hinges on the threshold behaviour introduced by Plumb and Hou and the dependence on boundary layer moist potential temperature θ_e formulated by Emmanuel. West Africa displays meridional gradients of θ_e that can vary interannually and intraseasonally. The existence of dry heat to the north and moist conditions to the south across a relatively confined range of latitudes means there are compensating factors at play that can influence the local Hadley circulation. The temperature gradient opposes the humidity gradient in the setting up of boundary layer θ_e. It is thus possible for this gradient to be either flat or steep depending on the surface conditions, in particular the SST in the Gulf of Guinea. A steep gradient of boundary layer θ_e easily sends the tropopause ζ to zero, which initiates a vigourous meridional circulation and a healthy monsoon (see fig. 5).[2] This happens when the Gulf of Guinea is cool. Conversely, when the gulf is warm, θ_e gradients are weak, the divergent circulation is not initiated and the monsoon fails. Eltahir and Gong contrast a wet year (1958) with a dry year (1960). Their deductions about the gradient of boundary layer equivalent potential temperature being a good indicator of monsoon circulation are borne out by these two examples. The wet year has a steep θ_e gradient, a vigourous meridional circulation and their diagnostics show strong signals in relative vorticity which are able to cancel the planetary vorticity at the tropopause to excite the angular momentum conserving regime. The dry year has weaker θ_e gradients, a weak meridional circulation and the absolute vorticity at the tropopause is closer to the planetary value. A statistical analysis by Philippon and Fontaine [25] confirms these findings, and underlines the importance of soil moisture in determining the gradient of θ_e.

Since west Africa is relatively close to the equator, the planetary vorticity $2\Omega \sin\phi$ is small and its gradient is large. Eltahir and Gong show that because of this parameter regime, the west African region is particularly sensitive to changes in surface θ_e gradients (in fact they show that it is fifteen times more sensitive than the Indian monsoon region !). Small thermal changes can create large differences in the meridional circulation. Changes in surface conditions can therefore have a considerable impact, whether they are changes in the SST, in land ocean contrast or in the position of the desert boundary. Some interesting sensitivity studies were carried out by Zeng and Eltahir [26] to establish where the greatest sensitivity lies. Using a Hadley cell model similar to those described above attached to a land surface scheme, they suppressed selected bands of vegetation, allowing the Sahara to grow southwards or deforesting the coast. They found the circulation relatively insensitive to an extended Sahara, but more sensitive to perturbations further south. This is consistent with the findings of Plumb and Hou and Hou and Lindzen, who showed that the meridional circulation is much more sensitive to redistributions of heating in the dominant southern cell. The work of Zeng and Eltahir amounts to a strong critique of Charney's classic hypothesis [2]. The desert feedback mechanism depends on sensitivity to destruction

---

[2] It should be noted that some authors refer to 'moist static energy' $= c_p T + \Phi + Lq$, which has the dimensions of energy per unit mass. Its use as a field variable in these discussions is equivalent to the use of θ_e.

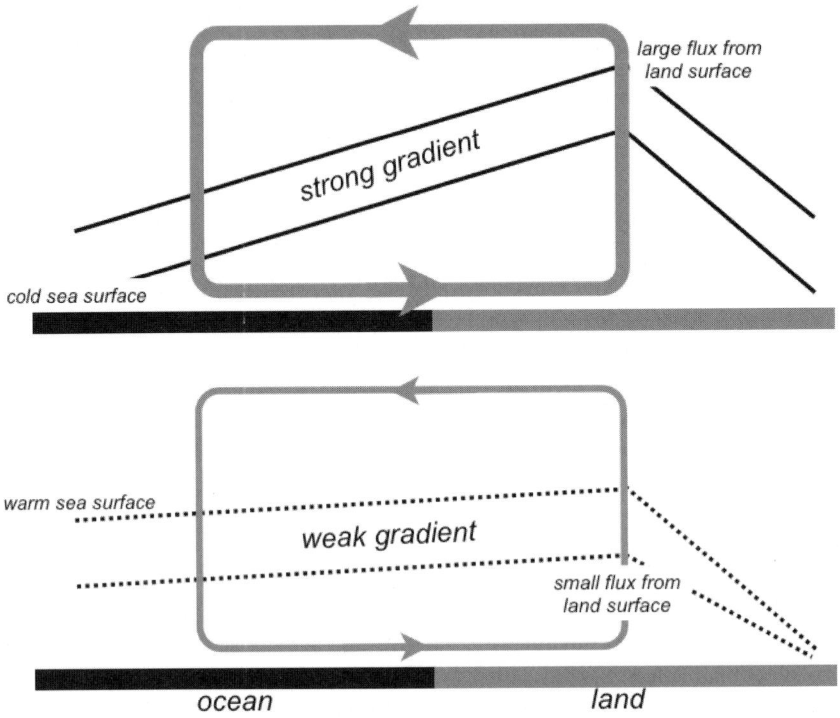

Figure 5. Strong (top) and weak (bottom) monsoon circulations as a consequence of gradients of boundary layer θ_e arising from SSTs and moist entropy flux over the land surface. Adapted from Eltahir and Gong [24].

of vegetation in the Sahel, which appears to be absent, and on a memory of it from one year to the next, which is unlikely since grassland vegetation can re-grow quickly. Removal of the coastal forest, on the other hand, appears to have more dynamical impact and is harder to restore.

A number of questions remain concerning the detailed structure and dynamics of the meridional circulation and the interplay between the observed features of the WAM listed in section 2. Thus far most studies of the mean meridional circulation have either used a dry framework or have considered a moist atmosphere with simplified physics. To make further progress, we must also consider the interactions between moist and dry processes, and further physical processes need to be taken into account (aerosols, SSTs, radiation) in order to quantify their relative importance.

A zonally-symmetric model designed to simulate WAM dynamics over a wide range of scales and processes was developed by Peyrillé et al ([27] hereafter PLR). Their model consists of a vertical-meridional plane from 30^0S to 40^0N with wall lateral boundary conditions to preclude interactions with the mid-latitudes. The SSTs are prescribed but the humidity and temperature of the continental surface are free to evolve in time. Surface properties such as vegetation and albedo are fixed. The atmospheric model has a complete physical package including a convection scheme, a turbulence scheme and a radiative scheme. The model is started with a dry atmosphere at rest, which is moistened during the 10-15 day spin-up period and is then run in perpetual 15 July mode to investigate sensitivities independently of the seasonal cycle.

The theory based on the meridional profile of θ_e is useful to understand the WAM at large scales but one must keep in mind that regional processes, such as dry air capping the monsoon flow, or the interaction of aerosols with radiation, can also be important. This is illustrated in fig. 6, which shows the humidity budget in the ITCZ and heat low regions. The water vapor budget in a given volume depends on advection, convection, turbulence and friction. For upper levels in the ITCZ the balance is between

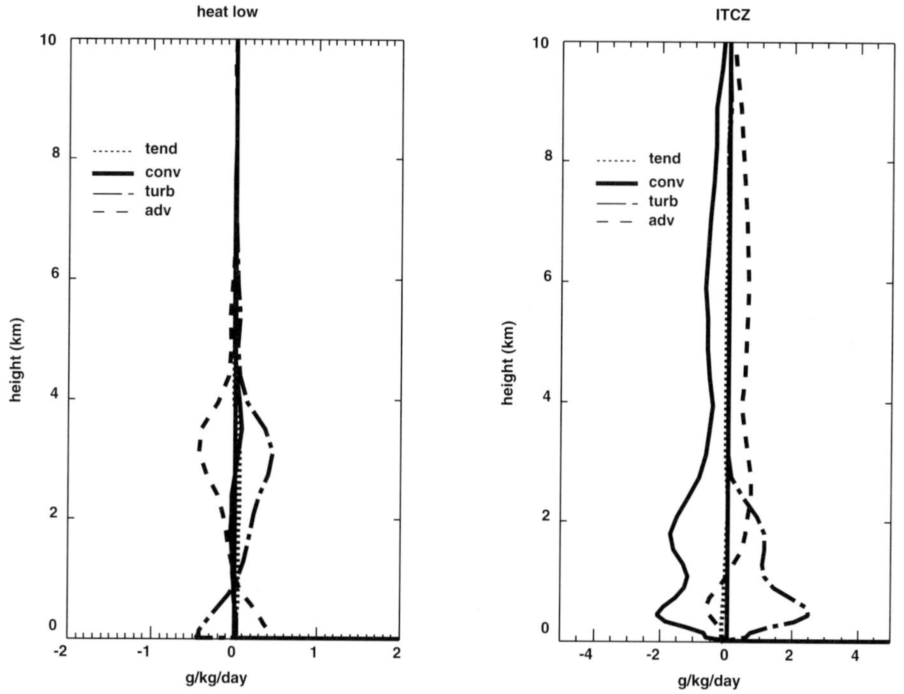

Figure 6. 20-day mean humidity budget from an idealised 2-D monsoon simulation. (a) in the heat low. (b) in the ITCZ. The humidity terms are in g kg^{-1} day^{-1}. From Peyrillé and Lafore [28].

convection, which acts as a moisture sink, and advection (mainly vertical), which supplies moisture. The balance at low levels is more complicated, with the drying effect of convection countered by turbulent fluxes. The net effect is a slight moistening of the low levels. There is also a drying by advection around 2 km, corresponding the the dry inflow from the Sahara shown in fig. 1. The monsoon flux is thus dried out by convection and kept moist by turbulence. The equilibrium is of course very different in the Sahara as mainly dry convection occurs (there are no clouds but moisture is still present). The total advection moistens the lower levels (from the south and north) and dries the 2-4 km layer over the Sahara. This layer corresponds to the level where there was the moistening by advection in the ITCZ to the south. The turbulent fluxes thus have the opposite effect in the Sahara, as they dry the low levels and moisten the top of the boundary layer. This was observed during the JET2000 field campaign [29]. A maximum of relative humidity is found at 5 km height over the Sahara. We thus have two regions where both low level advection and turbulence play a key role but with opposite effects. The turbulence moistens the monsoon layer whereas it dries the Saharan boundary layer. Now that we have a better idea of how these two regions are balanced, we can test the impact of several processes on the θ_e profile and then on these equilibria. PLR considered the effect of changes in SST in the Mediterranean and the Gulf of Guinea, along with other factors, in modifying the action of thermal forcing on these budgets. Some useful diagnostics to illustrate the sensitivity to these factors are the meridional distributions of θ, θ_e and rainfall at the surface. These are shown in Fig. 7. The meridional profiles of SST tested correspond to May, June, and July in the Gulf of Guinea and in the Mediterranean Sea separately. PLR found that a warmer Atlantic ocean in May helps to increase θ_e over the ocean (through increased latent heat flux) and two rain bands are generated around the equator and at 9^0N. In June, when the SSTs in the the Gulf of Guinea are 2^0C lower, the maximum in θ_e shifts 4 degrees northwards. The θ_e of the monsoon flow decreases (through reduced moisture in the boundary layer) and the two rain bands move to the coast and to 11^0N. Finally the lowest SSTs in the Gulf of Guinea, occurring in July, fix

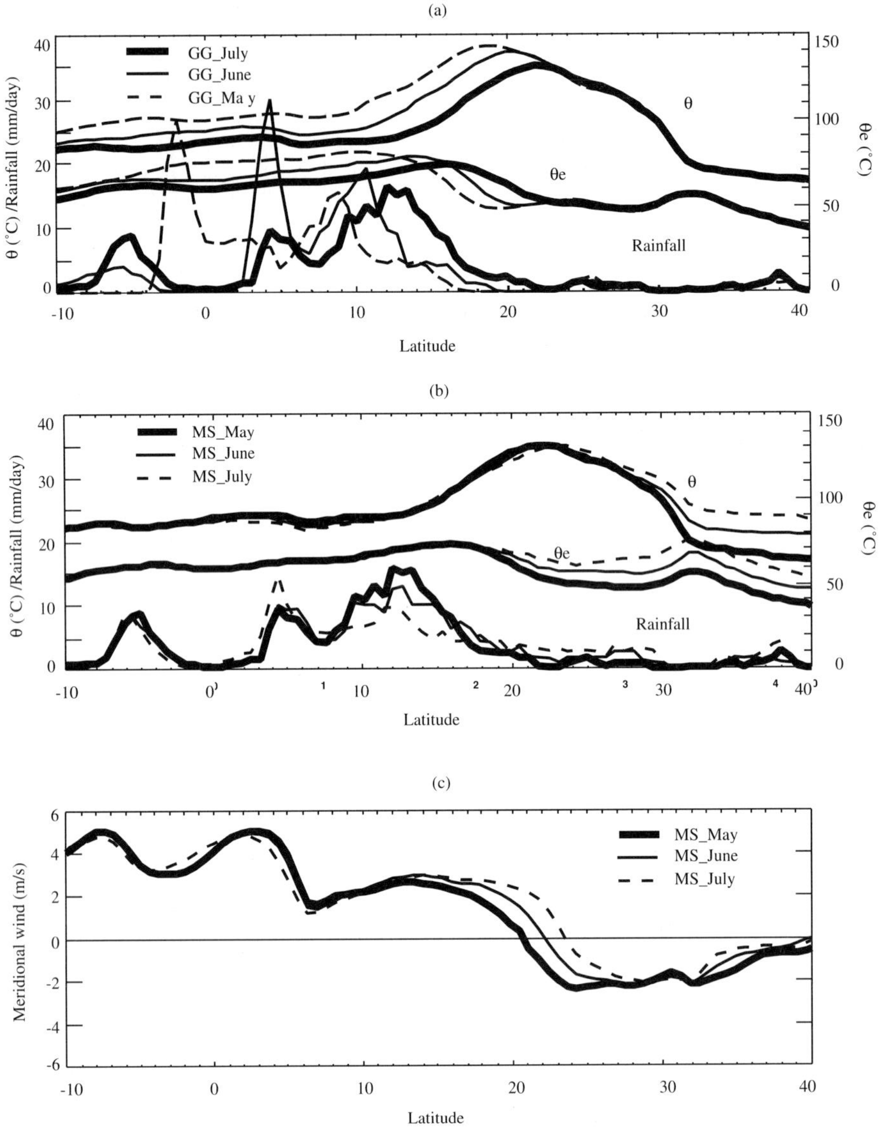

Figure 7. Meridional profiles of θ, θ_e and rainfall for sensitivity experiments to a) Gulf of Guinea SST and b) Mediterranean SST. c) Meridional wind for the different Mediterranean SSTs shown in b). From PLR [27].

the rain band at around 13^0N. Altogether the cooling of the Gulf of Guinea between June and July can alone explains a northward shift of the WAM by about 6 degrees of latitude. The rapid cooling of the Gulf of Guinea between April and June therefore contributes to inland penetration of the monsoon. The gulf is the main source of moisture for West Africa [30] [31], and determines the θ_e gradient between the ocean and the continent. Since the thermal contrast is greater in July, the meridional advection into the ITCZ is enhanced but its moisture content is reduced. When a higher (May) SST is imposed, the moisture content is greater but the southerly flow is weaker, so the ITCZ does not go very far north. For interannual variations a warm Atlantic ocean is associated with an ITCZ located further south [32]. The same appears to be true here for the seasonal cycle.

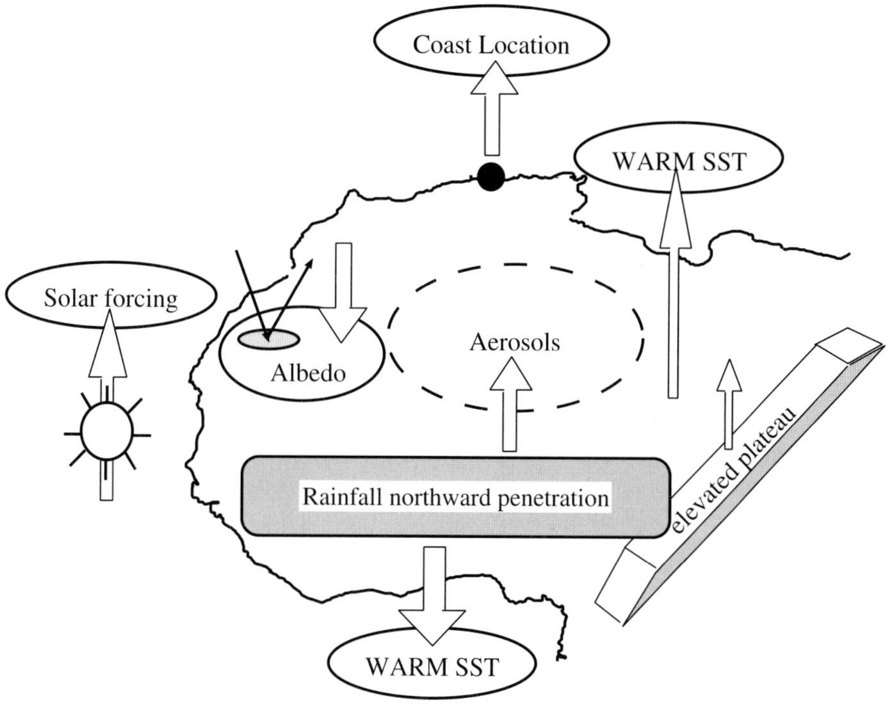

Figure 8. Schematic diagram of the mechanisms that affect the northward penetration of rainfall over West Africa. The arrows indicate the rainfall displacement induced by each mechanism. Their widths are proportional to the intensity of the rainfall shift. From PLR [27].

The Mediterranean Sea also contributes moisture to sustain the WAM through an increase in southward transport into to the ITCZ [31] [33]. The role of the Mediterranean in WAM dynamics is less well documented than that of the Gulf of Guinea, especially at the seasonal timescales. Fig. 7b illustrates the response of the 2-d model to the 5^0C warming observed in the Mediterranean Sea between May and July. This results in a warming and a moistening above the Mediterranean Sea at low levels. The magnitude and position of the heat low are slightly affected, and humidity increases over the desert, leading to the development of light precipitation over the Sahara in the 2-D experiment. So a warmer Mediterranean Sea favours the northward penetration of the WAM (in fact, other mechanisms not represented in this model maintain a dry Sahara, and they will be discussed in the next section). The change in the distribution of meridional wind shown in fig. 7c gives further insight into the processes involved in this response. The magnitude of the northerlies and their extension decrease with Mediterranean Sea warming. The strength of the Harmattan and its southward penetration depend therefore on the thermal contrast between the African continent and the Mediterranean Sea. Two mechanisms favouring the northward penetration of the monsoon have thus been identified by PLR. First, the increase of moisture supply by this basin helps moisten of the lower layers of the Sahara through low level advection. Second, a warm Mediterranean Sea decreases the thermal contrast with the Saharan heat low, resulting in a decrease in Harmattan flux.

To summarise, a warm Gulf of Guinea reduces the northward extension of the WAM by reducing the meridional temperature gradient. A warm Mediterranean sea, as observed in July and August, favors penetration of the WAM and precipitation over the Sahel. These mechanisms are shown schematically in fig. 8. Further sensitivities identified by PLR are as follows. The change in insolation from spring to summer forces the northward propagation of the WAM through differential heating of the land and sea. Mineral dust aerosols have the same effect but with greater efficiency. As no coupling between aerosols and atmospheric dynamics is allowed in the model, only the radiative effect of aerosols is exhibited here. When aerosols are present, the Saharan heat low becomes warmer because they absorb solar radiation. The mean elevation of West Africa is about 400 m. Surface pressure therefore lowers over the continent and the resulting pressure gradient force is greater than the case without elevated plateau, favouring monsoon penetration.

It is also interesting to note that the latitude of the Mediterranean coast varies from 30 to 35^0N over West Africa. This also affects the penetration of the WAM. The northern limit of the Sahel precipitation band may therefore be conditioned by the latitudinal extent of the Mediterranean Sea, or in other words by the northern extent of the Sahara. PLR found that the bigger the Sahara, the greater the northern extent of rainfall. This is because the advection of cool moist air from the Mediterranean is not enough to trigger convection, but instead serves to reduce the θ_e gradient, and weaken the overall monsoon circulation by the mechanism discussed above.

This series of experiments shows that the meridional gradients of θ and θ_e are important not only south of the ITF [24] but also north of it. The last experiment also shows the limits of the purely zonally symmetric approach. The Eastern part of West Africa experiences a significant import of moist and cold air from the Mediterranean Sea that breaks the zonal homogeneity over the continent. Mechanisms associated with zonal asymmetries of the WAM are the subject of the next section.

4. ZONAL VARIATIONS

Although the shape of the continent lends itself to a zonally uniform analysis, the WAM is not just a mini-Hadley circulation. Remote climatic factors can also influence the local circulation, particularly those arising from SST anomalies in the equatorial Atlantic and Pacific and associated changes in convection, and from the onset of the Indian monsoon. Zonally varying aspects of the global circulation may be important on a range of timescales, including the monsoon seasonal cycle, variation within the seasonal cycle, interannual variability, and even the maintenance of surface features such as vegetation and albedo. Remote influences are transmitted to the West African region by mean flow transport: the divergent 'Walker' circulation; by wave motion: equatorial Kelvin waves propagating to the east and sub-equatorial Rossby waves to the west, and by complex traveling disturbances like the Madden-Julian oscillation [34]. A thorough phenomenological review of remote factors influencing the WAM is not in the spirit of this article, see Rowel et al [35], Fontaine and Janicot [32]. Instead we highlight a few dynamical mechanisms.

We start with a fundamental question: why does the Sahara desert exist in its current location? As discussed above, the Sahara plays an important role in the WAM. Charney's albedo feedback mechanism may contribute to its maintenance, but neither this nor any other local feedback mechanism can account for its location. Deserts do not exist at all longitudes at 30^0N, and the traditional explanation that the Hadley cell has its descending branch here only works for the wintertime anyway. In the summer, when the Sahara is hottest and driest, the Hadley cell descends south of the equator. All these points were made by Rodwell and Hoskins [36] [37], who put forward a global theory of interacting monsoon circulations connected by planetary waves. Localised deep heating on the equator gives rise to a baroclinic response that satisfies hydrostatic balance and vorticity conservation. The form of this response at upper levels is a quadrupole in streamfunction with equatorial westerlies to the east and an anticyclone to the northwest [38] [39]. This triggers a Kelvin wave that propagates eastwards along the equator and a pair of sub-equatorial Rossby waves that propagate westwards. If the heating is moved northwards and placed in a realistic zonal flow, the northern hemisphere Rossby wave response is strengthened and the localised subtropical heating can influence vertical motion at remote locations. Rodwell and Hoskins placed a deep convective heat source at 25^0N, 90^0E in an idealised simulation of the Indian monsoon onset. They showed that the associated Rossby wave induces descent in the southern Mediterranean / Saharan region. They argue that this suppresses summer convection north of the Sahel, limiting the penetration of the WAM and maintaining the Sahara. It is important to note that this is a wave effect and not the result of a modified Walker circulation in which the Indian monsoon outflow descends over the Sahara. In fact the descent induced in this manner draws in air from the midlatitude westerlies. This air has low θ_e and is warmed adiabatically by descent. Scale analysis of the thermodynamic equation [40] suggests a vertical equilibrium near the equator in which vertical advection of potential temperature is balanced by diabatic terms. In the midlatitudes horizontal advection dominates. But between the two, in the subtropics, both processes can be important. So in the Sahara, midlatitude air can flow down isentropes contributing to the descent, and further Charney-style radiative cooling may also serve to augment this descent and maintain local thermodynamic equilibrium.

The local vertical nature of the thermodynamic budget in the tropics is of course one of the cornerstones of the zonal theories of the WAM reviewed in the last section, allowing the link to be made between surface θ_e and the upper level circulation. The extension to three dimensions has been pursued by Neelin and co-workers. The point of departure is the idea of quasi-equilibrium, in which the local vertical temperature structure is always close to a moist neutral profile (the basis of 'convective adjustment' schemes used in GCMs). This is a large-scale approximation to the behaviour of an ensemble of convective systems. Rather than reasoning in terms of heating and response, it is recognised that the two are coupled and a model is constructed in terms of departures from a reference profile. The departures are also based on realistic vertical structures which are furnished by analytical solutions for wind and temperature under various simplifying assumptions such as height independence of pressure gradients and temperature advection. Using these profiles as basis functions, an alternative to having vertical levels, numerical solutions are then found that do not depend on the simplifying assumptions used to define the basis functions. The result is a model tailored to tropical thermodynamic processes [41], which has been used [42] [43] [44] to deconstruct various influences on the northward penetration and east-west asymmetry of monsoonal flows, including the WAM. In a study with an idealised continent symmetric about the equator, experiments were performed to isolate the effects of the following mechanisms: soil moisture feedback; ocean heat transport; horizontal advection of temperature and humidity, and the Rossby wave mechanism discussed above.

The former two considerations are in opposition: limiting surface moisture availability dries the continent, but ocean heat flux (away from the equator) is vital in allowing the ocean surface to stay cool, displacing convection onto the continent. In reality, ocean heat fluxes can be considered given, but surface soil moisture is interactive and so its distribution over the continent can influence the shape of the climatological monsoon (see for example Cook [45], for a dynamical analysis of the response to soil moisture distributions over Africa in a GCM).

The latter two considerations, horizontal advection and Rossby waves, both have the effect of limiting the northward penetration of the monsoon rainfall. Subtropical westerly winds cool and dry the continent to the west and clear warm moist air from the east, concentrating rainfall on the east coast. This effect has been termed 'ventilation'. East coast convection establishes a source of Rossby waves that induce descent to the west, further disfavouring convection. The relative importance of ventilation and Rossby waves has been measured by Chou et al [43] for a handful of monsoons around the world by carefully modifying certain terms in the dynamical equations of their model. They find that the Rossby wave mechanism is more important for the Asian and American monsoons than for the WAM, where ventilation is the most important effect. However, it is the prescribed albedo (high for the Sahara) that makes the difference, by controlling the local thermodynamics. Indeed the only ways to make the WAM penetrate north of the Sahel are to saturate the Sahara or drop its albedo. So the question of ventilation vs the Rossby wave mechanism seems to depend on whether we are asking the fundamental question of why the Sahara exists, or whether we are trying to understand variability in the monsoon system where the Sahara is given.

Much remains to be understood about the seasonal cycle of the WAM, and how it is modulated. As well as a variable onset date and an apparent rapid onset, variability on subseasonal timescales has also been identified [46] with periods around 15 and 40 days. Statistical studies of outgoing long-wave radiation (OLR) [47] link the organisation of convection on these two timescales with propagation

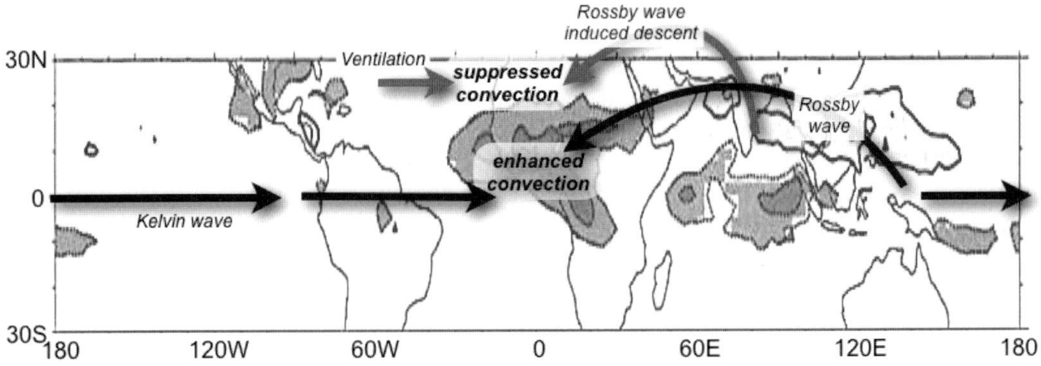

Figure 9. Regression of OLR onto the first principal component of >20-day OLR variability in the African region. Contour interval 4 W m^{-2}, negative values (denoting enhanced convection) shaded dark, zero contour omitted. Also shown are arrows signifying remote influences on the WAM circulation and convection in the WAM region. Adapted from Matthews [48].

through the region in the form of a westward propagating Rossby-wave mode that splits the Sahel into an east-west dipole, and a quasi-stationary mode, that affects convection near the Guinea coast and is of Kelvin-wave form. As we have already seen, these types of disturbance can have remote tropical origins and Matthews [48] presents evidence that intraseasonal modulation of the WAM can emanate from variations in convective heating over the Indonesian warm pool. Using filtered OLR data, lag correlations between longitudes show that when convection is suppressed in the Indonesian source region the resulting anomalous Rossby and Kelvin waves propagate westwards and eastwards respectively at unequal speeds. After circumnavigating the globe they encounter one another again 20 days later to enhance convection over the WAM region. This is shown in fig. 9 in terms of a regression map of OLR onto an active phase over the region. The associated dynamical mechanism involves changes in static stability, monsoon flux and moisture convergence. Increased shear in the AEJ is even posited in [48] as a cause of increased African easterly wave activity and a corresponding increase in convection. The relationship between AEWs and convection is still quite poorly understood and at such short timescales it becomes increasingly difficult to separate the ‘dynamical’ signal from the noise (or self-organisation of convective systems). There is a wealth of literature on AEWs, starting with the analysis of GATE data [49] [50], but synoptic variations such as these, interesting though they may be, are not the focus of this article. For a recent observational study of AEWs see Kiladis et al [51] and for the state of the art in modeling see Hall et al [52]. Superimposed on fig. 9 is a set of arrows that depict the various remote influences on the WAM circulation discussed on this section.

5. DISCUSSION

In this article we have given an overview of the fundamental mechanisms of the West African Monsoon, concentrating on large scale atmospheric dynamics. The WAM is geographically unique in that the lower boundary condition essentially follows lines of latitude. So we can make a lot of progress in understanding the WAM using a two-dimensional framework, and theories developed to explain the zonal mean Hadley circulation are also useful for explaining the WAM. Starting with the laws of angular momentum conservation, force balance and an energy budget, then incorporating the moist thermodynamics of deep convective equilibrium, a relationship emerges whereby the overturning circulation, that includes the onshore monsoon flux, is very sensitive to the large scale near-surface gradient of θ_e (or alternatively moist static energy). Strong or weak gradients can account for vigorous or feeble monsoon states, translating into rain or drought in some regions. Further studies with more complete models refine this view and also show the importance of other factors such as aerosols over the Sahara or the SST in the Mediterranean. We have also sampled a handful of ideas on east-west asymmetry in the WAM. Land-sea contrasts and tropical waves help to explain the monsoon's northern limit, and may give further clues to variations in onset date and the existence of wet and dry years.

As if this picture were not already complicated enough, at this point we must admit that we have not covered everything, far from it. We have restricted our attention to the large scales, neglected clouds and mostly ignored the land surface, vegetation and hydrology. The land surface differs from the ocean in that beyond the diurnal cycle, the net surface energy flux must be essentially zero. How this zero flux condition is met depends on vegetation and soil moisture, which are both interactive. These in turn determine albedo, and thus, along with clouds, the radiative balance of the entire atmospheric column. They also determine the low level temperature and humidity, which condition the local thermodynamic budget and, in turn, the dynamics. Even the simplest vegetation schemes contain threshold effects which introduce further nonlinearities into the system. Our understanding of the monsoon state may ultimately depend on assessing whether natural variability and weather noise can coalesce the multiple equilibria brought on by these nonlinearities.

What we describe as 'weather noise', of course consists of the many traveling organised convective systems that make up the monsoon. The ITCZ is nothing more than an average of these entities, and the way they interact with the large scale flow needs to be properly addressed. We see, therefore, that although a foundation of understanding can be built on large scale atmospheric dynamics alone, to go further, and hope to make useful predictions, a multi-disciplinary, multi-scale approach is needed.

The AMMA campaign [1], mentioned in the introduction, is based on an approach that spans scales and disciplines. An ambitious observational program has been launched, which aims to supply much needed data in a region where the synoptic network is historically poor. Aircraft data and surface measurements will help provide hitherto missing information on the state and progression of the monsoon system. The observations are organised into a long observing period (2002-2010) in which the existing network is reinforced, and a three year 'enhanced' observing period, which includes an intensive 'special' observing period that is in full swing at the time of writing (summer 2006). In this way it is hoped that some continuity can be provided between observations at different timescales. The full range of spatial scales are also covered, with nested systems of observing sites in locations that are mainly arranged along a north-south transect, in recognition of the sharp north-south gradients that characterise the WAM. The measurements are designed to cover the range of phenomena discussed in section 2, from the Saharan heat low and aerosol layer, to the inter-tropical front, the African easterly jet, easterly waves, mesoscale convective systems, the inter-tropical convergence zone, the coastal waters of the northern Gulf of Guinea and the equatorial Atlantic Ocean. Observations are currently being made using a large array of equipment, including satellites, six research aircraft, dropsondes, balloon-borne soundings, three research vessels, rain radars and ground-based instrument platforms. These are being operated simultaneously by French, British, German, American and African scientists. The onset of the monsoon was late this year, and so far the season has proved to be relatively dry, but not without a rich selection of cases to be studied. It will take time to fully synthesise and exploit the considerable amount of data generated by the campaign. Ultimately it is to be hoped that a more detailed description of the system will help to further refine the theories and models that have been introduced in this article, and will improve our capacity to make predictions of the West African Monsoon.

Acknowledgements

Figures and adaptations of figures reproduced with the kind permission of the Royal Meteorological Society (1,2), the American Geophysical Union (3), Cambridge University Press (4) and the American Meteorological Society (5-9)

References

[1] http://amma.mediasfrance.org/
[2] Charney, J., 1975: *J. Atmos. Sci.*, **32**, 193-202.
[3] Mohr, K. I. and C. D. Thorncroft, 2006: *Q. J. R. Met. Soc.*, **132**, 163-176.
[4] Parker. D. J., C. D. Thorncroft, R. R. Burton and A. Diongue-Niang, 2005: *Q. J. R. Meteorol. Soc.*, **131**, 1461-1482.
[5] Smith, E., 1986: *Mon. Wea. Rev.*, **114**, 1084-1102.
[6] Haywood, J. P. et al, 2003: *J. Geophys.Res.*, **108**, 8577.
[7] Mathon, V. and H. Laurent, 2002: *J. Appl. Meteorol.*, **41**, 1081-1092.
[8] Rocca, R., J-P. Lafore, C. Piriou and J-L. Redelsperger, 2005: *J. Atmos. Sci.*, **62**, 390-407.
[9] Koteswaram, P., 1958: *Tellus*, **10**, 43-57.
[10] Hastenrath, S., 1995: *Climate dynamics of the tropics,* Kluwer academic press.
[11] Thorncroft, C. D. and M. Blackburn, 1999: *Q. J. R. Meteorol. Soc.*, **125**, 763-786.
[12] Sultan, B. and S. Janicot, 2003: *J. Climate*, **16**, 3407-3427.
[13] Lebel, T., A. Diedhiou and H. Laurent, 2003: *J. Geophys. Res.*, **108**, doi: 10.1029/2001JD001580.
[14] Persson, A., 1998: *Bull. Am. Met. Soc.*, **79**, 1373-1385.
[15] Hide, R., 1969: *J. Atmos. Sci.*, **26**, 841-853.
[16] Schneider, E. K., 1977: *J. Atmos. Sci.*, **34**, 280-297.
[17] Held, I. M., and A. Y. Hou, 1980: *J. Atmos. Sci.*, **37**, 515-533.
[18] Lindzen, R. S., and A. Y. Hou, 1988: *J. Atmos. Sci.*, **45**, 2416-2427.
[19] James, I. A., 1994: Introduction to circulating atmospheres. *Cambridge University Press.*
[20] Hou, A. Y., and R. S. Lindzen, 1992: *J. Atmos. Sci.*, **49**, 1233-1241.
[21] Plumb, R. A., and A. Y. Hou, 1992: *J. Atmos. Sci.*, **49**, 1790-1799.
[22] Emmanuel, K. A., 1995: *J. Atmos. Sci.*, **52**, 1529-1534.
[23] Xu, K-M., and K. A. Emmanuel, 1989: *Mon. Wea. Rev.*, **117**, 1471-1479.
[24] Eltahir, E. A. B., and C. Gong, 1996: *J. Climate*, **9**, 1030-1042.
[25] Philippon, N., and B. Fontaine, 2002: *Ann. Geophys.*, **20**, 575-582.
[26] Zeng, X., and E. A. B. Eltahir, 1998: *J. Climate*, **11**, 2078-2096.
[27] Peyrillé, P., J.-P. Lafore and J.-L. Redelsper, 2006: *J. Atmos. Sci.*, accepted.
[28] Peyrillé, P., and J.-P. Lafore, 2006: *J. Atmos. Sci.*, submitted.
[29] Thorncroft, C. D. et al, 2003: *Bull. Am. Met. Soc.*, **84**, 337-351.
[30] Cadet, D., and O. Nnoli, 1987: Q. J. R. Meteorlol. Soc., **113**, 581-602.
[31] Fontaine, B., N. Philippon, S. Trzaska and P. Roucou, 2002: *J. Geophys. Res.*, **107**, 834.
[32] Fontaine, B., and S. Janicot, 1996: *J. Climate*, **9**, 2935-2940.
[33] Rowell, D. P., 2003: *J. Climate*, **16**, 849-862.
[34] Madden, R. A., and P. R. Julian, 1994: *Mon. Wea. Rev.*, **122**, 814-837.
[35] Rowell, D. P., C. K. Folland, K. Maskell and M. Ward, 1995: *Q. J. R. Meteorol. Soc.*, **125**, 225-252.
[36] Rodwell, M. J., and B. J. Hoskins, 1996: *Q. J. R. Meteorol. Soc.*, **122**, 1385-1404.
[37] Rodwell, M. J., and B. J. Hoskins, 2001: *J. Climate*, **14**, 3192-3211.
[38] Gill, A. E., 1980: *Q. J. R. Meteorol. Soc.*, **106**, 447-462.
[39] Simmons, A. J., 1982: *Q. J. R. Meteorol. Soc.*, **108**, 503-534.
[40] Hoskins, B. J., 1986: Pp. 57-73 in *Atmospheric and Oceanic variability.* Ed. H. Cattle., R. Meteorol. Soc.
[41] Neelin, J. D., and N. Zeng, 2000: *J. Atmos. Sci.*, **57**, 1741-1766.
[42] Zeng, N., J. D. Neelin and C. Chou, 2000: *J. Atmos. Sci.*, **57**, 1767-1796.
[43] Chou, C., J. D. Neelin and H. Su, 2001: *Q. J. R. Meteorol. Soc.*, **127**, 1869-1891.
[44] Chou, C., and J. D. Neelin, 2003: *J. Climate*, **16**, 406-425.
[45] Cook, K. H., 1999: *J. Climate*, **12**, 1165-1184.

[46] Sultan, B., S. Janicot and A. Diedhiou, 2003: *J. Climate*, **16**, 3389-3406.
[47] Mounier, F., and S. Janicot, 2004: *Geophys. Res. Lett.*, **31**, doi:10.1029/2004GL020665.
[48] Matthews, A. J., 2004: *J. Climate*, **17**, 2427-2440.
[49] Burpee, R. W., 1975: *Mon. Wea. Rev.*, **103**, 921-925.
[50] Reed, R. J., D. C. Norquist and E. E. Recker, 1971: *J. Atmos. Sci.* **28**, 317-333.
[51] Kiladis, G. N., C. D. Thorncroft and N. M. J. Hall, 2006: *J. Atmos. Sci.*, **63**, 2212-2230.
[52] Hall, N. M. J., G. N. Kiladis and C. D. Thorncroft, 2006: *J. Atmos. Sci.*, **63**, 2231-2245.

J. Phys. IV France **139** (2006) 101–118
© EDP Sciences, Les Ulis
DOI: 10.1051/jp4:2006139008

Regional climate modeling: Status and perspectives

F. Giorgi[1]

[1] *Abdus Salam International Centre for Theoretical Physics, Trieste, Italy*

Abstract. This paper is presents a concise review of regional climate modeling, from its ensuing stages in the late 1980s to the most recent developments. A tremendous progress has been achieved in improving the performance of regional climate models, which are currently used by a growing research community for a wide range of applications, from process studies to paleoclimate and future climate simulations. Basic concepts underlying the nested modeling technique, along with the current debate on outstanding issues in regional climate modeling, are discussed. Finally, perspectives of future developments in this rapidly evolving research area are briefly outlined. An extensive reference list is provided to support the discussion.

1. INTRODUCTION

The field of regional climate modeling originated in the late 1980s in response to the need to produce fine scale regional climate information for impact assessment studies, and since then it has undergone a tremendous development. Today most major laboratories worldwide and a large number of universities and other academic institutions maintain, develop and routinely use regional climate models, or RCMs. In addition, the advent of low cost, high power desktop Personal Computer (PC) technology has further enlarged the community of RCM users. A recent Science editorial (Huntingford and Gash 2005) indeed recognizes RCMs as a fundamental resource to engage scientists from developing countries into climate modeling research.

To date RCMs have been applied to virtually all land areas and many ocean areas of the world for a wide variety of applications, from process studies to regional climate change and paleoclimate simulations. Especially in the last decade, the quality of RCM experiments has improved dramatically and an increased understanding of the basic advantages, limitations and technical aspects of regional modeling has been achieved. This is also due to the inception of international projects involving a relatively large number of RCMs in coordinated experiments.

This paper is presents an overview of major developments in regional climate modeling during the last decade or so along with a discussion of its future perspectives. We begin (section 2) with a summary of the basic concepts underlying the nested regional modeling technique followed by a brief historical overview of this field of research, including major research landmarks (section 3). A review of the ongoing discussion of outstanding issues related to regional climate modeling is then presented (section 4) along with an overview of the current range of RCM applications (section 5). Finally (section 6) the paper discusses possible future directions in RCM research. Note that, although an extensive literature list is given here to support the discussion, this paper is not intended to provide a comprehensive review, but rather an introduction, to this area of research.

2. REGIONAL CLIMATE MODELING: BASIC CONCEPTS

The climate of a given region is characterized by the sequence of weather events that affect the region, which is in turn determined by large scale features of the general circulation of the atmosphere. For example, the climate of mid-latitude regions is affected by the seasonal migration of major storm tracks and related interactions with tropical and extratropical modes of variability. The general circulation of the atmosphere is driven by large scale climatic forcings (where large scale here refers to scales from a few hundred km to global), such as due to the input of solar radiation, the concentration of greenhouse gases and long-lived stratospheric aerosols, the distribution of continents and oceans and the presence of large topographical complexes (e.g. the Himalayas).

Superimposed to these large scale climatic features are the effects of regional scale climatic forcings, where regional scale here refers to scales from a few hundred km to local. Examples of regional scale forcings are those due to complex topography and coastlines, surface vegetation distribution, inland bodies of water, and short-lived tropospheric aerosols (Giorgi and Mearns 1991). In addition, weather systems are often characterized by small scale circulation features, e.g. as found during the development and evolution of mesoscale convective systems (Pielke 1984).

It is thus clear that the successful simulation of regional climates requires the representation of processes that occur on a wide range of spatial scales, from the global and large scale to the regional and local scale. The most advanced numerical tools today available to carry out climate (and climate change) simulations are coupled atmosphere-ocean general circulation models, or AOGCMs (e.g. Washington and Parkinson 1986). However, because of limitations in computational resources, the horizontal resolution of present day AOGCMs is still of the order of a few hundred km. Therefore, while AOGCMs can provide a good description of the general circulation of the atmosphere and its response to large scale forcings (e.g. greenhouse as concentration) (McAvaney et al. 2001), they cannot capture processes occurring at finer scales.

The "nested" regional climate modeling technique was thus developed to circumvent this problem and enhance the climate information of AOGCMs at the regional scale. This technique (Figure 1) consists of using a limited area climate model, i.e. an RCM, over a given region of interest. Because the RCM covers only a limited area, it can reach a much higher resolution than the AOGCM and thus it can describe the effects of fine scale, sub-AOGCM (or more generally, GCM) grid scale processes. In order to be run, the RCM requires the provision of initial and lateral meteorological boundary conditions (typically wind components, temperature, water vapor and cloud variables, surface pressure, chemical tracers if needed) and surface boundary conditions (sea surface temperature) (Giorgi and Mearns 1999). In the nesting technique these are provided from a GCM simulation.

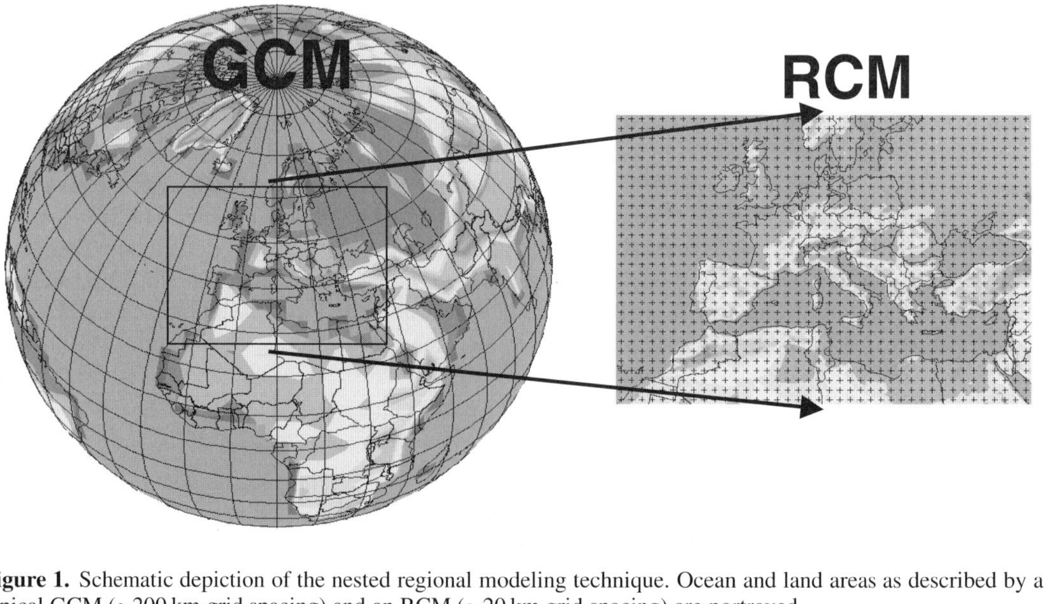

Figure 1. Schematic depiction of the nested regional modeling technique. Ocean and land areas as described by a typical GCM (~200 km grid spacing) and an RCM (~20 km grid spacing) are portrayed.

The nesting procedure described above can be implemented either in one-way or two-way mode. In the one-way nesting approach the GCM information drives the RCM at the lateral boundaries but the RCM information does not feed back into the GCM, while in the two-way nesting approach the RCM solution feeds back into the GCM. To date, the vast majority of RCM studies have adopted the one-way nesting technique, which is much easier to implement, and only recently some two-way nesting studies have been published (e.g. Lorenz and Jacob 2005).

The basic strategy underlying the RCM nesting approach is that the GCM provides the response of the general circulation to large scale forcings. From this information, the RCM describes the effects of sub-GCM grid scale processes and provides fine scale regional information (Giorgi and Mearns 1991; 1999). Especially in the one-way nesting approach, the RCM is thus not intended to strongly modify the large scale circulations of the GCM but mostly to add regional detail in response to the regional scale forcings (e.g. topography). For this reason, the RCM nesting has also been referred to as "dynamical downscaling". This also implies that if the large scale GCM fields are characterized by large errors (for example large displacements of storm tracks) the nested RCM is generally not intended to correct these errors (although in practice some correction might occur, Giorgi et al. 1998). As a consequence, it is extremely important to evaluate the large scale fields used for RCM nesting before proceeding to the RCM simulation.

Another important aspect of the nesting technique is that different types of large scale fields can be used to "drive" an RCM at the lateral boundaries depending on the specific application. These include fields from coupled AOGCMs (e.g. when performing a climate change simulation), fields from global atmospheric models (AGCMs) or fields from global analyses of observations. Experiments for the latter case are usually referred to as "perfect boundary condition (PBC)" experiments because analyses of observations, although produced with models characterized by systematic biases, also include assimilation of observations and thus can be expected to provide the best available forcing large scale fields. PBC experiments can be directly compared with available observations for actual periods, and thus are optimal to test and optimize the RCM performance.

Finally, it is important to emphasize that RCM nesting is not the only technique today available to regionally enhance the AOGCM information. Other such techniques include variable resolution AGCMs (Deque and Piedelievre 1995) and empirical/statistical downscaling methods (Hewitson and Crane 1996). Each technique is characterized by its own advantages and limitations and is possibly preferable depending on the specific applications (Giorgi et al. 2001). It is beyond the purpose of this paper to address other "regionalization" techniques, but it is useful to summarize here the primary advantages and limitations of the one-way nesting approach.

The main advantage of RCM nesting is that it is a physically based downscaling approach, which implies on the one hand that all climate variables are calculated in a physically consistent way and on the other hand that RCMs are general tools applicable to a wide variety of studies. The main limitation of RCMs resides in the one-way nesting approach, which implies, first, that major errors in the driving GCMs are not corrected by the RCM and, second, that regional to global feedbacks are not accounted for. These problems are somewhat ameliorated by the two-way nesting approach. In addition, as will be seen in section 4, the use of RCMs requires the careful consideration of a number of technical issues.

Perhaps the fundamental question of RCM nesting is that usually referred to as "added value", that is: what is the added value of using an RCM compared to a GCM? This question does not have an obvious answer and depends on the specific problem of interest. Section 4 deals extensively with this question, but it is important to emphasize that the "added value" of the use of an RCM needs to be carefully considered before embarking in any RCM-based study.

3. A BRIEF HISTORY OF REGIONAL CLIMATE MODELING

Regional climate modeling first originated in the late 1980s. Limited area model nesting was by no means a new approach, as it had been used for many years in numerical weather prediction (NWP). In this sense, nested regional modeling was essentially borrowed from NWP. On the other hand, limited area models had been run only for periods of up to five days and the NWP community was very skeptical about their use in long term climate simulations. This is because it was felt that errors generated at the lateral boundaries would swamp and contaminate the model solution at long temporal scales. For this reason, the original approach proposed by Dickinson et al (1989) for regional climate simulation was to carry out multiple short experiments and build a regional climatology from the statistics of the ensemble of short runs.

Indeed, the extension of available limited area models to long term climate simulations was not trivial in that these models did not include processes that are secondary at short time scales (few days) but are critical for long temporal scales (decades to centuries). Examples of such processes are atmospheric radiative forcing and interactive soil moisture. The first RCM was developed by a small research team at the National Center for Atmospheric Research (NCAR) via the augmentation of the Penn State NCAR mesoscale model MM4 in the areas of radiative forcing and land surface processes (Dickinson et al. 1989). This model was used for the first simulations in "climate mode", i.e. in simulations of length greater than a few days (more specifically in month-long simulations) by Giorgi and Bates (1989) in PBC experiments and Giorgi (1990) in GCM-driven experiments. It is important to realize that a huge conceptual step was taken when it was shown that limited area models could be successfully run for long integration times, something that today is taken essentially for granted. The importance of a long simulation time for climate application is twofold. First, it allows a model to exceed the atmospheric spin-up time (typically a few days) during which the information from the lateral boundaries reaches a dynamical equilibrium with the internal model physics. The resulting climatology is thus more fully representative of the model behavior. Second, it allows the model to equilibrate with its surface components (e.g. soil moisture and snow), thereby providing a more internally consistent model climatology.

After the first month-long simulations demonstrated the applicability of RCMs to long numerical experiments, the first multi-year RCM simulations were completed by Giorgi et al. (1993a) in the PBC mode and Giorgi et al. (1994) in a GCM-driven regional climate change experiment for the continental U.S. Later, the first decadal simulations were completed by Jones et al. (1995, 1997) and McGregor et al. (1995). Today multi-decadal simulations have become the norm (e.g. Machenhauer et al. 1998; Deque et al. 2005) and even some 140 year long transient climate change experiments have been carried out (Mcgregor et al. 1999).

In terms of horizontal resolution, the first RCM experiments employed grid intervals in the range of 50-125 km. The resolution of RCM simulations has increased over the years, and to date a number of RCM experiments have been published with grid intervals in the range of 15-25 km (Marinucci et al. 1995; Rotach et al. 1997; Kauker 1998; Machenhauer et al. 1998; Christensen et al. 1998; Kato et al. 1999; Christensen and Kuhry 2000; Leung and Qian 2003; Diffenbaugh et al. 2005; Gao et al. 2006; Im et al. 2006). Many of these experiments employ the so-called double nesting procedure, in which an intermediate resolution RCM simulation is used to generate lateral boundary conditions for further nested very fine scale RCM experiments.

Important developments also occurred during the last decade concerning the regions of RCM application. In the early days of regional modeling, most RCMs were developed for mid-latitude regions such as North America, Europe, East Asia and Australia. Today, virtually all land regions and many ocean regions of the world have been covered by RCM experiments. It is however important to mention that the extension of an RCM from mid-latitude to tropical regions is not trivial, as convective processes and their representation in regional models can be very different across latitudinal zones. Indeed, one of the potential advantages of RCMs is the possibility of using different convection schemes in different domains.

In this regard, great progress in the understanding of RCM behavior has been achieved through a number of regional model intercomparison projects: The Project to Intercompare Regional Climate Simulations (PIRCS, Takle et al. 1999), the North American Climate Change Application Project

(NARCCAP), and the North American Monsoon model Assessment Project (NAMAP, Gutzler et al. 2006) for North America; The Prediction of Regional scenarios and Uncertainties for Defining EuropeaN Climate change risks and Effects (PRUDENCE, Christensen et al. 2002) and the NEWBALTIC I and II (Jacob et al. 2001) for Europe; The Arctic Regional Climate Model Intercomparison Project (ARCMIP; Rinke et al. 2006); The Regional Climate Model Intercomparison Project for East Asia (RMIP, Fu et al. 2005); The IRI/ARC Intercomparison project for South America (Roads et al. 2003). More recently, Takle et al. (2006) launched a Model Transferability Project to evaluate the sensitivity of RCMs to their use in different regional domains. It is also worth mentioning that the European modeling community has been especially active in RCM research through a series of EU-sponsored projects culminating in the projects PRUDENCE and ENSEMBLES, which have represented major landmarks in regional climate modeling.

An important recent evolution of RCMs is the coupling of the atmospheric model component with other components of the climate system towards the development of regional Earth System Models (Giorgi 1995). A number of efforts in this direction started in the mid and late 1990s and are continuing. For example, Lynch et al. (1995) and Bailey and Lynch (2000) coupled regional atmosphere and ocean/sea ice models for the Polar regions, Hostetler et al. (1993) and Small et al. (1999b) developed coupled atmosphere-lake models; Leung et al. (1996) coupled an RCM with a basin hydrology model; Qian and Giorgi (1999), Giorgi et al. (2003), Ekman and Rodhe (2003), Solmon et al. (2006) and Zakey et al. (2006) developed interactively coupled regional chemistry/aerosol and climate models. Efforts are also ongoing to couple atmosphere and biosphere model components (Lu et al. 2001).

Finally, throughout the years the development of RCM research has been documented in a number of excellent review papers, which the readers are encouraged to consult: Giorgi and Mearns (1991), McGregor (1997); Giorgi and Mearns (1999); Giorgi et al. (2001); Leung et al. (2003); Wang et al. (2004). Furthermore, large collections of RCM-based research papers, showing in particular the wide range of applications of such models, can be found in four journal special issues: The April 1999 issue of the Journal of Geophysical Research-Atmosphere (Giorgi and Mearns 1999), the December 2004 issue of the Journal of the Meteorological Society of Japan (Wang et al. 2004) and upcoming special issues of Theoretical and Applied Climatology (Giorgi et al. 2006) and Climatic Change (Christensen et al. 2006).

4. OUTSTANDING ISSUES IN REGIONAL CLIMATE MODELING

This section summarizes the current debate concerning outstanding issues in regional climate modeling. As already mentioned, the most fundamental one is the issue of "added value". The use of a nested RCM is justified to the extent that it adds useful and improved information compared to the driving GCM, which is not necessarily always the case. A clear situation in which RCMs provide substantial added value is in the presence of complex fine scale topographical features, which are well known to affect precipitation. An example of this added value is shown in Figure 2, which clearly demonstrates the improvement in the simulation of precipitation by an RCM compared to the driving GCM. Similar conclusions have been obtained consistently when using RCMs in mountainous regions (e.g. Giorgi and Mearns 1999). In fact, various studies have shown that the surface climate change signal is strongly dependent on fine scale topographical features (Giorgi et al. 1994, 1997; Jones et al. 1997; Gao et al. 2006).

Another instance of clear added value is in the simulation of extreme events, which often occur at fine spatial and temporal scales. Various studies (Christensen et al. 1998; Huntingford et al. 2003; Frei et al. 2003) have shown that RCMs can describe better than GCMs the frequency of occurrence of high intensity daily precipitation events (e.g. see Figure 3). Similarly, high resolution is a critical requirement for the simulation of realistic tropical storms and hurricanes (e.g. Knutson et al. 2000).

Areas of complex coastlines, such as in the presence of narrow islands and peninsulas (e.g. the Italian peninsula or the Japan Islands) are not even described at the resolution of many current AOGCMs, and

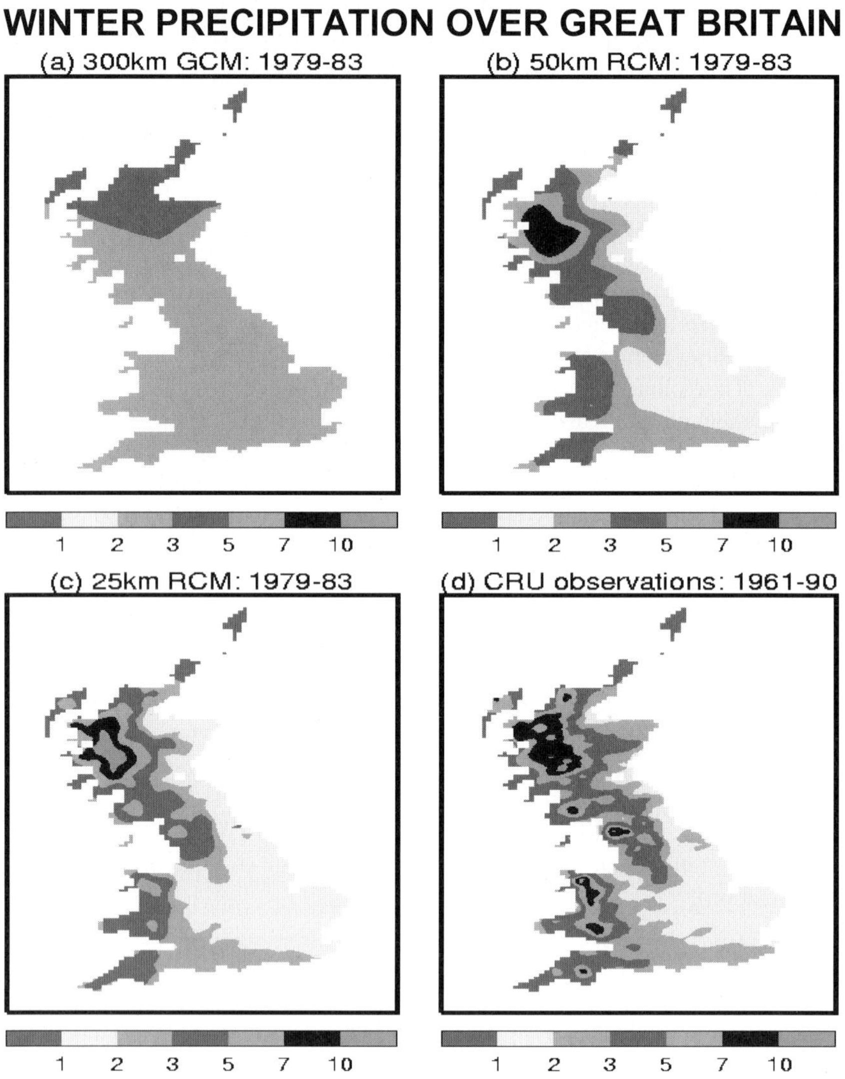

Figure 2. Winter precipitation (mm/day) over Great Britain as simulated by a GCM (a) and an RCM at 50 km (b) and 25 km (c) grid spacing. Corresponding observations are shown in (d). The improved agreement with observations at the higher RCM resolutions is evident. (Courtesy of R.G. Jones).

this clearly calls for the use of fine scale RCMs. In addition, the fine resolution of nested RCMs is necessary to simulate the effect of surface heterogeneity-driven organized mesoscale circulations (such as the sea-breeze), which are well known to affect local climate (e.g. Pielke 1974; Giorgi and Avissar 1997). Regions where tropospheric aerosol distributions are highly variable at fine scales may also require the use of high resolution models (e.g. Giorgi et al. 2003). All these examples clearly illustrate the added value of RCMs, however they also show that the question of added value needs to be carefully addressed before carrying out any RCM application, especially because the performance of an RCM (or more generally a climate model) does not necessarily increase with increased resolution.

Another issue which has been often debated within the regional climate modeling community is that of the physics representations of an RCM with respect to those of the driving GCM. Traditionally, one line of thought has supported the approach that the physics packages of an RCM should be the

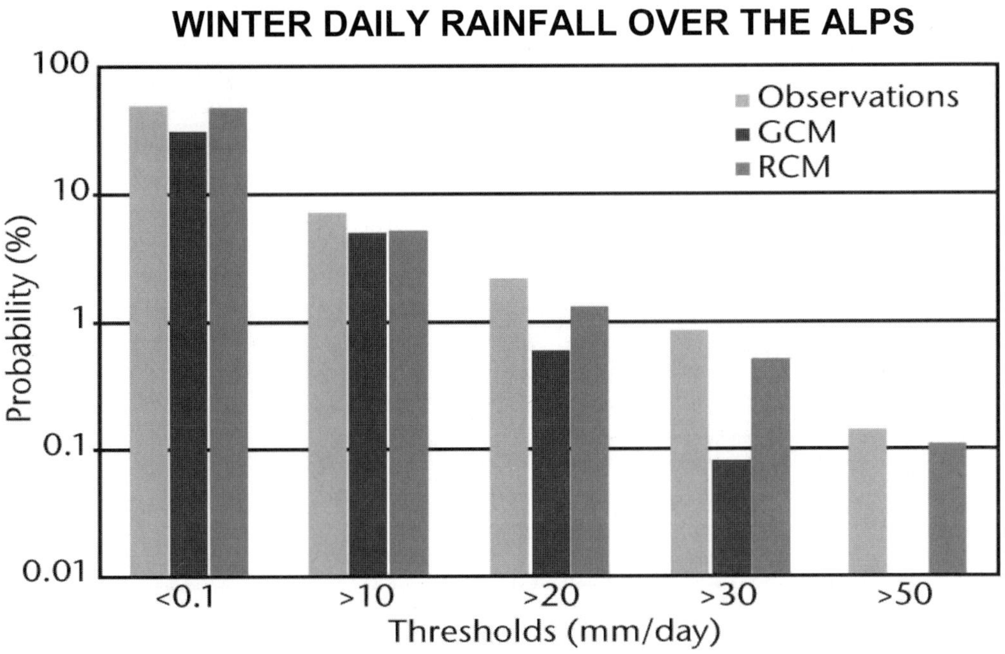

Figure 3. Winter daily rainfall probability of occurrence as a function of intensity thresholds over the Alps as observed and as produced by a GCM and a nested RCM. Only the RCM is capable of producing high intensity events in agreement with observations. (Courtesy of R.G. Jones).

same as that of the driving GCM in order to maximize the consistency between the two models. Another approach has been to use different physics schemes in the global and regional models, with the underlying assumption that each model uses the physics representations most appropriate at the respective resolution. Indeed, the same physics scheme can behave quite differently at different resolutions (Giorgi and Marinucci 1996). The PRUDENCE project has addressed this issue by virtue of sets of simulations in which global and regional models had both the same and different physics schemes. This exercise showed that there was not a clear difference in performance between models with and without the same physics as in the driving GCM (Deque et al. 2006; Jacob et al. 2006). Therefore, the issue of GCM-RCM physics consistency does not appear to be a critical one, although having the same physics in the driving and nested models might indeed help in the interpretation of differences between the model results.

The issue of the technique used to provide lateral boundary conditions has been recently raised by von Storch et al. (2000). The traditional approach has been to provide boundary forcing in a lateral buffer zone whose width can be selected by the model user (Figure 4) using a relaxation procedure (essentially a Newtonian relaxation term) that drives the model prognostic variables towards the boundary conditions (Davies and Turner 1977). The resulting model solution is thus determined by a dynamical equilibrium (obtained after a sin-up time of several days) between the lateral boundary information pervading the interior of the domain and the internal model physics and dynamics. In this regard, different functional forms can be used to provide a smooth blending of model solution and boundary fields in the buffer zone (Giorgi et al. 1993b).

An alternate approach, usually referred to as spectral nesting or nudging (Kida et al. 1991, von Storch et al. 2000), is to provide large scale forcing to the model fields throughout the entire model domain, but only in the long wave spectrum of the solution. In other words, the large scale component of the solution is forced throughout the domain, while the RCM only solves for the small scale spectrum

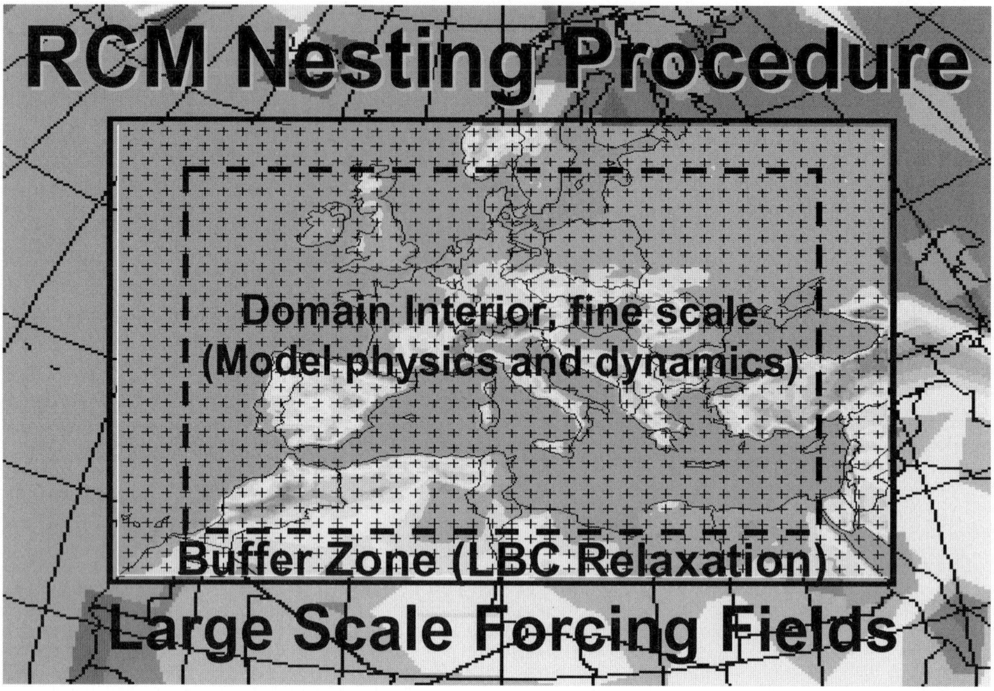

Figure 4. Depiction of the standard procedure use to provide lateral boundary conditions to an RCM. A Newtonian relaxation term is applied to the lateral buffer zone in order to smoothly blend the model solution towards the large scale forcing fields.

of the solution. The two approaches are characterized by a different philosophical interpretation of the nesting method. In the spectral nudging the RCM is purely seen as a dynamical downscaling tool that adds regional detail to a large scale field entirely determined by the global model. The advantage of spectral nudging is that it provides a strong consistency between large scale forcing and model fields. The advantage of the standard relaxation method is that it leaves the RCM more freedom to generate its own circulations in the interior of the domain, especially when the domain is large. This, particularly when using GCM forcing fields characterized by larger errors than analyses of observations, might lead to a partial improvement of the simulation also at the large scale (e.g. Giorgi et al. 1998). To date, it has been difficult to rigorously evaluate which of the two approaches is better, and often the choice of approach is determined by the particular application being considered and by a subjective decision by the individual modeler.

As already mentioned, several technical issues need to be carefully considered in the design of an RCM experiment. One is the choice of domain size and location in relation to the model resolution. The domain should be large enough to include all forcing areas that are relevant for the problem under consideration and to allow the RCM to develop its own circulations (the smaller the domain the greater the relative contribution of the lateral boundary forcing). By the same token, the model resolution should be sufficient to capture relevant forcings for the problem at hand (e.g. the structure of mountainous systems). On the other hand, recent experiments using the so-called big-brother approach (Denis et al. 2002; Dimitrijevic and Laprise 2005) have indicated that the ratio in resolution between the large scale forcing fields and the RCM should not exceed a value of 8-10, above which the climatology of the nested model becomes inconsistent with that of the driving fields. Therefore the issues of model domain, boundary fields and resolution are interconnected.

The placement of the domain boundaries is also important. For example, the domain boundaries should not cross major mountain systems, since there the mismatch between large scale forcing fields and fine scale model fields is maximum and could lead to the generation of spurious model behavior. In addition, the location of the lateral boundary might affect the model sensitivity to internal parameters (e.g. Seth and Giorgi 1998). Often, spurious precipitation is produced at the downwind boundary of an RCM domain because differences in the boundary and model wind fields can create spurious areas of convergence there (Giorgi and Mearns 1999). More generally, the effects of the lateral boundary conditions tend to incrementally affect the model solution as we move towards the lateral boundaries. It is therefore best to place the lateral boundaries of an RCM domain as far away as possible from the area of interest.

Both, larger model domain and increased model resolution lead to more computationally intensive simulations and hence, in practice, often the decision on model domain size and resolution is reached on a trial and error basis also taking into account the availability of computational resources. It should be recognized, however, that in general an RCM simulation depends on the location of the model lateral boundaries, and this dependency should be reduced as much as possible through the testing of different model domains. Typically, RCM domains have varied from a few hundred km size to sizes sufficient to cover entire continents and adjacent ocean areas.

5. RCM APPLICATIONS

During the last 15 years, RCMs have been applied to a wide variety of studies. Most of the first applications were carried out for the European (e.g. Jones et al. 1995, 1997), North America (e.g. Giorgi et al. 1994), East Asia (e.g. Hirakuchi and Giorgi 1995) and Australia (McGregor et al. 1995) regions, but more recently RCMs have been increasingly applied over Africa (e.g. Sun et al. 1999), south, southeast and central Asia (e.g. Hassell and Jones, 1999; Francisco et al. 2006; Small et al. 1999), and South America (e.g. Seth and Rojas 2003; Rojas and Seth 2003).

Much of the RCM work goes into the evaluation, testing and improvement of the model performance. This can be done either in PBC mode or in GCM-driven mode (Giorgi et al. 2001). In this regard, it is important to note that the RCM performance has been considerably enhanced over time. In the early years of regional modeling, climatological biases exceeding several degrees for temperature and 50% for precipitation were common. Today, state-of-the-art RCMs are expected to show temperature biases of less than 1-2 degrees and precipitation biases not exceeding 25% (Giorgi and Mearns 1999). In addition, when driven by good quality boundary conditions, RCMs have shown good performance in reproducing interannual variability (e.g. Giorgi and Shields 1999; Sun et al. 1999) as well as extreme events (Pal and Eltahir 2003; Huntingford et al. 2003; Frei et al. 2003). The improvement found in RCM simulations is due to both better models and better quality GCM and reanalysis fields available to produce the lateral boundary conditions.

Many studies, also related to the model intercomparison projects mentioned in section 2, aimed at a better understanding of the behavior of RCMs. Among these we find studies of the RCM sensitivity to domain size and location (e.g. Seth and Giorgi 1998; Noguer et al. 1998; Landman et al. 2005), use of different physics schemes (e.g. Ratnam and Kumar 2005), assimilation of lateral boundary conditions (e.g. von Storch et al. 2000; Vidale et al. 2003; Marbaix et al. 2003; Wu et al. 2005), resolution of the driving lateral boundary data (e.g. Denis et al. 2002) and initialization of land surface variables (e.g. Christensen et al. 2001). A few studies have addressed the issue of internal model variability (e.g. Giorgi and Bi 2000; Christensen et al. 2001; Rinke et al. 2004), which is today recognized to be an important factor in evaluating the RCM response to external forcings, and the dependency of model physics parameterizations on model resolution (e.g. Giorgi and Marinucci 1996).

RCMs can also be used for process studies. Among these applications we can highlight studies of the regional climatic effects of land-use change (e.g. Copeland et al. 1996; Fu 2003) and atmospheric aerosols (e.g. Giorgi et al. 2003; Ekman and Rodhe 2003), and studies of soil moisture – atmosphere

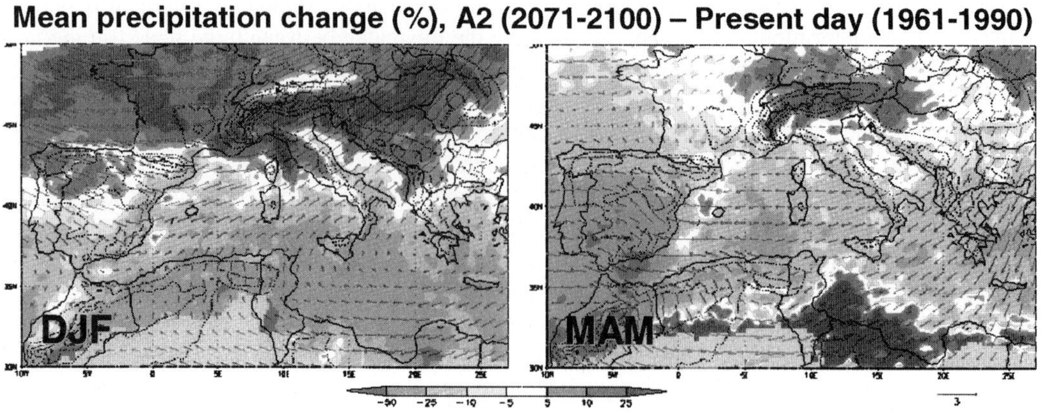

Figure 5. Mean precipitation change (winter in the left panel, spring in the right panel) calculated in a high resolution (20 km grid spacing) simulation over the Mediterranean region for the A2 IPCC emission scenario compared to present day. The arrows show the change in low level wind. The effect of topography in modulating the precipitation change signal is for example evident along the Italian Peninsula and over the Alps (Adapted from Gao et al. 2006).

feedbacks (e.g. Pal and Eltahir 2003). These types of processes are markedly regional in nature and RCMs have proven to be extremely useful high resolution modeling tools to explore them.

Perhaps the primary application of nested RCMs has been in the development of climate change scenarios due to increased greenhouse gas concentration over different regions of the world (see for example the reference lists in Giorgi et al. 2001; Christensen et al. 2001; Deque et al. 2006). In these applications the length of RCM simulations has been from multi-year to multi decadal and the model grid interval has varied in the range of ~20 to 125 km. In particular, RCMs have been especially useful in capturing and understanding the effects of fine scale topography on the climate change signal (Figure 5), which can often even locally reverse the sign of the signal (e.g. for precipitation) compared to the driving GCM (e.g. Giorgi et al. 1994; Jones et al 1997). Because of the better topographical representation in the RCM it can be argued that the RCM-produced change signal is in this case more credible. Of particular relevance within the context of climate change studies is also the use of RCMs to simulate changes in extreme events (Gao et al. 2002; Christensen and Christensen 2003; Huntingford et al. 2003; Bell et al. 2004).

An important conclusion recently reached from the PRUDENCE project (Deque et al. 2006) is that the uncertainty in a regional climate change projection due to the use of different RCMs can be as large as the uncertainty due to different GCMs (e.g for summer precipitation, see Figure 6). This somewhat surprising result, which indicates that the internal model physics can be dominant over the lateral boundary forcing for some climatic variables, implies the need to use ensembles of RCMs in the development of regional climate change scenarios.

A substantial development in RCM research is that RCM-based regional climate change scenarios are now considered to be of sufficient quality and resolution to be used in impact assessment studies. A range of such studies is today available in the literature. Examples of these include regional climatic impacts on agriculture (e.g. Mearns et al. 2001; Tvetsinskaya et al. 2003), water resources (e.g. Stone et al. 2003; Kleinn et al. 2005), human health (e.g. Diffenbaugh et al. 2006) and regional economy (e.g. Adams et al. 2003).

The application of RCMs to paleoclimate studies has been so far rather limited. The first RCM paleoclimate simulations were reported by Hostetler et al. (1991, 1994). Since then, additional studies were carried out for example by Barron and Pollard (2000), Hostetler et al. (2000) and Diffenbaugh et al. (2004). However the use of RCMs in paleoclimate research still needs to be better explored

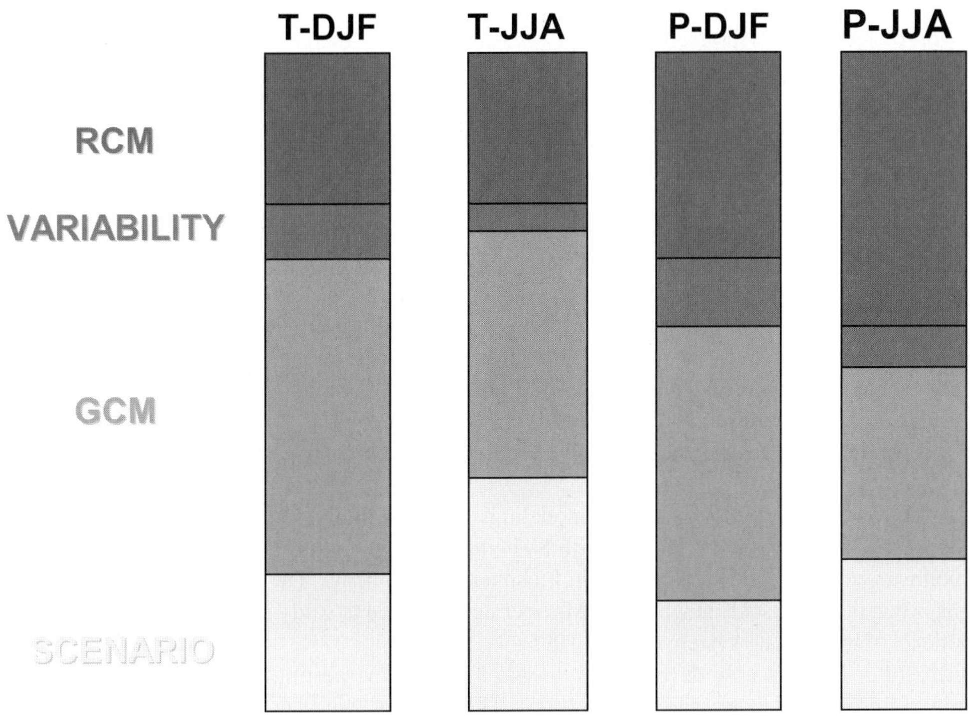

Figure 6. Relative contribution to the uncertainty in the simulation of climate change over Europe originating from various sources: 1) Use of different RCMs (8 models); 2) Internal variability of GCMs; 3) Use of different GCMs (4 models); Use of different scenarios (2 scenarios covering about half the IPCC scenario range). T is temperature, P is precipitation, DJF is winter, JJA is summer. Uncertainty due to GCMs and scenarios (when accounting for the reduced range considered) generally dominates except for summer precipitation, when uncertainty due to RCMs is of comparable magnitude.

and exploited. An excellent discussion of the use of RCMs in paleoclimate research is provided by Sloan (2006).

Finally, a recent research thrust has been directed towards the use of RCMs for regional downscaling of seasonal forecasts (Wang et al. 2004). This application was first explored by Fennessy and Shukla (2000), but this exploratory study was not followed by a sustained research effort. More recent applications of RCMs to seasonal prediction can be found in Cocke and Larow (2000), Druyan et al. (2002) and Rauscher et al. (2006). This relatively unexplored application is today considered to be a very promising area of research in regional modeling (Wang et al. 2004).

6. SUMMARY AND FUTURE PERSPECTIVES

It is clear from this brief review that since its inception in the late eighties, the field of regional climate modeling has steadily grown and matured. The behavior of RCMs and the advantages and limitations of the (one-way) nesting technique are much better understood, the quality of state-of-the-art RCMs as well as that of the GCM fields used to provide lateral boundary conditions has noticeably improved (and will likely continue to do so), and the RCM user community as well as the range of RCM applications has consistently broadened. As rapidly increasing computing resources allow the GCM modeling community to increase the horizontal resolution of global models, the regional modeling community is confronted with the task of planning for the future.

First, along with the increase in the GCM resolution, we will certainly see a rapid increase in the resolution of RCMs in response to the ever pressing need for increasingly higher resolution climate information usable in impact assessment studies. Until the early 2000s, the typical grid spacing of RCMs was ~50 km, while today the target grid spacing is ~20 km. RCM simulations at grid spacing of a few km is the target for the coming decade and a new generation of RCMs. This however will present a number of problems, not only of computational nature. For resolutions higher than about 10 km non-hydrostatic effects become important. As a result, most current RCM systems (which are still hydrostatic) are being upgraded to include non-hydrostatic dynamics. In addition, most physics parameterizations in current RCMs, most noticeably for precipitation and cumulus convection, were designed for model resolutions at which the scale separation assumption underlying many physics parameterizations is robust. At scales of a few km this assumption is not valid any more and a new set of physics schemes may be required in RCMs to describe more explicitly precipitation, cloud, deep convection and boundary layer processes. Furthermore, as the model resolution increases, higher resolution observation datasets will be needed to evaluate the model performance, and this may prove to be a formidable task, especially in regions where data are sparse (e.g. mountainous and remote regions).

Also following the trend in global modeling, we will likely see increased efforts toward the development of fully coupled regional Earth System Models including coupled atmosphere, ocean-sea ice, chemistry-aerosol, land hydrology and biosphere components. The importance of such modeling systems is that often the scale at which RCMs operate is optimal to study interactions across the different components of the climate system (Giorgi 1995).

Different groups are now developing two-way GCM-RCM nesting capability, encouraged by the preliminary results of Lorenz and Jacob (2005). The issue of two-way nesting, however, needs to be carefully approached. One of the most attractive aspects of the one-way nesting procedure is the relative simplicity of its implementation. Executing a two-way nested simulation implies running at the same time and with a complex interfacing procedure two very complex modeling systems (the GCM and RCM). This may pose severe technical problems. It is thus important to carefully evaluate weather the added value of using two-way nesting overweighs the added complexity of implementing it. This likely depends on the region and problem under consideration.

One of the greatest advantages of regional modeling is that these can be effectively run on inexpensive computing platforms, such as desktop PCs or PC clusters. As already emphasized, this has tremendously enlarged the scientific community directly involved in climate modeling, especially from the developing countries (Giorgi et al. 2006; Pal et al. 2006), which has important consequences. First, with more scientists using the models, the models themselves become better understood and improved. Second, a wider range of scientists can acquire hands-on experience and know-how in climate modeling and application and thus RCMs can provide a critical capacity building tool for the scientific community (Huntingford and Gash 2005). This has the important implication that scientists in developing countries do not have to only rely on the information deriving from models run in major laboratories to address global change issues. Rather, they can produce their own information. Third, RCMs can be applied to scientific problems specifically tied to local and regional needs. Not only this meets a local demand for climate information but it also allows the overall modeling community to benefit from the knowledge of locally specific problems by the local scientists.

With a bigger scientific community being involved in regional modeling, the need will be pressing that this community finds a way to organize its activities in a coordinated fashion. In the past, most RCM efforts have been rather individual and applied to specific problems. It has now become clear that a better understanding of global change requires the compounded information of large ensembles of different model simulations (Giorgi 2005), which requires a strong coordination of modeling efforts. This coordination has happened effectively in the global modeling community, but it is still lagging in the regional modeling one, perhaps due to the more "local" nature of regional modeling. The RCM intercomparison projects of section 2 have been important steps towards a greater coordination of RCM work. A recent effort to develop a Regional Climate research NETwork (RegCNET; Giorgi et al. 2006, Pal et al. 2006) is another step in this direction. It is important that such coordinated international programs are further developed and strengthened in the future to provide means of better integration within the RCM research community.

References

Adams, R.M., B.A. McCarl, and L.O. Mearns, 2003: The effects of spatial scales of climate scenarios on Economic Assessments: An example from U.S. agriculture. Climatic Change, 60, 131-148.

Barron, E.J. and D. Pollard, 2000: High-resolution climate simulations of Oxygen Isotope Stage 3 in Europe, Quaternary Research, 58, 296-309.

Bailey, D.A. and A.H. Lynch, 2000: Development of an Antarctic regional climate system model. Part I: Sea ice and large scale circulations. J. Climate, 13, 1337-1350.

Bell, J.L., L.C. Sloan and M.A. Snyder, 2004: Regional changes in extreme climatic events: A future climate scenario. J. Climate, 17, 81-87.

Christensen, J.H. and P. Kuhry, 2000: High resolution regional climate model validation and permafrost simulation for the East European – Russian arctic. J. Geophys. Res., 105, 29647-29658.

Christensen, J.H. and O.B. Christensen, 2003: Climate modeling: Severe summertime flooding in Europe. Nature, 421, 805-806.

Christensen O.B., J.H. Christensen, B. Machenhauer and M. Botzet, 1998: Very high-resolution regional climate simulations over Scandinavia – Present climate. J. Climate, 11, 3204-3229.

Christensen, J.H. et al., 2001: Synthesis of regional climate change simulations: A Scandinavian perspective. Geophys. Res. Lett., 28, 1003-1006.

Christensen, O.B., M.A. Gaertner, J.A. Prego and J. Polcher, 2001: Internal variability of regional climate models. Clim. Dyn., 17, 875-887.

Christensen, J.H., T.R. Carter, and F. Giorgi, 2002: PRUDENCE employs new methods to assess European climate change. EOS, 83, 147.

Christensen, J.H., T.R. Carter, and M.R. Rummukainen, 2006: Evaluating the performance and utility of regional climate models: The PRUDENCE project. Climatic Change, in press.

Cocke, S.D. and T.E. LaRow, 2000: Seasonal Prediction using a Regional Spectral Model embedded within a Coupled Ocean-Atmosphere Model. Mon. Wea. Rev., 128, 689-708.

Copeland, J.H., R.A. Pielke and T.G.F. Kittel, 1996: Potential climatic impacts of vegetation change: A regional modeling study. J. Geophys. Res., 101, 7409-7418.

Davies, H.C., and R.E. Turner, 1977: Updating prediction models by dynamical relaxation: An examination of the technique. Quart. J. Roy. Met. Soc., 103, 225-245.

Denis, B. et al., 2002: Downscaling ability of one-way nested regional climate models: The Big Brother Experiment. Clim. Dyn., 18, 627-648.

Deque, M. and J.P. Piedelievre, 1995: High resolution climate simulation over Europe. Clim. Dyn., 11, 321-339.

Deque, M. et al., 2005: Global high resolution vs. regional climate model climate change scenarios over Europe: Quantifying confidence level from PRUDENCE results. Clim. Dyn., 25, 653-670.

Deque, M. et al., 2006: An intercomparison of regional climate simulations for Europe. Assessing uncertainties in model projections. Climatic Change, in press.

Dickinson, R.E., R.M. Errico, F. Giorgi, and G.T. Bates, 1989: A regional climate model for the western United States, Climatic Change, 15, 383-422.

Diffenbaugh, N.S. and L.C. Sloan, 2004: Mid-Holocene orbital forcing of regional-scale climate: A case study of Western North America using a high-resolution RCM. J. Climate, 17, 2927-2937.

Diffenbaugh, N.S., J.S. Pal, R.J. Trapp and F. Giorgi, 2005: Fine scale processes regulate the response of extreme events to global climate change. Proc. Nat. Acad. Sci., 102, 15774-15778.
Diffenbaugh, N.S., J.S. Pal, F. Giorgi and X. Gao, 2006: Coastal amplification of heat stress risk in the Mediterranean hotspot. Submitted to Nature.
Dimitrijevic, M. and R. Laprise, 2005: validation of the nesting technique in a regional climate model and sensitivity tests to the resolution of the lateral boundary conditions during summer. Clim. Dyn., 25, 555-580.
Druyan, L.M., M. Fulakeza and P. Lonergan, 2002: Dynamic downscaling of seasonal climate predictions over Brazil. J. Climate, 15, 3411-3426.
Ekman, A.M.L. and H. Rodhe, 2003: Regional temperature response due to indirect sulfate aerosol forcing: impact of model resolution. Clim. Dyn., 21, 1-10.
Fennessy, M.J. and J. Shukla, 2000: Seasonal prediction over North America with a regional model nested in a global model. J. Climate, 13, 2605-2627.
Francisco, R.V. et al., 2006: Regional model simulation of summer rainfall over the Philippines. Theor. Appl. Climatol., 86, 211-224.
Frei C. et al. 2003. Daily precipitation statistics in regional climate models: Evaluation and intercomparison for the European Alps. J. Geophys. Res., 108, 4124-4136.
Fu, C., 2003: Potential impacts of human induced land cover change on East Asia monsoon. Glob. Planet. Change., 37, 219-229.
Fu, C. et al., 2005: Regional Climate Model Intercomparison Project for Asia. Bull. Amer. Met. Soc., 86, 257-266.
Gao, X., Z.-C. Zhao and F. Giorgi, 2002: Changes in extreme events in regional climate simulations over East Asia. Adv. Atmos. Sci, 19, 927-942.
Gao, X., J.S. Pal and F. Giorgi, 2006: Projected changes in mean and extreme precipitation over the Mediterranean region from high resolution double nested RCM simulations, Geophys. Res. Lett., 33, L03706, doi: 10.1029/2005GL024954.
Giorgi F., 1990: Simulation of regional climate using a limited area model nested in a general circulation model. J. Climate, 3, 941-963.
Giorgi, F., 1995: Perspectives regional Earth System modeling. Glob. Planet. Change, 10, 23-42.
Giorgi, F., 2005: Climate change prediction. Climatic Change, 73, 239-275.
Giorgi, F. and G. T. Bates, 1989: The climatological skill of a regional model over complex terrain, Mon. Wea. Rev., 117, 2325-2347.
Giorgi, F. and L.O. Mearns, 1991: Approaches to the simulation of regional climate change: a review. Rev. Geophys., 29, 191-216.
Giorgi, F. and M.R. Marinucci, 1996: An investigation of the sensitivity of simulated precipitation to model resolution and its implications for climate studies. Mon. Wea. Rev., 124, 148-166.
Giorgi, F. and R. Avissar, 1997: The representation of heterogeneity effects in Earth System modeling: Experience from land surface modeling. Rev. Geophys., 35, 413-438.
Giorgi, F. and L. O. Mearns, 1999: Introduction to special section: Regional climate modeling revisited. J. Geophys. Res., 104, 6335- 6352.
Giorgi, F. and C. Shields, 1999: Tests of precipitation parameterizations available in the latest version of the NCAR regional climate model (RegCM) over the continental United States. J. Geophys. Res., 104, 6353-6375.
Giorgi, F. and X. Bi, 2000: A study of internal variability of a regional climate model. J. Geophys. Res., 105, 29503-29516.
Giorgi, F., G.T. Bates, and S.J. Nieman, 1993a: The multi-year surface climatology of a regional atmospheric model over the western United States. J. Climate, 6, 75-95.
Giorgi, F. M.R. Marinucci, G.T. Bates and G. DeCanio, 1993b: Development of a second generation regional climate model (RegCM2). Part II: Convective processes and assimilation of lateral boundary conditions. Mon. Wea. Rev., 121, 2814-2832.

Giorgi, F., C.S. Brodeur and G.T. Bates, 1994: Regional climate change scenarios over the United States produced with a nested regional climate model. J. Climate, **7**, 375-399.

Giorgi, F., J.W. Hurrell, M.R. Marinucci and M. Beniston, 1997: Elevation signal in surface climate change: A model study. J. Climate, 10, 288-296.

Giorgi, F., L.O. Mearns, C. Shields and L. McDaniel, 1998: Regional nested model simulations of present day and 2XCO2 climate over the Central Plains of the U.S. Climatic Change, 40, 457-493.

Giorgi, F., B. Hewitson, J.H. Christensen, M. Hulme, H. von Storch, P. Whetton, R. Jones, L.O. Mearns and C. Fu, 2001: Regional Climate Information – Evaluation and Projections. Chapter 10 of Intergovernmental Panel on Climate Change (IPCC), 2001: Climate Change 2001: The Scientific Basis. Contribution of Working Group I to the Third Assessment Report of the Intergovernmental Panel on Climate Change [Houghton, J.T.,Y. Ding, D.J. Griggs, M. Noguer, P.J. van der Linden, X. Dai, K. Maskell, and C.A. Johnson (eds.)]. Cambridge University Press, Cambridge, United Kingdom and New York, NY, USA, 583-638.

Giorgi, F, X. Bi and Y. Qian, 2003: Indirect vs. direct effects of anthropogenic sulfate on the climate of East Asia as simulated with a regional coupled climate-chemistry/aerosol model. Climatic Change, 58, 345-376.

Giorgi, F, et al., 2006: Introduction to the TAC special issue: The RegCNET network. Theor. Appl. Climatol., 86, 1-4.

Gutzler, D.S. et al., 2006: The North American Monsoon Model Assessment Project. Bull. Amer. Meteor. Soc. (submitted).

Hassell, D. and R.G. Jones, 1999: Simulating climatic change of the southern Asia monsoon using a nested regional climate model (HadRM2). HCTN 8, Hadley Centre for Climate Prediction and Research, London Road, Bracknell, U.K.

Hirakuchi, H. and F. Giorgi, 1995: Multi year present day and 2XCO2 simulations of monsoon-dominated climate over Eastern Asia and Japan with a regional climate model nested in a general circulation model. J. Geophys. Res., 100, 21,105-21,126.

Hewitson, B.C. and R.G. Crane, 1996: Climate downscaling: Techniques and application. Clim. Res., 7, 85-95.

Hostetler, S.W. and F. Giorgi, 1991: Use of a regional atmospheric model to simulate lake-atmosphere feedbacks associated with Pleistocene Lakes Lahontan and Bonneville. Clim. Dyn., 7, 39-44.

Hostetler, S.W., G.T. Bates, and F. Giorgi, 1993: Interactive nesting of a lake thermal model within a regional climate model for climate change studies. Geophys. Res. Lett., 98, 5045-5057.

Hostetler, S.W., F. Giorgi, G.T. Bates and P.J. Bartlein, 1994: The role of lake-atmosphere feedbacks in sustaining paleolakes Bonneville and Lahontan 18,000 years ago. Science, 263, 665-668.

Hostetler, S.W., P.J. Bartlein, P.U. Clark, E. Small, and A. Solomon, 2000: Simulated influences of Lake Agassiz on the climate of central North America 11,000 years ago. Nature, 405, 334-337.

Huntingford, C. and J. Gash, 2005: Climate equity for all. Science, 309, 1789.

Huntingford, C. et al., 2003: regional climate model predictions of extreme rainfall for a changing climate. Quart. J. Roy. Meteor. Soc., 123, 265-292.

Im, E.-S., E.-H. Park, W.-T. Kwon and F. Giorgi, 2006: Present climate simulation over Korea with a regional climate model using a one-way double-nested system. Theor. Appl. Climatol., 86 183-196.

Jacob, D. et al., 2001: A comprehensive model intercomparison investigating the water budget during the BALTEX-PIDCAP period. Meteor. Atmos. Phys., 77, 19-43.

Jacob, D. et al., 2006: An intercomparison of regional climate models for Europe: Design of the experiments and model performance. Climatic Change, in press.

Jones, R.G., J.M. Murphy and M. Noguer, 1995: Simulations of climate change over Europe using a nested regional climate model. I: Assessment of control climate, including sensitivity to location of lateral boundaries. Quart. J. Roy. Meteor. Soc., 121, 1413-1449.
Jones, R.G., J.M. Murphy, M. Noguer and A.B. Keen, 1997: Simulation of climate change over Europe using a nested regional climate model. II: Comparison of driving and regional model responses to a doubling of carbon dioxide. Quart. J. Roy. Meteor. Soc., 123, 265-292.
Kato, H., H. Hirakuchi, K. Nishizawa and F. Giorgi, 1999: Performance of the NCAR RegCM in the simulations of June and January climates over eastern Asia and the high-resolution effect of the model. J. Geophys. Res., 104, 6455-6476.
Kauker, F., 1998: Regionalization of climate model results for the North Sea. PhD Thesis, University of Hamburg, 109 pp.
Kida, H., T. Koide, H. Sasaki and M. Chiba, 1991: A new approach to coupling a limited area model with a GCM for regional climate simulations. J. Meteor. Soc. Japan., 69, 723-728.
Kleinn, J., 2005: Hydrologic simulations in the Rhine basin driven by a regional climate model. J. Geophys. Res., 110, D05108.
Knutson, T.R., R.E. Tuleya, W. Shen, and I. Ginis, 2000: Impact of 2000-induced warming on hurricane intensities as simulated in a hurricane model with ocean coupling. J. Climate, 14, 2458-2468.
Landman, W.A., A. Seth and S.J. Camargo, 2005: The effect of regional climate model domain choice on the simulation of tropical cyclone-like vortices in the southwestern Indian Ocean. J. Climate, 18, 1263-1274.
Leung, L.R. and Y. Qian, 2003: The sensitivity of precipitation and snow-pack simulations to model resolution via nesting in regions of complex terrain. J. Hydrometeorology, 8, 145-167.
Leung, L.R., M.S. Wigmosta, S.J. Ghan, D.J. Epstein, and L.W. Veil, 1996: Application of a subgrid orographic precipitation/surface hydrology scheme to a mountain watershed. J. Geophys. Res., 101, 12803-12817.
Leung, L.R., L.O. Mearns, F. Giorgi and R.L. Wilby, 2003: Regional climate research: Needs and opportunities. Bull. Am. Meteor. Soc., 82, 89-95.
Lorenz, P. and D. Jacob, 2005: Influence of regional scale information on the global circulation: A two-way nesting climate simulation. Geophys. Res. Lett., 32, L14826.
Lu, L. et al., 2001: Implementation of a two-way interactive atmospheric and ecological model and its application to the central United States. J. Climate, 4, 900-919.
Lynch, A.H., W.L. Chapman, J.E. Walsh and G. Weller, 1995: Development of a regional climate model of the western Arctic. J. Climate, 8, 1555-1570.
Machenhauer, B. et. al., 1998: Validation and analysis of regional present-day climate and climate change simulations over Europe. MPI Report No.275, MPI, Hamburg, Germany.
Marbaix, P., H. Gallee, O. Brasseur, and J.P. Van Ypersele, 2003: Lateral boundary conditions in regional climate models: A detailed study of the relaxation procedure. Mon. Wea. Rev., 131, 461-479.
Marinucci, M.R. et al., 1995: High resolution simulations of January and July climate over the western Alpine region with a nested regional modeling system. Theor. Appl. Climatol., 51, 119-138.
McAvaney, B.J., C. Covey, S. Joussaume, V. Kattsov, A. Kitoh, W. Ogana, A.J. Pitman, A.J. Weaver, R.A. Wood, and Z.-C. Zhao, 2001: Model Evaluation. Chapter 8 of Intergovernmental Panel on Climate Change (IPCC), 2001: Climate Change 2001: The Scientific Basis. Contribution of Working Group I to the Third Assessment Report of the Intergovernmental Panel on Climate Change [Houghton, J.T.,Y. Ding, D.J. Griggs, M. Noguer, P.J. van der Linden, X. Dai, K. Maskell, and C.A. Johnson (eds.)]. Cambridge University Press, Cambridge, United Kingdom and New York, NY, USA, 583-638.
McGregor, J.L., 1997: Regional climate modeling. Meteorol. Atm. Phys., 63, 105-117.

McGregor, J.L., J.J. Katzfey and K.C. Nguyen, 1995: Seasonally varying nested climate simulations over the Australian region. Third Int. Conference on Modeling of Global Climate Change and Variability, Hamburg, Germany, 4-8 Sept.
McGregor, J.L., J.J. Katzfey and K.C. Nguyen, 1999: Recent regional climate modelling experiments at CSIRO. In: Research Activities in Atmospheric and Oceanic Modelling. H. Ritchie (ed). (CAS/JSC Working Group on Numerical Experimentation Report; 28; WMO/TD – no. 942) [Geneva]: WMO. P. 7.37-7.38.
Mearns, L.O., W. Easterling, C. Hays and D. Marx, 2001: Comparison of agricultural impacts of climate change calculated from high and low resolution climate change scenarios. Part I: The uncertainty due to spatial scale. Climatic Change, 51, 131-172.
Noguer M., R.G. Jones and J.M. Murphy, 1998: Sources of systematic errors in the climatology of a nested regional climate model (RCM) over Europe. Clim. Dyn., 14, 691–712.
Pal, J.S. and E.A.B. Eltahir, 2003: A feedback mechanism between soil moisture distribution and storm tracks. Quart. J. Roy. Meteor. Soc., 129, 2279-2297.
Pal, J.S. et al. 2006: The ICTP RegCM3 and RegCNET: Regional climate modeling for the developing World. Submitted to Bull. Amer. Meteor. Soc.
Pielke, R.A., 1974: A three-dimensional numerical model of the sea breeze over South Florida. Mon. Wea. Rev., 102, 115-139.
Pielke, R.A., 1984: Mesoscale Meteorological Modeling, Academic Press, Orlando, Fl. 612 pp.
Qian, Y. and F. Giorgi, 1999: Interactive coupling of regional climate and sulfate aerosol models over East Asia. J. Geophys. Res., 104, 6501-6514.
Ratnam, J.V. and K.K. Kumar, 2005: Sensitivity of the simulated monsoons of 1987 and 1988 to convective parameterization schemes in MM5. J. Climate, 18, 2724-2743.
Rauscher, S.A., A. Seth J.-H. Qian and S.J. Camargo, 2006: Domain choice in an experimental nested modeling prediction system for South America. Theor. Appl. Climatol., 86, 225-242.
Rinke, A., P. Marbaix and K. Dethloff, 2004: Internal variability in Arctic regional climate simulations: Case study for the SHEBA year. Clim. Res., 27, 197-209.
Rinke, A. et al., 2006: Evaluation of an ensemble of Arctic regional climate models: Spatial patterns and height profiles. J. Climate (submitted).
Roads, J. et al., 2003: The IRI/ARCs regional model intercomparison over South America. J. Geophys. Res., 108. 4425, doi: 10.1029/2002JD003201.
Rojas, M. and A. Seth, 2003: Simulation and sensitivity in a nested modeling system for South America. Part II: GCM boundary forcing. J. Climate, 16, 2454-2471.
Rotach, M.W. et al., 1997: Nested regional simulation of climate change over the Alps for the scenario of doubled greenhouse gas forcing. Theor. Appl. Climatol., 57, 209-227.
Seth, A. and F. Giorgi, 1998: The effects of domain choice on summer precipitation simulation and sensitivity in a regional climate model. J. Climate, 11, 2698-2712.
Seth, A. and M. Rojas, 2003: Simulation and sensitivity in a nested modeling system for South America. Part I: Reanalyses boundary forcing. J. Climate, 16, 2437-2453.
Sloan, L.C., 2006: A framework for regional modeling of past climates. Theor. Appl. Climatol. 86, 267-277.
Small, E.E., F. Giorgi and L. C. Sloan, 1999a: regional climate model simulation of precipitation in Central Asia: Mean and interannual variability. J. Geophys. Res., 104, 6563-6582.
Small, E.E., L. C. Sloan, S.H. Hostetler and F. Giorgi, 1999b: Simulating the water balance of the aral sea with a coupled regional climate lake model, J. Geophys. Res., 104, 6583-6602.
Solmon, F., F. Giorgi and K. Liousse, 2006: Development of a regional anthropogenic aerosol model for climate studies: Application and validation over a European/African domain. Tellus B, 58, 51-72.

Stone, M.C., R.H. Hotchkiss and L.O. Mearns, 2003: Water yield response to high and low spatial resolution climate change scenarios in the Missouri River Basin. Geophys. Res. Lett., 30, 1186.

Sun, L., F.H.M. Semazzi, F. Giorgi and L. Ogallo, 1999: Application of the NCAR regional climate model to eastern Africa. Part II: Simulation of interannual variability of the short rains. J. Geophys. Res., 104, 6549-6565.

Takle, E.S. et al., 1999: Project to Intercompare Regional Climate Simulations (PIRCS): Description and initial results. J. Geophys. Res., 104, 19,443-19,462.

Takle, E.S. et al., 2006: Transferability Intercomparison: An opportunity for new insight on the global water cycle and energy budget. Bull. Amer. Meteor. Soc., Submitted.

Tvetvinskaya, E.A. et al. 2003: The effect of spatial scale of climatic change scenario on simulated maize, winter wheat and rice production in the southeastern United States. Climatic Change, 60, 37-71.

Vidale, P.L. et al., 2003: Predictability and uncertainty in a regional climate model. J. Geophys. Res., 108, 4586.

von Storch, H., H. Langenberg and F. Feser, 2000: A spectral nudging technique for dynamical downscaling purposes. Mon. Wea. Rev., 128, 3664-3673.

Wang, Y. et al., 2004: Regional climate modeling: progress challenges and prospects. J. Meteor. Soc. Japan, 82, 1599-1628.

Washington, W.M. and C.L. Parkinson, 1986: An Introduction to Three Dimensional Climate Modeling, University Science Books, Mills Valley, CA, 388 pp.

Wu, W.L., A.H. Lynch and A. Rivers, 2005: Estimating the uncertainty in a regional climate model related to initial and lateral boundary conditions. J. Climate, 18, 917-933.

Zakey, A.S., F. Solmon and F. Giorgi, 2006: Development and testing of a desert dust module in a regional climate model. Submitted to Atmospheric Chemistry and Physics.

J. Phys. IV France **139** (2006) 119–142
© EDP Sciences, Les Ulis
DOI: 10.1051/jp4:2006139009

Forcings and feedbacks by land ecosystem changes on climate change

R.A. Betts[1]

[1] *Met Office, Hadley Centre for Climate Prediction and Research, Fitzroy Road, Exeter EX1 3PB, UK*
e-mail: richard.betts@metoffice.gov.uk

Abstract. Vegetation change is involved in climate change through both forcing and feedback processes. Emissions of CO_2 from past net deforestation are estimated to have contributed approximately 0.22 – 0.51 Wm^{-2} to the overall 1.46 Wm^{-2} radiative forcing by anthropogenic increases in CO_2 up to the year 2000. Deforestation-induced increases in global mean surface albedo are estimated to exert a radiative forcing of 0 to -0.2 Wm^{-2}, and dust emissions from land use may exert a radiative forcing of between approximately +0.1 and -0.2 Wm^{-2}. Changes in the fluxes of latent and sensible heat due to tropical deforestation are simulated to have exerted other local warming effects which cannot be quantified in terms of a Wm^{-2} radiative forcing, with the potential for remote effects through changes in atmospheric circulation. With tropical deforestation continuing rapidly, radiative forcing by surface albedo change may become less useful as a measure of the forcing of climate change by changes in the physical properties of the land surface. Although net global deforestation is continuing, future scenarios used for climate change prediction suggest that fossil fuel emissions of CO_2 may continue to increase at a greater rate than land use emissions and therefore continue to increase in dominance as the main radiative forcing. The CO_2 rise may be accelerated by up to 66% by feedbacks arising from global soil carbon loss and forest dieback in Amazonia as a consequence of climate change, and Amazon forest dieback may also exert feedbacks through changes in the local water cycle and increases in dust emissions.

1. INTRODUCTION

1.1 Climate forcings and feedbacks

The global patterns and functioning of vegetation are now changing, both as a direct result of human land use activities and as a consequence of climate change. Since vegetation itself affects climate, both the above processes play a role in climate change. Land use change directly caused by human activity can lead to a number of "forcings" of climate change, where a "forcing" refers to a direct driver of change. Changes in vegetation cover which results from climate change and then exerts a further effect on climate constitutes a "feedback". This paper discusses the roles of vegetation in both these processes, in both past and future anthropogenic climate change.

1.2 Processes through which vegetation affects climate

A key contribution to climate change comes from changes in the atmospheric concentration of carbon dioxide, which, as a greenhouse gas (GHG), increases the absorption and re-emission of terrestrial longwave radiation by the atmosphere and hence is identified as a significant warming influence on climate [1]. Plants take in carbon dioxide from the atmosphere through the process of photosythesis, and although they return some to the atmosphere through autotrophic respiration, there is a net uptake of carbon enabling them to grow. Some carbon is transferred from plants to the soil through litter fall, leading to a buildup of carbon in the soil which itself can be returned to the atmosphere through heterotrophic respiration as part of the process of decay. The uptake of carbon by photosynthesis is generally observed to increase when plants are exposed to higher CO_2 concentrations, although this rate of increase saturates at higher concentrations. The release of carbon by autotrophic and heterotrophic respiration is generally observed to increase with temperature, with exponential increases being commonly reported.

The global vegetation cover currently stores an estimated 500 gigatonnes of carbon (GtC), and the world's soils store an estimated 1500 GtC [1]. For comparison, the atmosphere currently holds approximately 800 GtC. Tropical forests can store over 120 tonnes of carbon per hectare ($tC^{-1}ha$) in above-ground biomass, and temperate forests about 50 $tC^{-1}ha$ [2]. Grasslands hold approximately 7 $tCha^{-1}$ in above-ground biomass, and croplands about 2 $tCha^{-1}$.

Large-scale changes in forest cover can therefore affect climate change by providing sources or sinks of carbon dioxide. Changes in the rates of photosynthesis and respiration as a result of changes in CO_2 concentration or temperature can also modify the nature of vegetation as either a sink or a source of carbon.

Another aspect of atmospheric composition affected by vegetation is the concentration of mineral dust aerosol particles. Dust can exert significant effects by modifying the fractions of incoming solar radiation which are absorbed and reflected by the Earth, and by altering the absorption and transmission of outgoing terrestrial longwave radiation. Dry landscapes can be significant sources of dust, which be carried across continents and oceans and hence affect climate over large fractions of the Earth's surface. Dust escape from the soil surface can be affected by the extent and nature of vegetation cover, with the exposed soil of a less vegetated landscape releasing in more dust to the atmosphere.

Vegetation also affects climate through its influence on the physical properties of the land surface. Correlations between the global patterns of climate and vegetation have long been recognised [3], and climate provides the dominant influence on the density of vegetation cover in different parts of the world [4]. However, vegetation itself influences climate through the surface fluxes of radiation, heat, moisture and momentum [5-7]. If the character of the vegetation cover is modified, changes to the climate can result.

The nature of the vegetation cover exerts a strong influence on the albedo of the land surface. Forests are generally darker than open land, particularly when snow is lying because trees generally remain exposed while cultivated land can become entirely snow-covered [8]. Snow-free foliage is considerably darker than snow, but even if large quantities of snow are held on the canopy, multiple reflections within the canopy scatter rather than reflect shortwave radiation, which also reduces the landscape albedo [8]. Models suggest that the resulting low surface albedo causes boreal and cool-temperate forests to exert a warming influence on climate relative to non-forested land [9, 10].

Relative to bare soil, vegetation can enhance the evaporative flux of moisture to the atmosphere through the extraction of moisture deep in the soil by plant roots for transpiration. Furthermore, the vegetation canopy can capture a greater fraction of precipitation which is then re-evaporated back to the atmosphere, compared to bare soil which holds less water on the surface before runoff and infiltration. Also, the higher aerodynamic roughness of a vegetated land surface can promote the flux of moisture to the atmosphere through enhanced turbulence. Changes in the nature of vegetation cover, particularly from forest to non-forest, can therefore significantly alter the surface moisture budget and exert further effects on the surface energy budget. Deforestation can reduce evaporation, causing a greater proportion of the available energy at the land surface to flow to the atmosphere in the form of sensible heat rather than latent heat; this exerts a warming influence on the near-surface air temperature. Reduced evaporation also reduces the flux of moisture to the atmosphere, potentially decreasing the quantity of moisture available for precipitation.

The present-day patterns of climate across the globe are therefore influenced by the presence and character of vegetation [11, 12]. Model results suggest that while tropical forests exert a cooling effect on their regional climates, forests in cold regions exert a warming influence through their large impact on surface albedo which outweighs the influence of transpiration [11]. Continental precipitation and evaporation are generally simulated as higher when the land is vegetated, but runoff from the land and convergence of moisture from the oceans are both decreased indicating that the vegetation allows more local 'recycling' of water over land masses [11].

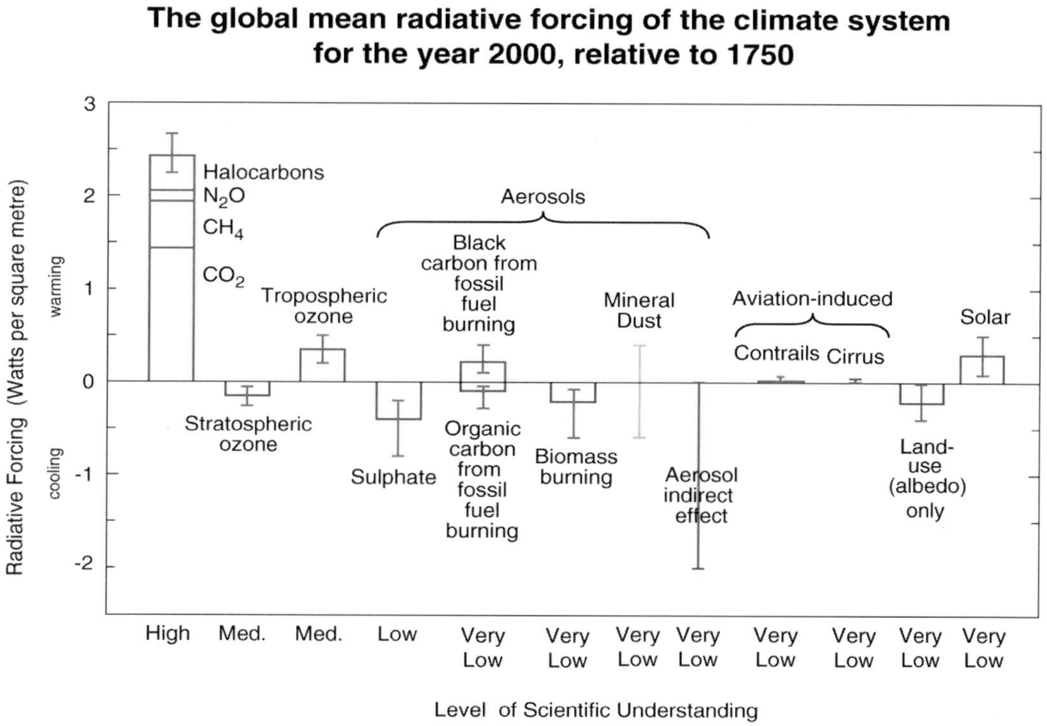

Figure 1. Global mean radiative forcing due to natural and anthropogenic causes, at 2000 relative to 1750, as assessed in the IPCC Third Assessment Report [1]. Copyright IPCC.

1.3 Roles of land cover change in climate change

The relative importance of different drivers of climate change is generally quantified in terms of radiative forcing, which is a measure of the perturbation to the Earth;s radiation budget. Radiative forcing is routinely used by the Intergovernmental Panel on Climate Change (IPCC) [1] to compare the influence of historical changes in different greenhouse gases, aerosols, land cover and solar activity on climate change relative to the pre-industrial era (Figure 1). The Millennium Ecosystem Assessment [13] noted that ecosystem processes were involved in many of these radiative forcings (Figure 2).

Future climate predictions and attribution of past changes generally focus on the radiative forcing of climate by greenhouse gases and aerosols (Figure 3a), and land use change (mainly deforestation and afforestation) is involved in these through emissions or uptake of carbon dioxide, other greenhouse gases and some aerosol species as a result of deforestation or afforestation. Land use change can also exert a radiative forcing through changes in surface albedo, which directly perturbs the planetary radiation budget and constitutes a radiative forcing which does not act via changes in atmospheric composition. (Figure 3b).

Moreover, land use change can also drive climate change through processes which do not directly perturb the radiation budget, such as changes in the surface moisture flux and associated changes in the partitioning of the available energy into sensible and latent heat. In many cases the responses to these forcings will be predominantly local, mainly affecting the environment of the ecosystem providing the feedback. In other cases, land surface perturbations can exert significant remote effects through teleconnections, sometimes extending across the globe. Such processes are not quantifiable in terms of radiative forcing, but nevertheless can be considered as drivers of climate change and fall within the

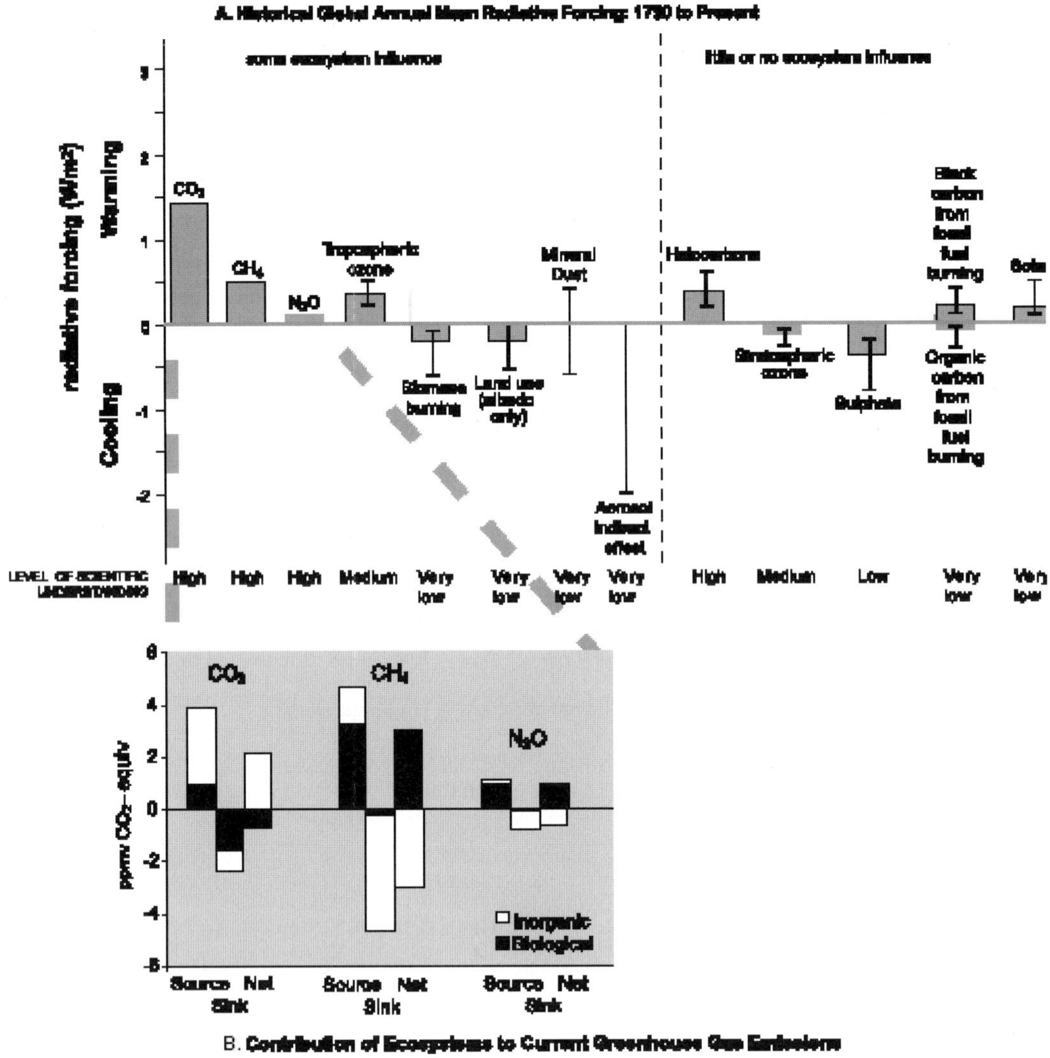

Figure 2. (A) Global mean radiative forcing due to natural and anthropogenic causes, at 2000 relative to 1750, from IPCC Third Assessment Report [1] and grouped according to the level of ecosystem influence. (B) Contribution of ecosystems to current greenhouse gas (GHG) emissions, expressed in terms of "equivalent CO_2" based on the 100-year Global Warming Potential (the mean radiative forcing of each GHG relative to the radiative forcing of the same mass of CO_2). Copyright Millennium Ecosystem Assessment [13].

"forcing" component of the "forcing – feedback – response" conceptual model (Figure 3c). The term "non-radiative forcing" has been proposed for these processes [14].

Similarly, land cover changes resulting from climate change, such as expansion or die-back of forest cover, can exert feedbacks through changes in greenhouse gas and aerosol concentrations (Figure 3d) and changes in the physical properties of the land surface (Figure 3e). Land cover change resulting either directly from human activity or indirectly from climate change is therefore involved in the climate change process through a number of routes, including the traditionally-central GHG-radiative forcing pathway (Figure 3a), non-GHG radiative forcing through surface albedo (Figure 3b), non-radiative

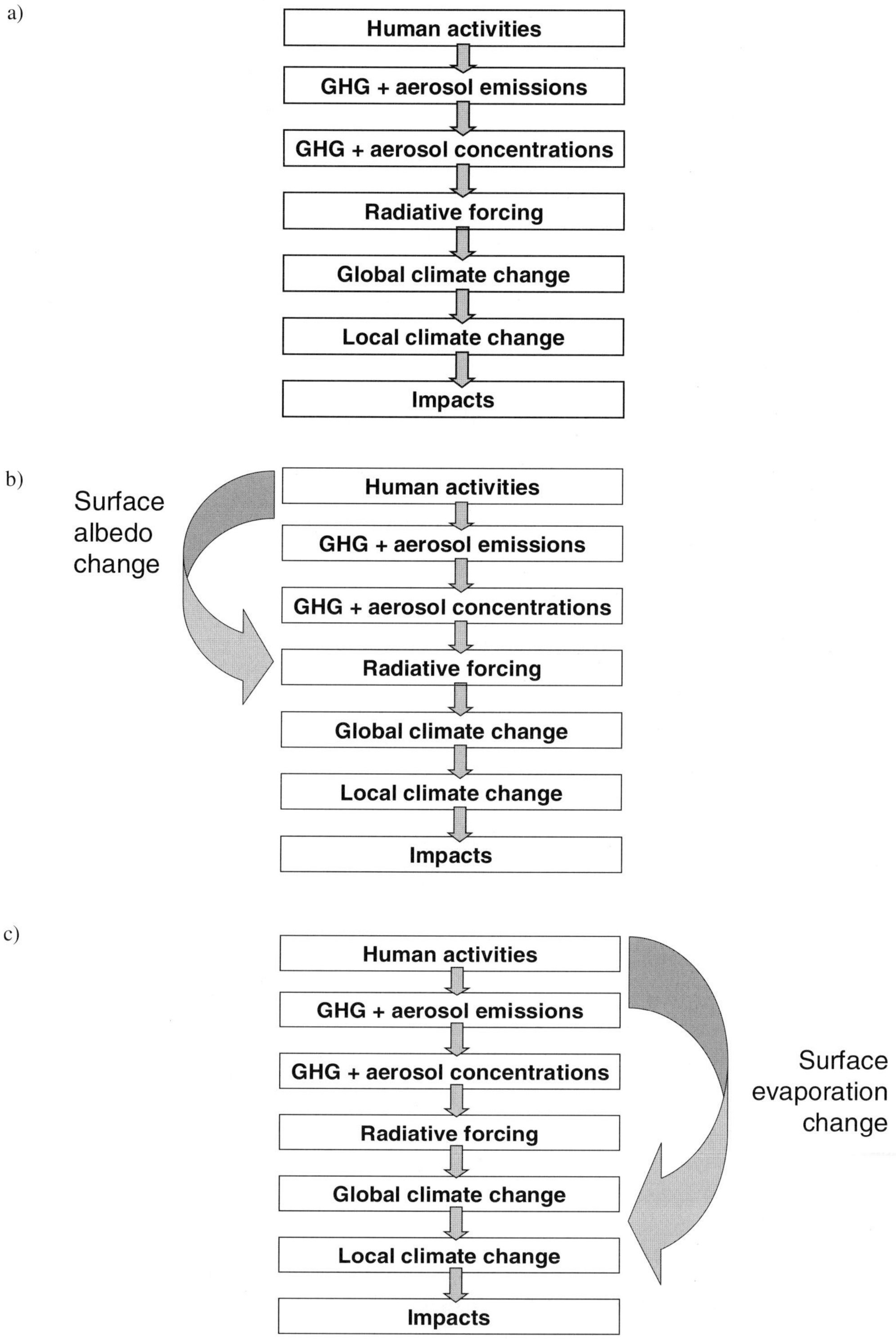
a)
Human activities
GHG + aerosol emissions
GHG + aerosol concentrations
Radiative forcing
Global climate change
Local climate change
Impacts
b)
Surface albedo change
Human activities
GHG + aerosol emissions
GHG + aerosol concentrations
Radiative forcing
Global climate change
Local climate change
Impacts
c)
Human activities
GHG + aerosol emissions
GHG + aerosol concentrations
Radiative forcing
Global climate change
Local climate change
Impacts
Surface evaporation change

d)

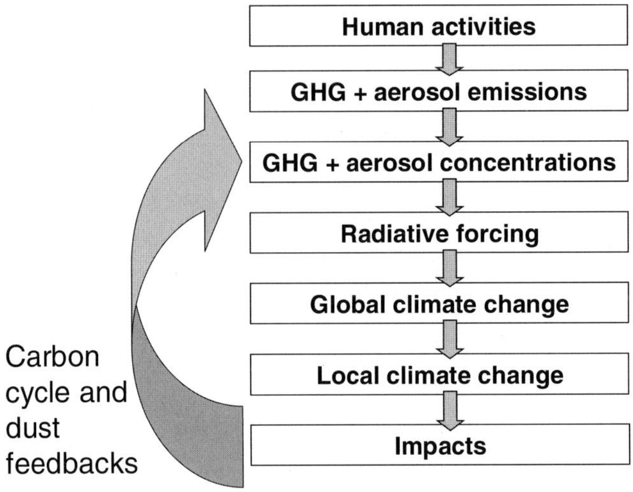

e)

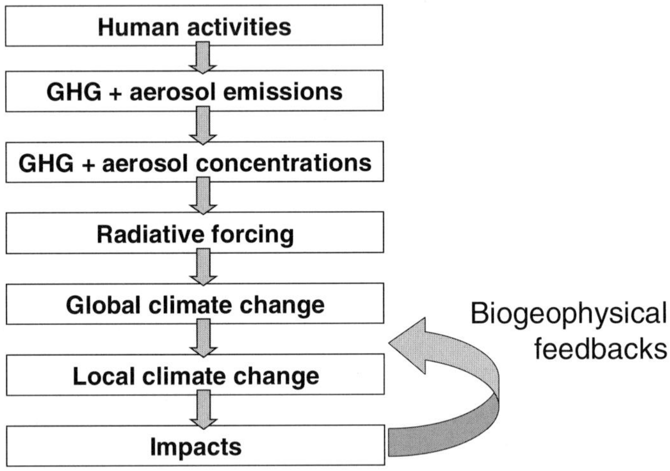

Figure 3. Roles of vegetation change in the linear cause-effect chain conceptual model of climate change. (a) General view of 1st order effects, with human activities affecting climate through radiative forcing by GHG and aerosol emissions. (b) Radiative forcing by land use through surface albedo change. (c) Additional forcing of climate by land use through non-radiative processes. (d) Biogeochemical feedbacks on GHG and aerosol concentrations by vegetation responses to climate change. (e) Biogeophysical feedbacks on global and local climate change through changes in the physical properties of the land surface.

forcings (Figure 3c), biogeochemical feedbacks (Figure 3d) and biogeophysical feedbacks (Figure 3e). Key examples of these forcing and feedback pathways are discussed in this paper.

2. FORCING OF CLIMATE CHANGE BY ANTHROPOGENIC LAND COVER CHANGE

2.1 Anthropogenic land cover change: Past changes and current trends

If global forests were able to reach their potential natural extent, they would cover an estimated 52–59 million km^2 [15, 16]. However, humans have been removing parts of the forest cover of their local landscape, either on a temporary or permanent basis, for thousands of years. Temporary, small-scale land cover change may have begun with the use of fire approximately 500,000 years ago, as means of opening up the landscape for hunting and possibly also as a tool for hunting itself. Land transformation became more widespread and permanent with the advent of agriculture approximately 10,000 years ago, which required fertile (and therefore often forested) land to be cleared for cultivation. Over the last millennium, increasing populations and advancements in agricultural practices led to increasing demands for land particularly in Europe, the Indo-Gangetic Plains and China. By the time of the industrial revolution in the 17th Century, global forest cover had decreased by about 5 – 7%, mostly in Europe and Asia.

In 1750, when industrial activity was beginning, deforestation in North America was relatively small-scale and confined to the eastern coastal regions because European settlers had arrived only relatively recently and major agricultural expansion had yet to take place [15, 16]. By 1850, however, deforestation had taken place over a wider area, reflecting the westward expansion of European settlement. By 1900 North American agriculture had undergone major expansion, and the landscape was subject to some degree of human modification in much of the eastern half of the present-day conterminous USA.

Meanwhile, in Europe, India and China, deforestation continued to increase. In these regions, the continental-scale extent of the disturbed areas remained largely unchanged from 1850 to 1900, although local-scale intensification of deforestation continued.

Between 1900 and 1950, deforestation continued to intensify by a similar degree all the main centres of global agriculture. North American agriculture continued to expand westwards, widening the area of deforestation. In contrast, however, forest regrowth took place in the north-east USA, as a result of cropland abandonment as the agricultural activity moved westwards.

Since 1950, cropland areas in parts of Europe, China and North America have begun to decrease and some areas of forest have regrown. However, this period has also seen very rapid expansion of deforestation in the tropics. Overall, global forest cover has decreased by approximately 11 – 17 million km^2 (20 - 30% of potential natural forest cover) since human deforestation began [15, 16]. Approximately 75% of this deforestation has occurred since the industrial revolution.

Currently, the trend is still towards net global deforestation, with tropical deforestation continuing at about eight times the rate of reforestation in the temperate regions [17].

2.2 Contributions of land use change to the radiative forcing of climate change by increasing CO_2 concentrations

From historical inventories of land use types and carbon stocks, it has been estimated that the net deforestation between 1850 and 2000 resulted in net emissions of 156 GtC [18]. House *et al* [19] estimated the mean C emissions from land use over the 1990s to be 1.4 – 3.0 Gt C y^{-1}, compared to 0.9 – 2.8 Gt C y^{-1} over the 1980s. However, annual C emissions from land use are estimated to have peaked in the early 1990s with a decline over the rest of the decade [18].

The IPCC Third Assessment Report [1] noted apparent inconsistencies between inventory-based emissions estimates [20] and emissions estimates obtained by applying historical land use change reconstructions to terrestrial biosphere models [21]. The models gave a decrease in land use emissions since the 1950s, in contrast with the ongoing rise until 1990 suggested by the inventory method [18, 20]. House *et al* [19] attribute this inconsistency to the use of different estimates of land clearance, especially in the tropics. The rising emissions estimate of Houghton [18] used tropical deforestation rates from the Food and Agriculture Organisation (FAO) Forest Resources Analysis [17] whereas the falling estimate of McGuire et al [21] used the cropland areas of the FAO-STAT database [22] as used by Ramankutty and Foley [15].

Brovkin et al [23] applied different reconstructions of historical land use change [15, 16] or historical land use change emissions [24], to a coupled climate-vegetation-carbon cycle model incorporating 2 alternative terrestrial models, to estimate the relative contribution of land use and fossil fuel emissions [25] to the atmospheric CO_2 growth rate. The simulated land-atmosphere and land-ocean

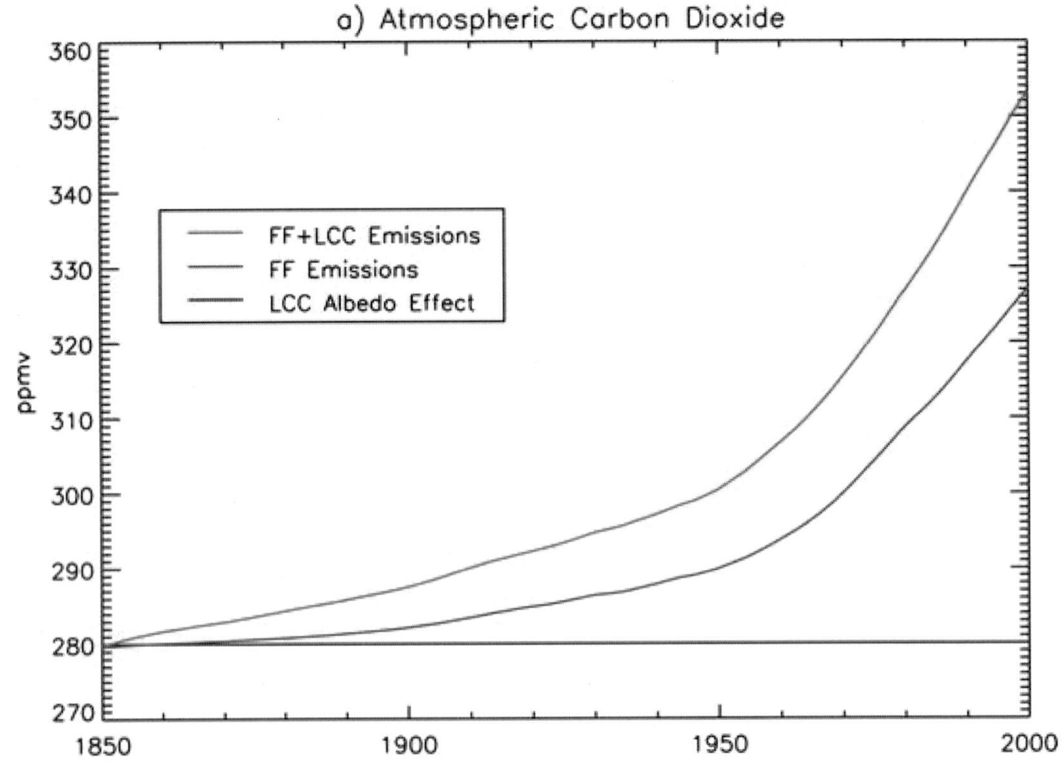

Figure 4. CO_2 concentrations from 1850 to 2000 simulated with past fossil fuel emissions only [25] (purple line) and additionally with past emissions from land use change [24] (red line), and (b) resulting simulated temperature change. Reproduced from Matthews *et al* [33].

carbon fluxes were close to the central estimates of ranges cited by the IPCC [1] and by more recent studies [26, 27] for the 1980s and 1990s, and the simulated atmospheric CO_2 concentrations reproduced the changes seen in measurements/reconstructions from 1850 to 2000 [28, 29] to within 3 ppmv when land use change reconstructions [15, 16] were used. When an emissions reconstruction [24] was used, the CO_2 rise simulated from 1950 to 2000 was overestimated by 16ppmv. Similarly, the HadCM3LC coupled-climate carbon cycle model [30] reproduced the observed CO_2 rise when using 70% of the Houghton [20] land use emissions [31], as suggested in the IPCC Second Assessment Report [32] to account for forest regrowth.

By performing simulations with and without land use emissions, Brovkin et al [22] estimated land use emissions to have contributed 15% - 35% of the overall CO_2 rise between 1850 and 2000 depending on which of the 3 land use change/emissions reconstructions was used [15, 16, 24]. With similar techniques, Matthews et al [33] estimated land use emissions [24] to have contributed 35% of the overall CO_2 rise (Figure 4). The overall radiative forcing due to CO_2 in 2000 relative to pre-industrial has been estimated as 1.46 Wm^{-2} [1], so these modelling studies suggest that emissions of CO_2 from past land use change have contributed approximately 0.22 – 0.51 Wm^{-2} to the overall current radiative forcing by anthropogenic increases in CO_2.

The global total CO_2 emissions in 2003 were estimated as 7.3 GtC [34]. As fossil fuel emissions have increased over the last century much more rapidly than land use emissions, the relative contribution of land use change to the rate of increase in atmospheric CO_2 concentrations (the "CO_2 growth rate") has become smaller. The land use contributions to the CO_2 growth rate between 1850 and 1900 were estimated as 42%-68%, and this fell to 5%-38% in the 1990s [22]. Fossil fuel emissions are therefore the dominant cause of the current CO_2 rise. Nevertheless, past land use emissions appear to have made a significant contribution to anthropogenic climate change over the 20th Century.

2.3 Radiative forcing of climate change by land use through emissions of mineral dust aerosol

The IPCC [1] noted large uncertainties in the measurements and model estimates of radiative forcing due to mineral dust aerosol, and cited a range of +0.4 to -0.6 Wm-2 for the total global radiative forcing by dust from both anthropogenic and natural sources. The net forcing depends on the relative offset of positive and negative forcings, with the global mean longwave forcing being positive while the global mean shortwave forcing is negative [35]. The shortwave forcing can vary spatially from positive to negative depending on the albedo of the underlying land or ocean surface [35]. Models of dust emissions have been used to estimate the anthropogenic contributions to this through land use change, based on estimates of the extent of anthropogenic desertification [36] or by comparing observed dust emissions with emissions simulated with a model in which land cove has been assigned to the potential natural vegetation state [37]. These suggest that land use has contributed 0 -30% of the total global atmospheric dust load, so the radiative forcing by land use-related increases in dust may be either positive or negative and range between approximately +0.1 and -0.2 Wm^{-2} relative to potential natural vegetation. However, this may not capture the full range of uncertainty. The radiative forcing relative to pre-industrial vegetation is likely to be smaller than that relative to potential natural vegetation.

2.4 Radiative forcing of climate change by land use through changes in surface albedo

From a global perspective, the dominant aspect of historical land cover change has been deforestation in temperate regions, and a number of modelling studies [38-46] have suggested that this has exerted a radiative forcing which is most likely negative (ie: a cooling effect) as a result of increased surface albedo especially in winter and spring when snow is lying.

Betts et al [47] used reconstructions of land cover at 1750 [15, 16] applied to the HadAM3 climate model [48] to simulate the radiative forcing due to land use-induced surface albedo change at the present-day relative to the pre-industrial state. The radiative forcing was simulated by performing the calculations of surface albedo and the shortwave radiation budget twice on each model timestep, once with spatial patterns of albedo parameters appropriate to present-day vegetation, and again with albedo parameter patterns appropriate to 1750 vegetation. All other quantities in the climate model were identical in the two sets of calculations, so any feedbacks such as changes in snow or cloud cover were excluded. This provided a pure estimate of the initial perturbation to the Earth's radiation budget, ie: the radiative forcing.

The global mean radiative forcing at 2000 due to surface albedo change relative to 1750 was simulated as -0.18 Wm^{-2}, within the range of -0.2$Wm^{-2}$$\pm$ 0.2Wm^{-2} estimated by the IPCC [1] and other studies [38-46]. This global forcing is small in comparison to the positive radiative forcing of approximately 3 Wm^{-2} estimated to arise from increased GHG concentrations (CO_2, CH_4, N_2O and the halocarbons), but it should be noted that the forcing due to GHGs is spread relatively evenly across the globe whereas that due to anthropogenic surface albedo change is likely to be concentrated in temperate agricultural regions (Figure 5). Regional radiative forcings due to anthropogenic surface albedo change were estimated to be -5Wm^{-2} or greater [47], leading to a regional cooling of 1-2 K in the absence of any other influences on climate [39]. This implies that the climate warming observed in these regions may be smaller than that which would have occurred as a result of increased GHG concentrations alone, with the GHG warming having been partly offset by anthropogenic land cover change in a similar manner to that associated with increased aerosol concentrations.

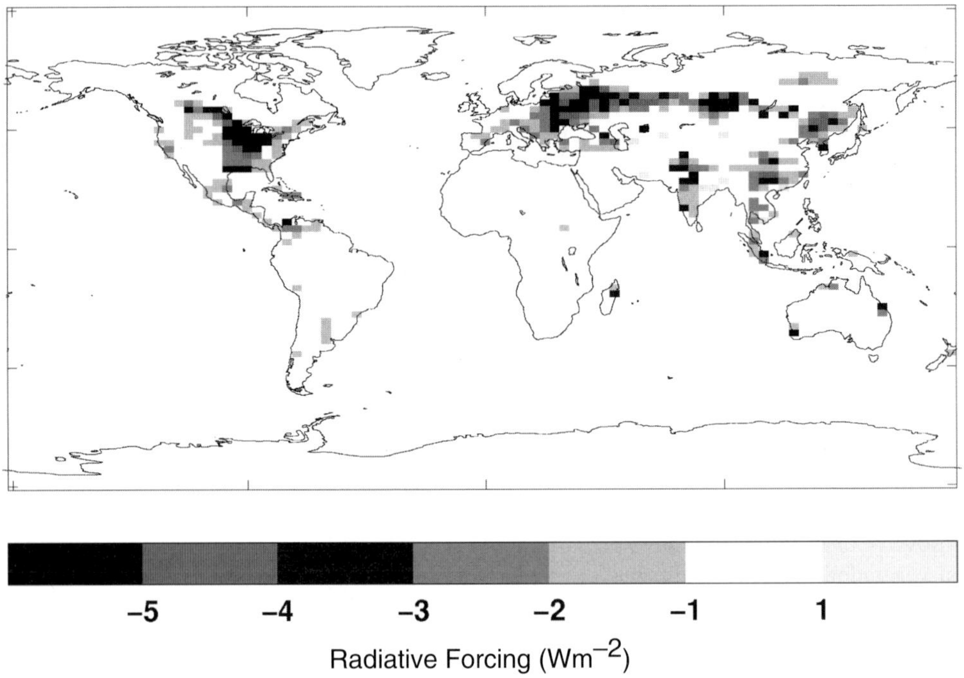

Figure 5. Radiative forcing at 1990 relative to 1750 due to surface albedo change arising from land use change, simulated with the HadAM3 climate model [47].

2.5 Forcing of climate change by land use through non-radiative processes

Land cover change can also modify the surface energy budget through changes in the fluxes of latent and sensible heat. For example, latent heat fluxes are observed to be smaller over tropical pasture land than over nearby forests [49] and complete deforestation is typically simulated to reduce evaporation and transpiration, providing a warming influence on near-surface air temperatures and potentially reducing precipitation [50]. The warming effect of reduced evaporation can therefore offset the cooling effect of increased surface albedo in areas with a plentiful supply of soil moisture for evaporation and if the albedo of the non-forested land is not greatly higher than that of the forested landscape. Reconstructions of past surface temperature changes from below-ground temperatures in boreholes suggest that removal of forest cover led to local warming in the temperate forests of Vancouver Island [51]. However, in regions where snow cover is present for long periods of the year, the warming effect of changes in evaporation in summer may be outweighed by the cooling effect of large increases in surface albedo in winter and spring [11].

Large-scale tropical deforestation is therefore likely to exert a warming effect through decreased evaporation [50] in addition to the warming effect caused by the emission of CO_2 and resulting increases in radiative forcing. While changes in the surface moisture budget act as anthropogenic perturbations to the climate system, they do not involve direct perturbations to the Earth's radiation budget and therefore cannot be considered to be radiative forcing. The term "non-radiative forcing" has been proposed [14], but this cannot be quantified in terms comparable with radiative forcing so at the present time the effects of these processes are often discussed in terms of temperature changes. Model results suggest that the combined effects of past tropical deforestation may have exerted regional warmings of approximately 0.2 K, and total clearance of continental-scale tropical forest areas may lead to warming of 1K or more [52]. Alternative metrics, focussing on impacts such as Net Primary Productivity (NPP), have also been suggested [53].

Although model results suggest that complete deforestation of tropical regions such as Amazonia would lead to decreased precipitation through reduced water recycling [50], the partial deforestation which has already occurred may be more complex. High-resolution mesoscale models which resolve the patterns of partial deforestation suggest that the fragmentation of the forest cover may induce small-scale atmospheric circulations which can increase atmospheric ascent and therefore enhance convection [54]. Therefore, partial deforestation may actually increase precipitation if the supply of moisture from advection remains unchanged, and this is consistent with observed patterns of precipitation change over Amazonia [55]. With current computing limitations, atmospheric models with resolution high enough to capture the effects of fine-scale forest fragmentation can only be applied over relatively small regions. Since actual patterns of deforestation often consist of fine-scale partial clearance over large areas, there is a need to investigate the interactions between mesoscale circulations and large-scale advection and evaporative recycling. There is therefore a requirement for large-domain modelling studies with either higher resolutions or adequate parametrizations of fine-scale circulations.

Large-scale deforestation in the tropics may also exert more far-reaching effects, with changes in the near-surface energy balance modifying the large cells of ascending and descending motion near the equator (the Hadley-Walker circulation). Changes to these may cause shifts in the atmospheric circulation which propagate across the globe. Gedney and Valdes [56] found that complete deforestation of Amazonia could lead to increased precipitation in Europe through this mechanism. Deforestation may already have perturbed the global atmospheric circulation, affecting regional climates remote from the land cover change [57, 58, 59]. However, these mechanisms have yet been investigated in studies for the detection and attribution of past climate change.

Feedbacks from the oceans may enhance the effects of land cover change on climate. For example, Delire *et al* [60] found that in the Indonesian Archipelago, the impacts of deforestation on wind speeds may be sufficient to modify ocean up-welling and cause warming over the surrounding ocean surfaces, in addition to the warming caused over land by reduced evaporation. This amplified warming may impact global-scale atmospheric circulations through changes in the Hadley-Walker circulation.

Many scenarios of future GHG emissions include contributions from land cover change, but the biophysical effects of such changes are not usually included as inputs to climate models in climate change prediction studies. Feddema *et al* [61] examined the importance of land cover changes implied by two of the IPCC SRES emissions scenarios, and demonstrated that the effects on regional climate change are significant in comparison with those of GHG-induced climate change (Figure 6). For example, the SRES B1 scenario implies reforestation in mid-latitudes and relatively little tropical deforestation (Figure 6b), whereas the A2 scenario implies less mid-latitude reforestation but extensive tropical deforestation (Figure 6c). Feddema *et al* [61] found that these differences in projected land cover led to significant variations in the predicted climate change at regional scales. For example, in the A2 scenario, greater warming occurred in Amazonia as a result of deforestation reducing evaporation while less warming occurred in northern China due to deforestation increasing the surface albedo (Figure 6e). In contract, the B1 scenario included no further Amazon deforestation so no additional climate change occurred in that region, but greater warming occurred in both northern and southern China as a result of reforestation decreasing the surface albedo (Figure 6d). Feddema *et al* [61] also reported impacts of land use change on the Asian monsoon, the Hadley-Walker circulation and the positioning of the near-equatorial band of major atmospheric ascent known as the Inter-Tropical Convergence Zone.

Plans and strategies for adaptation to climate change require specific, local detail of climate change. Although this can be provided by regional climate models, these models are typically only used for down-scaling of radiatively-forced global climate change. Climate change adaptation plans may therefore be inappropriate if based on projections which ignore land use change.

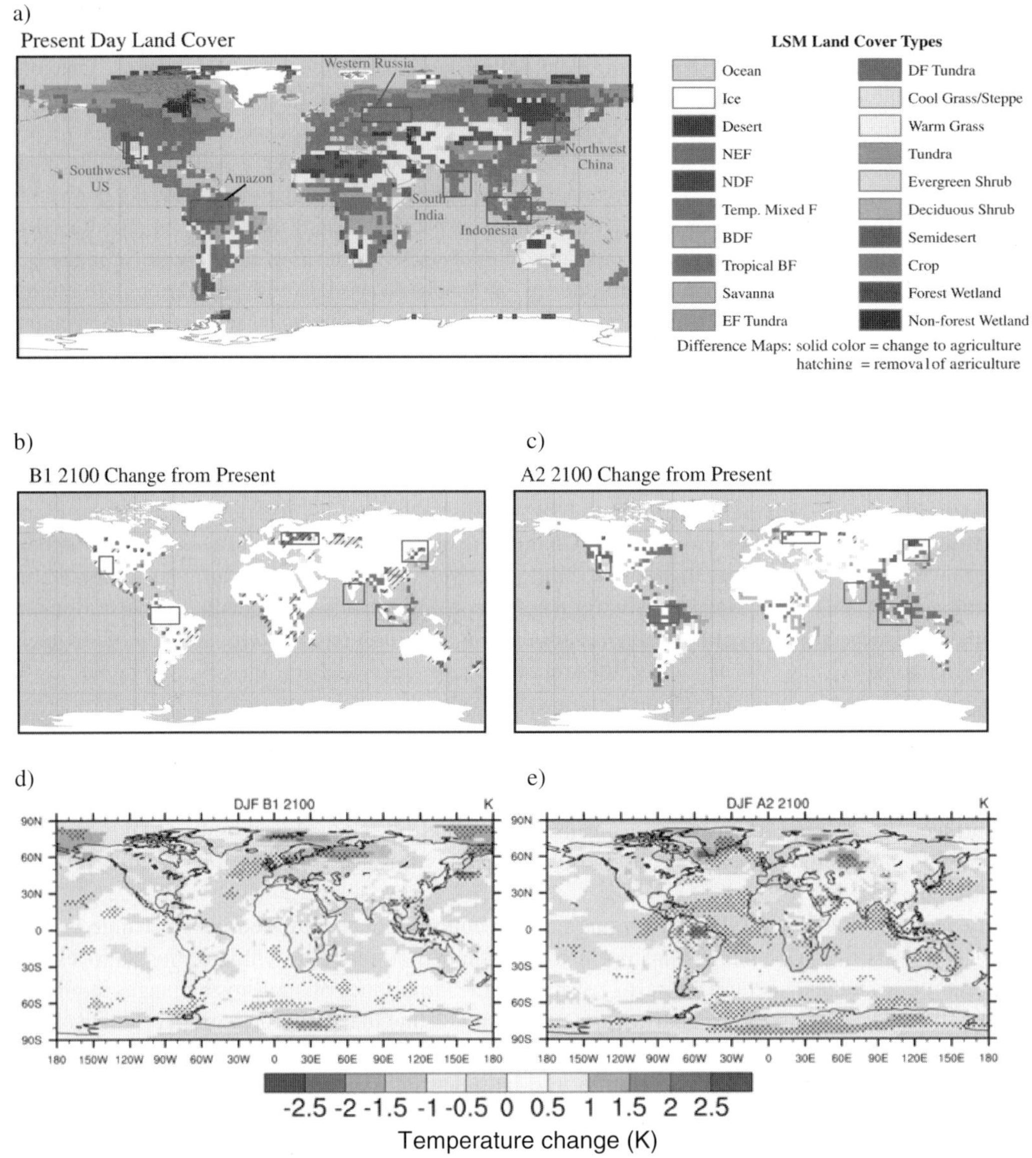

Figure 6. a) Present day land cover as represented in the DOE-PCM climate model. (b) Land use change by 2100 consistent with the IPCC SRES B1 emissions scenario. (c) As (b) for the IPCC SRES A2 emissions scenario. (d) Temperature changes in December-January-February (DJF) in addition the greenhouse forced changes, due to B1land use changes. (f) as (e) for A2 land use changes. Feddema *et al* [61].

3. FEEDBACKS ON CLIMATE CHANGE FROM RESPONSES OF GLOBAL VEGETATION

3.1 Responses of global vegetation to climate change

A number of climate models now include sub-models of ecosystem dynamics, allowing feedbacks on climate change to be examined. In the Hadley Centre coupled climate-carbon cycle model HadCM3LC [30], major changes in global vegetation patterns were simulated as a result of GHG emissions following IPCC "business-as-usual" scenario, IS92a. The climate was simulated to warm as a result of the increased radiative forcing by greenhouse gases, with more rapid warming over the land surface than over the oceans due to the large heat capacity of water (Figure 7a). The model also projected more rapid warming at high latitudes than at low latitudes, due to positive feedbacks on the warming arising from the melting of snow and ice. While snow and ice reflect a large proportion of solar radiation back to space, the underlying ground and ocean surface is generally darker and absorbs a greater fraction of the radiation. Melting of snow and ice therefore exposes the darker underlying surface, increasing the fraction of solar radiation absorbed and causing a further warming of these regions.

Global mean precipitation was generally simulated to increase, as a result of an increased supply of moisture to the atmosphere by evaporation from the oceans and continents in a warmer world (Figure 7b). However, significant decreases in precipitation were simulated in some regions, particularly Amazonia (Figure 7b). This was associated with more rapid warming of the ocean surface in the equatorial east Pacific and North Atlantic [62, 63, 64], which modified the Hadley-Walker circulation and position of the Inter-Tropical Convergence Zone and hence affected the positioning and rate of convection in the South American equatorial region.

As a result of these climatic changes, the simulations featured significant changes in the character and distribution of global ecosystems (Figure 7c). As the climate warmed in the cold regions, shrub cover expanded in the Artic tundra and on the Tibetan plateau. The boreal forests became more dense, and also spread towards the pole. In gridboxes on northern edge of boreal forest, tree cover became more dense, implying a northward movement of the treeline. A number of dry regions such as central Asia showed a greening trend, consistent with increased water-use efficiency but also partly a result of increased rainfall. However, other dry regions such as South-west Africa lost tree and shrub cover due to a decrease in rainfall. The forests of South-East Asia and central Africa increased in density in the absence of future deforestation. However, the forests of Amazonia showed a very large reduction in tree cover as a result of decreased rainfall (Figure 7c).

Some signs of the beginning of Amazon forest die-back were already simulated by 2000, with broadleaf tree cover reducing in the north-east of Amazonia in response to a drier climate than that simulated for 1860. The reduction in rainfall spread towards the south-west through the 21st Century, and the tree cover reduced until it was less than 1% in the north-east quarter of Amazonia by 2100. Almost all of the Amazon basin lost at least 50% of its tree cover by the end of the simulation, to be replaced mainly by C4 grass but also with large areas of bare soil. The general character of the region is therefore fundamentally changes from dense evergreen broadleaf forest to savanna, grassland or even semi-desert.

3.2 Terrestrial ecosystem feedbacks on climate change via the carbon cycle

In the simulations described in section 3.1 with the HadCM3LC coupled climate-carbon cycle GCM, climate-carbon cycle feedbacks enhanced the rate of CO_2 rise by 75% compared to CO_2 concentration projections which neglected the effect of climate change on the carbon cycle [30]. In a simulation neglecting the effects of climate change on the carbon cycle, the terrestrial biosphere was projected to take up 620 Gigatonnes of carbon (GtC) by 2100 (Figure 8a), with global soil carbon storage increasing by 400 GtC (Figure 9b) and global vegetation carbon increasing by 220 GtC (Figure 9a). In contrast, a simulation including the effects of climate change on the carbon cycle projected an overall loss of 90 GtC from the terrestrial biosphere (Figure 8b, due the soil losing 150 GtC (Figure 9b) and vegetation carbon only increasing by 60 GtC (Figure 9a). The soil carbon content was reduced worldwide as a result of increased heterotrophic respiration induced by the temperature rise (Figure 9b). While some vegetation areas, particularly the boreal forests, still took up carbon from the atmosphere and therefore provided a negative feedback, the Amazon forest became a large source of CO_2 as a result of the forest dieback induced by the drying climate in that region as described in section 3.1 (Figures 7c and 9a). The smaller net increase in vegetation carbon was therefore primarily a result of the loss of carbon from Amazonia.

a) b) c)

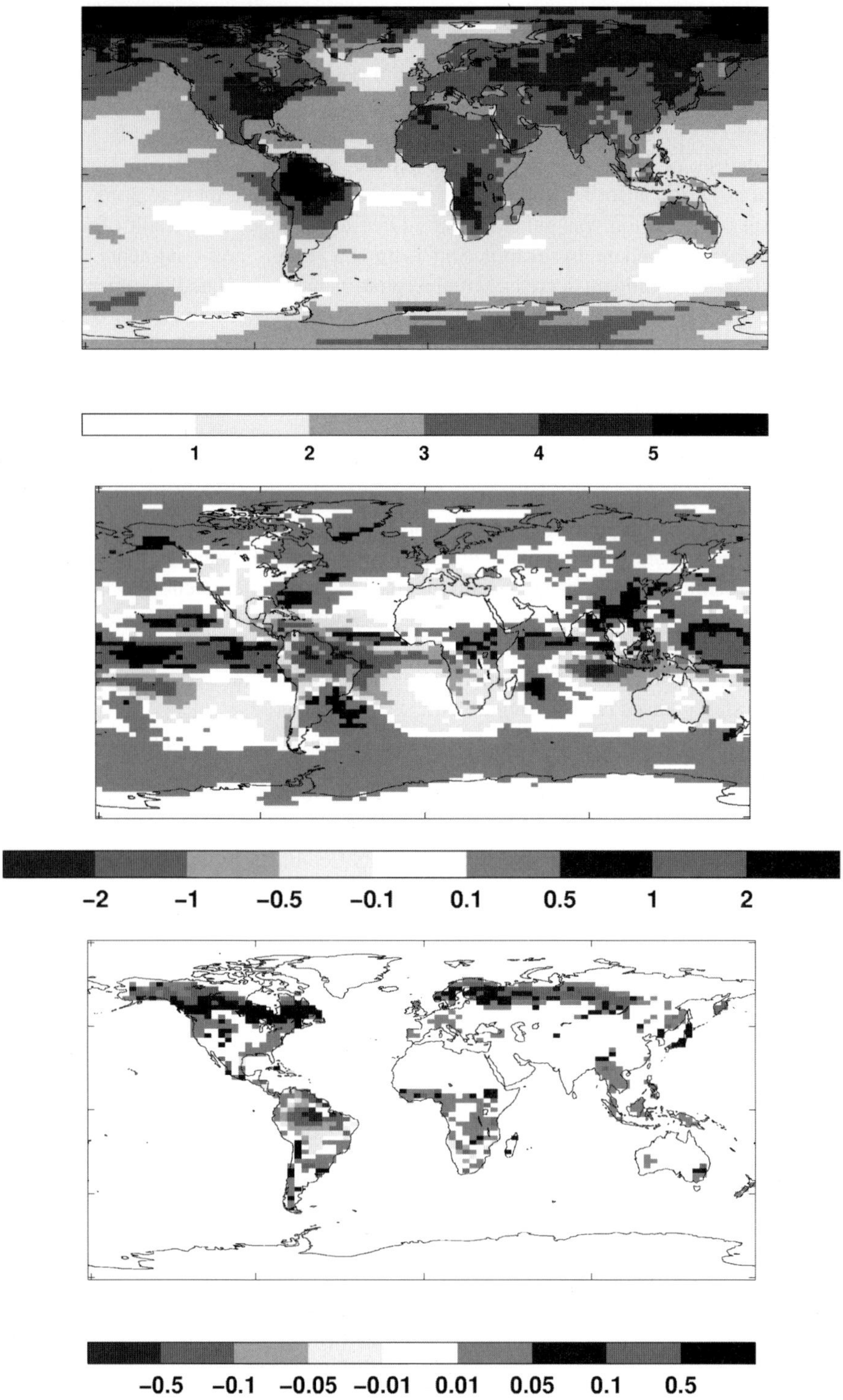

Figure 7. Changes in (a) annual mean near-surface air temperature (K), (b) annual mean precipitation (mm day^{-1}) and (c) forest cover fraction, relative to 2000 simulated by HadCM3LC. 30-year mean centred around 2080.

Climate change therefore induced a deficit of 710 GtC in the terrestrial biosphere in comparison with the changes projected without the effects of climate change. Approximately 10% of the deficit arose from the dieback of the Amazon forest [65], with most of the rest of the deficit arising from increased soil respiration. The ocean carbon cycle also responded to climate change, with the net uptake of CO_2 by the oceans increasing by approximately 100 GtC. The overall feedback resulted in CO_2 concentrations rising by 1500 GtC between 1850 and 2100 (Figure 8b), rather than approximately 900 GtC when carbon cycle feedbacks were neglected (Figure 8a). The CO_2 rise was therefore accelerated by 66%. The atmospheric CO_2 concentration, initially at 280ppmv in 1850, rose to 1000ppmv in 2100 rather than 700ppmv as projected by both this model and the standard IPCC IS92a scenario without the feedback.

This feedback was found to significantly increase the rate of global warming. Without the feedback, with a CO_2 rise of only 700ppmv, the global mean temperature rise over land was 5K, whereas in with the carbon cycle feedback included the warming over land was 8K. The inclusion of climate-carbon cycle feedbacks was therefore to enhance the rate of global land warming by 50%. Global mean temperatures rose by 5.5K with carbon cycle feedbacks and 4K without, so the carbon cycle feedbacks accelerated global warming by 37%.

Positive feedbacks on the CO_2 rise and climate change have also been found to emerge in all other coupled climate-carbon cycle models in the Coupled Climate-Carbon Cycle Model Intercomparison Project ("C4MIP" [68]). The strength of the feedback varies significantly between models, with HadCM3LC showing the largest feedback (Figure 10). There is therefore a significant uncertainty in the translation of CO_2 emissions scenarios into scenarios of CO_2 concentrations. However, this uncertainty is often neglected, and the scenarios of CO_2 *concentrations* used to drive climate models are routinely described as "*emissions* scenarios". For example, climate model simulations are almost always described as having been driven by a particular IPCC SRES emissions scenario, when in fact they have been driven by a scenario of CO_2 concentrations which neglected climate-carbon cycle feedbacks. The results of C4MIP clearly imply that it is incorrect to use emissions scenario names to label such simulations. With climate-carbon cycle feedbacks being of potentially great significance for projections of future global climate change, it is important to find a more informative system of identifying the assumptions made in climate projections.

3.3 Terrestrial vegetation feedbacks on climate change through mineral dust aerosol emissions

Changes in vegetation cover in climates which are sometimes dry may lead to changes in the emission of mineral dust aerosol from the land surface. This can affect climate by exerting radiative forcings in both the shortwave and longwave – the net effect is complex, and depends on other factors such as the albedo of the underlying surface. Woodward *et al* [69] used a dust emissions and transport model in the HadAM3 climate model [48] to simulate the changes in atmospheric dust load as a consequence of a drying climate and forest die-back in Amazonia as simulated by HadCM3LC [30, 63, 65]. By 2100, Amazonia was simulated to become a source of dust equivalent to that of the present-day Sahara desert (Figure 11a,b), as a result of dried soil, exposure of soil by lost forest cover, and increased windspeed which was a consequence of the reduced aerodynamic roughness of the landscape. This resulted in a positive radiative forcing at the top of the atmosphere of over 10 Wm^{-2} over the Amazon region (Figure 11c,d), as a result of increased absorption of outgoing longwave radiation by the airborne dust.

3.4 Terrestrial ecosystem feedbacks on climate change through changes in the physical properties of the land surface

Changes in terrestrial ecosystems may also exert some significant feedbacks on regional climates through changes in the physical properties of the land surface. Betts *et al* [65] examined these feedbacks on 21st-Century climate change with two simulations with HadCM3LC with CO_2 concentrations prescribed to the standard IS92a scenario neglecting climate-carbon cycle feedbacks, with one

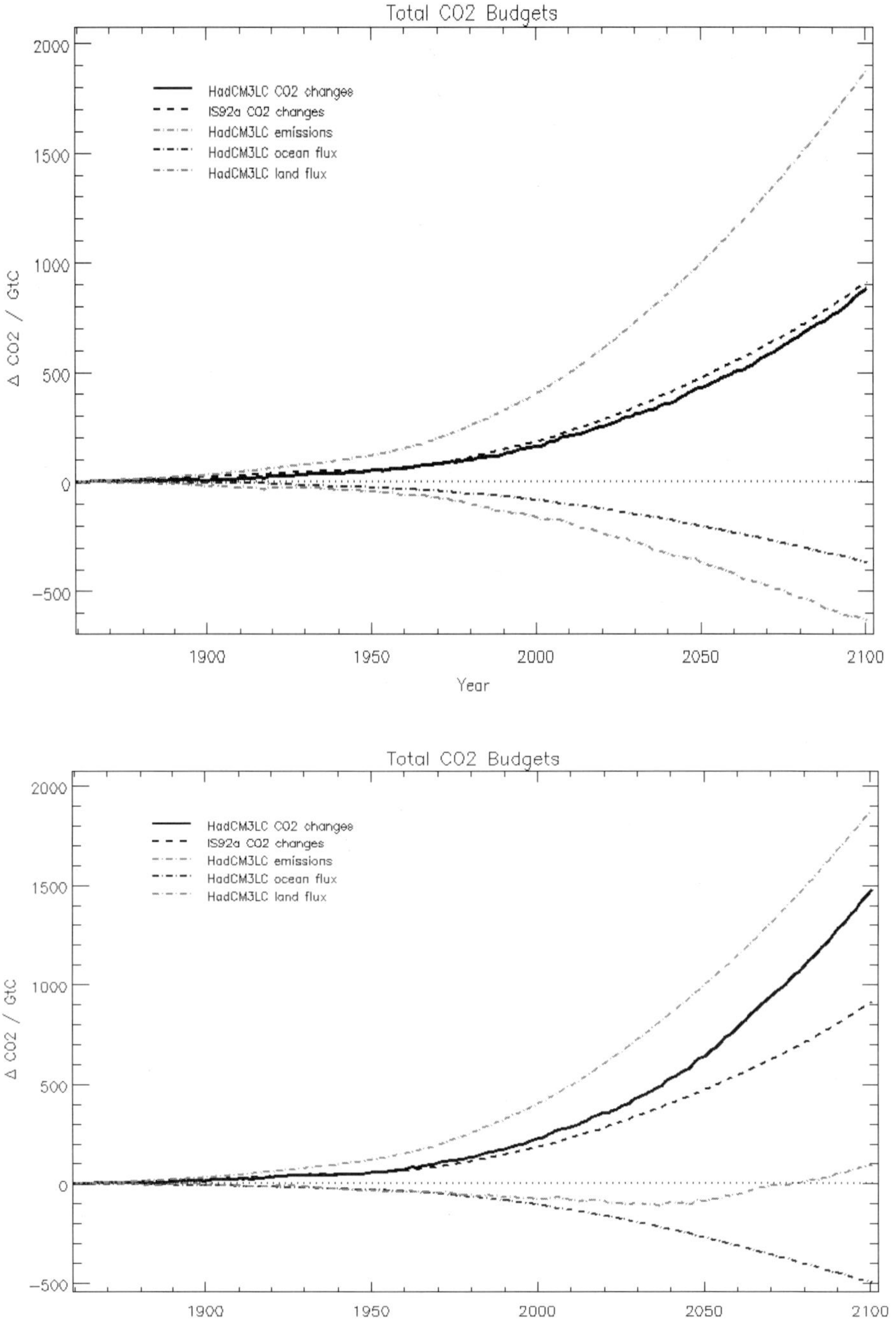

Figure 8. Cumulative carbon budgets (GtC) simulated by HadCM3LC [30] without (a) and with (b) the effects of climate change on the terrestrial and ocean carbon cycles. Red dotted line: anthropogenic emissions from CDIAC database (1850 - 1990s) [66] and IPCC IS92a emissions scenario (1990s – 2100) [67]. Blue dotted line: ocean-atmosphere flux. Green dotted line: land-atmosphere flux. Black solid line: atmospheric CO_2 rise simulated by HadCM3LC. Black dotted line: atmospheric CO_2 emissions from ice core records (until 1958), flask measurements (1958-1990s) and standard IPCC concentrations scenario from IS92a emissions scenario (1990s – 2100). In the atmosphere, 715 GtC = 337 ppmv.

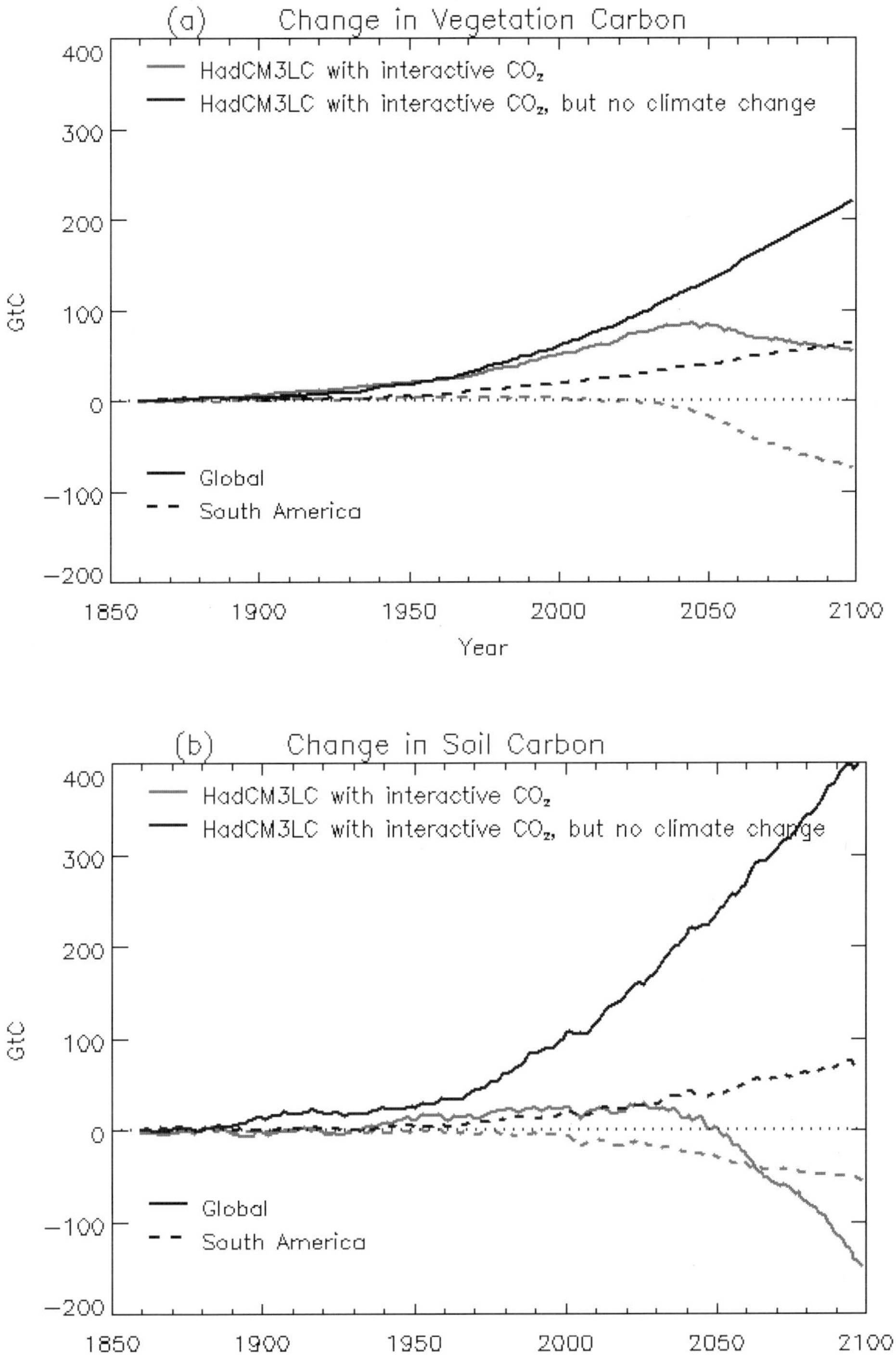

Figure 9. Changes in vegetation (a) and soil (b) carbon contents (GtC) relative to 1850 simulated by HadCM3LC with climate change (dashed lines) and without (solid lines) [30]. Blue lines: global total. Red lines: Total for South America.

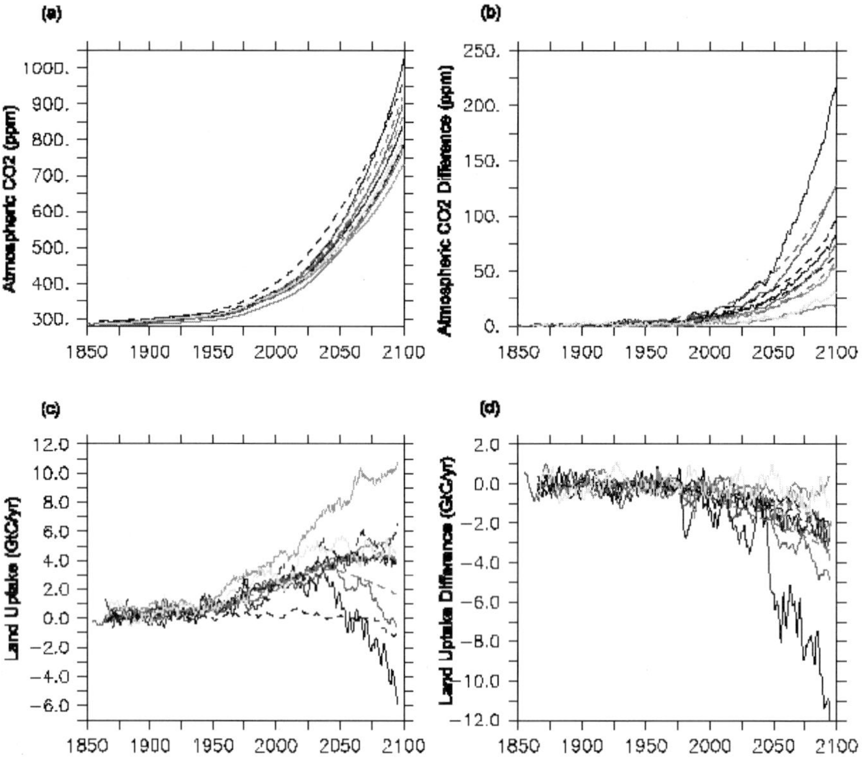

Figure 10. Uncertainties in CO_2 concentrations and land carbon uptake associated with one emissions scenario, as a result of climate-carbon cycle feedbacks. (a) CO_2 concentrations simulated by the 11 "C4MIP" coupled climate-carbon cycle models [68] driven by the SRES A2 emissions scenario, with climate-carbon cycle feedbacks included. (b) Differences in CO_2 concentrations between simulations including and excluding carbon cycle feedbacks, for each of the C4MIP models. (c), (d) as (a), (b) for global land carbon uptake. In each panel, results from the Hadley Centre model HadCM3LC [30] are shown by the solid black lines. Reproduced from Friedlingstein *et al* [68].

simulation including changes in vegetation cover simulated by the TRIFFID Dynamic Global Vegetation Model [70], and the other with vegetation cover fixed at the present-day state. Comparison between these therefore reveals the extent of biogeophysical feedbacks on climate change in the HadCM3LC model.

The general global patterns of climate change were similar in the two simulations, with almost all changes in temperature and precipitation being of the same sign irrespective of the inclusion of vegetation feedbacks. This implies that vegetation feedbacks are not a significant influence on atmospheric circulation in comparison with the greenhouse-gas forcing. However, some of the regional climate changes were significantly affected by vegetation feedbacks (Figure 12). In particular, the precipitation reduction over Amazonia was found to be enhanced by 25% by feedbacks from the loss of forest cover [65]. In the western part of the basin, the feedback was greater still, magnifying the precipitation reduction by over 30%. The larger precipitation decrease in western Amazonia was attributed to drought-induced die-back of the eastern forests contributing to further rainfall reductions in the west. The forest loss also increased surface albedo which reduced convection and moisture convergence, providing a further positive feedback on rainfall reduction [71].

Further biogeophysical feedbacks were seen at high latititudes, with the boreal forests expanding and thickening and tundra shrub cover also expanding. The more extensive vegetation cover resulted in a lower surface albedo, which increased the absorption of solar radiation. This gave a further 0.5 -1K

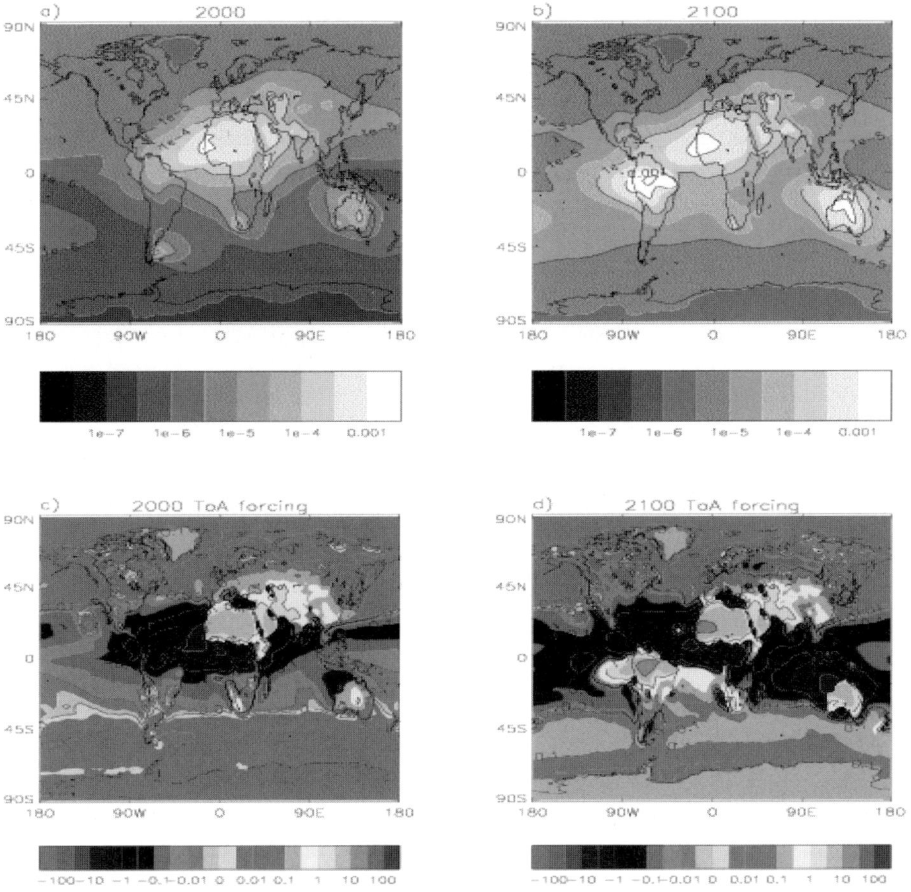

Figure 11. Effect of vegetation responses to climate change on atmospheric dust load and radiative forcing, simulated with the HadAM3 climate model [69]. (a) Atmospheric dust load at 2000. (b) Atmospheric dust load at 2100, including changes in climate and vegetation change from Cox *et al* [30]. (c) Net Top of Atmosphere (ToA) radiative forcing due to dust at 2000. (d) as (c) for 2100.

of warming over land north of 50°N, in addition to the 6K simulated with fixed vegetation cover (Figure 13).

4. CONCLUSIONS

Land cover has been directly modified by human activities over several millennia, resulting in an overall reduction in global forest cover of approximately 20-30% with most of this occurring in the last 250 years. Since such modifications are direct consequences of human action and not a result of climate change, their effects on climate constitute forcings of climate change as opposed to feedbacks. Some of these are radiative forcings, directly perturbing the radiation balance of the planet. Others are "non-radiative" forcings which act directly on the surface climate.

The net release of carbon dioxide from net deforestation is estimated to have contributed approximately 15 - 35% of the overall CO_2 rise since the industrial revolution. This "land-use CO_2" exerts a radiative forcing of approximately 0.22 – 0.51 Wm^{-2} which is approximately 10 – 20% of

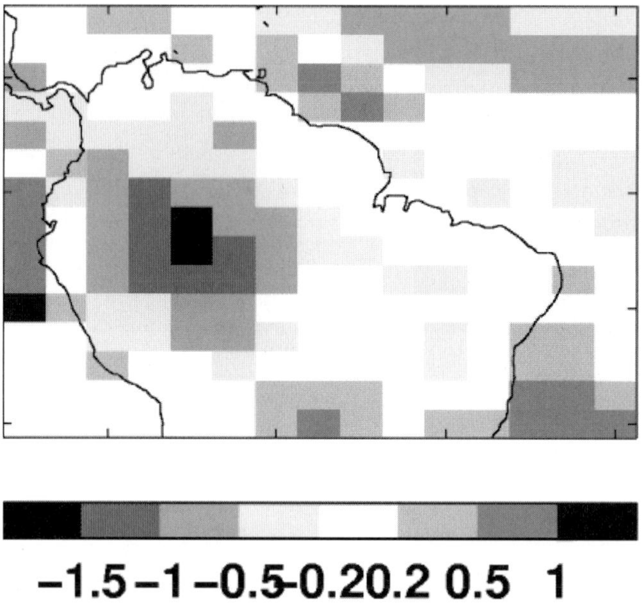

Figure 12. Effect of biogeophysical feedback on precipitation patterns over Amazonia in the HadCM3LC climate model. Difference in precipitation (mm day^{-1}) between simulations with forest cover dying-back due to climate change and with vegetation fixed at pre-industrial state. 30-year mean centred around 2080.

the overall warming influence of anthropogenic greenhouse gas increases (including carbon dioxide, methane, nitrous oxide and the halocarbons).

Since most deforestation has occurred in temperate regions, the main biogeophysical effect has been through an increase in surface albedo due to exposure of the brighter unforested land surface, especially when snow is lying. This is estimated to exert a radiative forcing of approximately 0 - -0.2 Wm^{-2}, constituting a cooling influence. Deforestation in the tropics may have exerted a small warming influence through reduced evaporation, which cannot be quantified in terms of Wm^{-2} radiative forcing.

Exposure of soil by land use in dry regions has contributed to an increase in the quantity of dust in the atmosphere, which is estimated to have exerted a radiative forcing of approximately +0.1 to -0.2 Wm^{-2} with a high level of uncertainty.

Current trends are for some reforestation in the temperate regions and major deforestation in the tropics. However, although this will still contribute to the anthropogenic CO_2 rise, the IPCC SRES scenarios suggest that increases in fossil fuel emissions may continue to make increasingly greater contributions to rising CO_2 than ongoing emissions from deforestation.

Ongoing tropical deforestation would continue to warm the tropical land surfaces through reduced evaporation in addition to contributing to global warming through CO_2 emissions. If tropical deforestation becomes the dominant form of land use change, radiative forcing will no longer be a suitable metric for comparing its biogeophysical effects on climate with the effects of greenhouse gases and aerosols.

Reforestation in temperate regions may also exert warming effects through decreases in surface albedo. Future climate is therefore likely to be affected by anthropogenic land cover change as well as by the enhanced greenhouse effect, at least at the regional scale.

Climate change itself may lead to changes in vegetation which exert further effects on climate as feedbacks. The drying of the climate in Amazonia seen in some climate models may lead to a die-back of forests and a release of CO_2 to the atmosphere. This may add to worldwide net releases of

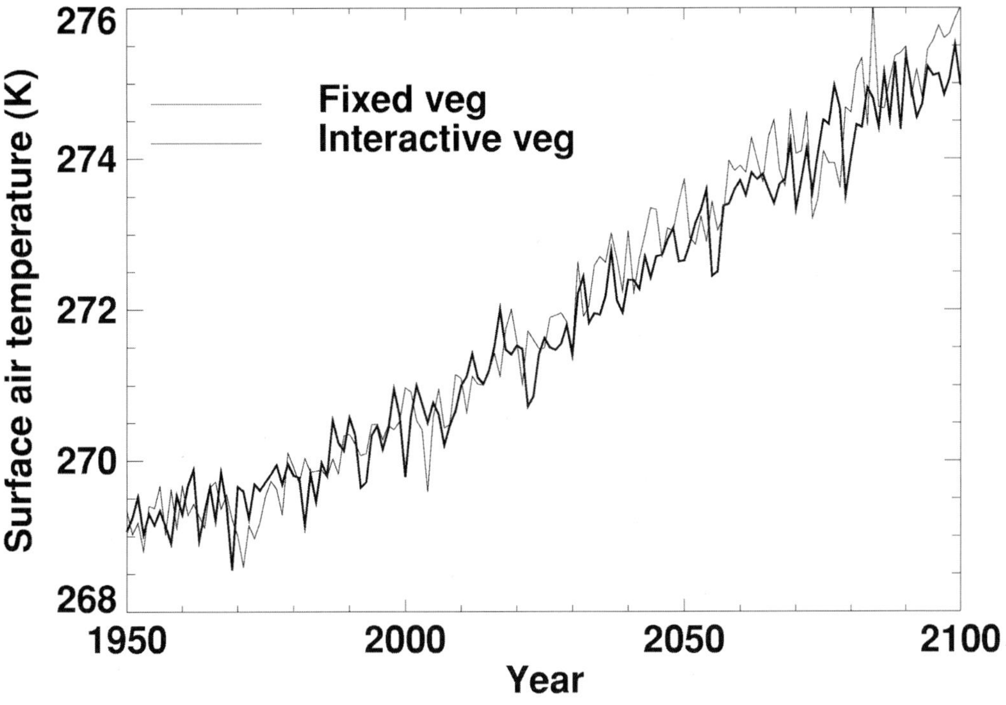

Figure 13. Timeseries of mean near-surface annual mean air temperature above 50°N simulated by the HadCM3LC climate model [30], with forest and tundra expanding due to climate change (blue line) and with vegetation fixed at the pre-industrial state (black line).

CO_2 from soils as heterotrophic respiration increases with warming, and overall the rate of global warming may be accelerated by up to 37% (but with most models showing more modest accelerations). With large uncertainties in the rate of change of CO_2 concentrations arising from a given emissions scenario, it is incorrect to use emissions scenario names to identify scenarios of changes in CO_2 concentration.

Forest dieback in Amazonia would accelerate the local drying by reducing the evaporative re-cycling of rainfall. Should drying and forest dieback occur as simulated by HadCM3LC, this could cause Amazonia to become a major new global source of dust comparable with the Sahara.

Some climate models used for climate change prediction are now beginning to include some vegetation-related forcings and feedbacks on climate change. However, this still not routine, and often only global-scale feedbacks (such as through the carbon cycle) are considered. However, predictions at the regional scale are needed to inform strategies for adaptation. Well-validated biogeophysical effects of human land use and vegetation responses to climate change need to be included in global and regional climate models used for assessments of future climate change impacts.

Acknowledgements

The author is grateful to V. Brovkin, P. Cox, C. Jones, P. Falloon, J. Feddema, P. Friedlingstein, K. Klein Goldewijk, D. Lowe, D. Matthews, N. Ramankutty, and S. Woodward for valuable collaborations, discussion and figures. This work forms part of the Climate Prediction Programme of the U.K. Department of the Environment, Food and Rural Affairs (contract PECD 7/12/37).

References

[1] IPCC, Climate Change 2001: The Scientific Basis. Contribution of Working Group I to the Third Assessment Report of the Intergovernmental Panel on Climate Change (Cambridge University Press, 2001) 879pp.

[2] IPCC, Land Use, Land-Use Change and Forestry (Cambridge University Press, 2000) 377pp.

[3] Holdridge, L.R., *Science* 105 (1947) 367-368.

[4] Woodward, F.I., Climate and Plant Distribution (Cambridge University Press, 1987) 174pp.

[5] Charney, J.G., *Q. J. R. Meteorol. Soc.* (1975) 101:193-202.

[6] Potter, G.L., Ellsaesser, H.W., MacCracken, M.C., Luther, E.M., *Nature* (1975) 258:697-698.

[7] Shukla, J., and Mintz, Y., 1982. *Science* (1982) 215:1498-1501.

[8] Harding, R.J., and Pomeroy, J.W., *J. Clim.* (1996) 9:2778-2787.

[9] Bonan, G.B., Pollard, D., and Thompson, S.L., *Nature* (1992) 359:716-718.

[10] Thomas, G., and Rowntree, P.R., *Q. J. R. Meteorol. Soc.* (1992) 118:469-497.

[11] Betts, R.A., *Geophys. Res. Lett.* (1999) 26, 1457-1460.

[12] Betts, R.A., *Journal de Physique IV Proceedings* (2004), 121, 37-60.

[13] House, J., Brovkin, V., Betts., R., Costanza., B., Assuncao Silva Dias, M., Holland, B., Le Quere, C., Kim Phat, N., Riebesell, U., and Scholes, M., in Millennium Ecosystem Assessment: Ecosystems and Human Well-Being, (Island Press, 2005).

[14] Jacob, D.J., et al., Radiative forcing of climate change. The National Acadamies Press, Washington, D.C., (2005) 207pp.

[15] Ramankutty, N., and Foley, J.A., *Global Biogeochem. Cycles* (1999) 14, 997-1027.

[16] Klein Goldewijk, K., *Global Biogeochem. Cycles* (2001) 15, 417-433.

[17] FAO, Global Forest Resource Assessment 2000. Main Report, Food and Agriculture Organisation of the United Nations, Rome (2001).

[18] Houghton, R.A., *Tellus* (2003) 55B, 378-390.

[19] House, J.I., Prentice, I.C., Ramankutty, N., Houghton, R.A., and Heimann, M. Tellus (2003) 55B, 345-363.

[20] Houghton, R.A., *Tellus* (1999) 51B, 298-313.

[21] McGuire, A.D., Sitch, S., Clein, J.S., Dargaville, R., Esser, G. Foley, J., Heimann, M., Joos, F., Kaplan, J., Kicklighter, D.W., Meier, R.A., Melillo, J.M., Moore, B., Prentice, I.C., Ramakutty, N., Reichenau, T., Schloss, A., Tian, H., Williams, L.J., and Wittenberg, U., Global Biogeochem. Cycles (2001) 15, 183-206.

[22] FAO, Land use, FAO_STAT-PC, Food and Agriculture Organisation of the United Nations, Rome (1995).

[23] Brovkin, V.M. et al., 2004. *Global Change Biol.,* (2004) 10, 1253-1266.

[24] Houghton, R.A., and Hackler, J.L., in *Trends*, CDIAC, ORNL, Oak Ridge (2002).

[25] Marland, G., Boden, T.A., and Andres, R.J., in *Trends*, CDIAC, ORNL, Oak Ridge (2002).

[26] Plattner, G.K., Joos, F., and Stocker, T.F., *Global Biogeochem. Cycles* (2002), 16 doi:10.1029/2001GB001746.

[27] Bopp, L., Le Quéré, C., and Hemiann, M., et al., *Global Biogeochem. Cycles* (2002), 16, doi:10.1029/2001GB001445.

[28] Francey, R.J., Allisn, C.E., and Etheridge, D.M., et al., *Tellus* (1999), 51, 170-193.

[29] Keeling, C.D., Bacastow, R.B., and Carter, A.F., et al., *AGU Geophysical Monographs* (1989) 55, 165-236.

[30] Cox, P.M., Betts, R.A., Jones, C.D., Spall, S.A., and Totterdell, I.J., *Nature* (2000), 408, 184-187.

[31] Jones C.D., Cox, P.M., Essery, R.L.H., Roberts, D.L., and Woodage, M.J., *Geophys. Res. Lett.* (2003) 30, doi:10.1029/2003GL016867.

[32] Schimel, D., et al., in Climate Change 1995: The Science of Climate Change, contribution of Working Group 1 to the Second Assessment Report of the Intergovernmental Panel on Climate Change (Cambridge University Press, New York, 1996) pp. 65–131.
[33] Matthews, H.D., Weaver, A.J., Meissner, K.J., Gillett, N.P., and Eby, M., *Clim. Dyn.*, (2004) 22, 461-479.
[34] Marland, G., Andres, R.J., and Boden, T.A., in *Trends*, CDIAC, ORNL, Oak Ridge (2003).
[35] Woodward, S., *J. Geophys. Res.* (2001) 106, 18155-18166.
[36] Sokolik, I.N., and Toon, O.B., *Nature* (1996) 381, 681-683.
[37] Tegen, I., Werner, M., Harrison, S.P., and Kohfeld, K.E., *Geophys. Res. Lett.* (2004) 31, L05105, doi:10.1029/2003GL019216.
[38] Hansen, J., Sato, M., and Ruedy, R., *Geophys. Res.*, (1997) 102, 6831-6864.
[39] Betts R.A., *Atmos. Sci. Lett.* (2001) doi:10.1006/asle.2000.0023.
[40] Govindasamy, B., Duffy, P., and Caldeira, K., *Geophys. Res. Lett.* (2001) 28, 291-294.
[41] Myhre G., and Myhre, A., *J. Clim.* (2003) 16, 1511-1524.
[42] Matthews, H.D., Weaver, A.J., Eby, M., and Meissner, K.J., *Geophys. Res. Lett.*, (2003) 30, 271-274.
[43] Matthews, H.D., Weaver, A.J., Meissner, K.J., Gillett, N.P., and Eby, M., *Clim. Dyn.*, (2004) 22, 461-479.
[44] Myhre, G., Kvalevåg, M.M., and Schaaf, C.B., *Geophys. Res. Lett.*, (2005) 32, L21410.
[45] Hansen, J., et al., *J. Geophys. Res.* (2005) 110, D18104, doi:10.1029/2005JD005776.
[46] Brovkin, V.M., et al., *Clim. Dyn.* (2006) 26, 587-600.
[47] Betts, R.A., Falloon, P.D., Goldewijk, K.K., and Ramankutty, N., *Agricultural and Forest Meteorology* (in press).
[48] Pope, V.D., Gallani, M.L., Rowntree, P.R., and Stratton, R.A., *Climate Dynamics*, (1999) 16, 123-146.
[49] Priante-Filho, N., et al., *Global Change Biol.*, (2004) 10, 863-876.
[50] Lean, J., and Rowntree P.R., *J. Clim.* (1997) 10(6):1216-1235.
[51] Lewis, T., *Global and Planetary Change* (1998) 18, 1-13.
[52] Kleidon, A., and Heimann, M., Clim. Dyn. (2000) 16, 183-199.
[53] Kleidon, A., *Global and Planetary Change* (2006) in press.
[54] Roy, S.B., and Avissar, R., *J. Geophys. Res.* (Atmos) (2002) 107 (D20).
[55] Chagnon, F.J.F., and Bras, R.L., *Geophys. Res. Lett.*, (2005) 32: Art. No. L13703.
[56] Gedney, N., and Valdes, P.J., *Geophys. Res. Lett.* (2000) 27(19):3053-3056.
[57] Chase, T.N., Pielke, R.A., Kittel, T.G.F., Nemani, N., and Running, S.W., *Clim. Dyn.* (2000) 16, 93-105.
[58] Zhao, M., Pitman, A., and Chase, T.N., *Clim. Dyn. (2001)* 17, 467-477.
[59] Pielke Sr., R.A., Marland, G., Betts, R.A., Chase, T.N., Eastman, J.L., Niles, J.O., Niyogi, D.S., and Running, S.W., Philos. Trans. R. Soc. London Ser. A, (2002) 360, 1705-1719.
[60] Delire C., Behling P., Coe M.T., Foley J.A., Jacob R., Kutzbach J., Liu Z.Y., and Vavrus S., *Geophys. Res. Lett.* (2001) 28 (10):2081-2084.
[61] Feddema *et al.*, *Science* (2005), 310, 1674-1678.
[62] Cai, W., and Whetton P.H., *J. Clim* (2001) **14**, 3337-3355.
[63] P. M. Cox, Betts R.A., Collins M., Harris P.P., Huntingford C., and Jones C.D., *Theor. Appl. Climatol.* (2004) 78:137-156.
[64] Harris, P.P., Modelling South American Climate and Climate Change. (Ph.D. thesis, University of Reading, 2005).
[65] Betts, R.A., Cox P.M., Collins M., Harris P.P., Huntingford C., and Jones C.D., Theor. Appl. Climatol. (2004) 78:157-175.
[66] Keeling, C.D., and Whorf, T.P., in *Trends*, CDIAC, ORNL, Oak Ridge (2005).

[67] Leggett, J., Pepper, W.J., and Swart, R.J., in IPCC, Climate Change 1992: The Supplementary Report (Cambridge University Press, 1992) pp. 69-95.

[68] Friedlingstein, P., Cox, P., Betts, R., Bopp, L., Von Bloh, W., Brovkin, V., Cadule, P., Doney, S., Eby, M., Fung, I., Bala, G., John, J., Jones, C., Joos, F., Kato, T., Kawamiya, M., Knorr, W., Lindsay, K., Matthews, D., Raddatz, T., Rayner, P., Reick, C., Rieckner, E., Schnitzler, K.-G., Schnur, R., Strassmann, K., Weaver, A.J., Yoshikawa, C., and Zeng, N., *J. Clim.*, 19, 3337–3353.

[69] Woodward, S., Roberts, D.L., and Betts, R.A., *Geophys. Res. Lett.* (2005) 32, L18810, doi: 10.1029/2005GL023482.

[70] Cox, P.M., Description of the TRIFFID dynamic global vegetation model. Technical Note 24, Hadley Centre, Met Office (2001).

[71] Charney J.G., *Q. J. R. Meteorol. Soc.* (1975) 101:193-202.

J. Phys. IV France **139** (2006) 143–155
© EDP Sciences, Les Ulis
DOI: 10.1051/jp4:2006139010

Earth system analysis and the future of the biosphere

W. Lucht[1]

[1] *Potsdam Institute for Climate Impact Research, PO Box 601203, 14412 Potsdam, Germany*
e-mail: Wolfgang.Lucht@pik-potsdam.de

Abstract. The global biosphere has begun to change fundamentally as a consequence of human actions. This change can be understood as a consequence of a major transition in the evolution of life on Earth, the emergence of human language, which opened up new pathways of biological information transmission. The challenge facing the humans species now is to not just suffer the consequences of this change, but to develop a science of Earth system analysis that will allow the collective, globally networked reflective capacity of humans to chart paths into the future that are sustainable. Global observation and computer modelling are important elements of this process. Such models for the biosphere predict large-scale reorganisation of the functional composition of the terrestrial biosphere under strong anthropogenic climate change. Considering the advent of humans and of global change on the background of the past evolution of life on Earth, it is obvious that the co-evolution between geosphere and biosphere that has characterised Earth history in the past has expanded to include the anthroposphere as a third interacting element.

1. GLOBAL CHANGE AS A PLANETARY PHENOMENON AND A CHALLENGE TO HUMANS

Humans are changing the planet with incredible speed. Anthropogenic climate change, expanding human land use, chemical pollution of the environment, extinction and mixing of biological species, and, in the near future, genetic modification of organisms are putting the Earth's biosphere under multiple pressures [1]. As a consequence, it is likely that in a hundred years from now the environment in most regions of the world will be substantially different from what it is today.

The biosphere is the most complex product of planet Earth's evolution over the 4.5 billion years of its history. But only a tiny fraction of all life forms and hence ecosystems that have existed at one time or another are still extant : 99.9% of species are estimated to be extinct. The environment in which these organisms lived has also continuously seen change, though mostly on a time scale that spanned enormous numbers of generations. Many of these changes were caused by geologic events such as continental drift, volcanic activity and weathering, or by impacts of space debris and radiation. Others were a direct consequence of life itself, such as changes in atmospheric and oceanic chemical composition through metabolism, sedimentation, burrowing, grazing and other biotic processes.

Currently, it is one species, the human species, that is the cause of global environmental change of a magnitude and rate that is worrisome. The exact history of life still remains sketchy from the fossil and genetic records, but it is unlikely that ever before in the long run of evolution has a single species had such an impact in such a short time. It is a current phenomenon that cannot be attributed even to the existence of humans as such, for the impact has increased to a global scale only in the last several thousand years and humans have existed for several millions years already. But the roots of that development lie earlier. One surely has to think back to the taming of fire and the invention of language perhaps 250 000 years ago.

Today humans, at least in the materially affluent world, but increasingly also elsewhere, are finding themselves confronted with the knowledge of the increasingly planetary consequences of their collective actions. Modern science, observation, communication and politics have uncovered a vivid picture of what is happening in what has been called global change, of which global climate change is just one element. Though the intense societal debate about relative priorities tends to still lead to denial about the extent of imminent changes, and though it is far from clear how much change societies can tolerate, sometimes with surprising ease, it is likely that societies will be confronted more fundamentally than even now with the actual changes occurring. Life in a degraded environment has an element of misery, but on the other hand human capacity for life in misery is also large. Given such knowledge, even if the priorities were set differently, changing the course of action remains a difficult process because it means changing well-established power structures and cultural preferences.

2. FOUR KEY QUESTIONS ABOUT THE BIOSPHERE AND HUMANS

From the view point of considering the Earth as a system and the evolutionary history of life since its beginnings, a number of questions are interesting and relevant to the global change debate which this paper will briefly consider :

(1) Given that the human species is in the process of greatly altering the Earth's biosphere, can we better understand what the biosphere is as a planetary phenomenon?
(2) Given that change has always been a feature of the biosphere in its co-evolution with the planetary environment into with which it is merged, is its existence and course of development robust or fragile?
(3) Given that humans with their potential to change the world are a product of evolution as other species are, how can their planetary impact be understood from an evolutionary viewpoint?
(4) What (transformed, remnant, managed) nature do reflective modern societies want?

3. THE BIOSPHERE IN SPACE AND TIME

At least in Western thinking, Earth is a special place because it harbours life. Earth is a planet with a biosphere. And that biosphere is almost as old as the planet, it emerged at an early stage.

Was it then inevitable that life occurred on a planet with Earth's properties? Current knowledge does not provide an answer to this question. The largest limitation is that we still know only of one planet in the universe that has a biosphere, our own planet, though it is quite possible that continued advances in the search for extraterrestrial planets will eventually reveal other biospheres. As of today, it is unknown whether life is common in the universe or not.

Given that we have only Earth's biosphere to work with, can anything general be learned from this sample the size of one? The answer is yes only if a very fundamental assumption is made, which is that the Earth's biosphere is a typical example of a biosphere, at least for earth-like planets. Then one can try to cautiously deduce general properties from that only example at hand. If, however, Earth is an oddity in that it has life at all, or in that it is somewhere on the fringes of a probability distribution for earth-like planets and not typical of the average case at all, then there is little of general value that can be deduced from our own planet about the phenomenon of biospheres in general. All substantial arguments therefore have to rest on the assumption that Earth and its biosphere are typical of other biospheres in some way. Of course they may not be.

Assuming, then, that something can be deduced, what is the evidence? On the one hand, the appearance of life on Earth relatively soon after its formation, in fact almost as soon as it was possible, could be taken as an indication that life forms easily, and that, if conditions are right, it will form. Biospheres are then a normal feature at least of Earth-like planets.

On the other hand, as far as we know, and despite of the immense amount of time that has passed since those early days of Earth, life has never since formed again from dead matter, and is not now being formed. Great similarity in the genetic and biochemical structure of extant life forms make it reasonable that they derive from a common ancestor. All living beings have always originated and continue to originate from some form of parents. A possible conclusion from this observation that life only originated once on Earth is that formation of life is not easily achieved (and even if in the near future test-tube life should be assembled from dead ingredients by humans, it would still be a life form, humans, that will have caused the act by consciously arranging a suitable chemical sequence of events). The conclusion would be that a biosphere is not a usual feature of Earth-like planets but rather the product of very special circumstances that are unlikely to be repeated often.

A possible way out of explaining how the most complex chemical phenomenon on Earth originated so early is to assume that life did not in fact originate on Earth but was imported from somewhere else at that early stage [2]. The early timing, if it is not through pure luck, would then, however, indicate that the interstellar transport of life is wide-spread, which would again be in contradiction with the observation that there is no evidence that life has since been re-imported a second or third time (though one should never discount the possibility that future research will come up with some big surprises).

One way out of these dilemmas is to assume that the likelihood of life origination events, whether they are chemical reactions on Earth or imports from space, have a likelihood that is about once in the lifetime of a biosphere. In other words, it is likely to happen once at some point in the history of an earth-like planet, but less likely to happen several times. The early timing would then be luck, but not the phenomenon itself. This is an intellectual path down the middle of the problem. Another possibility is to argue for special conditions present only on early Earth that enabled the particular, still unknown chemical reactions that caused the first transition from abiotic to biotic forms [3]. Early Earth was a place that was very different from the planet at later stages because of radioactive heating, early geologic structuring of Earth, a primordial atmosphere and continued impacts of large and small bodies remaining from the formation of the solar system (as conserved in the structures of the moon).

There is one final ingredient in this mix of questions, and that is the observation that, once formed, the biosphere has shown a remarkable persistence. It has existed for most of the life span of our planet, and has never been eliminated despite the relatively narrow corridor of environmental conditions, particularly of temperature, which organic life as we know it depends on for survival. Several times it may have been close, as during almost global glaciations in Precambrian times [4]. It is the controversial argument of Gaia theory that life itself entered into the planetary system of feedbacks in such a manner as to maintain Earth within the habitable zone of conditions [5]. While Gaia theory puts a strong focus on positive regulation, neglecting somewhat the destructive forces equally at work in cosmic and planetary environments, it is not very convincing to assume that Earth has remained within the habitable corridor for 3.5 billion years simply by chemical chance. Much about the functioning of Earth as a planet with an interactive biosphere remains to be discovered. And again, the required assumption is that Earth is a typical planet with life. Otherwise we may simply be dealing with observer self-selection: our very existence as observers obviously requires that life did not perish earlier, preventing our appearance on the stage. It was Vernadsky, the great Russian biogeochemist, who pointed out that from the viewpoint of empirical generalisation, nothing can be said about the origin of life since not a single observable trace of such an event has ever been found from any period of time (whether modern biochemical and genetic analysis can provide some traces remains to be seen) [6].

Earth is a planet with a biosphere, and it is interesting to note how fundamentally the existence of the biosphere has shaped the planet. Since its origins, it has been a fundamental aspect of Earth system dynamics that life has been present and has co-evolved with its environment. Not only have that the physical and chemical properties of earth been shaped by life but these changes have in turn equally allowed life to evolve in new directions, many of which are now characteristic of the extant biosphere [7,8]. The chemical signature of the Earth's atmosphere, which could be detected from great distances in space by spectral analysis of photons scattered on Earth, reveals that Earth is far removed from the only mild chemical disequilibrium one would expect for a dead planet with Earth's physical and chemical properties [9]. As Lovelock and his co-workers have pointed out, an observer from space would notice the co-existence of methane and oxygen in the Earth's atmosphere, two elements that would quickly react and decay on an abiotic planet. The only possible conclusion would be that a continuous process re-supplies the methane, and would perhaps conclude that it is a metabolic, that is, a biological process. Earth is also notably different from its neighbour planets Mars and Venus in the unusually low concentration of carbon dioxide and high concentration of atmospheric oxygen, which also have biotic explanations (photosynthesis and biologically enhanced rock weathering). The replenishment of numerous environmental constituents by metabolic biological processes causes the presence of these constituents for periods far longer than their abiotic chemical residence times. Earth carries a strong signature of being a planet with life. At the same time, many of the properties of life on Earth, such as the consumption of oxygen by many organisms, cannot be understood without reference to this altered chemical state of the planet.

Overall, these two enigmas remain as we consider the sample of one provided by our own biosphere, that the most complex of chemical appearances on Earth, life, its genetic code and cellular structure have appeared seemingly easily and at a very early stage of planetary development. And that once formed, life has persisted ever since, and shaped its own environment. A complete theory would not just include an understanding of how evolution by selection and feedback controls on units from genes to proteins, cells, traits, organisms and groups of species interactively co-evolves with the environment on all levels, but would also answer such interesting questions as how and why variation arises and is maintained, why short life cycles of organisms are necessary (that is, why we all have to die relatively quickly), and what the functional role, if any, of diversity is within this planetary system of feedbacks, regulations and evolutions.

4. HUMANS A MAJOR TRANSITION

For most of its history, life on Earth consisted of small, single-cell organisms that lived in oceans. Only the most recent 1 billion years (or so) of the 3.5 billion years of life's evolution have been characterised by large multicellular organisms with cell differentiation. Around 630 million years ago, the Ediacaran fauna appeared and left the first large multicellular fossil traces of soft-bodied organisms with sometimes peculiar body plans (their exact interpretation remain a challenge) [10]. Not before 543 million years ago the Cambrian explosion began, a rapid radiation of a large number of macro-bodied species, including ancestors of most that are extant [11,12]. After that expansion, life continued to change in fundamental ways. Plants appeared on land in the Silurian some 475 million years ago, vascular plants about 430 million years ago, and flowering plants, so common today, perhaps 140 million years ago; the dinosaurs perished 65 million years ago. The average life span of a species is estimated to be about 10 million years, though with large variations (a few extant species, the living fossils, have existed in similar form for much longer periods). The extended single-celled biosphere continued to exist, however, though today it receives scant attention in discussions of the biosphere. The biomass of prokaryotes alone has been estimated to rival that of the world's vegetation [13].

The biosphere of Earth, then, was not always what we think of as the biosphere today. For much of its history it had a distinctly different appearance, quantity and quality. Evolution is a strong agent of change when considering geologic time intervals. Though it is far from clear whether the existence of life on a planet should necessarily lead to intelligent life forms (that is, not to ask whether life in the universe is common but whether intelligent life is), from the perspective of evolution the appearance of mentally reflective humans is not an inexplicable phenomenon: substantial innovations were repeatedly made during the course of evolution. For example, the genetic inheritance system was formed and paired with epigenetic inheritance pathways, often involving particular three-dimensional patterns of DNA, RNA and proteins, and the topographical pattern of their many riders [14]. On the other hand, the transition from largely genetic and epigenetic information transmission between generations to symbolic information flow through language in humans is a very large step. The transmission of information has opened new channels of inheritance, an innovation that has occurred only very recently. It is what has been called a major transition, the consequences of which are currently taking on the form of global change.

Major transitions are marked by the emergence of biological properties that were difficult to achieve, either because a number of statistically unlikely events had to occur (for example an unlikely chemical constellation, genetic mutation or environmental constellation), or because a number of uncommon events had to occur simultaneously or in a fixed sequence. A crude mathematical estimate can be derived of the number of major transitions that are required leading to the appearance of humans. Statistically very unlikely can be taken to mean that the transition in question occurs with an e-fold Poisson distribution rate constant of about the lifespan of the biosphere (if it were more frequent, humans should have evolved much earlier; if it were much less frequent, we should not be here, in a statistical sense). This life span is thought to be about 1-1.5 billion years more than the 3.5 billion years that have already passed (that is, most of evolution has already passed, unless life finds ways to leave Earth), after which time the Sun will become too hot for organic life on Earth [15]. One can then compute the expectance value for the number of transitions that have occurred under the (optimistic) assumption that they are independent of each other [16,17]. The surprising answer is that the number of these major transitions is found to have been small: only four. Or, taking into account the crude nature of the derivation, less than ten. Nobody would claim that such a crude estimation resembles the complexity of the actual biochemical and environmental processes of evolution. However, it gives a feeling for the order of magnitude that may be involved. That order of magnitude is not found to be 50, or 100, as one might have expected in view of the great complexity of higher animals and of humans.

Evolutionary biologists and geneticists have proposed a list of such major transitions. It is very interesting to note, that they independently arrive at a very compatible number of only a few major transitions. Smith and Szathmáry [18], for example, list eight. These are the transition from replicating molecules to collections of such molecules enclosed in protocells; the transition from independent replicators to chromosomes; the transition from using the RNA both as gene and as enzyme to using DNA for genes and proteins for enzymes; the transition from prokaryotes to eukaryotes, that is, the emergence of cells with nuclei and organelles; the transition from asexual clones to sexual populations (this transitions is perhaps the least understood of all); the transition from single-celled organisms to multi-celled organisms with cell differentiation (this transition requires a sophisticated network system of gene regulation during embryonic development and subsequent growth); the transition from solitary individuals to groups with social behaviour (as in some insect colonies and mammals ; in insect colonies the hallmark case is the existence of unreproductive casts); and finally, the transition from primate societies to human societies, characterised by the emergence of language.

From this perspective, if the emergence of intelligence and language is taken to be a major transition, it is one of a very small number of fundamental breakthroughs in evolution to something qualitatively new. It is not surprising then that each of these major transition has had a fundamental impact on the biosphere, bringing with it very far-reaching change. It would therefore also not be surprising that humans are now changing the biosphere, though the speed with which this has now begun is certainly, as far as is known, unprecedented (that is, it is also qualitatively new). The co-evolution of geosphere and biosphere that has characterised the past history of Earth has expanded into a triangle that now also includes the anthroposphere (figure 1).

Human history can also be interpreted in terms of a number of breakthroughs that have strongly affected the role of humans in their environment. These could be, for example, the following six transitions [19]: the use of fire, which has allowed forest clearing, cooking, heating and defence; the use of language, which has allowed the building of more complex societies and thought structures through communication; the invention of agriculture, which has allowed more stable food supplies and sedentary life styles, but also a more hierarchical structuring of larger societies; the formation of civilisations, with their cities, states, and political structures, use of mechanical devices and metals; the invention of long-distance travel, which launched globalisation through world-wide exchanges of people, plants and animals, diseases, minerals and other resources while also leading to a loss of indigenous cultures and colonial political systems; and the emergence of science and technology, with their important advances in the fields of hygiene and medicine, mechanised resource extraction and land clearing, travel and production of new materials. With respect to averting global change of a dangerous magnitude, the

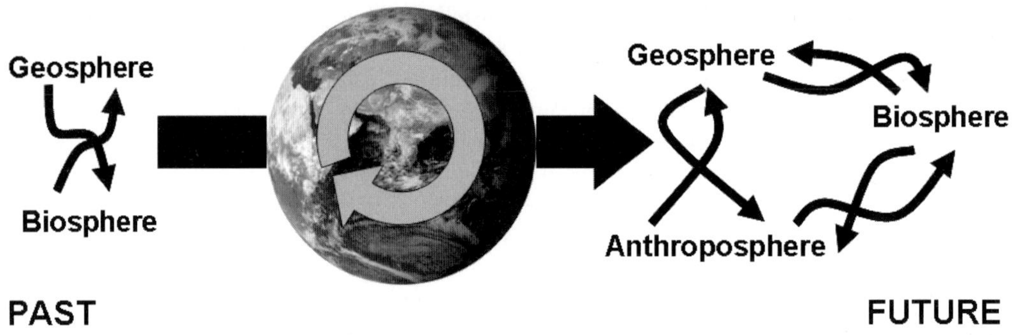

Figure 1. The co-evolution of geosphere and biosphere that has dominated Earth history has been expanded to include a third element, the anthroposphere, which is characterised by a dynamic derived from reflective mental activity.

challenge at hand is now to achieve another transition, one to sustainability. It will have to be the first of these transitions that does not increase the required material and energy use.

On the one hand, the concept of major transitions allows to explain how humans are deeply rooted in the evolution of life, and hence are as much simply a part of nature as are other living beings. On the other hand, it states a qualitative difference between humans and other life forms. The relationship of humans to their living environment has been a central and uneasy theme throughout the history of human thought. In practical culture (such as gardening), philosophy and theology the nature-culture divide, whether nature and culture can be separated and if so, how, and what the implications are for the position of humans in the world, have been a constant source of cultural tension and productivity. The three main flavours [20] are concepts of the controlling power of humans over nature (which can be adapted, rearranged and exploited), a friendly embedding of humans in nature (as in the underpinnings of many environmental movements), or fantasies about a detachment of humans from nature (as in common visions of dematerialisation, spaceships and robotic environments).

From the view point of major transitions, homo sapiens is both just another animal and, as the carrier of symbolic information transmission, a new phenomenon in the biosphere. Its relation to the environment is that of material dependence and of symbolic construction of its perceived niche. The disjunction between these two strands of the human condition, the tension between nature and culture in humans, is causing global change, which is a major blurring of these boundaries that have been constructed by humans in their culture (though not as much in magically and mythically based cultures as in rationalistic cultures). The world that is emerging from this last major transition is an inseparable hybrid of nature and culture.

5. EARTH SYSTEM ANALYSIS

Human societies around the world are affecting their environment, a situation that is now amounting to a significant disturbance of the flows in the global system. Among the causes are population growth, a consequence of improved hygiene, medicine and crop production, and the immense increase of human capabilities through manipulative technology.

Currently, human societies more or less are simply suffering the impacts of global environmental degradation. They strive to deal with the consequences as they occur. In some world regions, landscape planning is an important tool for directing change, though always contentious because of conflicting interests. In other regions landscape processes are less or not controlled, due to a local lack of planning capacity, change driven by dire need and social pressure, or because particular, often short-term interests dominate regional decision-making and a general dominance of purely economic considerations. On a continental and global scale, coordinated management of environmental change through regulatory regimes is still in its infancy and debates about adequate political, economic and social structures that support governance are still very open. A need for more coordinated planetary management, however, is increasingly being recognised and acknowledged.

An important dimension of the problem of global change is that the groups of people that cause environmental degradation are frequently not the ones which suffer the greatest impact. There is a geographically heterogeneous, only partially overlapping distribution of cause and effect. This remains true even though the transformation, for example, of much of the European landscape to industrialised agriculture and forestry is a change that also runs deep. Lovelock [21] called the pre-industrial world achingly beautiful, a term expressing a modern urbanised worldview more than that of a subsistence farmer, who typically has a different relationship to the land, but one that expresses very well the deep roots of the loss. Similar losses are developing everywhere around the world.

The challenge at hand is whether humankind will now be able to acquire some form of systems understanding that will permit the development of management instruments that allow to limit global change to levels that are compatible with continued future societal dependences on natural resources, general opportunities for all humans to not live in misery, and respect for ecosystems and the living beings that form them. In its weakest form, such global management would aim to avoid the most dramatic consequences of global change, for example ice sheet collapses with large sea level rise, substantial alterations of oceanic and atmospheric circulation (such as in the monsoon systems), or advanced deforestation of the tropical forests. In a stronger form, sustainability can be interpreted to include aspects of global equity, maintaining options for future generations, and minimising alterations of the Earth system's evolutionary path. An intermediate option would be one of controlled development with the aim of limiting the rates of change to levels that allow adaptation.

Schellnhuber [22], taking up earlier considerations by Vernadsky [6], Teilhard de Chardin [23] and others and applying them to modern scientific Earth system analysis, global-scale networking and observation, has described the required reflective capacity of humankind as the emergence of a collectively active global subject. The pre-modern world was governed by natural forces acting on nature and societies, and the modern world is characterised by both natural and human forces acting jointly on both the environment and societies. A future auto-evolving world in which human actions are not blind to the implied systemic pathways of the natural and societal subsystems of the Earth system would then be marked by the additional influence of this global subject, which selects trajectories of development from the catalogue of projected options, each qualified by associated uncertainties, on the natural and human systems. At issue is whether the nature-culture hybrid produced by human action is a hybrid of accident or a hybrid of conscious action.

The emergence of such a global reflective mental agent requires global observation, analysis and projection tools as well as methods of intercultural dialogue and mediation of action. These tools of Earth system analysis are global macroscopic observation systems, manipulative field experiments, and computer simulation models of intermediate complexity that allow integrated assessments [22]. Together they form the basis of the emerging science of Earth system analysis. Macroscopic observation systems have been or are being implemented, though beyond the classic geoscientific observations and socioeconomic statistics large gaps remain in ecological, political and cultural observation, which will have to be closed through advanced global sampling strategies and qualitative empirical social research to allow a discursive advancement of ecological and social theories and models [24]. Computer modelling faces the challenge of finding appropriate scales and methods of integration across diverse fields of interacting factors.

Scientific analysis is one mode of intercultural dialogue about global change that is widely accepted. However, understanding the Earth system does not automatically imply appropriate action. Rather, generalised knowledge from Earth system analysis has to be contextualised in concrete regional situations to be meaningful [25]. A step of reinsertion of global knowledge into local contexts is required. The resulting fields of meaning fuse with questions of identity and images of the self to produce the drivers of political representation and action. It is nothing less than putting the old slogan of thinking globally and acting locally into practice. Earth system analysis as a science plays an important role in this process.

6. THE FUTURE OF THE BIOSPHERE

The Earth's biosphere today is under the dual pressure of climate change and land use change. Both of these will cause large-scale reorganisations of its species, diversity and functioning. On the one hand climate change, caused by human emissions of gases with greenhouse effects into the atmosphere and by deforestation, affects the environmental conditions under which plants operate their physiological processes, causing complex systemic shifts ranging from changes in biochemical rates to alterations of ecosystemic dynamics. Fauna depends on the primary production of this vegetation through the trophic chains. On the other there is extensive human land use, which is still expanding and is converting forests into increasingly industrialised agriculture. In the process, remaining natural or semi-natural landscapes are being fragmented or simply replaced.

Numerical simulations of the climate system driven by human emissions of greenhouse gases into the atmosphere show that, unless these emissions are substantially reduced from current levels in the next two to three decades, it will not be possible to limit global warming to a temperature increase of less than 2°C. Temperatures may rise by 3 to 5°C or more if fossil fuel burning remains the dominant form of human energy production and is not replaced by different technologies of energy production. Such a temperature increase would likely translate to a warming of up to 10°C in some regions, particularly in the high northern latitudes, because warming over land is larger than over the oceans and is geographically not uniformly distributed.

What is more, water availability rather than temperature is the dominant control of most of the biosphere (the exception are the arctic and boreal zones). While current climate models agree that overall precipitation will increase in a warmer greenhouse future, differences between models are still uncomfortably large in important geographical regions, such as in Africa and Southern America, with respect to the spatial distributions of increase and decrease. These patterns depend on changes in atmospheric circulation and are affected by feedbacks from the land surface.

The consequences of climate change for the spatial organisation of the global biosphere in terms of functional differences in broad groups of vegetation types can be simulated with Dynamic Global Vegetation Models (DGVMs) [26]. The LPJ-DGVM [27] is a global biogeochemical-biogeographical simulation model that transforms time series of monthly temperature, precipitation, radiation data, and yearly atmospheric concentration of carbon dioxide, into spatially explicit estimates of land surface carbon and water fluxes, and the fractional composition of vegetation by type. Ten plant functional types are differentiated by the LPJ-DGVM according to their life form (woody, herbaceous), seasonality (evergreen, deciduous), leaf morphology (broadleaf, needleleaf), photosynthetic pathway (C3, C4) and climatic adaptation (boreal, temperature, tropical). Each functional type coexists at each global location with others depending on their competition for resources (space, light, water), growth (net primary production) and interaction with disturbance (fire frequency and extent). Plant responses to climate are simulated through numerical process descriptions of photosynthesis, stomatal coupling of plant carbon and water fluxes, and carbon allocation as a function of allometric and functional constraints. The model has been extensively validated in terms of large-scale vegetation distribution, variability, carbon and water fluxes.

In a climate change simulation with the HadCM3 climate model, in which the atmospheric concentration of carbon dioxide increases to 856 ppm by the year 2100 (IPCC SRES-A2 scenario), average global temperature over land increases by 5.3°C. The underlying emission scenario is one of an economically oriented world with regional differences in developments, close to a business-as-usual scenario. As a consequence of the simulated climate change, the spatial distribution of macro-level functional types of global vegetation as computed by the LPJ-DGVM are projected to undergo

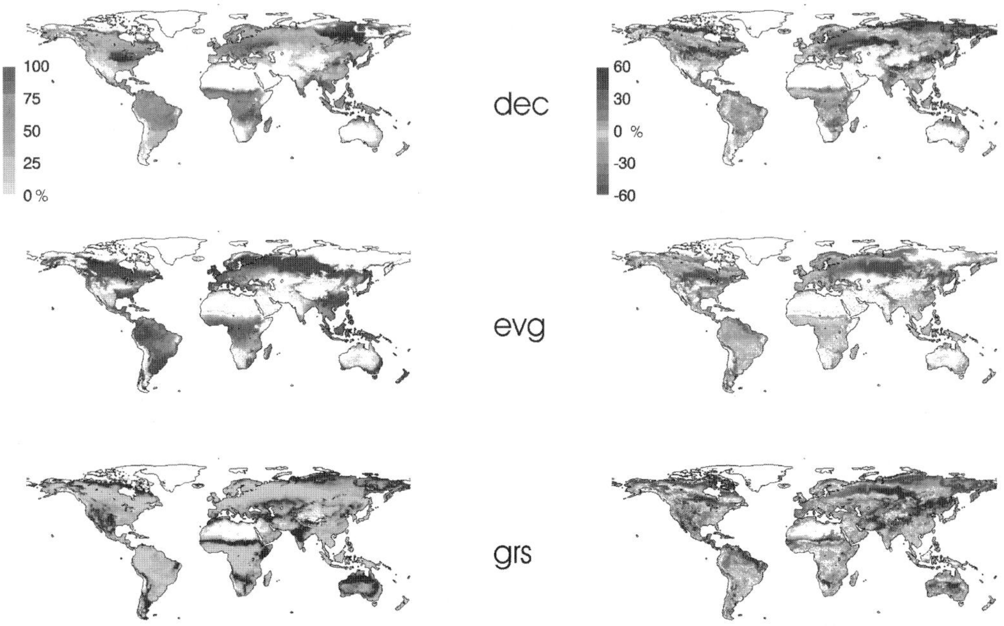

Figure 2. Composition of the terrestrial biosphere in the year 2000 in terms of major functional differences (deciduous woody, evergreen woody, herbaceous/grasses), left panel, and changes in that composition from the year 2000 to the year 2100, right panel. Simulations with the LPJ Dynamic Global Vegetation Model using a HadCM3 IPCC SRES A2 climate change scenario, in which the atmospheric concentration of carbon dioxide reaches 856 ppm in the year 2100 and the global mean temperature over land increases by 5.3°C.

substantial changes by the year 2100 (figure 2) [28]. In the high northern latitudes, tundra is replaced by woody vegetation as warming permits the treeline to move northward. Tundra expands into previously unvegetated northern Arctic fringes. In the boreal zone there is a noticeable shift away from evergreen to deciduous vegetation. In some boreal areas, forest dieback occurs and grasslands appears. The reason is increased mortality of trees due to heat stress, which also causes an interruption of reproduction. Boreal tree species are not replaced by temperate species because winter temperatures are still too low and temperate species suffer from frost damages.

In south America, the Amazonian rainforest does not collapse as predicted by some models but is under pressure, and suffers from increased fire activity in its eastern sections. In semiarid areas in the US, in the Sahel, Australia, the Middle East, Central Asia and elsewhere, increased water use efficiency of vegetation due to reduced stomatal opening under higher ambient carbon dioxide concentration causes increased vegetation cover. In Africa, a slight general shift toward more deciduousness is found in these simulations, and woody encroachment in some areas that currently are savannahs, particularly in the eastern parts of the continent. In the southeastern US and in eastern China, evergreen vegetation advances its fractional presence, while in India, deciduous woody vegetation is on the rise (though most land in India is agricultural so that these modelled shifts are not reflective of the actual landscape).

Overall, shifts in vegetation composition are projected on all continents. Such shifts are associated with substantial changes in the underlying ecosystems. The biosphere will undergo substantial transformations and spatial reorganisations as a consequence of strong global warming. Human action is the cause of this change, which through anthropogenic changes in the atmosphere reach all parts of global vegetation. The degree to which vegetation will be able to adapt to such change is insufficiently known at present, but close relationships between climate and vegetation composition has long been established in biogeography, so such shifts do not come as a surprise. Pollen records also show that vegetation composition at most sites has fluctuated in the last millennia, reflecting much smaller changes in climate. A factor is also the rapidity of this projected climate change, which will make it more difficult for ecosystems to adapt. Important transitional effects can be expected. A literature survey of expected climate change impacts on a wide range of ecosystems shows that some are predicted to already be in peril with just 1°C of global warming, and that most will be at 2-3°C of warming [29]. From this analysis, the ecological change will be profound if warming is not limited to 2°C. Current warming over pre-industrial levels is 0.8°C. The role of biodiversity in maintaining ecosystem stability and functioning has been discussed controversially, but is unlikely to be a marginal factor [30].

A brief look at the implications of these shifts for biogeochemical cycles in the Earth system is interesting. Studies show that stimulated growth of the terrestrial biosphere due to carbon fertilisation and slight warming presently is the main cause of net carbon sequestration from the atmosphere into the biosphere. However, this land carbon uptake, which currently contributes to a slowing of global warming (though it is being strongly countered by concurrent deforestation), will not increase in step with global warming and rising carbon dioxide concentration in the atmosphere. Rather, after an initial increase, the carbon sequestration potential of the terrestrial biosphere will flatten out or decline [31]. In some scenarios, the terrestrial biosphere switches from being a sink to being a source of atmospheric carbon. The main reason is an increase in the decomposition of organic matter in soils, particularly in temperate and boreal regions, where soil carbon stocks are high. This increase outpaces the concurrently observed increase in global net primary production. It is quite certain, therefore, that a positive feedback operates: global warming, through time-lagged effects on the terrestrial biosphere, will be enhanced due to shifts in the land carbon balance toward decreased carbon uptake or even net carbon emissions. Studies from coupled climate-biosphere models show that atmospheric carbon dioxide concentration will increase by between 20 and 200 ppm due to terrestrial feedbacks, causing an additional increase in temperature of between 0.1 K and 1.5 K [32].

Average water stress is not projected to increase substantially in many areas because increased transpirational water demand by vegetation due to higher temperatures in countered by increased water use efficiency due to increased atmospheric carbon dioxide concentration. However, changes in vegetation density, cover and composition also contribute to this overall hydrologic resilience [33].

The broad functional composition of global vegetation is here taken as a top-level indicator of change to be expected. The simulations show that change will be global. Two other cases of projected change in the biosphere serve to highlight anthropogenic biosphere change (a phrase that should come into use as much as the word climate change) from two other perspectives: that of fauna and that of the oceans. A study of extinction risk in terrestrial mammals due to human pressure shows high current extinction risk on the South American Pacific coast, in the Caribbean, in Central Asia and in Madagascar to expand greatly in the next decades to also include large parts of southeastern Asia, southern Africa and the Arctic, with potential threats from geographical shifts in human activity in Northern America and India [34]. Life in the oceans is threatened not only by overfishing, pollution and global warming, which will lead to increased stratification of the oceans, but also by the current increase in atmospheric carbon dioxide concentration because of its direct effects on ocean acidity and the ability of marine organisms such as plankton to form the calcium carbonate shells in which they depend. By 2100, undersaturation of southern oceans in aragonite are projected to lead to dissolution effects on important marine organisms, as was also demonstrated in laboratory experiments [35].

But the biosphere is not just in danger from climate change. Even without climate change, it would change considerably in the next decades due to non-climatic global change. The area of land used for agricultural production continues to expand as it has throughout the 20th century, during which it increased from less than 12 million km^2 to more than 18 million km^2. The high rates of global deforestation, particularly in the tropics (as temperate latitudes have already been deforested in earlier periods), lead to a substantial loss of terrestrial carbon and biodiversity. Slowly, the planet is turning into a planet dominated by grasses, as on pastures and in the form of crops on the fields. Land use change has been projected to be a more important driver of biodiversity than climate change [1]. Estimates of the current global human appropriation of net primary production amount to 22 percent, about half of which is due to changes in vegetation type due to human alteration (forest replaced by crops), and half by harvesting of the resulting biomass products [36].

Global vegetation modelling is just beginning to meet the challenges of also modelling global agricultural activity within the same frameworks in which natural vegetation is being modelled. The LPJ-DGVM, for example, has been expanded to include crop production and grazing, including effects such as harvesting, irrigation, residue use, multiple cropping or crop rotation (LPJmL model) [37]. Global forestry will subsequently also be implemented. One of the largest changes in the future dynamics of human land use is a potentially greatly expanded use of biomass, which as a carbon-neutral form of energy provision is seen by many as an important element in a global climate protection strategy. However, producing a projected 200-300 EJ of bioenergy per year will require to double current global harvest (present global primary energy consumption is about 400 EJ/year, of which 35-55 EJ/year are in the form of bioenergy [38]). Whether the global puzzle consisting of increased food production for an increased global population of 9-10 billion people, considerable biodiversity protection under the pressure of climate change, population growth and agricultural expansion, and such bioenergy production are possible, remains a topic that urgently requires a close quantitative assessment. The impacts of bioenergy production of the global biosphere could be as large as those of other human land use or climate change, given its impact on space, carbon and water fluxes. Such questions belong to the core of what a science of Earth system analysis will have to address.

In summary, the next century, which is about the life span of children born today, will see wide-ranging changes in the global biosphere. The human species is an agent of unprecedented biospheric change, which will leave its mark on the history of life and therefore of Earth as a planet. On an evolutionary time scale, the activity of humans is likely to be less of a factor. Based on paleoclimatic evidence, a strong global warming pulse is expected to decline over 100000 to 200000 years. Biodiversity is likely to recover on evolutionary time scales, as it has a number of times after large extinction events. The basic genetic inventions characterising current life on Earth are shared by very large groups of organisms that differ more in their particular expression of these basic principles than in their fundamental makeup. The genetic loss will therefore be limited, though biodiversity in itself is a value to humans that should not be treated lightly. Overall, it is the human species itself that stands to loose through biosphere change essential elements of its environment, with a danger of having to live in less attractive circumstances.

From an evolutionary perspective, the ultimate outcome of the current change cannot be foreseen. It is possible that humans will decline rather earlier than later, at least in the magnitude of their numbers, and perhaps as a consequence of environmental changes. It is equally possible that paths into the future will be found and that we currently are only at the beginning of a far larger development, continuing the theme of a major transition that has yet to reveal its full implication. The innovation of this transition is symbolic information inheritance through language. This is a form of information transition that is separate from the genetic and epigenetic streams of information that otherwise are dominant in the world. The detachment of information transmission from molecular matter would, if carried forward with consequence, imply that self-replicating machines forming networks of communication might be the next direction of evolution. Humans would have been the evolutionary agents in the emergence of such inforobots. It is interesting to note that the principles of natural selection would be fully functioning on such machines, that is, they would undergo evolutionary development. The public arena is today full of fantasies about hybrid beings half organism and half machine, of humans with integrated technical elements and of animated, half-alive robotic beings. On balance, though, it seems much more likely that the influence of human tampering with genes and proteins will alter the future course of evolution, though very possibly in unforeseen and unplanned ways.

Estimates are that about 6% of all individuals of homo sapiens that ever lived are alive, a very high fraction but far from the majority [39]. It is in the hands of these individuals to develop a programme of Earth system analysis that will allow out species to fit into the global system in some form that is sustainable, that is, not detrimental to our own well-being nor to that of the majority of other organisms that evolution has produced. The biosphere of planet Earth is unique, most certainly to ourselves.

Life is a process of continued internal differentiation of the Earth system, with an element of the rolling-up of life onto itself that Teilhard de Chardin observed [22]. What is required today, then, is an integrated view of humans and Earth.

Acknowledgments

I am deeply grateful for the support I have received and the many essential discussions I have had with John Schellnhuber, Ottmar Edenhofer, Tim Lenton, Wolfgang Cramer, Dieter Gerten, Tim Erbrecht and Carlo Jaeger. I thank the students who participated in my seminars on earth system analysis and geosphere-biosphere-anthroposphere co-evolution in the past several years at the Institute of Geoecology of Potsdam University, and the members of my research group at the Potsdam Institute for Climate Impact Research. Special thanks to Sibyll Schaphoff for LPJ calculations and the preparation of figure 2. I thank the ERCA winter school for having invited me to present this talk on a wintery night in Grenoble in January 2006. I acknowledge funding from the Leibniz Pakt für Forschung project on the biosphere and societies under global change.

References

[1] Sala O.E., Chapin F.S., Armesto J.J., Berlow E., Bloomfield J., Dirzo R., Huber-Sanwald E., Huenneke L.F., Jackson R.B., Kinzig A., Leemans R., Lodge D.M., Mooney H.A., Oesterheld M., Poff N.L., Sykes M.T., Walker B.H., Walker M., and Wall D.H., *Science* 287 (2000), 1170-1774.

[2] Napier W.M., *Mon. Not. Roy. Astr. Soc.* **348** (2004), 46-51.

[3] Szathmáry E., *Nature* **387** (1997) 662-663.

[4] Hoffman P.F. and Schrag D.P., *Terra Nova* **14** (2002), 129-155.

[5] Lovelock, J., Gaia: a new look at life on Earth (Oxford Univ. Press, Oxford, 1979).

[6] Vernadsky V.I., The biosphere (Springer, New York, 1991 (1926)).

[7] Lenton T.M., Schellnhuber H.-J. and Szathmáry E., *Nature* **431** (2004), 913.

[8] von Bloh W., Bounama C. and Franck S., *Geophys. Res. Lett.* **30** (2003), 1963 doi:10.1029/2003GL017928.

[9] Lenton T.M., *Nature* **394** (1998), 439-447.

[10] Narbonne G.M., *Ann. Rev. Earth Planet. Sci.* **33** (2005), 421-442.

[11] Valentine J.W., Jablonski D. and Erwin D.H., *Development* **126** (1999), 851-859.

[12] Morris S.C., *Current Biology* **7** (1997), R71-R74.

[13] Whitman W.B., Coleman D.C. and Wiebe W.J., *Proc. Natl. Acad. Sci. USA* **95** (1998), 6578-6583.

[14] Jablonka E. and Lamb M.J., Evolution in four dimensions (MIT Press, Cambridge MA, 2005).

[15] Franck S., Bounama C., and von Bloh, W., *Biogeosciences* **3** (2006), 85-92.

[16] Feoli A. and Rampone S., *arXi:* **gr-qc**/9812093 (1998).

[17] Carter B., *Phil. Trans. R. Soc. Lond. A* **310** (1983), 347-363.

[18] Smith J.M. and Szathmáry E., The major transitions in evolution (Oxford Univ. Press, Oxford, 1995).

[19] Takács-Sánta A., *Human Ecol. Rev.* **11** (2004), 51-66.

[20] Schellnhuber H.-J., Crutzen P.J., Clark W.C. and Hunt J., *Environment* **47** (2005), 10-25.

[21] Lovelock J, The revenge of Gaia (Penguin, London, 2006), p. 56.

[22] Schellnhuber H.-J., *Nature* **402 Suppl** (1999), C19-C23.

[23] Teilhard de Chardin P., The phenomenon of man (Collins, London, 1959 (1955)).

[24] Lucht, W. and Jaeger C.C., in: Contributions to Global Change Research: A Report by the German National Committee on Global Change Research (NKGCF, Bonn , 2001), 138-144.

[25] Lucht W. and Pachauri R.K., in: Earth System Analysis for Sustainability (MIT Press, Cambridge, MA, 2004), pp. 341-365.
[26] Prentice I.C., Bondeau A., Cramer W., Harrison S.P., Hickler T., Lucht W., Sitch S., Smith B. and Sykes M.T., in: Terrestrial Ecosystems in a Changing World, Canadell, J., Pataki D.E. and Pitelka L.F. Eds. (Springer, Berlin, 2006).
[27] Sitch S., Smith B., Prentice I.C., Arneth A., Bondeau A., Cramer W., Kaplan J.O., Levis S., Lucht W., Sykes M.T., Thonicke K. and Venevsky S., *Global Change Biol.* **9** (2003) 161-185.
[28] Lucht W., Schaphoff S., Erbrecht T., Heyder U. and Cramer W., *Carbon Balance and Management*, **1** (2006), 6, doi:10.1186/1750-0680-1-6.
[29] Hare W., Assessment of Knowledge on Impacts of Climate Change – Contribution to the Specification of Art. 2 of the UNFCCC (Scientifc Advisory Board on Global Environmental Change of the German Government WBGU, Berlin, 2003).
[30] Loreau M., Naeem S., Inchausti P., Bengtsson J., Grime J.P., Hector A., Hooper D.U., Huston M.A., Raffaelli D., Schmid B., Tilman D. and Wardle DA, *Science* **294** (2001), 804-808.
[31] Schaphoff S., Lucht W., Gerten D., Sitch S., Cramer W. and Prentice I.C., *Climatic Change* (2006) doi:10.1007/s10584-005-9002-5.
[32] Friedlingstein P., Cox P., Betts R., Bopp L., von Bloh W., Brovkin V., Cadule P., Doney S., Eby M., Fung I., Bala G., John J., Jones C., Joos F., Kato T., Kawamiya M., Knorr W., Lindsay K., Matthews H.D., Raddatz T., Rayner P., Reick C., Roeckner E., Schnitzler K.-G., Schnur R., Strassmann K., Weaver A.J., Yoshikawa C., Zeng N, *J. Clim* **19** (2006), 3337-3353.
[33] Gerten D., Lucht W., Schaphoff S., Cramer W., Hickler T. and Wagner W., *Geophys. Res. Lett.* **32** (2005) L21408, doi: 10.1029/2005GL024247.
[34] Cardillo M., Mace G.M., Gittleman J.L. and Purvis A., *Proc. Nat. Acad. Sci. USA* **103** (2006), doi: 10.1073/pnas.0510541103, 4157-4161.
[35] Orr J.C., Fabry V.J., Aumont O., Bopp L., Doney S.C., Feely R.A., Gnanadesikan A., Gruber N., Ishida A., Joos F., Key R.M., Lindsay K., Maier-Reimer E., Matear R., Monfray P., Mouchet A., Najjar R.G., Plattner G.-K., Rodgers K.B., Sabine C.L., Sarmiento J.L., Schlitzer R., Slater R.D., Totterdell1 I.J., Weirig M.-F., Yamanaka Y. and Yool A., *Nature* **437** (2005), 681-686.
[36] Haberl H., Erb K.-H., Krausmann F., Gaube V., Bondeau A., Plutzar C., Gingrich S., Lucht W., Fischer-Kowalski, M., *Proc. Nat. Acad. Sci. USA*, submitted.
[37] Bondeau A., Smith P., Zähle S., Schaphoff S., Lucht W., Cramer W., Gerten D., Lotze-Campen H., Müller C., Reichstein M. and B. Smith, *Global Change Biol.*, in press.
[38] Berndes G., Hoogwijk M. and van den Broek R., *Biomass and Bioenergy* **25** (2003), 1-28.
[39] Haub C., *Population Today* **Nov/Dec2002** (2002), 2-3.

J. Phys. IV France **139** (2006) 157–166
© EDP Sciences, Les Ulis
DOI: 10.1051/jp4:2006139011

Atmospheric electromagnetics and climate change

M. Fullekrug[1]

[1] *Center for Space, Atmospheric and Oceanic Science, University of Bath, Bath, UK*

Abstract. Clouds are the largest uncertainty in future projections of climate. This critically important uncertainty is addressed by studying the role of atmospheric electromagnetics in cloud initiation. Cloud droplets are produced by cloud condensation nuclei. However, the formation of cloud condensation nuclei ultimately relies on physical mechanisms on the atomic and molecular scale which influence the meta-stable phase transition between condensation and evaporation. At these small spatial scales, electrical forces are important. A small yet non-zero influence of electric fields and energetic charged particles on the formation of cloud condensation nuclei would have important consequences for climate change, as small changes in cloud initiation can lead to large changes in global temperature. The current knowledge on the role of atmospheric electromagnetics and energetic charged particles in cloud initiation and climate change is summarised and recommendations for future work are proposed.

1. INTRODUCTION

It is well beyond any doubt that the Earth's atmosphere is currently warming (Watson, 2001; Hansen & Lebedeff, 1987). However, the physical cause of this warming is contentious (Gray *et al.*, 2005). Climate change during the last century could result from an increase in greenhouse gases (Lindzen, 1990; Hansen *et al.*, 1981) or solar variability (Reid, 2000; Friis-Christensen, 2000). The most important greenhouse gas is water vapour which keeps the Earth's atmosphere at an average temperature of $\sim 14^\circ$ C. Another greenhouse gas is carbon dioxide the abundance of which is steadily increasing within the Earth's atmosphere as a result of fossil fuel burning by man kind (Lindzen, 1990; Hansen *et al.*, 1981). The Sun's variable short wave (UV) radiation is absorbed by stratospheric ozone and the resulting heat may be advected into the troposphere (Baldwin *et al.*, 2003; Haigh, 1996). The earthward flux of energetic charged particles from the Sun and other celestial sources may have an effect on cloud initiation through aerosols (Harrison & Stephenson, 2006; Marsh & Svensmark, 2000; Svensmark, 1998).

The Earth's atmosphere exhibits at a very high level of complexity in terms of composition and behaviour. It is therefore very likely that several of the above mentioned physical mechanisms or even yet unknown mechanisms simultaneously affect climate change within the Earth's atmosphere. The ranking of the relative importance of the various mechanisms is a major undertaking of the scientific community which requires a high level of understanding of the underlying physics before an educated assessment can be made (Watson, 2001). For example, the United Nations Intergovernmental Panel on Climate Change specifies that the level of scientific understanding of solar influences on the Earth's climate is very low. In fact, the solar electromagnetic (black body) radiation, i.e., the energy output from the Sun, is practically constant on climate time scales $\geq$ 1 month (Willson & Hudson, 1988), whilst numerous climate proxy records suggest an influence of solar variability on the atmosphere on decadal and centennial time scales (Lean & Rind, 1999; Lockwood *et al.*, 1999). This contribution summarises the current knowledge on the role of atmospheric electromagnetics and energetic charged particles in climate change since these physical mechanisms have the potential to mediate solar activity into climate change.

2. ELECTROMAGNETIC FIELDS IN THE ATMOSPHERE

2.1 Convection of the atmosphere

The Earth's orbit around the Sun causes differential heating between geographical latitudes caused by the constant solar radiation. This heat is dissipated to other forms of energy through conduction, radiation, and convection which drives the global circulation of the atmosphere. Fair weather convection lifts air parcels beyond the condensation level at some km height to form clouds. Deep convection lifts air parcels up to the tropopause, where $\sim$ 10 km height to form thunderstorms. The strong updrafts within convective thunderstorms transport water vapour up to the tropause, where the water vapour spreads out over large areas and subsequently modifies the natural water vapour greenhouse effect within a feedback loop affecting climate change (Price, 2000). Whether this feedback is positive and increases the average temperature of the atmosphere remains to be determined.

2.2 Charging of the atmosphere

In the mixed phase region of thunderclouds at $\sim$ 5 km height, the collisions of riming graupel and ice crystals result in a charge separation which produces strong electric fields (Saunders *et al.*, 1991; Takahashi, 1978). The negative charges are bound to the graupel which remains in the mixed phase region as a result of gravitational forces, while the ice crystals carry positive charge upwards to the amboss-shaped thunderstorm anvil (Rakov & Uman, 2003; MacGorman & Rust, 1998). The resulting charge configuration corresponds to an electric dipole which drives an electrical current through the weakly conductive atmosphere up to the ionosphere at $\sim$ 100 km height (Chalmers, 1967), where the current is modified by ionospheric variability resulting from space weather. The current subsequently maps down in the fair-weather regions of the Earth, i.e., the oceans and the polar regions. Finally, the current flows back along the Earth's surface towards the thunderstorm areas to form a closed current loop, denoted the global atmospheric electric circuit (Uman, 1974). The electromagnetic fields associated with the global atmospheric electric circuit are continuously maintained by the global thunderstorm activity and they are hence quasi-static in nature (Bering *et al.*, 1998; Hoffmann, 1923). The observed quasi-static electric fields from thunderstorms are closely linked to the electrodynamic fields resulting from intense lightning discharges (Füllekrug *et al.*, 1999; Holzer & Deal, 1956) which occur during numerous thunderstorms around the globe, i.e., the global lightning activity (Christian *et al.*, 2003; Füllekrug & Constable, 2000; Brooks, 1925). Yet, the direct effect of giant lightning discharges (super bolts) on the global atmospheric electric circuit remains to be discovered.

2.3 Fair-weather cloud initiation

The global atmospheric electric circuit is closely linked to climate change (Rycroft *et al.*, 2000; Bering, 1995; Williams, 1992; Markson, 1978; Markson, 1986; Markson & Muir, 1980). In the fair weather regions, the global circuit (air-Earth) current acts upon the condensation level of the atmosphere where fair-weather clouds form. Two different physical mechanisms have been proposed to assist fair-weather cloud formation. Both mechanisms require the presence of energetic charged particles to produce ions in the atmosphere, but only the latter mechanism requires the air-Earth current maintained by global thunderstorm activity.

2.3.1 Ion-aerosol clear-air mechanism

The first mechanism is the ion-aerosol clear-air mechanism caused by ion-growth (Tinsley & Yu, 2004; Yu & Turco, 2001). Energetic charged particles produce ions in the atmosphere. The presence of ions assists the growth of aerosol particles up to ultrafine condensation nuclei ($\sim$1-2 nm). A fraction of these stable particles further grow to cloud condensation nuclei ($\sim$100 nm), which finally form cloud droplets (~10-20 μm) (Gray *et al.*, 2005; Carslaw *et al.*, 2002). This clear-air mechanism requires the growth of aerosols over many orders of magnitude. Experimental observations of large cluster ions in the atmosphere are indicative of the presence of ultrafine condensation nuclei in the atmosphere (Eichkorn *et al.*, 2002). Yet, it remains to be seen if the ion growth over many orders of magnitudes can be observed in the laboratory and within the Earth's atmosphere.

2.3.2 Ion-aerosol near-cloud mechanism

The second mechanism is the ion-aerosol near-cloud mechanism caused by the air-Earth current of the global atmospheric electric circuit. The air-Earth current produces charge layers at the boundary of fair-weather clouds. The air-Earth current pulls the ions inside the fair-weather cloud, where they charge aerosols and subsequently act as cloud condensation nuclei (Gray *et al.*, 2005; Harrison & Carslaw, 2003; Carslaw *et al.*, 2002; Tinsley, 2000). When the charged aerosol particles contact supercooled droplets, spontaneous ice formation ('electro-freezing') occurs (Tripathi & Harrison, 2002; Tinsely *et al.*, 2000). The phase transition from liquid to ice modifies the radiative properties of clouds. The effect of these modified radiative properties of the clouds on climate change remains to be determined.

The time varying atmospheric electric field changes induced by short lived solar activity (solar bursts) and the cross polar-cap potential are on the order of ~ 10 V/m (Farrell & Desch, 2002; Corney *et al.*, 2003) and the associated changes of the air-Earth current have the potential to mediate solar activity into climate change. The quasi-static electric field resulting from the global thunderstorm activity is on the order of ~ 100 V/m. This fact means that variations in the global thunderstorm activity ~ 10 % can conceal the expected effect of ion-aerosol near-cloud assisted cloud initiation. It is therefore of critical importance to perform simultaneous control measurements of the global thunderstorm activity (ICSU, 1986; Holzworth & Volland, 1986) to ensure that no competing changes of the global thunderstorm activity did happen. Simultaneous measurements of the cloud content, the fair-weather air-Earth current and the global lightning activity are a prime target of future research.

3. ENERGETIC CHARGED PARTICLES IN THE ATMOSPHERE

The Earth's atmosphere is populated by a rich variety of energetic charged particles which can contribute to the ion-aerosol clear-air mechanism and the ion-aerosol near-cloud mechanism by ionising air molecules and/or by modifying the air-Earth current of the global atmospheric electric circuit. (1) Top down particles: Solar particles impinge on the top of the atmosphere and ionise the ambient air molecules. Soft particles travel down to ~70-80 km height whilst relativistic particles reach down to ~40-50 km height (e.g., Baker *et al.*, 1993, Sentman, 1990, and references therein). The particles ionise the air and thereby modify the sub-ionospheric propagation of radio waves. Magnetars are the source of strong γ-ray bursts, which produce short-lived ionospheric disturbances with oscillation periods of some seconds lasting for some minutes (Inan *et al.*, 1999; Fishman & Inan, 1988). Ultra-high energy cosmic rays produce cascades of particles in the atmosphere which produce bursts of electromagnetic radiation for tens of nanoseconds. (e.g., Falcke *et al.*, 2005, Huege & Falcke, 2004, and references therein). (2) Bottom up particles: Lightning discharges during thunderstorms occasionally produce X-rays, where small scale electric fields inside the thundercloud act as mini particle accelerators (Dwyer *et al.*, 2003; Dwyer, 2003). The large scale electric fields of lightning discharges act as maxi particle accelerators for electrons which propagate along the geomagnetic field lines into near-Earth space (Inan, 2005). The accelerated electrons may produce an exotic kind of lightning above thunderclouds termed gigantic jets, blue jets and sprites (Füllekrug *et al.*, 2006; Su *et al.*, 2003; Neubert, 2003; Pasko *et al.*, 2002; Fishman *et al.*, 1994; Sentman & Wescott, 1993; Franz *et al.*, 1990). Whether terrestrial γ-ray bursts observed in near-Earth space are associated with regular lightning discharges or sprites is the topic of ongoing research (Dwyer & Smith, 2005). (3) Strange particle: The photon has an impulse but seemingly no mass. Classical electromagnetism is described with Maxwell's equations with a zero rest mass of the photon. Proca's equations describe electromagnetism with a small, but finite rest mass of the photon (Tu *et al.*, 2005). Which of the two theories describes electromagnetism correctly is not known. The Earth's atmosphere is a giant laboratory which is ideally suited to distinguish between the two theories. The comparison of large scale electromagnetic fields in the Earth's atmosphere with their theoretical description offers a unique opportunity to place a new upper limit on the photon rest mass (Füllekrug, 2004) and the atmospheric photon gas may be used to monitor the thermal radiation of the Earth's atmosphere to monitor global climate change.

3.1 Top down

Protons with energies up to $\sim$100 MeV are accelerated in the solar corona and their occurrence is often associated with solar flares. Proton emissions can last for many days, while the bursts of optical and X-ray radiation from solar flares usually last for less than one hour. The high-energy proton flux is often associated with the emission of electrons with energies $>$1 MeV following the flare. Another class of high-energy electrons is accelerated in the co-rotating interaction regions of the solar wind. These high-energy electrons preferentially occur with a periodicity of 26-28 days during minimum solar activity. No proton flux is associated with these high-energy electrons (Schlegel & Füllekrug, 1999). The proton and electron flux with energies $>$1 MeV is continuously monitored on board the space environment system on the GOES satellites. The emitted particles travel through interplanetary space and penetrate deep into the Earth's atmosphere where they ionise the air molecules. This ionisation increases the conductivity of the atmosphere by an order of magnitude, superposed on the background ionisation from solar short wave (UV) radiation and soft (keV) particles accelerated in the Earth's magnetosphere (Baker *et al.*, 1993; Sentman, 1990). The increase of the atmospheric ionisation results in an absorption of radio waves which are produced by naturally occurring lightning discharges and man-made transmitters. The search for celestial sources of ionospheric disturbances resulted in the detection of an extremely intense hard X-ray/γ-ray flare from a soft gamma ray repeater located some thousands of light years away (Inan *et al.*, 1999; Fishman & Inan, 1988). The illumination of the terrestrial night side by this neutron star led to an increase of the atmospheric ionisation up to daytime levels at 60 km height. This ionisation exhibits a 5.16 s long cyclic variability which coincides exactly with the rotation period of the neutron star. Unknown celestial objects are the source of ultra-high energy cosmic rays with energies $> 10^{17}$ eV (Nagano & Watson, 2000). These cosmic rays propagate straight through the solar system, while low energy galactic cosmic rays $\sim 10^9$ eV are deflected by the heliosphere or guided by the magnetic lines of the Earth's magnetosphere to high latitudes as a result of their beta value and the energy dependence of their penumbra (Smart *et al.*, 2000). The largest cosmic ray energies observed to date are on the order of $\sim 10^{20}$ eV, i.e., some Joule, and occur sporadically, i.e., about one incidence per century per km^2. Using the Earth's atmosphere as a particle detector with an effective surface area of $\sim 500 \cdot 10^6$ km^2 results in $\sim$ 10 occurrences per minute on the global scale. Ultra-high energy cosmic rays produce cascades of electron-positron pairs which spiral around the magnetic field line and emit highly beamed electromagnetic radiation during the propagation of the air shower through the atmosphere (Huege & Falcke, 2005). These bursts of electromagnetic radiation can reach peak electric field amplitudes $\sim$ 1 V/m and decay away during some tens of nanoseconds such that they can be recorded with electric field antennas in the frequency range of some MHz (Falcke *et al.*, 2005). The radio signals are reflected by the highly conductive ground and the ionosphere and thus propagate to remote distances. The detection of cosmic rays over large distances at low frequencies poses a new challenge to radio remote sensing.

3.2 Bottom up

The fundamental problem of atmospheric electricity is that the conventional breakdown electric field in the atmosphere needs to exceed $\sim 3 \cdot 10^6$ V/m to initiate a lightning discharge, but these large electric fields are not observed inside thunderclouds. It was speculated that the required electric fields within the thundercloud are highly localised in space and time and therefore remain undetected. However, balloon measurements of electric fields in thunderclouds suggest that the breakdown electric field is on the order of $\sim 10^5$ V/m, i.e., one order of magnitude smaller than the conventional breakdown threshold (Marshall *et al.*, 1995). This observation is explained with the newly recognised phenomenon of relativistic breakdown in the atmosphere (Gurevich & Zybin, 2005; Gurevich *et al.*, 1992). The key point of relativistic breakdown is that the cross section of collisions between energetic electrons and neutral air molecules decreases during acceleration of the electrons by the electric field. In this way, the electrons 'run away' and subsequent avalanches of energetic electrons are produced. This new physical mechanism requires one energetic particle, e.g., a cosmic ray, $\sim 10^{16}$ eV to start the relativistic breakdown process (Gurevich & Zybin, 2005). Positive feedback of runaway breakdown results from energetic photons and positrons which travel in opposite direction to the propagation of the avalanche and subsequently initiate additional electron avalanches (Dwyer, 2003). The runaway breakdown process results in a population of particles with energies of some tens of MeV in the mixed phase region of the thundercloud at ~ 5 km height where the collisions of riming graupel and ice crystals result in the necessary charge separation to produce strong electric fields. About 50 % of the mass of the atmosphere is located below and above 5 km height such that the energetic radiation from runaway breakdown can be observed from the ground (Dwyer *et al.*, 2005; Dwyer *et al.*, 2003) and from space (Inan, 2005; Smith *et al.*, 2005). The observations of terrestrial γ-ray flashes on board of satellites are explained with a giant particle accelerator in the Earth's atmosphere which is produced by the lightning electric field above thunderclouds (Inan, 2005). In this picture, a particularly intense lightning discharge deposits a charge of ~ 100 C inside the thundercloud. The resulting monopole electric field decays slower with height than the higher order multipole electric fields, while the runaway breakdown threshold falls off exponentially with height since it scales with the neutral gas density. The result is that the lightning electric field exceeds the relativistic breakdown threshold at some height in the atmosphere and accelerates free electrons to some MeV of energy producing bursts of terrestrial γ-rays resulting from bremsstrahlung radiation of the energetic electrons (Lehtinen *et al.*, 1996). The electrons may produce a transient luminous (red-bluish) airglow above the thundercloud termed sprite (Yukhimuk *et al.*, 1998, Fishman *et al.*, 1994). The electrons propagate along the geomagnetic field lines into near-Earth space where they are trapped in the radiation belt and bounce back and forward between conjugate hemispheres while drifting eastward. The primary electron beam is highly localised in space ($\sim 10 - 100$ km) and time (~ 1 ms) such that the direct detection of the electrons with particle detectors on satellites is improbable. However, some of the injected electrons are predicted to form eastward drifting 'curtains' extending over $\sim 70°$ in latitude within a few minutes after injection (Lehtinen *et al.*, 2000). This spreading of the particles substantially increases the likelihood of detection before the particles dissipate by precipitating back into the Earth's atmosphere. A number of challenging space missions is now under way to simultaneously detect the electron beam, the emitted energetic radiation and the luminous manifestation of the relativistic particles in the form of sprites.

3.3 Strange

The photon is a strange particle. It carries an impulse but seemingly no mass. The current knowledge on the photon rest mass ultimately relies on experimental observations and their theoretical understanding putting an upper limit on the rest mass of the photon (Tu *et al.*, 2005; Gul'yel'mi & Pokhotelov, 1994; Goldhaber & Nieto, 1971). Maxwell's equations are a cornerstone of modern physics to describe electromagnetic phenomena. Proca's equations are an extension of Maxwell's equations which include the effects of a massive photon to describe electromagnetic phenomena (Morse & Feshbach, 1953). The mass of the photon $m_\gamma = \mu_\gamma \hbar / c$ is given by the characteristic length scale of the photon μ_γ, the Planck constant $\hbar$ and the speed of light c. It is not known if Maxwell's or Proca's equations describe electromagnetism correctly. The solutions of Proca's equations suggest a number of ways in which the effects of a non-zero photon rest mass might be measurable. The four most widely explored of these are: (i) a frequency dependence of the speed of electromagnetic radiation in vacuum, (ii) deviations from the inverse square law of electrostatics, (iii) changes to the form of the magnetic dipole field, and (iv) astrophysical measurements of magneto-hydrodynamic effects. The current upper limits for the photon rest mass derived from these four approaches are 3×10^{-49} kg, 8×10^{-51} kg, 8×10^{-52} kg and 1×10^{-52} kg respectively (Tu *et al.*, 2005). Progress on the knowledge of the photon's physical properties is monitored by the Particle Data Group at Lawrence Berkeley National Laboratory in the United States (Eidelman *et al.* (Particle Data Group), 2004). The effect of a non-zero rest mass becomes highly significant if the rest energy and the quantum energy of the photon become comparable, i.e., $mc^2 = \hbar\omega$. It is clear that the lower the frequency, the easier it becomes to observe the effects of a non-zero rest mass. Extremely low frequency radio waves are emitted by lightning discharges in the troposphere (Sentman *et al.*, 1995) and gigantic transient luminous events above thunderstorms (Boccippio *et al.*, 1995) termed sprites (Füllekrug *et al.*, 2006). These radio waves propagate over long distances, reflected between the highly conductive Earth and the lower ionosphere, which together form a giant natural spherical capacitor. They have wavelengths as long as the circumference of the Earth (40,000 km) and frequencies as low as 10 Hz (Füllekrug, 2005; Füllekrug & Constable, 2000). Using the above expression, it is easy to see that a photon mass on the order of 1×10^{-50} kg would have readily observable effects at 10 Hz. A more detailed analysis shows that the photon rest mass must be smaller than 4×10^{-52} kg to be consistent with observations, this upper limit being constrained by the natural variability of the conductivity within the Earth's ionised upper atmosphere (Füllekrug, 2004, Füllekrug *et al.*, 2002; Füllekrug, 2000). The ultimate upper limit for the photon rest mass is given by the characteristic length scale of the photon which must be smaller than the size of the universe. This ultimate limit is given by $m_\gamma \approx h/Tc^2 \approx 10^{-69}$ kg, where $T \approx 10^{10}$ years is the age of the universe (Goldhaber & Nieto, 1971). It is apparent that a substantial gap of 17 orders of magnitude remains between the current best upper limit and the ultimate upper limit for the photon rest mass which poses a challenge to all scientists using Maxwell's equations as a working hypothesis during their daily work.

4. SUMMARY

An understanding of climate change is essential for the future of civilisation. The United Nations Intergovernmental Panel on Climate Change states that the Sun was probably responsible for much of the temperature changes in the early part of the 20th century. Yet the level of scientific understanding of the causative processes is very low at present. The variable solar radiation and energetic charged particles ionise the upper atmosphere and thereby modulate the air-Earth current of the global atmospheric electric circuit which is maintained by global thunderstorm activity. In the lower atmosphere, the modified air-Earth current acts upon cloud microphysical processes which have an impact on the radiative properties of clouds and hence on global climate change. The global electric field may in this way mediate solar variability into climate. The proposed physical mechanisms are difficult to observe in the Earth's atmosphere as a result of the complex behaviour of the atmosphere. Simultaneous measurements of the cloud content, the air-Earth current and the global lightning activity are necessary to disentangle cause and effect in future scientific studies. The Earth's atmosphere is populated by a rich variety of energetic charged particles which ionise air molecules and thereby assist subsequent cloud initiation through microphysical processes such as the ion-aerosol clear-air mechanism and the ion-aerosol near-cloud mechanism. Top down particles from stellar and celestial sources penetrate deep into the Earth's atmosphere where the particles ionise air and emit electromagnetic radiation. The physical mechanisms by which these energetic charged particles affect cloud radiative properties needs to be determined. Bottom up particles are accelerated to relativistic energies by small scale electric fields inside thunderclouds and large scale lightning discharge electric fields above thunderclouds. The radio waves emitted from these relativistic runaway breakdown processes and the associated electron beams in near-Earth space remain to be discovered. The Earth's atmosphere hosts a peculiar long-wave photon gas which may help to determine the thermal radiation of the Earth's atmosphere for a more effective monitoring of global climate change in the future. The variable ionisation of the atmosphere places an upper limit on the rest mass of these long-wave photons. The current best upper limit for the photon rest mass leaves a whopping wiggle-room of 17 orders of magnitude for worries if Maxwell's or Proca's equations describe electromagnetism correctly, a charming potential which deserves to be fostered in the future.

Acknowledgements

I wish to thank Claude Brouton for giving me the unique opportunity to present my views on atmospheric electromagnetics and climate change to a thriving audience of young scientists during the European Research Course on Atmospheres in Grenoble, France. I also wish to thank the attendants of the sprite summer school in Corsica and numerous colleagues around the world for inspiring scientific discussions.

References

Baker, D.N., Mason, G.M., Figueroa, O., Colon, G., Watzin, J.G., & Aleman, R.M. 1993. An Overview of the solar, anomalous and magnetospheric particle explorer (SAMPEX) Mission. *IEEE Transactions on Geoscience and Remote Sensing*, **31**(3), 531–574.

Baldwin, M.P., Thompson, D.W.J., Shuckburgh, E.F., Norton, W.A., & Gillett, N.P. 2003. Weather from the stratosphere? *Science*, **301**, 317–319.

Bering, E.A. 1995. The global circuit: Global thermometer, weather by-product or climatic modulator ? *Reviews of Geophysics*, **33**(Suppl., U.S. National Report to Int. Union of Geodesy and Geophys. 1991-1994), 845–862.

Bering, E.A., Few, A.A., & Benbrook, J.R. 1998. The global electric circuit. *Physics Today*, **51**(10), 24–30.

Boccippio, D.J., Williams, E.R., Heckman, S.J., Lyons, W.A., Baker, I.T., & Boldi, R. 1995. Sprites, ELF transients, and positive ground strokes. *Science*, **269**, 1088.

Brooks, C.E.P. 1925. The distribution of thunderstorms over the globe. *Geophys. Mem.*, **III**(24), 147.

Carslaw, K.S., Harrison, R.G., & Kirkby, J. 2002. Cosmic rays, clouds and climate. *Science*, **298**, 1732–1737.

Chalmers, J.A. 1967. *Atmospheric Electricity*. Pergamon Press, Oxford.

Christian, H.J., Blakeslee, R.J., Boccippio, D.J., Boeck, W.L., Buechler, D.E., Driscoll, K.T., Goodman, S.J., Hall, J.M., Koshak, W.J., Mach, D.M., & Stewart, M.F. 2003. Global frequency and distribution of lightning as observed from space by the Optical Transient Detector. *Journal of Geophysical Research*, **108**(D1), 4.1–15.

Corney, R.C., Burns, G.B., Michael, K., Frank-Kamenetsky, A.V., Troshichev, O.A., Bering, E.A., Papitashvili, V.O., Breed, A.M., & Duldig, M.L. 2003. The influence of polar-cap convection on the geoelectric field at Vostok, Antarctica. *Journal of Atmospheric and Solar-Terrestrial Physics*, **65**, 345–354.

Dwyer, J.R. 2003. A fundamental limit on electric fields in air. *Geophysical Research Letters*, **30**(20), 2055–2058.

Dwyer, J.R., & Smith, D.M. 2005. A comparison between Monte Carlo simulations of runaway breakdown and terrestrial gamma-ray flash observations. *Geophysical Research Letters*, **32**(doi:10.1029/2005GL023848), 1–4.

Dwyer, J.R., Uman, M.A., Rassoul, H.K., Al-Dayeh, M., Caraway, L., Jerauld, J., Rakov, V.A., Jordan, D.M., Rambo, K.J., Corbin, V., & Wright, B. 2003. Energetic radiation produced during rocket-triggered lightning. *Science*, **299**, 694–697.

Dwyer, J.R., Rassoul, H.K., Al-Dayeh, M., Caraway, L., Chrest, A., Wright, B., Kozak, E., Jerauld, J., Uman, M.A., Rakov, V.A., Jordan, D.M., & Rambo, K.J. 2005. X-ray bursts

associated with leader steps in cloud-to-ground lightning. *Geophysical Research Letters*, **32**(doi:10.1029/2004GL021782), 1–4.
Eichkorn, S., Wilhelm, S., Aufmhoff, H., Wohlfrom, K.H., & Arnold, F. 2002. Cosmic ray-induced aerosol-formation: First observational evidence from aircraft-based ion mass spectrometer measurements in the upper troposphere. *Geophysical Research Letters*, **29**(10.1029/2002GL015044), 1–4.
Eidelman *et al.* (Particle Data Group), S. 2004. Review of particle physics. *Physics Letters B*, **592**, 1.
Falcke, H., Apel, W.D., & et al., A.F. Badea. 2005. Detection and imaging of atmospheric radio flashes from cosmic ray air showers. *Nature*, **435**(19), 313–316.
Farrell, W.M., & Desch, M.D. 2002. Solar proton events and the fair weather electric field at ground. *Geophysical Research Letters*, **29**(9), 37–1.
Fishman, G.J., & Inan, U.S. 1988. Observation of an ionospheric disturbance caused by a gamma-ray burst. *Nature*, **331**, 418–420.
Fishman, G.J., Bhat, P.N., Mallozzi, R., Horack, J.M., Koshut, T., Kouveliotou, C., Pendleton, G.N., Meegan, C.A., Wilson, R.B., Paciesas, W.S., Goodman, S.J., & Christian, H.J. 1994. Discovery of intense gamma-ray flashes of atmospheric origin. *Science*, **264**, 1313–1316.
Franz, R.C., Nemzek, R.J., & Winckler, J.R. 1990. Television image of a large upward electrical discharge above a thunderstorm system. *Science*, **249**, 48–51.
Friis-Christensen, E. 2000. Solar variability and climate. *Space Science Reviews*, **94**, 411–421.
Füllekrug, M. 2000. Dispersion relation for spherical electromagnetic resonances in the atmosphere. *Physics Letters A*, **275**, 80–89.
Füllekrug, M. 2004. Probing the speed of light with radio waves at extremely-low frequencies. *Physical Review Letters*, **93**(4), 043901.1–3.
Füllekrug, M. 2005. Detection of thirteen resonances of radio waves from particularly intense lightning discharges. *Geophysical Research Letters*, **32**(doi:10.1029/2005GL023028), 1–4.
Füllekrug, M., & Constable, S. 2000. Global triangulation of intense lightning discharges. *Geophysical Research Letters*, **27**(3), 333–336.
Füllekrug, M., Fraser-Smith, A.C., Bering, E.A., & Few, A.A. 1999. On the hourly contribution of global lightning to the atmospheric field in the Antarctic during December 1992. *Journal of Atmospheric and Solar-Terrestrial Physics*, **61**, 745–750.
Füllekrug, M., Fraser-Smith, A.C., & Schlegel, K. 2002. Global ionospheric D-layer height monitoring. *Europhysics Letters*, **59**(4), 626–632.
Füllekrug, M., Mareev, E.M., & Rycroft, M.J. (eds). 2006. *Sprites, elves and intense lightning discharges*. Springer, Dordrecht.
Goldhaber, A.S., & Nieto, M.M. 1971. Terrestrial and extraterrestrial limits on the photon mass. *Reviews on Modern Physics*, **43**(3), 277–296.
Gray, L.J., Haigh, J.D., & Harrison, R.G. 2005. *The influence of solar changes on the Earth's climate*. Hadley Centre, Technical note 62.
Gul'yel'mi, A.V., & Pokhotelov, O.A. 1994. Geophysical methods for estimating photon mass. *Physics of the solid Earth*, **29**(11), 1016–1019.
Gurevich, A., & Zybin, K.P. 2005. Runaway breakdown and the mysteries of lightning. *Physics Today*, **58**(4), 37–43.
Gurevich, A.V., Milikh, G.M., & Roussel-Dupre, R. 1992. Runaway electron mechanism of air breakdown and preconditioning during a thunderstorm. *Physics Letters A*, **165**, 463–468.
Haigh, J. 1996. The impact of solar variability on climate. *Science*, **272**, 981–984.
Hansen, J., & Lebedeff, S. 1987. Global trends of measured surface air temperature. *Journal of Geophysical Research*, **Vol. 92**(No. D11), p. 13345.
Hansen, J., Johnson, D., Lacis, A., Lebedeff, S., Lee, P., Rind, D., & Russell, G. 1981. Climate impact of increasing atmospheric carbon dioxide. *Science*, **Vol. 213**(No. 4511), p. 957.

Harrison, R.G., & Carslaw, K.S. 2003. Ion-aerosol-cloud processes in the lower atmosphere. *Reviews of Geophysics*, **41**(3), 2.1–26.

Harrison, R.G., & Stephenson, D.B. 2006. Empirical evidence for a nonlinear effect of galactic cosmic rays on clouds. *Proceeding of the Royal Society A*, **462**, 1221–1233.

Hoffmann, K. 1923. Bericht über die in Ebeltofthafen auf Spitzbergen (11°36′15″, 79°9′14″) in den Jahren 1913/14 durchgeführten luftelektrischen Messungen. *Beitr. Phys. Atmosph.*, **11**, 1–11.

Holzer, R. E., & Deal, D. E. 1956. Low audio frequency electromagnetic signals of natural origin. *Nature*, **177**, 536–537.

Holzworth, R., & Volland, H. 1986. Do we need a Geoelectric Index ? *EOS*, **67**(26), 545–548.

Huege, T., & Falcke, H. 2005. Radio emission from cosmic ray air showers. *Atronomy & Astrophysics*, **430**(3), 779–798.

ICSU. 1986. *A program for the study of the long-term behaviour of the upper atmosphere und near-space environment* STP Newsletters **1**, 1-20. International Council of Scientific Unions.

Inan, U.S. 2005. Gamma rays made on Earth. *Science*, **307**, 1054–1055.

Inan, U.S., Lehtinen, N.G., Lev-Tov, S.J., Johnson, M.P., & Bell, T.F. 1999. Ionization of the lower ionosphere by γ-rays from a magnetar: Detection of a low energy (3-10 keV) component. *Geophysical Research Letters*, **26**(22), 3357–3360.

Lean, J., & Rind, D. 1999. Evaluating sun-climate relationships since the little ice age. *Journal of Atmospheric and Solar-Terrestrial Physics*, **61**, 25–36.

Lehtinen, N.G., Walt, M., Inan, U.S., Bell, T.F., & Pasko, V.P. 1996. γ-ray emission produced by a relativistic beam of runaway electrons accelerated by quasi-electrostatic thundercloud fields. *Geophysical Research Letters*, **23**(19), 2645–2648.

Lehtinen, N.G., Inan, U.S., & Bell, T.F. 2000. Trapped energetic electron curtains produced by thunderstorm driven relativistic runaway electrons. *Geophysical Research Letters*, **27**(8), 1095–1098.

Lindzen, R.S. 1990. Some coolness concerning global warming. *Bull. Am. Meteor. Soc.*, **Vol. 71**, p. 288.

Lockwood, M., Stamper, R., & Wild, M.N. 1999. A doubling of the sun's coronal magnetic field during the past 100 years. *Nature*, **399**, 437–439.

MacGorman, D.R., & Rust, W.D. 1998. *The electrical nature of storms*. Oxford University Press, New York.

Markson, R. 1978. Solar modulation of atmospheric electrification and possible implications for the Sun-weather relationship. *Nature*, **273**, 103–109.

Markson, R. 1986. Tropical convection, ionospheric potentials and global circuit variation. *Nature*, **320**, 588–594.

Markson, R., & Muir, M. 1980. Solar wind control of the Earth's electric field. *Science*, **208**, 979–990.

Marsh, N.D., & Svensmark, H. 2000. Low cloud properties influenced by cosmic rays. *Physical Review Letters*, **85**(23), 5004–5007.

Marshall, T.C., McCarthy, M.P., & Rust, W.D. 1995. Electric field magnitudes and lightning inititaion in thunderstorms. *Journal of Geophysical Research*, **100**(D4), 7097–7103.

Morse, P.M., & Feshbach, H. 1953. *Methods of theoretical physics*. Mc Graw Hill, New York.

Nagano, M., & Watson, A.A. 2000. Observations and implications of the ultrahigh-energy cosmic rays. *Reviews on Modern Physics*, **72**(3), 689–732.

Neubert, T. 2003. Sprites and their exotic kin. *Science*, **300**, 747–749.

Pasko, V.P., Stanley, M.A., Mathews, J.D., Inan, U.S., & Wood, T.G. 2002. Electrical discharge from a thundercloud top to the lower ionosphere. *Nature*, **416**, 152–154.

Price, C. 2000. Evidence for a link between global lightning activity and upper tropospheric water vapour. *Nature*, **406**, 290–293.

Rakov, V.A., & Uman, M.A. 2003. *Lightning, physics and effects*. Cambridge University Press, Cambridge.

Reid, G.C. 2000. Solar variability and the Earth's climate: Introduction and overview. *Space Science Reviews*, **94**, 1–11.

Rycroft, M.J., Israelsson, S., & Price, C. 2000. The global atmospheric electric circuit, solar activity and climate change. *Journal of Atmospheric and Solar-Terrestrial Physics*, **62**, 1563–1576.

Saunders, C.P.R., Keith, W.D., & Mitzeva, R.P. 1991. The effect of liquid water content on thunderstorm charging. *Journal of Geophysical Research*, **96**, 11007–11017.

Schlegel, K., & Füllekrug, M. 1999. Schumann resonance parameter changes during high-energy particle precipitation. *Journal of Geophysical Research*, **104**(A5), 10111.

Sentman, D.D. 1990. Approximate Schumann resonance parameters for a two scale-height ionosphere. *Journal of Atmospheric and Terrestrial Physics*, **52**(1), 35–46.

Sentman, D.D., & Wescott, E.M. 1993. Observations of upper atmospheric optical flashes recorded from an aircraft. *Geophysical Research Letters*, **20**(24), 2857–2860.

Sentman, D.D., Wescott, E.M., Osborne, D.L., Hampton, D.L., & Heavner, M.J. 1995. Preliminary results from the Sprites94 aircraft campaign: 1. Red sprites. *Geophysical Research Letters*, **22**(10), 1205–1208.

Smart, D.F., Shea, M.A., & Flückiger, E.O. 2000. Magnetospheric models and trajectory computations. *Space Science Reviews*, **93**, 305–333.

Smith, D.M., Lopez, L.I., Lin, R.P., & Barrington-Leigh, C.P. 2005. Terrestrial gamma-ray flashes observed up to 20 MeV. *Science*, **307**, 1085–1088.

Su, H.T., Su, R.R., Chen, A.B., Wang, Y.C., Hsiao, W.S., Lai, W.C., Lee, L.C., Sato, M., & Fukunishi, H. 2003. Gigantic jets between a thundercloud and the ionosphere. *Nature*, **423**, 974–976.

Svensmark, H. 1998. Influence of cosmic rays on Earth's climate. *Physical Review Letters*, **81**(22), 5027–5030.

Takahashi, T. 1978. Riming electrification as a charge generation mechanism in thunderstorms. *Journal of the Atmospheric Sciences*, **35**, 1536–1548.

Tinsley, B., & Yu, F. 2004. *Atmospheric ionization and clouds as links between solar activity and climate*. AGU monograph (in press).

Tinsley, B.A. 2000. Influence of solar wind on the global electric circuit, and inferred effects on cloud microphysics, temperature, and dynamics in the troposphere. *Space Science Reviews*, **94**, 231–258.

Tinsley, B.A., Rohrbaugh, R.P., Hei, M., & Beard, K.V. 2000. Effects of image charges on the scavenging of aerosol particles by cloud droplets, and on droplet charging and possible ice nucleation processes. *Journal of the Atmospheric Sciences*, **57**, 2118–2134.

Tripathi, S.N., & Harrison, R.G. 2002. Enhancement of contact nucleation by savenging of charged aerosol. *Atmospheric Research*, **62**, 57–70.

Tu, L.C., Luo, J., & Gillies, G.T. 2005. The mass of the photon. *Reports on Progress in Physics*, **68**, 77–130.

Uman, M.A. 1974. The Earth and its atmosphere as a leaky spherical capacitor. *American Journal of Physics*, **42**, 1033–1035.

Watson, R.T. (ed). 2001. *Climate Change 2001: Synthesis Report*. Cambridge University Press.

Williams, E.R. 1992. The Schumann resonance: A global tropical thermometer. *Science*, **256**, 1184–1187.

Willson, R.C., & Hudson, H.S. 1988. Solar luminosity variations in solar cycle 12. *Nature*, **332**, 810–812.

Yu, F.Q., & Turco, R.P. 2001. From molecular clusters to nanoparticles: Role of ambient ionization in tropospheric aerosol formation. *Journal of Geophysical Research*, **106**(D5), 4797–4814.

Yukhimuk, V., Roussel-Dupre, R.A., Symbalisty, E.M.D., & Taranenko, Y. 1998. Optical characteristics of blue jets produced by runaway air breakdown, simulation results. *Geophysical Research Letters*, **25**(17), 3289–3292.

J. Phys. IV France **139** (2006) 167–174
© EDP Sciences, Les Ulis
DOI: 10.1051/jp4:2006139012

Wavelet reconstructions of solar magnetic activity

H. Lundstedt[1]

[1] *Swedish Institute of Space Physics, Lund, Sweden*

Abstract. Wavelet methods have become a very much common tool for exploring and analyzing solar data. Many methods have been developed. I will briefly describe three different methods: Wavelet Power Spectra (WPS), Ampligrams, and Multi-Resolution Analysis (MRA). I will then give examples of how the wavelet methods have been applied to indicators of solar activity. Interestingly these studies raise fundamental questions about what we mean by solar activity and how good the indicators are.

1. INTRODUCTION

Since the solar activity can be described as a non-linear chaotic dynamic system Mundt *et al.* [22] methods such as neural networks and wavelet methods should be very suitable (Lundstedt [15]; Lundstedt [16]; Lundstedt [18]). Many have used wavelet techniques for studying solar activity. Frick *et al.* [8] investigated the wavelet transform of monthly sunspot group number from 1610 to 1994. Two pronounced peaks were found corresponding to the Schwabe (11 years) and Gleissberg (90-100 years) cycles. The wavelet analysis also showed the Maunder and Dalton minima. Knaack and Stenflo [11] first calculated the spherical coefficients of the radial magnetic field from Kitt Peak Observatory maps. They then applied wavelet analysis to deduce the temporal variation. The wavelet power spectra of the spherical coefficients showed power for 22 years, 6-7 years, and for 2-3 years.

In Boberg *et al.* [4] we used both the daily solar mean magnetic field (SMMF) observed at the Wilcox Solar Observatory (Stanford) and one-minute resolution measurements of SMMF by MDI onboard SOHO. Peaks were found in the power for periods of 11 years (the solar cycle), 1-2 years (related to variations in internal rotation), 80-200 days (related to evolution of active regions) and 13/26 days (related to solar rotation). Using one-minute resolution SMMF we managed to detect peaks in power around 90 minutes. Further studies suggested that the 90 minutes oscillation could be associated with occurrence of CMEs. Polygiannakis *et al.* [27] used the sunspot number R_z and Krivova and Solanki [12] used the sunspot area (SA) as an indicator of the solar activity in a similar wavelet study. Polygiannakis et al., tried to explain the found periodicities as a result of two dynamos with periods of 11 and 2 years. Krinova and Solanki on the other hand suggest that the 1.3 year and 156 days periods are harmonics of the solar activity cycle. Richardsson *et al.* [28] found 1.3 years periodicities in the solar wind variation.

Oliver and Ballester [26] found 158 days periodicities in the number of X-ray flares, sunspot number, sunspot area and the MWSI. However, they noticed the amplitude varies from cycle to cycle.

The long-term solar activity of periods over 205 years and as long as 2300-2500 have been suggested from studies of proxies such as the C14 production rate and Be10 (Muscheler *et al.* [23], Muscheler *et al.* [24], Muscheler *et al.* [25], Lundstedt *et al.* [20]).

2. WAVELET METHODS

Today using wavelet techniques has become a common method of analyzing solar-terrestrial data. Good introductions to the use of wavelet transforms are given by Kumar and Foufoula-Georgiou [13], Torrence and Compo [32], Mallat [21] and Addison [1]. Wavelet analysis is a powerful tool both to find the dominant mode of variation and also to study how it varies with time, by decomposing a non-linear time series into time-frequency space.

2.1 Wavelet power spectra

Assume that we have a time series, x_n, with equal time spacing δt and $n = 0....N-1$. We assume also a wavelet function $\Psi(\eta)$ that depends on a nondimensional "time" parameter η. Torrence and Compo [32] then derive the continuous wavelet transform of the discrete sequence x_n as the convolution of x_n with the scaled and translated version of $\Psi(\eta)$:

$$W_n(s) = \sum_{n'=0}^{N-1} x_{n'} \Psi^* \left[\frac{(n'-n)\delta t}{s}\right], \tag{2.1}$$

where $(*)$ is the complex conjugate and Ψ the Morlet wavelet.

In the Fourier space the above expression becomes:

$$W_n(s) = \sum_{k=0}^{N-1} \hat{x}_k \hat{\Psi}^*(s\omega_k) e^{i\omega_k n \delta t}, \tag{2.2}$$

where the angular frequency is defined as

$$\omega_k = \begin{Bmatrix} \dfrac{2\pi k}{N\delta t}, & k \leq \dfrac{N}{2} \\ -\dfrac{2\pi k}{N\delta t}, & k > \dfrac{N}{2} \end{Bmatrix} \tag{2.3}$$

and the discrete Fourier transform.

$$\hat{x}_k = \frac{1}{N} \sum_{n=0}^{N-1} x_n e^{-\frac{2\pi i k n}{N}}, \tag{2.4}$$

Finally the wavelet power spectra is defined as $|W_n(s)|^2$.

2.2 Ampligrams

The wavelet transform of a function y(t) is given by

$$w(a,b) = a^{-1/2} \int_{-\infty}^{+\infty} y(t) \Psi^* \left(\frac{t-b}{a}\right) dt \tag{2.5}$$

where a is the scale dilation, Ψ is a Morlet wavelet, b is the translation parameter.

If the wavelet is admissible, then the function can be reconstructed from the wavelet coefficients.

$$y(t) = \frac{1}{C_\Psi} \int_0^{+\infty} \frac{1}{a^2} da \int_{-\infty}^{+\infty} w(a,b) \Psi_{ab} dt, \tag{2.6}$$

where the admissibilty condition is

$$C_\Psi = \int_{-\infty}^{+\infty} |\Psi(\omega)|^2 \frac{d\omega}{|\omega|} < \infty, \tag{2.7}$$

$\Psi_{ab} = a^{-1/2} \Psi((t-b)/a)$ and $\Psi(\omega)$ is the Fourier transform of $\Psi(t)$. To analyze a discrete signal $y(t_i)$ we need to sample the continuous wavelet transform on a grid in the time-scale plane (b,a). By setting a = j and b = k the wavelet coefficients $w_{j,k}$ are

$$w_{j,k} = j^{-1/2} \int_{-\infty}^{+\infty} y(t) \Psi^* \left(\frac{t-k}{j}\right) dt \tag{2.8}$$

When the wavelet coefficient magnitudes (WCM) are plotted for the scale and the elapsed time, a so called scalogram is produced. Skeleton spectrum (Polygiannakis *et al.* [27]) can be derived from scalograms. The scale maximal wavelet skeleton spectrum keeps only those wavelet components that are locally of maximum amplitude at any given time-scale. The instantly maximal wavelet skeleton spectrum keeps only those wavelet components that are locally of maximum amplitude at any given time.

Ampligrams and time-scale spectra Liszka [14] can be looked upon as a band pass filtering in the WCM domain analogues to Fourier analysis is in frequency domain. They can be used to separate independent components of the signal, assuming that the different components are characterized by different wavelet coefficient magnitudes (spectral densities). Ampligrams are constructed in the following way: The maximum magnitude (W) among the wavelet coefficients is first found. L magnitude intervals are then defined,

$I_l = [(l-1)\Delta w, l\Delta w], l = 1, 2..L$ with $\Delta w = \frac{|W|}{L}$. From that we construct L matrices W_l, 1 = 1, 2...L such that

$$[W_l]_{j,k} = \begin{cases} w_{j,k} & \text{if } | w_{j,k} | \in I_l \\ 0 & otherwise \end{cases} \tag{2.9}$$

Inverse the wavelet transform to get a new time-signal $y_l(t_i), l = 1, ..L$. Each $y_l(t_i)$ is what the signal should have looked like if only a narrow range of wavelet coefficient amplitude would be present in the signal. After that construct L x N matrix Y with $y_l(t_i)$ as rows. This matrix Y is the ampligram of the original time-signal $y_l(t_i)$. Each row can of course further be analyzed by e.g. a scalogram. By adding all the rows then the original signal is reconstructed.

Finally we may also construct time-scale spectra if each row of the ampligram matrix Y is first wavelet transformed, which results in L matrices. We then time-average these matrices (average along rows) leading to L arrays $w_{,l,ave}$ with J elements. Finally a matrix L x J is constructed Y_{ave}, with $w_{,l,ave}$ as rows. This matrix Y_{ave} is the time scale spectrum of the ampligram.

This set of wavelet methods developed by Liszka [14] and Wernik and Grzesiak [33], and applied in Lundstedt *et al.* [19] are summarized in Figure 1.

2.3 Multi-Resolution Analysis

The idea behind Multi-Resolution Analysis (MRA) (Mallat [21]) is to separate the information to be analyzed into a "principal" (low pass) and a "residul" (high pass) part. The process of decomposition can then be applied again to both parts. Mathematically simplified, it can be described by the following equations

$$s = A_J + \sum_{j \le J} D_j \tag{2.10}$$

where s is the signal, A_J the approximation (principal part) at resolution level J and D_j the detail (residul part) at level j. From the previous formula, it is seen that the approximations are related to one another by:

$$A_{J-1} = A_J + D_J \tag{2.11}$$

$$D_j = \sum_{k \in Z} C(j,k)\Psi_{j,k}(t) \tag{2.12}$$

where D_j is the detail at level j, $C(j,k)$ the wavelet coefficient and $\Psi_{j,k}(t)$ the wavelet function. The wavelet used in this study was a Daubechies of order six. The dyadic scale is $a = 2^j \le 2^J$ for level j. The resolution is given by $1/a$ or 2^{-j}. The dyadic translation is given by $b = ka$.

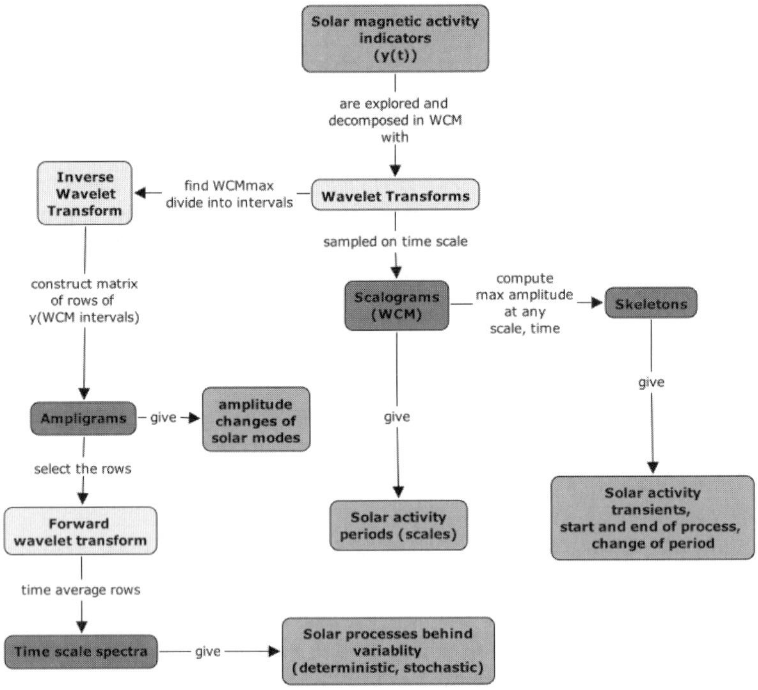

Figure 1. The set of developed wavelet methods described in Lundstedt *et al.* [19].

3. WAVELET METHODS APPLIED

I will now give examples of the three methods applied to solar activity data.

3.1 Wavelet Power Spectra

The first example shows a wavelet power spectrum of the daily mean solar magnetic field data (Scherrer *et al.* [30]), observed at Wilcox Solar Observatory (WSO) between May 1975 and March 13, 2001. The wavelet power spectrum, as described in section 2.1, shows variations with a period of 1-2 years. These variations may be connected to the rotational changes discovered at base of the convective zone (Howe *et al.* [9]). Oscillations with periods of 80-200 days are also seen. They maybe related to the 152 periodicity (Bogart and Bai [5]).

The peaks about 13 and 26 days are related to the solar rotation.

3.2 Ampligram

Figure 3 shows an ampligram of the group sunspot number (1610-1950) and the sunspot number 1995-2005.

The Maunder and Dalton minima are clearly seen as well as the high solar activity, which already started in early 1900. We also see a decreasing solar activity after about 1990 (Lundstedt *et al.* [19]). It can be seen in the ampligram that the strongest components start to decrease earlier than the weaker. The very weak components below 20 % (lower panel) interestingly show some semiperiodic structure even if they are noisier.

If the decreasing trend holds, then we would expect a weak solar cycle 24.

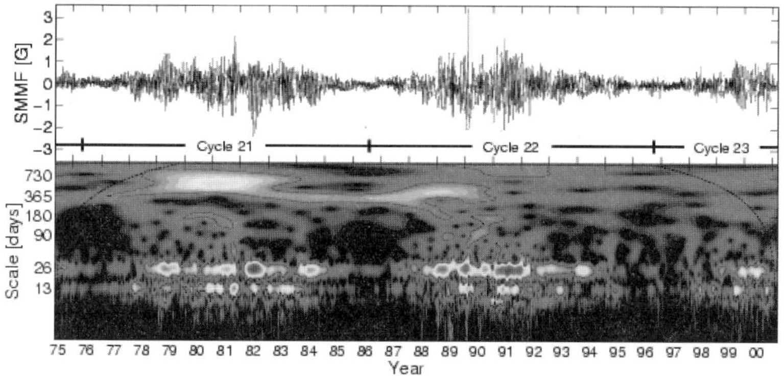

Figure 2. Daily WSO mean magnetic field observations between May 16 1975 and March 13, 2001 are presented in the top panel. The wavelet power spectrum in the bottom panel shows a number of oscillations.

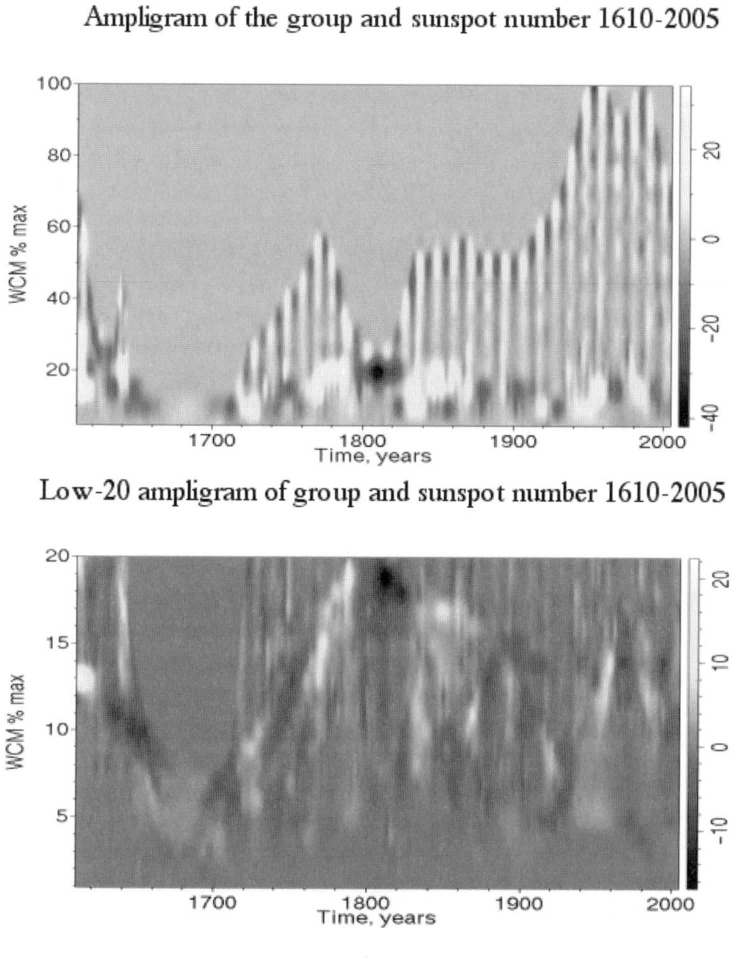

Figure 3. Ampligram of the group sunspot number 1610-1995 and the sunspot number 1995-2005. In lower panel only the weak signal, below 20% of WCM maximum, is shown.

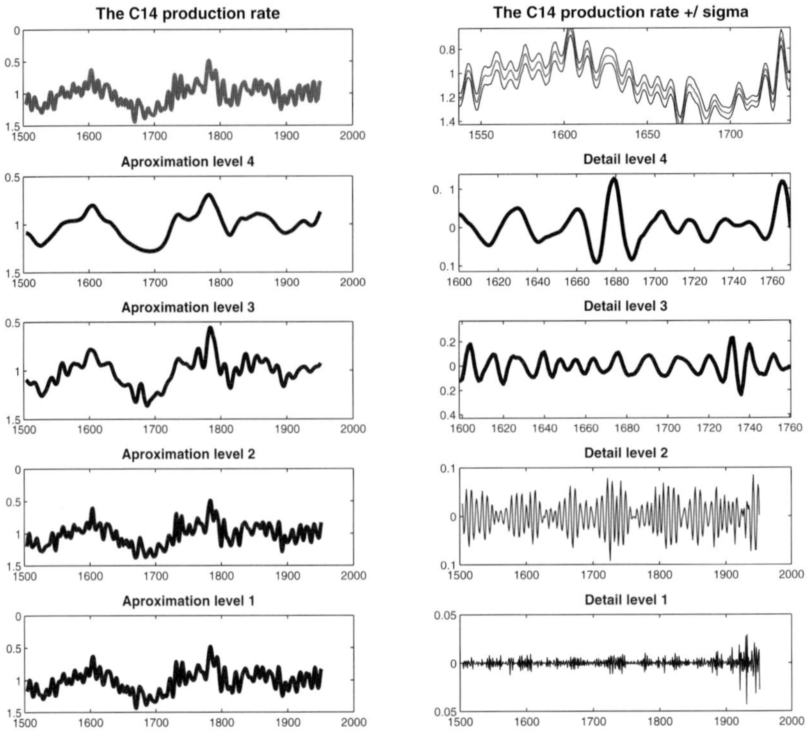

Figure 4. Multiresolution analysis of the ^{14}C production rate. The 22-year cycle is clearly present during the Maunder Minimum (zoomed in period) at detail level 4. The 11-year cycle is clearly present during the Maunder Minimum (zoomed in period) at detail level 3.

Scientists have already predicted the amplitude of the next solar cycle 24: Badalyan [2] predict a weak maximum, Svalgaard *et al.* [31] $R_z = 75 \pm 10$, Duhau [7] $R_z = 87.5 \pm 23.5$, Kane [10] $R_z = 105$, Schatten [29] $R_z = 100 \pm 30$ and Dikpati *et al.* [6] predict a large maximum.

Most expect a rather small cycle 24, i.e. in accordance with the trend. However, Dikpati *et al.* [6] claim that cycle 24 will become strong.

3.3 Multi-Resolution Analysis

In order to further study the cyclicity, revealing maximum and minimum, during the Maunder Minimum we carried out a Multi-Resolution Analysis of the annual ^{14}C production rate (Figure 4). The variation is studied at different resolutions. At level four in the detailed part the 22-years cycle is clearly present during the Maunder Minimum. The 11-years cycle is also clearly seen at level three in the detailed part. At the lowest resolution (approximation level four) several peaks appear. The strongest occur around 1600 and about 1780.

The 11-years cycle is not seen during the Maunder Minimum (Lundstedt *et al.* [20]). That is explained by that the toroidal magnetic field was too weak to produce sunspots (Beer *et al.* [3]).

4. CONCLUSIONS AND DISCUSSIONS

The three applications in section 3 exemplified: Detection of periods of changing intensity. Trends in long-term solar activity. Separation of solar magnetic field components; the toroidal magnetic field (indicated by the group sunspot number) and the poloidal magnetic field (indicated by the C14 production rate).

Wavelet methods are important for studies of the: Scalable structures of information and local changes, transients and trends, i.e. very much characteristic of solar activity.

The solar activity is driven by the changes of the solar magnetic field. However, the indicators of solar activity show large differences (Lundstedt *et al.* [19]). The choice of indicators is therefore very important.

We need a new looking glass at solar activity: Multi Resolution Analysis has shown to be very promising.

Acknowledgements

We are grateful to the following providers of data; ESA/NASA SOHO/MDI team, Wilcox Solar Observatory, Stanford, Mount Wilson Observatory, UCLA, RWC Belgium, Hoyt and Schatten for Rg, NOAA and SEC.

References

[1] Addison, P.: The Illustrated Wavelet Transform Handbook, Introductory Theory and Applications in Science, Engineering, Medicine and Finance, Institute of Physics Publishing, Bristol, 2002.

[2] Badalyan, O., Obridko, V., and Sykora, N., J.: Brightness of the coronal green line and prediction for activity cycles 23 and 24, Solar Physics, 199, 421, 2001.

[3] Beer, J., Tobias, S., and Weiss, N.: An Active Sun Throughout the Maunder Minimum, Solar Phys., 181, 237-249, 1998.

[4] Boberg, F., Lundstedt, H., Hoeksema, J. T., Scherrer, P. H., and Lui, W.: Solar mean magnetic field variability: A wavelet approach to WSO and SOHO/MDI observations, J. Geophys. Res., Vol. 107, No A10, 15-1–15-7, 2002.

[5] Bogart, R., and Bai, T., Confirmation of a 152 day periodicity in the occurrence of solar flares inferred from microwave data, Astrophys. J., 299, 51-55, 1985.

[6] Dikpati, M., de Toma, G., and Gilman, P.A.: Predicting the strength of solar cycle 24 using a flux-transport dynamo-based tool, Geophys., Res. Lett., Vol. 33, L05102, doi:10.1029/2005GL025221, 2006.

[7] Duhau, S.: An Early Prediction of Maximum Sunspot Number in Solar Cycle 24, Solar Physics, 213 (1), 203-212, 2003.

[8] Frick, P., Galyagin, D., Hoyt, D. V., Nesme-Ribes, E., Schatten, K.H., Sokoloff, D., and Zakharov, V.: Wavelet analysis of solar activity recorded by sunspot groups, Astron. Astrophys., 328, 670-681, 1997.

[9] Howe, R., Christensen-Dalsgaard, J., Hill, F., Komm, R.W., Larsen, R.M., Schou, J., Thompson, M.J., and J. Toomre, Dynamic variations at the base of the solar convection zone, Science, 287, 2456-2460, 2000.

[10] Kane, R.P.: Prediction of solar activity: Role of long-term variations, J. Geophys Res., 107, 3-1–3-3, 2002.

[11] Knaack, R., and Stenflo, J.O.: Harmonic Analysis of Solar Magnetic Fields, in ESA SP-505: SOLMAG 2002, *Proceedings of the Magnetic Coupling of the Solar Atmopshere Euroconference*, 453-456, 2002.

[12] Krivova, N.A., and Solanki, S.K.: The 1.3 year and 156-day periodicities in sunspot data: Wavelet analysis suggests a common origin, Astron. Astrophys., 394, 701-706, 2002.

[13] Kumar, P., and Foufoula-Georgiou, E.: Wavelet analysis for geophysical applications, Rev. Geophys. 35(4), 385-412, 1997.

[14] Liszka, L.: Cognitive Information Processing in Space Physics and Astrophysics, Pachart Publishing House, Tuscon, 2003.

[15] Lundstedt, H.: AI Techniques in Geomagnetic Storm Forecasting, in Magnetic Storm Geophysical Mono- graph 98, AGU, 1997.
[16] Lundstedt, H.: Solar Activity Predicted with Artificial Intelligence, in Space Weather Geophysical Monograph 125, AGU, 2001.
[17] Lundstedt, H.: Progress in Space Weather Predictions and Applications, in Adv. in Space Res., 36, 2516-22523, 2005.
[18] Lundstedt, H.: Solar Activity Modelled and Forecasted: A New Approach, Adv. in Space Res., doi: 10.1016/j.asr.2006.03.041, 2006.
[19] Lundstedt, H., Liszka, L., and Lundin, R.: Solar activity explored with new wavelet methods, Annal. Geophysicae, 23, 1505-1511, 2005.
[20] Lundstedt, H., Liszka, L., Lundin, R., and R. Muscheler.: Long-term solar activity explored with wavelets, Annal. Geophysicae, 2005.
[21] Mallat, S.: A wavelet tour of signal processing, Academic Press, 1998.
[22] Mundt, M.D., Maguire II, W.B., and Chase, R.P.: Chaos in sunspot cycle: Analysis and prediction, J. Geophys., Res. Vol. 96, 1705-1716, Februari1, 1991.
[23] Muscheler, R., Beer, J., and Kubik, P.W.: Long-Term Solar Variability and Climate Change Based on Radionuclide Data From Ice Cores, in Solar Variability and its Effect on the Earth's Atmospheric and Climate System, AGU Geophysical Monograph series (ed. J. Pap, P. F.) 221-235, 2004.
[24] Muscheler, R., Joos, F., Muller, S.A., and Snowball, I.: Not so unusual - today's solar activity, accepted for Nature, 2005.
[25] Muscheler, R., Beer, J., Kubik, P.W., and Synal, H.-A.: Geomagnetic field intensity during the last 60,000 years based on 10Be and 36Cl from the Summit ice cores and 14C. Quat. Sci. Rev., 10.1016/j.quascirev.2005.01.012, 2005.
[26] Oliver, R., and Ballester, J.L.: The 158-Day Periodicity in the Rising Phase of Cycle 23, in Proc. SOLSPA: The Second Solar Cycle and Space Weather Euro conference, Vico Equense, Italy, 24-29 September 2001, ESA SP-477, February 2002.
[27] Polygiannakis, J., Preka-Papadema, P., and Moussas, X.: On signal-noise decomposition of timeseries using the continuous wavelet transform: application to sunspot index, Mon. Not. R. Astron. Soc. 343, 725-734, 2003.
[28] Richardsson, J.D., Paularena, K.I., and Belcher, J.W.A., J. Geophys. Res. Lett., 21, 1559, 1994.
[29] Schatten, K.: Solar activity prediction: Timing predictors and cycle 24, J. Geophys., Res., 107, 15-1–15-7, 2002.
[30] Scherrer, P.H. Wilcox, J.M., Svalgaard, L., Duvall, T.L., Dittmer, P.H., and Gustafson, E.K.: The mean magnetic field of the sun: Observations at Stanford, Solar Physics, 54, 353-361, 1977.
[31] Svalgaard, L., Cliver, E. W., and Kamide, Y.: Sunspot cycle 24: Smallest cycle in 100 years?, Geophys. Res. Lett., 32, L01104, doi:10.1029/2004GL021664, 2005.
[32] Torrence, C., and Compo, G.P.: A practical guide to wavelet analysis, Bull. Am. Meteorol. Soc., 79, 61-78, 1998.
[33] Wernik, A.W., and Grzesiak, M.: Analysis of ionospheric plasma turbulence with the wavelet transform, in Sadowski, M., and H. Rothkaehl, (eds.), Proc. Int. Symp. Plasma 97, Jarnoltowek, Space Research Center, Polish Academy of Sciences, Vol. 1, p. 391, 1997.

J. Phys. IV France **139** (2006) 175–183
© EDP Sciences, Les Ulis
DOI: 10.1051/jp4:2006139013

Towards quantifying the contribution of the Antarctic ice sheet to global sea level change

M.R. van den Broeke[1]

[1] *Institute for Marine and Atmospheric research, Utrecht University, PO Box 8005, 3508 TA Utrecht, The Netherlands*

Abstract. At present, the mass balance of the Antarctic Ice Sheet (AIS) and its contribution to global sea level change are poorly known. Current methods to determine AIS mass balance as well as the inherent uncertainties are discussed. Special emphasis is placed on the increasingly important role of regional atmospheric climate models, which can reduce the uncertainties in surface accumulation, the correction for the firn layer depth and density in ice thickness calculations and moreover help in interpreting surface elevation changes in terms of accumulation and firn density variability. Some recent advances in these fields of research are presented.

1. INTRODUCTION

The Antarctic ice sheet (AIS) is the largest freshwater source on Earth, larger by an order of magnitude compared to the Greenland ice sheet and even by two orders of magnitude compared to all other glaciers and ice caps combined. The Antarctic climate is too cold to allow for significant melt and runoff, and model studies suggest that the first order effect of a warmer climate on AIS mass balance would be to increase snowfall on the ice sheet and hence lead to sea level lowering [1]. But recent results from both Antarctica and Greenland strongly suggest that fringing ice shelves play a central role in modulating the dynamic behaviour of the inland ice [2-6]. In turn, these ice shelves have proven to be very sensitive to changes in atmospheric/oceanic temperature [7, 8]. Ice shelf disintegration through enhanced meltwater ponding and basal melt may, in the future, be responsible for unexpectedly rapid changes in mass balance of the entire AIS.

Presently, three methods are used to estimate AIS mass balance: i) repeated weighing of the ice sheet using remotely sensed gravity anomalies, ii) equating mass input (accumulation) and output (solid ice flux) for individual ice drainage basins and iii) remotely sensed elevation changes in time (radar/laser altimetry). All three methods have their advantages and deficiencies, while methods ii) and iii) are partly complementary. Several recent studies used these techniques to estimate the mass balance of the East Antarctic ice sheet (EAIS), the West Antarctic ice sheet (WAIS) and their sum (Table 1) [9-12]. Despite three of the studies employing the same data set and the rapid, unambiguous thinning of some coastal glaciers in West Antarctica [6, 13], there is clearly no consensus over the contribution of the AIS to global sea level change. Although the three elevation change studies suggest that the mass of the EAIS increases (Table 1), other work based on ice cores and regional climate modelling indicates no significant increase in Antarctic precipitation over the last 25-50 years [14-16]. Results of studies 1-3 are also in conflict with recent (and particularly since 1999) sea level rise and ocean freshening trends [17].

This lack of consensus is due to the paucity of ice dynamical and climate data from Antarctica. For example, the average area density of accumulation observations on the Antarctic ice sheet is $1/6,000\,\mathrm{km}^2$, sixty times less than what is deemed necessary to capture precipitation variability at middle latitudes. This gap cannot be closed simply by increasing the number of *in-situ* observations, for these are expensive and time-consuming: a smart combination of new observations, modelling and remote sensing techniques must be used. In this paper we highlight the role of regional atmospheric climate modelling to narrow down the uncertainty margins in AIS mass balance estimates.

Table 1. Estimates of Antarctic ice sheet mass balance in 10^{12} kg per year. EAIS = East Antarctic Ice Sheet, WAIS = Wast Antarctic Ice Sheet, AIS = Antarctic Ice Sheet.

Study	*Ref.*	*Method*	*Period*	*EAIS*	*WAIS*	*AIS*
Davis and others (2005)	[9]	Radar altimetry	1992-02	$+45 \pm 7$	-	-
Zwally and others (2005)	[10]	Radar altimetry	1992-02	$+16 \pm 11$	-47 ± 4	-31 ± 12
Wingham and others (2006)	[11]	Radar altimetry	1992-02	$> +27$	< 0	$+27 \pm 29$
Velicogna and Wahr (2006)	[12]	Gravity data	2002-05	0 ± 56	-148 ± 21	-152 ± 80

2. METHODS AND UNCERTAINTIES

2.1 Repeated weighing of the ice sheet

This method uses the Gravity Recovery And Climate Experiment (GRACE) satellite, and is completely independent of the other two methods. AIS mass balance estimates obtained from the GRACE satellite cover both the EAIS and WAIS, but only for a short period of time (2002-2005, Table 1). The results imply a dramatic mass loss for the WAIS (-148 ± 21 km^3, equivalent to 0.4 mm of sea level rise per year) but approximate balance for the EAIS [12]. Apart from the uncertainties deriving from the short time period over which the trends have been calculated (3 years), several corrections must be applied to GRACE data, each introducing additional uncertainties. An example is the correction for upward motion of the Earth's crust after the last glacial maximum (postglacial rebound), which is poorly constrained over Antarctica. Another example is the atmospheric mass correction, for which the air pressure at the surface of the ice sheet must be known at better than 1 hPa precision. Given the sparse network of meteorological stations in Antarctica one must resort to modelled meteorological fields, which are known to be unreliable. This method will not be further discussed here.

2.2 Equating mass input and output

This method relies on estimating the difference between snowfall over a catchment area (mass input) and the outgoing ice flux along the catchment gate at the ice sheet grounding line (mass output). It requires techniques from various disciplines: i) meteorological modelling and in-situ observations to obtain the surface accumulation distribution, ii) satellite radar interferometry to obtain the flow speed of the narrow glaciers through the flux gates at the ice sheet grounding line and iii) satellite altimetry to accurately delineate the ice drainage basins as well as to obtain the elevation/thickness of the floating glacier. Because the resulting mass imbalance represents the difference between two large terms, this method is sensitive to uncertainties in the individual components. That is why until now, this method has mainly been applied to individual AIS drainage basins with sufficient data coverage [18, 19]. However, using remotely sensed ice velocities/thickness and elevation data and an ever-improving surface accumulation distribution, an AIS-wide assessment of the mass balance will soon be feasible.

Another problem is created by using a multitude of data sources covering different time intervals: local variations in mass balance may be driven by short or long term changes in ice velocity and/or accumulation that may or may not be related to recent climatic forcing. Combining this method with remotely sensed elevation change measurements (method iii, see below) can help elucidate whether an imbalance has a dynamic or snowfall driven origin and whether it is a long-term, secular trend or a short-term perturbation.

2.3 Remotely sensed elevation changes in time

This method assumes a direct relation between changes in ice sheet elevation (as observed by satellite radar/laser altimeters) and changes in ice volume. Datacoverage is not perfect: no data exist for the area south of 81.5°, while in the steeply sloping coastal margins of the ice sheet the signal to noise ratio is too small to obtain accurate height changes. This is especially problematic for the narrow, fast flowing outlet glaciers, which are not adequately resolved by present-day sensors but which are expected to react most rapidly to environmental changes. With the loss of CryoSat this problem will not be resolved for at least another 5-7 years.

Another major uncertainty is introduced by the fact that the altimeter time series are too short to infer the underlying cause of elevation changes (ice dynamics vs. snowfall). There are indications that decadal changes in accumulation and firn densification are a major source for the observed elevation changes [6]. Firn densification rate is a strong non-linear function of firn temperature; Zwally and others [10] corrected the observed elevation trends for densification changes using a firn densification model driven by AVHRR-derived surface temperatures [19], but owing to a lack of data they had to assume a constant accumulation rate. The three recent studies using radar altimetry listed in Table 1 suggest that over the 11-year period 1992-2002, the EAIS is gaining mass, while the WAIS looses a similar amount. These results are somewhat in conflict with recently observed salinity changes, which would require a freshwater source in the southern hemisphere [17].

3. REDUCING THE UNCERTAINTIES USING A REGIONAL ATMOSPHERE MODEL

Regional atmospheric climate models run at higher resolution than global models, and are therefore well suited to narrow down the uncertainties in AIS mass balance calculations. Below we describe three applications using output of the regional Antarctic climate model RACMO2/ANT at 55 km horizontal resolution [21], namely i) the surface accumulation distribution, ii) the firn depth correction and iii) elevation changes owing to firn densification and accumulation variability.

3.1 Surface accumulation

Errors in surface accumulation compilations based on interpolation of *in-situ* observations may differ by as much as 20% on the basin scale [22-24]. To make matters worse, the quality of many 'older' (pre-1980) *in-situ* observations has recently been questioned (C. Genthon, personal communication 2006), which affects *all* existing observation-based accumulation compilations.

A more observation-independent approach is to nudge output of a (regional) meteorological model towards fewer, but strict quality-controlled observations. This ensures that data sparse regions still get a physically correct accumulation magnitude [15], while potentially remaining erroneous measurements do not negatively impact large areas of the domain. Van den Berg and others [21] showed that even without any nudging, modelled accumulation from RACMO2/ANT compares very well to 1900 *in situ* quality-controlled observations over the AIS (R = 0.82). This is remarkable if one keeps in mind that the accumulation periods are different and that the model is not nudged towards observations. Fig. 1 shows the accumulation distribution after nudging in 500 m elevation intervals. The surface accumulation field shows detail on scales much smaller than the irregularly distributed observations, confirming the added value of this approach. Van den Broeke and others [25] showed that processes that are not yet explicitly modelled in RACMO2/ANT, such as the erosion by and sublimation of drifting snow, could account for the remaining differences between modelled and observed accumulation. Modelling these processes explicitly and further increasing the model resolutions is a line of future research.

Fig. 1 compares the modelled and observed accumulation. The model captures the steep accumulation gradient between the wet coastal zone and the dry interior. It also reliably reproduces the accumulation gradients over the flat ice shelves. New accumulation data should confirm the band with very high accumulation rates (>2000 kg m^{-2} yr^{-1}) along coastal West Antarctica and the west coast of the Antarctic Peninsula.

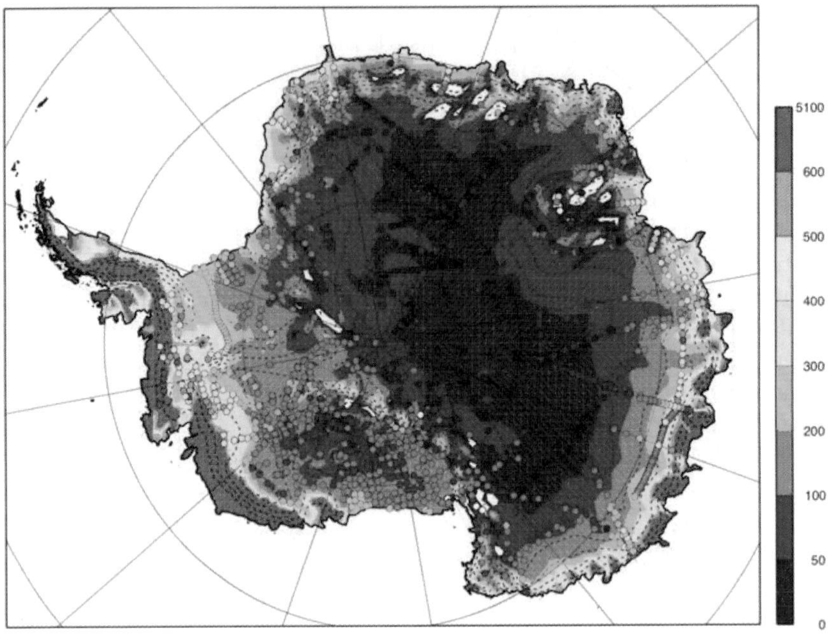

Figure 1. Observed (circles) and modelled accumulation in kg m $^{-2}$ year^{-1}.

3.2 Ice thickness and firn correction depth

For most Antarctic glaciers, ice thickness at the grounding line has not been directly measured and must be calculated from accurate satellite elevation observations and a floatation criterion. The floatation criterion equates the mass of the displaced seawater with the added weight of the combined ice/firn column:

$$(h_i + h_f - h_{asl})\,\rho_{sw} = h_f\rho_f + h_i\rho_i \equiv \rho_i H_i \tag{1}$$

where h_i, h_f and ρ_i, ρ_f are the thicknesses and average densities of the ice and firn layers, respectively, h_{asl} is the elevation of the floating glacier surface above sea level and ρ_{sw} is the density of seawater which for this application can be assumed constant. The equivalent ice thickness (H_i) is obtained by compressing the firn layer until it has the density of glacier ice (917 kg m^{-3}). An expression for H_i in terms of h_{asl} is obtained by eliminating h_i:

$$H_i = \frac{(h_{asl} - \Delta h)\rho_{sw}}{\rho_{sw} - \rho_i}; \quad \Delta h = h_f(1 - \rho_f/\rho_i) \tag{2}$$

where Δh is the *firn depth correction*, defined as the difference between the combined ice/firn column and the equivalent ice thickness. It is clear from (2) that to convert h_{asl} to H_i, Δh must be known at the grounding line, i.e. the firn depth and average firn density must be known.

Assessing the depth and average density of the firn layer requires medium-deep firn cores that reach the firn-ice transition at 50 – 150 m depth. These cores are rare in Antarctica and seldom drilled at the grounding line (Fig. 2). Therefore, the spatial distribution of Δh along the grounding line is poorly known at present, and Δh is often assumed constant in ice flux calculations. For a typical ice thickness of 500 m, ($h_{asl} \approx 54$ m), an uncertainty in Δh of 4 m introduces a 7.4% uncertainty in the ice flux estimate, which is a large number compared to uncertainties in the other components of the solid ice flux calculation, e.g. surface elevation from satellite laser altimetry (uncertainty <0.5 m or <1% for this example, if GLAS sensor onboard ICESAT can be used, but <5 – 10 m or 10 – 19% for this example

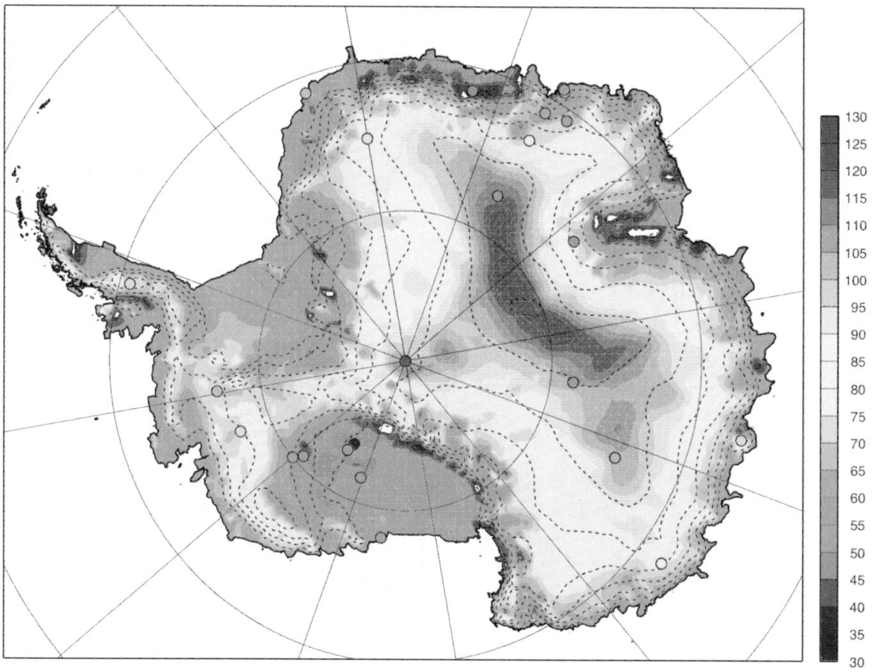

Figure 2. Observed (circles) and modelled depth of the firn layer h_f in m, taken to be the 830 kg m^{-3} density level.

if ERS1/2 data must used, [26]) and glacier velocity from radar interferometry (<5% uncertainty in column ice speed, [13]).

In the absence of significant melting, the steady-state firn densification rate depends mainly on temperature, accumulation rate and wind speed. Using a steady-state firn densification model [27] in combination with annual mean accumulation, surface temperature and near surface wind speed (1980-2004) from RACMO2/ANT we calculated the vertical density profile in Antarctica. As an example, Fig. 2 shows the resulting depth of the firn layer. Owing to the wide range of (near) surface climate conditions over the AIS, firn depth shows large spatial variability [10, 28, 29], which are well reproduced by the model. In the calm, dry and cold interior, densification is slow and the firn-layer thickness exceeds 100 m. In the windier, wetter and milder coastal zone, densification is more rapid and the firn layer shallower, typically 40 – 60 m. In regions with active katabatic winds and low precipitation rate, the firn layer may have been completely removed by snowdrift erosion and/or sublimation, exposing the glacier ice at the surface [25, 30].

From these results, a first-order estimate of the firn depth correction along the grounding line can be obtained, for use in solid ice flux calculations.

3.3 Elevation changes owing to temperature and accumulation variability

We used the same firn densification model, but in a time-dependent fashion, forced by 6-hourly temperature and accumulation rate from the regional atmospheric climate model RACMO2/ANT over the period 1980-2005. To minimize spin-up effects, the model was run for this 25-year period twice, but only the last 25 years were considered for analysis. As an example, Fig. 3 shows the results for an arbitrary location in coastal East Antarctica. This site has no melting in summer, moderately low temperatures (annual mean surface temperature 249.7 K), relatively high accumulation (522 kgm^{-2} year^{-1}) and strong near-surface katabatic winds (annual mean 10 m wind speed 13.1 m s^{-1}). Fig. 3a

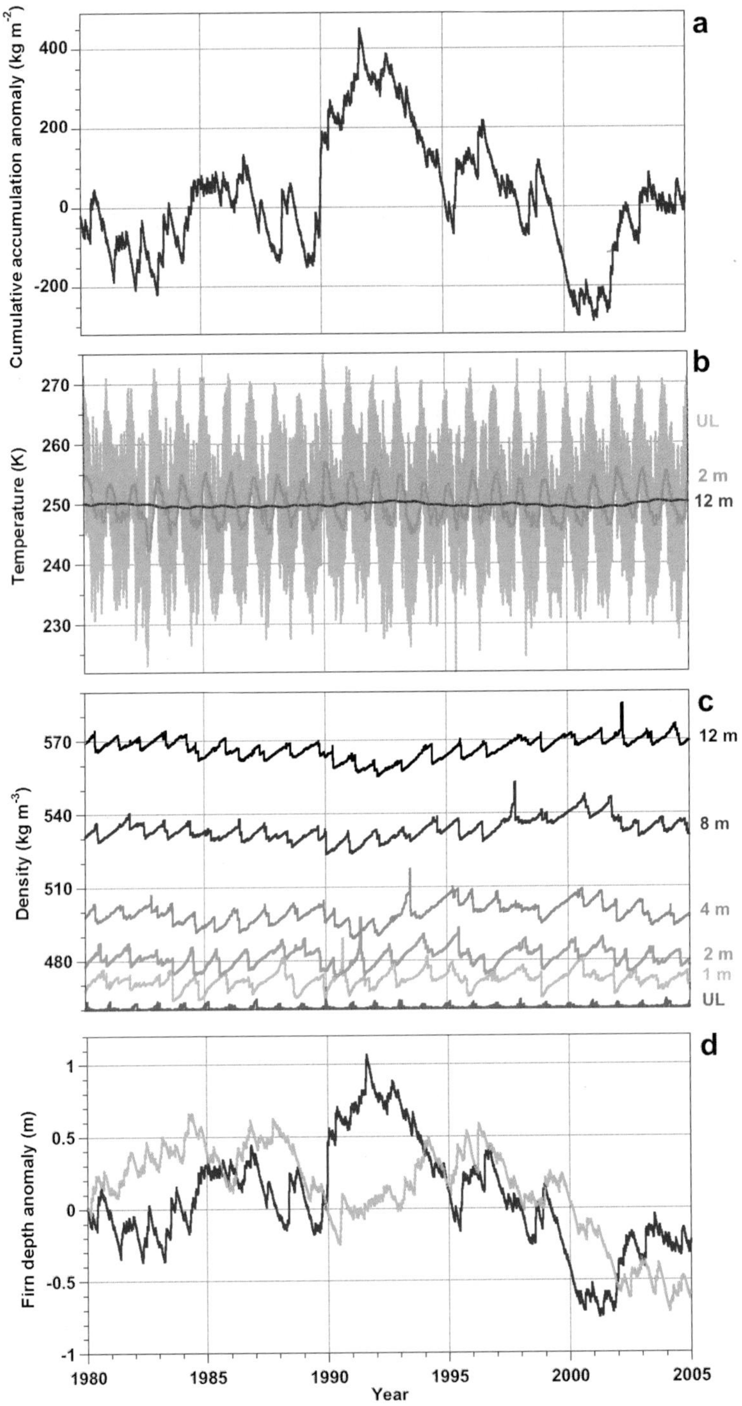

Figure 3. Modelled cumulative accumulation anomaly (a), Upper Layer (UL), 2 m and 12 m snow temperature (b), UL, 1 m, 2 m, 4 m, 8 m and 12 m snow density (c) and firn depth anomaly (d, blue line) for an arbitrary coastal points in East Antarctica. Yellow line in d) represents another arbitrary location in East Antarctica.

shows the cumulative accumulation deviation from the mean. Clearly, decadal accumulation variability is present, causing cumulative deviations as large as 80% of the annual accumulation. Fig. 3b shows that seasonal temperature variations are only important in the upper 15 m of the firn. In response to this annual temperature wave the density also shows a clear annual cycle, with more rapid densification in summer (Fig. 3c). Interannual temperature variations also have an effect, especially warm summers. For instance, the anomalously high summer surface temperatures in 1989/90 (Fig. 3b) caused a sharp density peak that is seen to slowly travel downward in the firn, reaching a depth of 12 m in 2002.

Fig. 3d shows surface elevation change resulting from non-steady densification and surface accumulation variability. *If we assume a constant vertical motion at the firn-ice and ice-bedrock boundaries, this signal represents the observed vertical displacements at the ice sheet surface.* From comparing Fig. 3a and Fig. 3d it is clear that at this site, temporal accumulation variability is the dominant mechanism explaining variations in surface height, the influence of temperature effects being only of second order. Elevation trends in Fig. 3d are as large as 10 cm per year, at the high end of the range of values reported from remote sensing studies [10-12]. Curiously, the period that is characterized by a pronounced surface lowering coincides more or less exactly with the time frame in which the ERS-1 satellite was active and on which the estimates of studies 1-3 in Table 1 are based (1992-2002). Note that this trend is not representative for the full period (1980-2005), nor for a larger area: the yellow line in Fig. 3d represents another arbitrary site in East Antarctica, showing entirely different characteristics.

These preliminary calculations show that estimating AIS mass balance from satellite altimetry alone is undesirable; it should be combined with method ii). Once the above calculations have been applied over the whole ice sheet (>4000 gridpoints for the present configuration), the effects of firn densification and accumulation variability can be isolated and subtracted from the observed vertical velocity, yielding the real mass imbalance. This can then be compared to results of methods i) and ii).

4. SUMMARY AND CONCLUSIONS

The balance state of the Antarctic Ice Sheet (AIS) and its contribution to global sea level change are poorly known at present. Three methods are currently used to determine the AIS mass balance, two of which involve surface accumulation and the (time-space) structure of the firn layer. Major uncertainties in these methods are caused by i) the poorly known present-day surface accumulation distribution, ii) the poor constraint on the firn layer depth and density at the grounding line to convert elevation to ice thickness and iii) the poor understanding of accumulation variability and time-dependent firn densification in the interpretation of remotely-sensed surface elevation changes. Recently, some progress has been made in minimizing these uncertainties, using a mix of atmospheric and firn densification modelling. Several of these efforts have been presented in this paper.

References

[1] Church, J.A., J.M. Gregory, P. Huybrechts, M. Kuhn, K. Lambeck, M.T. Nhuan, D. Qin and P.L. Woodworth, 2001: Changes in sea level, in J. T. Houghton, Y. Ding, D. J Griggs, M. Noguer, P. J. Van der Linden and D. Xiaou (eds.), *Climate Change 2001: The Scientific Basis*, Cambridge University Press, Cambridge and New York, 639-694.

[2] Shepherd, A., D. Wingham, T. Payne, Skvarca, 2003: Larsen Ice Shelf has progressively thinned, *Science* **302**, 856-859.

[3] De Angelis, H. and P. Skvarça, 2003: Glacier surge after ice shelf collapse, *Science* **299**, 1560-1562.

[4] Shepherd, A., D. Wingham and E. Rignot, 2004: Warm ocean is eroding West Antarctic Ice Sheet, *Geophysical Research Letters* **31**, L23402, doi:10.1029/2004GL021106.

[5] Rignot, E., G. Casassa, P. Gogineni, W. Krabill, A. Rivera, and R. Thomas. 2004. Accelerated ice discharge from the Antarctic Peninsula following the collapse of Larsen B ice shelf, *Geophysical Research Letters* **31**, L18401, doi:10.1029/2004GL020697.

[6] Thomas, R., E. Rignot, G. Casassa, P. Kanagaratnam, C. Acuña, T. Akins, H. Brecher, E. Frederick, P. Gogineni, W. Krabill, S. Manizade, H. Ramamoorthy, A. Rivera, R. Russell, J. Sonntag, R. Swift, J. Yungel, and J. Zwally, 2004: Accelerated Sea-Level Rise from West Antarctica, *Science* **306**, 255-258.

[7] Scambos, T., C. Hulbe, and M. Fahnestock, 2003: Climate-induced ice shelf disintegration in the Antarctic Peninsula, in: *Antarctic Peninsula Climate Variability: Historical and Paleoenvironmental Perspectives*, E. Domack et al. (eds.), *Antarctic Research Series* **79**, 77-92.

[8] Van den Broeke, M. R., 2005: Strong surface melting preceded collapse of Antarctic Peninsula ice shelf. *Geophysical Research Letters* **32**, L12815, doi:10.1029/2005GL023247.

[9] Davis, C. H., Y. Li, J. R. McConnell, M. M. Frey and E. Hanna, 2005: Snowfall-driven growth in East Antarctic ice sheet mitigates recent sea-level rise, Sciencexpress report, 10.1126/science.1110662.

[10] Zwally, H. J., M. B. Giovinetto, J. Li, H. G. Cornejo, M. A. Beckley, A. C. Brenner, J. L. Saba and D. Yi, 2005: Mass changes of the Greenland and Antarctic ice sheets and shelves and contributions to sea-level rise: 1992–2002, *Journal of Glaciology* **51(175)**, 509-527.

[11] Wingham D. J., A. Shepherd, A. Muir and G. J. Marshall, 2006: Mass balance of the Antarctic ice sheet, *Philosophical Transactions A Math Phys Eng Sci.* **364(1844)**, 1627-35.

[12] Velicogna, I. and J. Wahr, 2006: Measurements of time-variable gravity show mass loss in Antarctica, *Sciencexpress*: 10.1126science.1123785.

[13] Rignot, E., C. Davis, Y. Li, J. Bamber, R. Arthern, M. R. van den Broeke, W. J. van de Berg and E. van Meijgaard, 2006: A net source of ocean mass from the coastal thinning of Antarctic glaciers.

[14] Van de Berg, W. J., M. R. van den Broeke, C. H. Reijmer and E. van Meijgaard, 2005: Characteristics of the Antarctic surface mass balance, 1958-2002, using a Regional Atmospheric Climate Model, *Annals of Glaciology* **41**, 97-104.

[15] Van den Broeke, M. R., W. J. van de Berg and E. van Meijgaard, 2006: Snowfall in coastal West Antarctica much greater than previously assumed, *Geophysical Research Letters* **33**, L02505, doi:10.1029/2005GL025239.

[16] Monaghan, A. J., D. H. Bromwich, R. L. Fogt, S-H Wang, P. A. Mayewski, D. A. Dixon, A. Ekaykin, M. Frezzotti, I Goodwin, E. Isaksson, S. D. Kaspari, V. I. Morgan, H. Oerter, T. D. Van Ommen, C. J. Van der Veen, J. and Wen, 2006: Insignificant change in Antarctic snowfall since the International Geophysical Year, *Science* **313**, 827-831.

[17] Munk, W., 2003: Ocean Freshening, Sea Level Rising, *Science* **300**, 2041-2043.

[18] Joughin, I. and S. Tulaczyk, 2002: Positive Mass Balance of the Ross Ice Streams, West Antarctica, *Science* **295**, pp. 476 – 480, DOI: 10.1126/science.1066875.

[19] Rignot, E., 2002: East Antarctic Glaciers and Ice Shelves Mass Balance from Satellite Data, *Annals of Glaciology* **34**, 217-227.

[20] Comiso, J. C., 2000: Variability and trends in Antarctic surface temperatures from in situ and satellite infrared measurements, *Journal of Climate* **13**, 1674-1696.

[21] Van de Berg, W. J., M. R. van den Broeke, E. van Meijgaard and C. H. Reijmer, 2006: Reassessment of the Antarctic surface mass balance using calibrated output of a regional atmospheric climate model *Journal of Geophysical Research* **111**, D11104, doi:10.1029/2005JD006495.

[22] Vaughan, D. G., J. L. Bamber, M. Giovinetto, J. Russell and A. P. R. Cooper, 1999: Reassessment of net surface mass balance in Antarctica, *Journal of Climate* **12**, 933-946

[23] Giovinetto, M. B., and H. J. Zwally, 2000: Spatial distribution of net surface accumulation on the Antarctic ice sheet, *Annals of Glaciology* **31**, 171-178.

[24] Arthern, R. J., D. P. Winebrenner, and D. G. Vaughan, 2006: Antarctic snow accumulation mapped using polarization of 4.3-cm wavelength microwave emission, *Journal of Geophysical Research* **111**, D06107, doi:10.1029/2004JD005667.

[25] Van den Broeke, M. R., W. J. van de Berg, E. van Meijgaard and C. H. Reijmer, 2006: Identification of Antarctic ablation areas using a regional atmospheric climate model, *Journal of Geophysical Research* in press.

[26] Bamber, J. and J.-L. Gomez-Dans, 2005: The accuracy of digital elevation models of the Antarctic continent, *Earth Planetary Science Letters* **237**, 516-523.

[27] Barnola, J.-M., P. Pimienta, D. Raynaud, and Y. S. Korotkevich, 1991: CO2 -climate relationship as deduced from the Vostok ice core: a re-evaluation of the air dating, *Tellus* **43(B)**, 83-90.

[28] Kaspers, K. A., R. S. W. van de Wal, M. R. van den Broeke, N. P. M. van Lipzig and C. A. M. Brenninkmeijer, 2004: Model calculations of the age of firn air across the Antarctic continent, *Atmospheric Chemistry and Physics* **4**, 1817-1853.

[29] Li, J. and H. J. Zwally, 2004: Modeling the density variation in the shallow firn layer, *Annals of Glaciology* **38**, 303-313.

[30] Winther, J. -G., M. Nørman Jespersen and G. E. Liston, 2001: Blue-ice areas in Antarctica derived from NOAA AVHRR satellite data, *Journal of Glaciology* **47(157)**, 325-334.

J. Phys. IV France **139** (2006) 185–196
© EDP Sciences, Les Ulis
DOI: 10.1051/jp4:2006139014

The challenge from ice cores: Understanding the climate and atmospheric composition of the late Quaternary

E.W. Wolff[1]

[1] *British Antarctic Survey, Natural Environment Research Council, High Cross, Madingley Road, Cambridge CB3 0ET, UK*
e-mail: ewwo@bas.ac.uk

Abstract. The Quaternary period is a critical one for understanding the working of the Earth System because it shows a wide range of climates under a geography similar to the present. Ice cores are an important palaeorecord because they record aspects of the atmosphere (including trace gas concentrations) rather directly. This paper takes advantage of recently published results from ice cores completed since an earlier ERCA chapter was published. These extend the ice core record from Antarctica back towards 800,000 years, confirming the close relationship between different parameters (particularly CO_2 and Antarctic temperature), but showing a different behaviour (with cooler interglacials) in the period preceding 450,000 years before present compared to the later period. New records of the last glacial cycle have documented the entire suite of rapid climate warmings (Dansgaard-Oeschger events) in this period, shown the behaviour of Antarctica during these events, and given us a first clear view of Greenland climate in the later parts of the last interglacial. Taken together these results present a compilation of the behaviour of the Earth that challenges palaeoclimatologists and Earth System modellers towards better understanding of the system.

1. INTRODUCTION

In order to assess the future trajectory of the Earth's atmospheric composition and climate, under pressure from anthropogenic changes, we have to understand the natural evolution of the system. Apart from providing the context of present and future changes, such studies also allow us to see numerous examples of natural processes in action: processes that may mitigate or exaggerate changes in composition, or determine the detailed climatic responses that can occur. However, the instrumental era is typically a century or less for meteorological parameters, and just a few decades for measurements of atmospheric chemical composition. Beyond the instrumental era, we are forced to use palaeoclimatic and palaeoenvironmental archives to understand how the Earth has behaved.

While such archives can go back deep into Earth history, the Quaternary period has a particular importance. Beginning around 2.6 million years ago (the exact definition of the Quaternary is currently under discussion), it is the period during which significant northern hemisphere ice sheets have appeared. Crucially the topography of the Earth – the positions of land masses, ocean gateways and mountain chains – has been similar to today throughout at least the latter half of the Quaternary. It therefore represents a simplified set of conditions in which the Earth has exhibited a wide range of behaviours within similar boundary conditions, and with external forcings that can be estimated quite accurately.

A wide range of palaeoarchives allows access to the Quaternary. Some of them, such as tree ring inventories, cover only the last few thousand years, but at very high time resolution, while others, such as marine sediments, cover mush longer periods at rather low resolution. Ice cores hold a unique place in the spectrum of available archives because of the range of parameters they describe. They form a particularly direct archive of atmospheric composition, both the trace gases held in air bubbles in the ice, and aerosol and soluble gases trapped in surface snow. With recent work that I will describe below, the ice core record now reaches to 800 kyr BP (800,000 years before present), well into the mid-Quaternary.

The basic principles behind the ice core record, and how it is obtained, are already described in a chapter in an earlier ERCA volume [1], and will not be repeated here. Ice core records describing the last few hundreds to thousands of years in good detail are also summarised in the earlier chapter. In this paper, I will update the information obtained from deep Greenland and Antarctic ice cores in the light of recent major ice drilling projects, notably NorthGRIP [2, 3] in Greenland, and Dome Fuji [4, 5] and EPICA [6, 7] in Antarctica. The aim will be to describe the general nature of late Quaternary climate change, with particular emphasis on new information from ice cores, and to discuss the challenges the new data pose to our understanding of how the Earth System functions.

2. NEW ICE CORE DRILLINGS

Since the previous ERCA paper on the ice core record [1], several new deep ice cores have been drilled. In Greenland, the NorthGRIP project, led by Danish scientists but involving 9 nations, was completed. Its aim was to extend the GRIP record through the last interglacial. Although this goal was not fully realised, NorthGRIP did penetrate through the younger part of the last interglacial, to 123 kyr BP. It therefore provided the first clear record of the last glacial inception from Greenland, including the early part of the last glacial period. The bed of the ice was found to be at the melting point, so that large quantities of refrozen meltwater from the bed of the ice sheet could be collected.

Fig. 1 shows the locations of major Antarctic ice cores, including the ones discussed in this paper. The Japanese Dome Fuji project reached a depth of 2503.5 m (equivalent to 330 kyr BP) in 1996. This core therefore replicates a large part of the Vostok era but in a completely different sector of the Antarctic. A new project to drill through to the bed at the same place has been conducted recently, and has reached 3028.5 m, just above the bed, with the ice at the base expected to be significantly older than the oldest Vostok ice.

While a number of other cores have been drilled, covering the most recent part of the Quaternary, the most prominent European effort of recent years has been the European Project for Ice Coring in Antarctica (EPICA). This project has involved numerous scientists from 10 European nations. Its goal

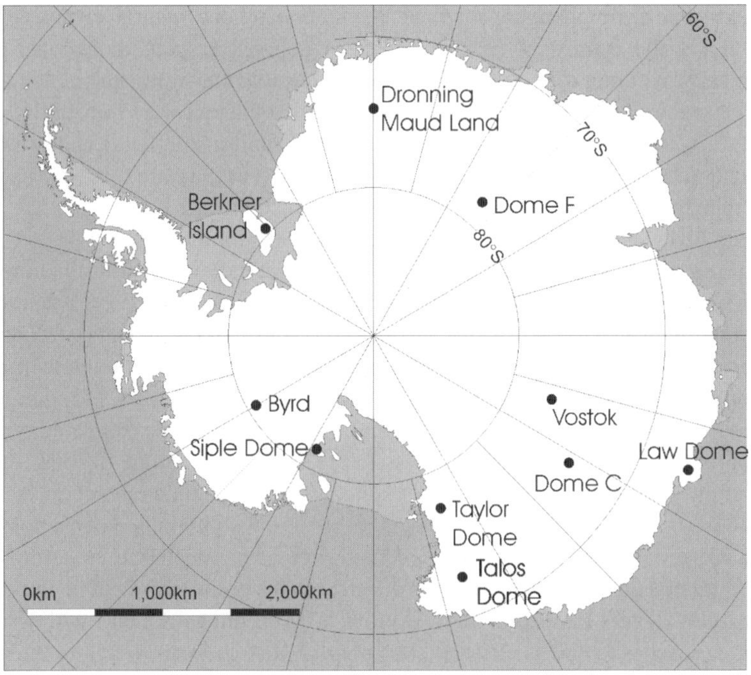

Figure 1. Map of Antarctica, showing the principal ice core drilling sites.

was to drill two cores to bedrock: one at Dome C, the other at a site in Dronning Maud Land (DML). The Dome C drilling ended at a depth of 3259 m, a few metres above the bed, in December 2004. The climate record in the core extends to just over 800 kyr BP, while the deepest 70 m of ice shows a very flat signal in most components suggesting that a flow disturbance has occurred. The deepest part will not be discussed further in this paper.

A second core was drilled by EPICA at DML (Kohnen station). There the bed was reached at 2774.15 m in January 2006, and again refrozen meltwater was obtained from the base. Ice from the last glacial cycle will be discussed in this paper although the core certainly contains ice extending further than that.

3. GLACIAL-INTERGLACIAL CYCLES IN ANTARCTIC ICE CORES

The general nature of late Quaternary climate is already well-known from marine archives [8] and, for the last 420 kyr, from the Vostok ice core [9]. The main contribution of the new cores has been to confirm the broad-scale significance of the Vostok results and to extend our knowledge of the parameters only available from ice out to 800 kyr. The climate of this period is dominated by the existence of glacial-interglacial cycles with a period of roughly 100,000 years. During the cold periods, large ice sheets existed over North America and northern Europe, sea level was substantially lower than at present (by 120 m at the Last Glacial Maximum (LGM)). The 100 kyr periodicity appears to be controlled by the period of the eccentricity of the Earth's orbit around the Sun, although it remains uncertain why this rather weak radiative forcing should dominate the record.

3.1 Spatial pattern of climate change across Antarctica

The Dome Fuji deuterium record [4] is extremely similar to that of Vostok over their common period; deuterium is taken as a good proxy for site temperature [10]. Apparent differences in timing of events certainly reflect issues regarding the age scales of one or both cores. Small differences in the amplitudes of change likely reflect genuine climatic differences as well as possible changes in the relative altitude of the two sites as the ice sheet grew and shrunk. However, the overwhelming commonality of the climate signal at the two sites from opposite sides of the continent shows that central East Antarctic climate as a whole is well-described by the record from each site.

3.2 Antarctic climate before 420 kyr BP

A similarly common signal is seen also in the Dome C deuterium record (Fig. 2) during the last 420 kyr (the "Vostok era"). However, the pattern of climate variability is very different in the previous 3 climatic cycles, back to 740 kyr BP (Fig. 3) [7]. The most obvious difference is that the interglacials were substantially cooler than recent ones, and also substantially longer (i.e. a more even distribution between glacial and interglacial time periods). However, a caveat must be added to the latter statement. The records here are shown on a glaciological timescale known as EDC2, derived by using an ice flow and snow accumulation model with uncertain parameters fixed by assigning time windows to a few fixed events. At the time of writing, a new timescale (EDC3) has been developed, in which many additional constraints have been introduced, and this will be used for papers extending the record through the full 800 kyr. In EDC3, at least some of the early interglacials are likely to be shorter than they appear with EDC2, although the general pattern will remain the same.

We have no clear explanation at present for the change in amplitude occurring around 450 kyr BP. The dominant period appears to remain at 100 kyr, which is quite surprising since we know from the marine record that periods of 40 kyr (corresponding to the period of obliquity, tilt of Earth's axis) tend to dominate in the period just before the base of the Dome C record. In any case, the change in style clearly indicates that the pattern we have seen in the last 4 glacial cycles is not the only pattern that can

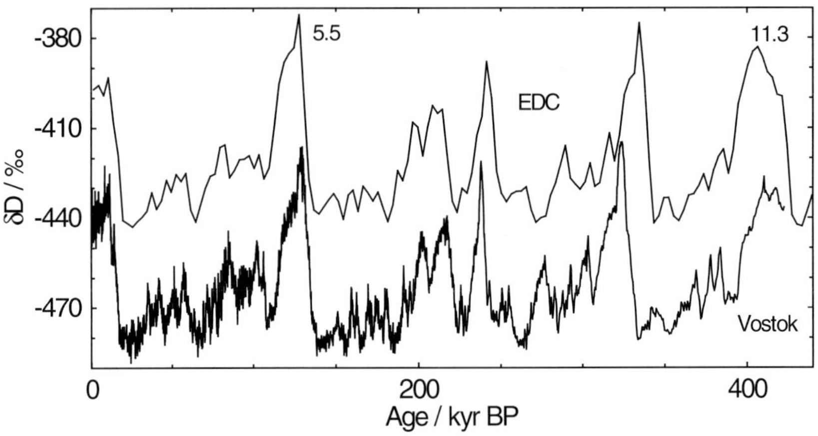

Figure 2. Comparison of the deuterium (temperature proxy) records from Vostok [9] and EPICA Dome C [7] (EDC, 3 kyr averages). Differences in timing result from differences in the construction of the respective timescales (GT4 and EDC2 respectively). Vostok has lower deuterium values because it is a colder site. Marine isotope stages (MIS) 5.5 and 11.3 are marked.

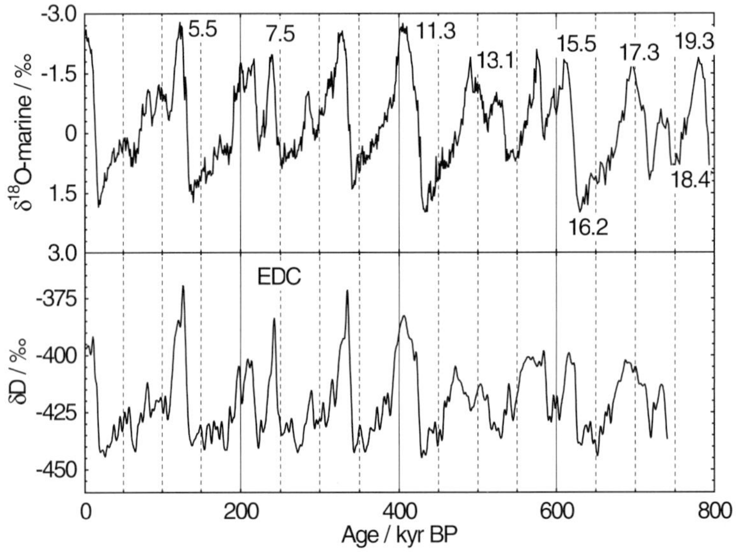

Figure 3. Deuterium (temperature proxy) [7] at the EPICA Dome C site (EDC2 timescale) over the last 740 kyr. The oxygen isotope record from a combined marine stack [8] is also shown for reference on its own timescale: this record represents some combination of global ice volume and deep sea temperature.

occur, and an obvious question is to determine what happened to other parameters describing aspects of the Earth system in the earlier period with weak interglacials.

3.3 The trace gas record

Carbon dioxide (CO_2), methane (CH_4) and nitrous oxide (N_2O) concentrations have all been measured in the Dome C core between 390 and 650 kyr BP, as well as in more recent ice. Vostok data, extending to 414 kyr, have also been used to compile a complete dataset to 650 kyr BP. This combination is

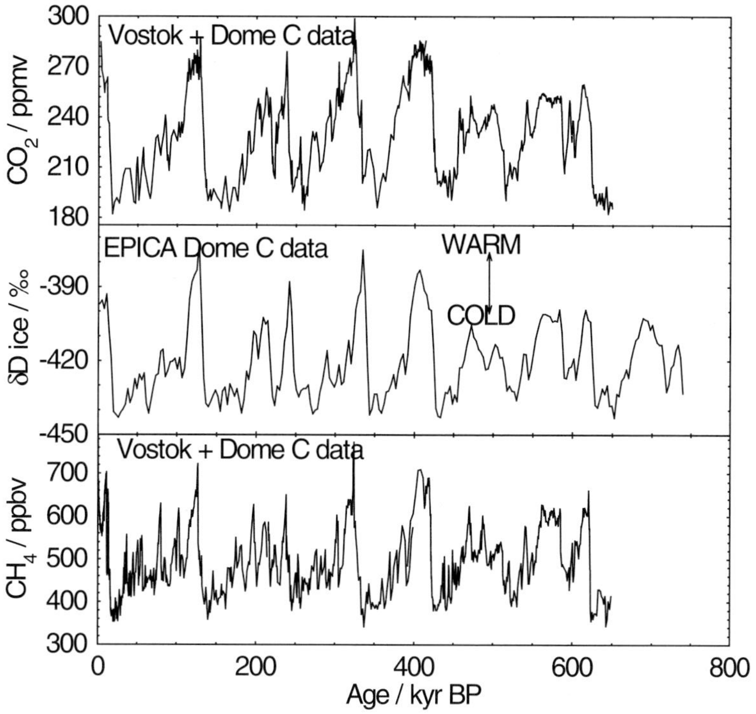

Figure 4. CO_2 [11], deuterium, and CH_4 [12] for the last 650 kyr. Deuterium is from EPICA Dome C, while the trace gases are a composite from Vostok [9] and Dome C.

justified because trace gases have rather long lifetimes, and are quite well-mixed in the atmosphere. Concentrations at Vostok and Dome C for a given time should therefore be identical.

The CO_2 record (Fig. 4) [11] shows clearly that weak interglacials also have lower CO_2 concentrations than later, stronger interglacials. Typical values in marine isotope stage (MIS) 13 and 15 are 240-250 ppmv, rather than the 280-300 ppmv that is typical of later warm periods. There is therefore no specific fixed concentration to be expected in an interglacial; rather CO_2 concentration and temperature (deuterium) scale linearly between climate periods. In fact they track each other remarkably closely; before the new record was published, a challenge was issued to the modelling community to predict the form of the CO_2 record over the pre-Vostok era. While a number of (mainly) conceptual and correlative models of different complexities were proposed [13], there is little scope for improvement on one that models CO_2 concentration simply as a function of deuterium at Dome C. This close relationship can certainly be attributed to a number of processes, but in any case implies that the most important determinants of CO_2 concentration over this timescale reside in the Southern Ocean. This impression is reinforced by the highly-detailed record of CO_2 change that has been produced from Dome C for the transition from the last glacial period, between about 18 and 11 kyr BP [14]. Here (Fig. 5) CO_2 and Antarctic temperature again track each other over several thousand years (with a best estimate that temperature leads by a few hundred years). The CO_2 increases in two major phases, just as Antarctic temperature, punctuated by the Antarctic Cold Reversal, does. Two episodes of sharp CO_2 increase by about 10 ppmv are contemporaneous with the rapid Greenland warmings (and methane increases) at the start of the Allerod-Bolling warm period and the end of the Younger Dryas cold period, and therefore may be related to northern hemisphere processes. The overall pattern of the increase is consistent with the idea that the temperature and CO_2 are operating as an amplifier in positive feedback!

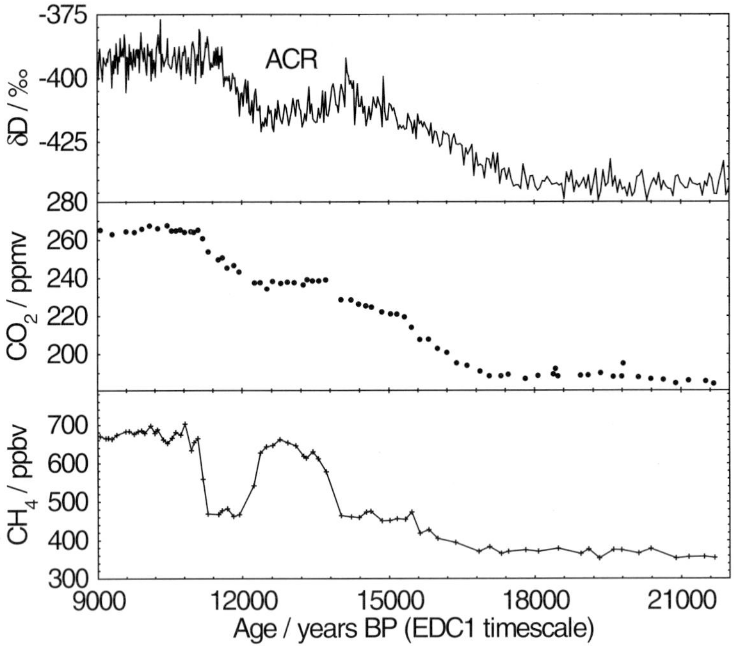

Figure 5. Deuterium, CO_2 and CH_4 across Termination I [14], the transition from the last glacial period to the Holocene. The Antarctic Cold Reversal (ACR) is marked. Methane acts here as a proxy for Greenland temperature change.

Methane (Fig. 4) also shows strong similarities with deuterium [12] over the longer timescale, including lower values in weak interglacials than in strong ones. However there is, even at low resolution, an obviously greater variability in CH_4 during glacial periods than exists in either CO_2 or deuterium. In the most recent glacial cycle, we know that this variability corresponds to rapid climate changes (Dansgaard-Oeschger (D-O) events) [15] seen in the Greenland isotopic (temperature) record, and it seems reasonable to assume that this may also be the case in earlier glacials, and that CH_4 concentrations may actually resemble more closely northern hemisphere temperature rather than Antarctic temperature. N_2O (not shown here) [12] also shows glacial-interglacial variability, but there is no obvious change in concentration between weak and strong interglacials.

It is worth noting that in the last two centuries, both CO_2 and CH_4 emerged well above the level encountered at any time in the last 650 kyr. CO_2 remained in the range 180-300 ppmv until the early 20th century [16]: the concentration in 2006 is over 380 ppmv. CH_4 was in the range 340-780 ppbv until the late 19th century [17], but has now reached 1780 ppmv. The rate of change is also unprecedented in the last 650 kyr.

3.4 The ion chemistry record of aerosol components

Along with the extended time period covered by the ion chemistry records from the Dome C core have come some new (although in some cases still controversial) interpretations. The focus has so far been on new data related to components of terrestrial dust and sea salt, and sulfur compounds [18].

3.4.1 Terrestrial dust

Components of terrestrial dust (including dust concentration itself [7], non-sea-salt Ca, and Fe [18]) show, as already observed in shorter cores, greatly enhanced concentrations and fluxes during peak

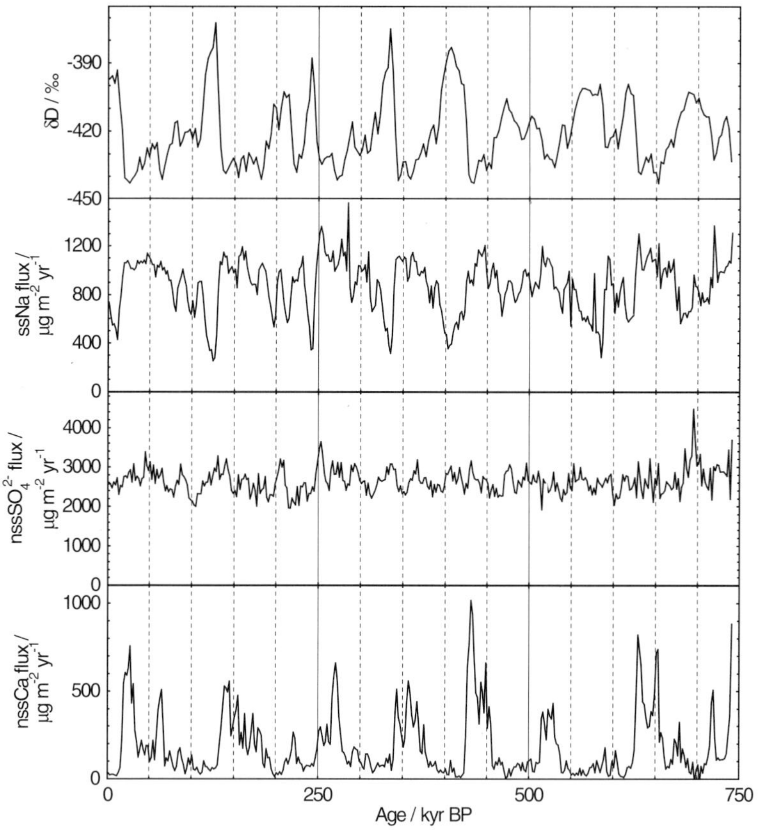

Figure 6. Chemical fluxes (on EDC2 timescale) in ice from Dome C over the last 740 kyr [18]. The deuterium record is also shown. Fe is not shown, but has similar variability to nssCa.

glacial periods (Fig. 6). At a site with very low snow accumulation rate, where dry deposition should dominate, the flux, rather than the concentration, is a good representative of changing atmospheric composition and of the input of material to Antarctica. The flux at glacial maxima is more than 10 times higher than it is in interglacials. This increase has to be a result of changing source strength, modulated by changes in transport and in the strength of the loss processes between the source and Dome C. The source, at least during glacials, is well-established from geochemical measurements as being mainly the Patagonian region of South America [19]. In the past, the effect of changes in transport (greater meridional transport) has been emphasised [9]. However, modelling studies for the LGM [20, 21] suggest that changes in the transport strength and atmospheric residence time between South America and Dome C were small, despite the changes in precipitation scavenging and atmospheric circulation that must have occurred. Assuming this applies to the whole record, it suggests that the huge changes in dust flux must originate mainly in changes at the source.

The source area in Patagonia was larger in glacial times because the lowered sea level exposed large areas of continental shelf to the east of Patagonia. However, the timing of sea level change in the last termination is too late for this to be the main factor affecting the dust concentration [18]. It seems therefore that climate conditions in Patagonia – either changes in wind strength, aridity, or source creation through glacial action – must be responsible. Whatever the cause of the dust changes, the fact that material reaching Antarctica must pass over the Southern Ocean means that the dust profile through the period of the Dome C record represents the input function for dust to the Southern Ocean. This is important because it has been suggested that at least a part of the change in CO_2 concentration across glacial-interglacial transitions arises from changes in iron fertilization of the ocean [22], induced by the changing input of Fe in atmospheric dust. While this process probably only accounts for a portion of the change [23], nonetheless the dust profile, modified by possible changes in the solubility of iron, will be an important input to estimates of this process.

3.4.2 Sea salt

Sea-salt Na (sodium) flux, as a marker of sea salt, also shows higher values in glacial periods than in interglacials [18], although only by a factor 2. Again, this must be a reflection of changes in source and transport. It used to be assumed that the source of sea salt to Antarctica, as elsewhere in the world, was bubble bursting from open sea water. Since we know that there must have been more sea ice in glacial periods, the open water source became more distant from the continent during cold periods. The increase in sea salt in the LGM had therefore to be ascribed to increased production (due to greater wind speeds over the sea surface) or stronger transport [9, 24]. However, recently the importance of an additional salt source from the surface of young sea ice has been demonstrated [25]. Such a source (coming from the salts expelled as sea ice forms, and occurring as brine slush, frost flowers, and brine-soaked snow) has been shown to be depleted in sulfate, because of the precipitation at low temperature of mirabilite. Sea salt similarly depleted in sulfate is found in coastal Antarctic aerosol and snow, and more recently in inland aerosol [26]. This and other evidence has led to the suggestion that the sea ice source may be sufficiently important that increasing sea ice extent (or strictly an increased annual production of sea ice) may be the principal cause of the glacial increase in salt flux to Antarctica [27]. Indeed, models cannot reproduce an increase in salt flux to central Antarctica using only an open water source [28, 29]; a sea ice source improves the situation although it is not yet known how to parameterise the source strength. The outcome of all this is that the 740 kyr sea-salt Na record from Dome C has been taken as a proxy for sea ice extent over that period [18], with (unsurprisingly) more sea ice during cold periods. However, more work will be needed to make a quantitative proxy for sea ice from the ice core data. Marine sediment data provide a complementary source of information about past sea ice extent [30], and it will be particularly important to reconcile the findings from the two archives.

3.4.3 Sulfur compounds

Two compounds (methanesulfonic acid (MSA) and non-sea-salt sulfate (nss sulfate)) measured in ice cores originate as oxidation products of dimethylsulfide (DMS), which is produced by various species of diatom in the Southern Ocean. The sulfate in particular is one of the main components of small aerosol particles that act as cloud condensation nuclei (CCN), and it has previously been proposed that feedbacks might occur whereby changes in climate could lead to changing marine biological productivity and hence, via DMS, to changes in aerosol and cloud, that in turn would affect climate [31]. In the past MSA in ice cores has been taken as a proxy for DMS production; it shows much higher concentrations in glacial times [32], and it was suggested that this might be an example of such a feedback in action. However we now know that, at sites with very low accumulation rate, MSA in snow is strongly influenced by postdepositional losses [33, 34]; under the dustier conditions of cold climates, it appears to be much better preserved in the ice, and the MSA record at sites such as Vostok and Dome C cannot be taken as indicating changes in DMS production. Nss sulfate is also a DMS oxidation product (assuming that pollution and volcanic input play a minor role) and does not suffer from postdepositional loss. The surprise in the Dome C nss sulfate flux [18] is that it is almost unchanged across 740 kyr (within about +/-20%). This could arise from almost unchanged DMS production, along with no important differences in transport, or could be a coincidence of changes in several factors. In either case, the result

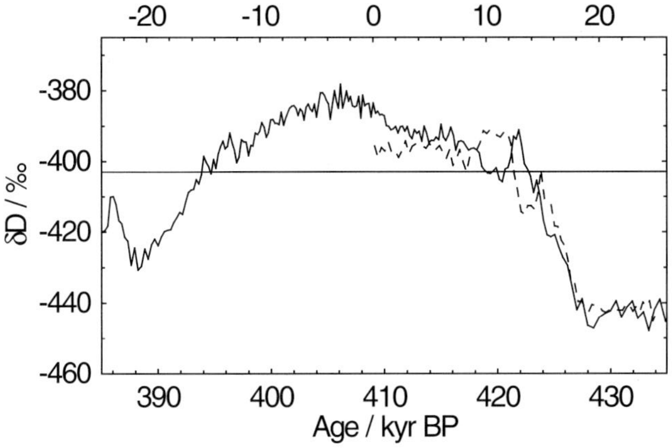

Figure 7. Comparison of two interglacials: MIS11 (solid line and lower x-axis) and the Holocene (dashed line and upper x-axis) [7]. The two records are aligned on the glacial termination, and not at the point of greatest similarity in orbital parameters.

is an unchanged flux of nss sulfate, implying that in this case (representing probably the Southern Ocean south of the polar front in the Indian Ocean sector), no climate feedback can be occurring.

3.5 Marine isotope stage 11

MIS11 is the interglacial that occurred some 400 kyr before our present interglacial. It has often been presented as an analogue for the present and near future [35], because the status of Earth's orbit around the Sun is most similar in the two periods: eccentricity is low, which implies that only small changes in precession occur. It is therefore very interesting to study how MIS11 evolved.

There are certainly many similarities between MIS11 and MIS1 (Holocene), with similar values of many parameters in the Dome C ice core before and after the termination that commenced the interglacial, and a similar rate of change during the transition. MIS11 maintained values (of Antarctic temperature, greenhouse gas concentrations, Antarctic aerosol chemistry) typical for an interglacial (Fig. 7) (depending how this is defined) for 28 kyr [7]; the Holocene is so far only 12 kyr long, a length similar to that of the interglacials in between MIS11 and the Holocene. However it has been pointed out that there is no direct analogue for insolation (incorporating all aspects of orbital cycles) that allows one to line up the two interglacials against each other [36]. One can only make the generalised comment that, at a time of low eccentricity, there is no strong impetus for the system to escape from an interglacial, and that this is consistent with the long MIS11.

4. RAPID CLIMATE CHANGE IN THE LAST GLACIAL PERIOD

Although this paper has mainly focussed on changes occurring at orbital timescales, there has also been progress at the millennial timescale. It has been well-known for some time that the climate variability within the last glacial period is dominated by the millennial scale Dansgaard-Oeschger (D-O) events, which are particularly prominent in the Greenland ice core record [37]. It has also been recognised that the larger D-O events have Antarctic counterparts [38]: smaller changes in Antarctic temperature (and indeed in CO_2 concentration [39]) that are not in phase with the Greenland warmings. Rather, it appears that Antarctica warms while Greenland is cold, and vice-versa, consistent with some mechanisms related to changes in the style of Atlantic thermohaline circulation [40].

The GRIP and GISP2 ice cores from Summit, Greenland provided a reliable record of climate only for just over 100 kyr BP: below that, flow disturbances had altered the climate signal. The new

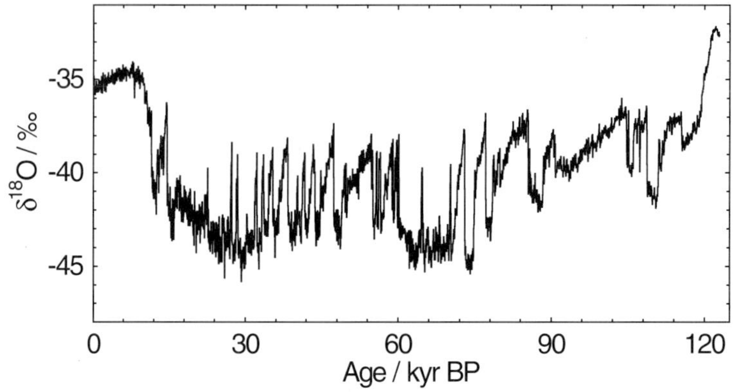

Figure 8. Climate signal of Greenland from oxygen isotopes in the NorthGRIP core, extending 123 kyr into the past [3], showing the D-O events of the last glacial, and the latter parts of the last interglacial.

NorthGRIP core has provided a detailed record back to 123 kyr BP, through the entire glacial, and the late part of the last interglacial. It shows the existence of D-O events right back to the time of glacial inception (Fig. 8) [3]. It also shows that for at least part of the last interglacial, the NorthGRIP site was substantially warmer than it currently is.

There is also new detail about the Antarctic counterparts. Analysis of the two EPICA cores (Dome C and Dronning Maud Land (EDML)) shows that not just the large events, but every D-O event, has an Antarctic counterpart [41]. The uncertainties in the relative age scales for the NorthGRIP and EDML cores make it impossible to state with certainty the phase relationship between the climate signals, but again it could be consistent with the oceanographic explanation [40].

5. CONCLUSIONS AND FUTURE PLANS

The last few years have seen very significant new datasets from the ice core community. At both orbital and millennial scale, the behaviour of different components of the Earth System has been exposed, and this provides a major challenge to the modelling community. The models they use for climate prediction need also to be able to replicate the behaviours we can observe in the palaeorecord.

A number of new plans have been proposed for future years, particularly under the auspices of IPICS (International Partnerships in Ice Core Sciences), which has a web site at: (http://www.pages.unibe.ch/science/initiatives/ipics/index.html). As well as plans for technical developments, IPICS has laid out an agenda of four major projects for the ice core community in the next decade or more.

Firstly, noting that marine data suggest that the era of 100 kyr dominance began just before the earliest date represented in the EPICA Dome C ice core record, the community wishes to find a place in Antarctica where they can obtain a longer record, extending into the 40 kyr era. This implies finding, if possible, a site with ice as old as 1.5 Myr. A major programme of geophysical survey will be required to find if such a site exists. The record from such a core would give valuable clues to the reasons for the change from 40 kyr to 100 kyr periodicity: which in turn is required to understand why we have the climate we currently enjoy.

A second project would aim to drill a core in Greenland that covers at least the whole of the last interglacial. This would allow us to assess the status of the Greenland ice sheet under conditions apparently somewhat warmer than those of today. A third project aims to obtain a network of cores in both polar regions covering at least 40 kyr. This time period encompasses major climate changes: the last termination, and D-O events. A key aim would be to assess the spatial pattern of climate change during these periods. A number of new relevant cores have recently been, or currently are being, drilled (e.g. Berkner Island and Talos Dome in Antarctica). The new inland West Antarctic drilling being started by US investigators would add a crucial new, high-resolution, core to the network.

Finally, reconstructions of natural climate variability, such as those used by the Intergovernmental Panel on Climate Change [42] tend to concentrate on a period of 2000 years, and are poorly represented in the polar regions, and by ice cores. The fourth IPICS project will aim to involve all nations with capability in an effort to construct a network of ice cores (polar and non-polar), of 2000 year length, to aid these reconstructions.

Acknowledgments

I am grateful to all my colleagues in the ice core community, but particularly from the EPICA project, who have generously shared data and ideas.

References

[1] Wolff, E. W., in ERCA - Volume 4 - From weather forecasting to exploring the solar system, edited by Boutron C. (EDP Sciences, Les Ulis, France, 2000), p. 147-177.

[2] Dahl-Jensen, D., Gundestrup, N. S., Miller, H., Watanabe, O., Johnsen, S. J., Steffensen, J. P., Clausen, H. B., Svensson, A. and Larsen, L. B., Ann. Glaciol. 35, 1-4 (2002).

[3] North Greenland Ice Core Project Members, Nature 431, 147-151 (2004).

[4] Watanabe, O., Jouzel, J., Johnsen, S., Parrenin, F., Shoji, H. and Yoshida, N., Nature 422, 509-512 (2003).

[5] Watanabe, O., Kamiyama, K., Motoyama, H., Fujii, Y., Igarashi, M., Furukawa, T., Goto-Azuma, K., Saito, T., Kanamori, S., Kanamori, N., Yoshida, N. and Uemura, R., Memoirs of National Institute of Polar Research, Special Issue 57, 1-24 (2003).

[6] Wolff, E. W. and Miller, H., EGGS Newsletter 2005, 12-15 (2005).

[7] EPICA Community Members, Nature 429, 623-628 (2004).

[8] Lisiecki, L. E. and Raymo, M. E., Paleoceanography 20, PA1003; doi:10.1029/2004PA001071 (2005).

[9] Petit, J. R., Jouzel, J., Raynaud, D., Barkov, N. I., Barnola, J. M., Basile, I., Bender, M., Chappellaz, J., Davis, M., Delaygue, G., Delmotte, M., Kotlyakov, V. M., Legrand, M., Lipenkov, V. Y., Lorius, C., Pepin, L., Ritz, C., Saltzman, E. and Stievenard, M., Nature 399, 429-436 (1999).

[10] Jouzel, J., Vimeux, F., Caillon, N., Delaygue, G., Hoffmann, G., Masson-Delmotte, V. and Parrenin, F., J. Geophys. Res.-Atmos. 108, 4361, doi:10.1029/2002JD002677 (2003).

[11] Siegenthaler, U., Stocker, T. F., Monnin, E., Luthi, D., Schwander, J., Stauffer, B., Raynaud, D., Barnola, J. M., Fischer, H., Masson-Delmotte, V. and Jouzel, J., Science 310, 1313-1317 (2005).

[12] Spahni, R., Chappellaz, J., Stocker, T. F., Loulergue, L., Hausammann, G., Kawamura, K., Fluckiger, J., Schwander, J., Raynaud, D., Masson-Delmotte, V. and Jouzel, J., Science 310, 1317-1321 (2005).

[13] Wolff, E. W., Kull, C., Chappellaz, J., Fischer, H., Miller, H., Stocker, T. F., Watson, A. J., Flower, B. P., Joos, F., Kohler, P., Matsumoto, K., Monnin, E., Mudelsee, M., Paillard, D. and Shackleton, N. J., EOS Trans. 86, 341,345 (2005).

[14] Monnin, E., Indermuhle, A., Dallenbach, A., Fluckiger, J., Stauffer, B., Stocker, T. F., Raynaud, D. and Barnola, J. M., Science 291, 112-114 (2001).

[15] Brook, E. J., Sowers, T. and Orchardo, J., Science 273, 1087-1091 (1996).

[16] Etheridge, D. M., Steele, L. P., Langenfelds, R. L., Francey, R. J., Barnola, J.-M. and Morgan, V. I., J. Geophys. Res. 101, 4115-4118 (1996).

[17] Etheridge, D. M., Steele, L. P., Francey, R. J. and Langenfelds, R. L., J. Geophys. Res. 103, 15979-15993 (1998).

[18] Wolff, E. W., Fischer, H., Fundel, F., Ruth, U., Twarloh, B., Littot, G. C., Mulvaney, R., Rothlisberger, R., de Angelis, M., Boutron, C. F., Hansson, M., Jonsell, U., Hutterli, M. A., Bigler, M., Lambeck, K., Kaufmann, P., Stauffer, B., Stocker, T. F., Steffensen, J. P., Siggaard-Andersen, M. L., Udisti, R., Becagli, S., Castellano, E., Severi, M., Wagenbach, D., Barbante, C., Gabrielli, P. and Gaspari, V., Nature 440, 491-496 (2006).
[19] Basile, I., Grousset, F. E., Revel, M., Petit, J. R., Biscaye, P. E. and Barkov, N. I., Earth planet. Sci. Lett. 146, 573-589 (1997).
[20] Lunt, D. J. and Valdes, P. J., Geophys. Res. Lett. 28, 295-298 (2001).
[21] Krinner, G. and Genthon, C., Tellus Ser. B-Chem. Phys. Meteorol. 55, 54-70 (2003).
[22] Martin, J., Paleoceanography 5, 1-13 (1990).
[23] Röthlisberger, R., Bigler, M., Wolff, E. W., Joos, F., Monnin, E. and Hutterli, M., Geophys. Res. Lett. 31, L16207, doi:10.1029/2004GL020338 (2004).
[24] Petit, J. R., Briat, M. and Royer, A., Nature 293, 391-394 (1981).
[25] Rankin, A. M., Wolff, E. W. and Martin, S., J. Geophys. Res. 107, 4683, doi:10.1029/2002JD002492 (2002).
[26] Hara, K., Osada, K., Kido, M., Hayashi, M., Matsunaga, K., Iwasaka, Y., Yamanouchi, T., Hashida, G. and Fukatsu, T., J. Geophys. Res. 109, D20208, doi:10.1029/2004JD004713 (2004).
[27] Wolff, E. W., Rankin, A. M. and Rothlisberger, R., Geophys. Res. Lett. 30, 2158, doi:10.1029/2003GL018454 (2003).
[28] Reader, M. C. and McFarlane, N., J. Geophys. Res. 108, 4253, doi:10.1029/2002JD002063 (2003).
[29] Mahowald, N., Lamarque, J.-F., Tie, X. X. and Wolff, E. W., J. Geophys. Res. 111, D05303, doi:10.1029/2005JD006459 (2006).
[30] Gersonde, R., Crosta, X., Abelmann, A. and Armand, L., Quat. Sci. Rev. 24, 869-896 (2005).
[31] Charlson, R. J., Lovelock, J. E., Andreae, M. O. and Warren, S. G., Nature 326, 655-661 (1987).
[32] Legrand, M., Feniet-Saigne, C., Saltzman, E. S., Germain, C., Barkov, N. I. and Petrov, V. N., Nature 350, 144-146 (1991).
[33] Wagnon, P., Delmas, R. J. and Legrand, M., J. Geophys. Res. 104, 3423-3431 (1999).
[34] Weller, R., Traufetter, F., Fischer, H., Oerter, H., Piel, C. and Miller, H., J. Geophys. Res. 109, doi:10.1029/2003JD004189 (2004).
[35] Loutre, M. F., Earth planet. Sci. Lett. 212, 213-224 (2003).
[36] Crucifix, M. and Berger, A., EOS Trans. 87, 352-353 (2006).
[37] Johnsen, S. J., Clausen, H. B., Dansgaard, W., Fuhrer, K., Gundestrup, N., Hammer, C. U., Iversen, P., Jouzel, J., Stauffer, B. and Steffensen, J. P., Nature 359, 311-313 (1992).
[38] Blunier, T. and Brook, E. J., Science 291, 109-112 (2001).
[39] Indermuhle, A., Monnin, E., Stauffer, B., Stocker, T. F. and Wahlen, M., Geophys. Res. Lett. 27, 735-738 (2000).
[40] Stocker, T. F. and Johnsen, S. J., Paleoceanography 18, art. no.-1087 (2003).
[41] EPICA Community Members, Nature 444, 195-198 (2006).
[42] IPCC, IPCC Third Assessment Report: Climate Change 2001: The Scientific Basis (Cambridge University Press, Cambridge, 2001) pp. 944.

J. Phys. IV France **139** (2006) 197–209
© EDP Sciences, Les Ulis
DOI: 10.1051/jp4:2006139015

Biogeochemical processes in the ocean and at the ocean-atmosphere interface

A. Saliot[1]

[1] *LOCEAN, IPSL/UPMC/CNRS/MNHN/IRD, Université Pierre et Marie Curie, Case 100, 75252 Paris Cedex 05, France*
e-mail: saliot@ccr.jussieu.fr

Abstract. The ocean can be considered as a chemical reactor, whose energy sources are the various matter inputs originating from the continent and the ocean. Among various elements, carbon plays a key role as it is involved in both inorganic form as CO_2 and organic forms such as compounds synthesized through photosynthesis. Thus, the ocean is presently an active actor in climate change and ocean-atmosphere exchange processes. This review will present some insights into: 1) schematic representations of the carbon cycle, with emphasis on CO_2 exchange between the ocean and the atmosphere and to the organic parts of this cycle, 2) concepts relative to the biological pump of CO_2, with a detailed view on photosynthesis, 3) concepts leading to the existence of oceanic provinces and associated productivity for open sea and coastal areas, 4) addressing the question: what is the net efficiency of the biological pump of CO_2in terms of exportation of organic carbon and sequestration in sediments and 5) specific aspects on biogeochemical processes occurring at the boundary between the ocean and the atmosphere.

1. INTRODUCTION

The scientific study of the chemistry of the sea was commenced during the seventeenth and eighteenth centuries with the observations of R. Boyle and the publication of "Observations and Experiments on the Saltness of the Sea" in 1670, and A. Lavoisier, who attempted to relate the observed abundances of salts in the rivers to the geology of the districts in which they flooded. Important developments of marine chemistry occurred on the knowledge on the abundances of the major components of sea water during the nineteenth century. Of note one can cite the works by J. Murray, A.M. Marcet, J.L. Gay-Lussac and G. Forchhammer. This latest scientist, over a period of twenty years, analysed several hundreds of sea water samples for calcium, magnesium, chloride, potassium. Forchhammer used differences in the salinities for tracing the movements of water masses in the Atlantic Ocean and in the Baltic Sea.

The expedition of the H.M.S. Challenger between 1872 and 1876 added greatly to the advancement of all branches of Oceanography. J.Y. Buchanan was the chemist of the expedition. He supervised the collection at each station of samples at various depths down to 800 m. Specific gravity and carbon dioxide were determined on board. Samples were treated for land-based measurement such as for dissolved gases and major elements conducted by Dittmar. Dittmar's results are a model of precision, compare very favourably with modern figures and confirmed Forchhammer's results: major elements are present in any oceanic area in constant ratios. At the suggestion of an International Commission, Forch, Knudsen and Sørensen thoroughly investigated the interrelationship between chlorinity, salinity and density of sea water. Modern Marine Chemistry was born and many points concerning the composition of sea water were further developed in the twentieth century: dissolved gases, carbon dioxide and the buffer mechanism of sea water, trace elements, biologically important trace nutrients, organic biogeochemistry [1].

In the 1980's, scientists first tried to describe and understand the complex set of interlocking processes involved in global change in a pluridisciplinary approach. New questions were raised: understanding the role of the ocean in the regulation of the climate, in the cycling of natural elements, in the cycling of pollutants and in the management of marine resources. The term biogeochemical cycles became familiar. Today it is a powerful discipline, with measured rates, reservoirs and mechanisms, innovative experiments, complex models, and vigorous testing of ideas.

The role of the ocean in regulating the climate is major owing to its surface, approximately 4/5 of the earth's surface, and as it is one of the major reservoir of an element, connected to the greenhouse effect, the carbon. There is presently a consensus to take into account in the ocean-atmosphere exchanges, carbon fluxes associated to photosynthesis and fluxes involved in the functioning to the pelagic and benthic food webs. This implies to know the functioning of all coastal and open-sea areas, and particularly a deep knowledge of water masses circulation, chemistry, biology and ocean-atmosphere interactions. These reflections lead to the creation of the JGOFS program (Joint Global Ocean Flux Studies) in 1987 [2].

The JGOFS Science plan was developed through 1989-1990 [3] and, together with the implementation plan [4], formed the basis of the JGOFS strategy.

The preface for the first JGOFS workshop held in Woods Hole in 1984 emphasized the need for study of "the physical, chemical and biological processes governing the production and fate of biogenic materials in the sea ... well enough to predict their influences on, and responses to, global scale perturbations, whether natural or anthropogenic ...".

In 1984, at the time of the first JGOFS meeting, atmospheric CO_2 levels were 344 ppm, or 64 ppm above the pre-industrial baseline. Today there are 378 ppm, or 98 ppm above the pre-industrial levels – an important increase while we have been planning and carrying our research.

2. THE OCEAN, AN ACTIVE CHEMICAL REACTOR AND ITS ROLE IN CLIMATE CHANGE

2.1 Carbon cycle and the role of the ocean in CO_2 pumping

A schematic diagram of the carbon cycle in reservoir-flux form is shown in Fig.1. The stock of carbon present in the atmosphere is presently estimated about 750 gigatonnes (1GtC = 10^{15} gC). It is mainly composed of CO_2(more than 99%), CH_4, with traces of CO, $(CH_3)_2S$ or DMS and organic compounds such as hydrocarbons. Present emissions of fossil $CO_2$are in the range of 6.3 ± 0.4 GtC/year [5] and increase annually by 9%. The annual increase of CO_2 in the atmosphere is estimated at 3.2 ± 0.1 GtC/year and thus represents about half of fossil emissions. The other half is fixed by the land vegetation and dissolved in the ocean. The present sink for atmospheric CO_2 in the oceans is the net imbalance between the ingoing and outgoing fluxes and is of order 2 GtC/year [6].

The main reservoir of oceanic carbon is dissolved inorganic carbon (35 000 GtC) (Fig. 1). This reservoir is composed of CO_2, HCO_3^- and CO_3^{2-}, three species in equilibrium known as the "carbonate system". The ocean exchanges CO_2 with the atmosphere, schematically from the ocean into the atmosphere in warm oceanic zones and from the atmosphere into the ocean in cold and productive ocean (the ocean is then considered as a pump for atmospheric CO_2). There is an excess of this pump, which can be decomposed into a physical pump (dissolution and transport into the deep ocean in zones of formation of deep waters (Greenland and Norwegian Seas in the Arctic, Labrador Sea in the Atlantic, Weddell Sea in the Antarctic).

The ocean is characterized by an original organic carbon cycle, in interaction with the huge dissolved inorganic carbon pool (Fig. 1) [8]. By decreasing order of importance are: i) the dissolved (and colloidal) organic carbon pool with a stock of 1 000 GtC, which is the most important reservoir of organic carbon at the earth's surface. This organic carbon is relatively chemically inert, with a long residence time in deep waters of the Atlantic and the Pacific, ~3000 years. But in surface waters it is biologically labile and constitutes a large reservoir of energy for heterotrophic organisms [10]; ii) the particulate organic pool (POC), with 30 GtC, major actor in sedimentation processes, slow with small-size settling particles and rapid with large-size (>50-60 μm) sinking particles and iii) the biomass carbon, which integrates all living organisms, from the virus to large mammals, with a modest pool of 3 GtC. Thus the size

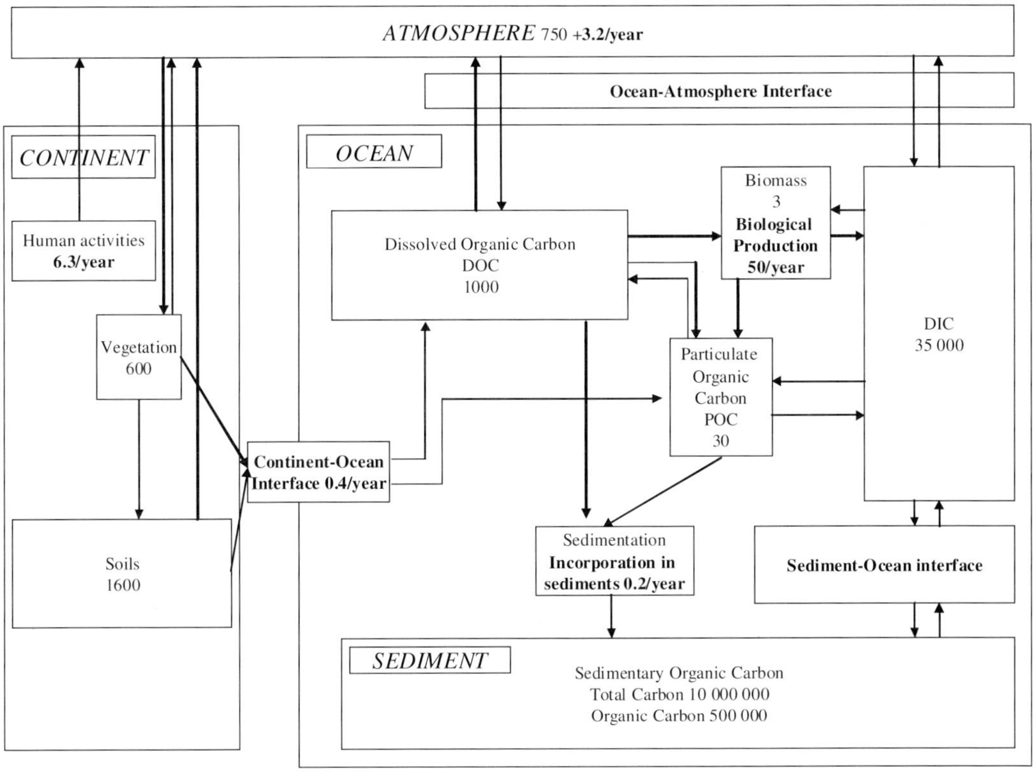

Figure 1. Schematic diagram of the carbon cycle. Sizes of reservoirs are in GtC; fluxes (in bold) are in GtC/year. From : Hedges [7] Saliot [8], and IPCC [9].

of marine biota carbon reservoir is much smaller than the fluxes in and out the reservoir. This implies that any changes in the activity of this reservoir can mean substantial changes in the fluxes to related reservoirs, such as the burial of particulate organic carbon in marine sediments.

Carbon fluxes are estimated in Fig.1. The major flux, 50 GtC/year, is from the oceanic net biological production. A small part of this net flux, 0.2 GtC/year will be incorporated in sediment, after complex sedimentation processes, which will be discussed in section 2.4. Natural inputs from the continent carried out by rivers and steams are in the range of 0.4 GtC/year, whereas vegetation emits organic material, both in gas phase and in small submicronic particles which a part, estimated in the range 1-1.5 GtC/year, is introduced in the ocean by direct deposition and washout of the atmosphere.

Figure 2 shows a representation of the organic carbon cycle, underlining the complexity of the oceanic machine or reactor, which is driven by carbon sources, both marine and terrigenous and biological and chemical processes [8]. Of note one can consider in any area oceanic area that various sources of carbon are the basis of this reactor. Terrestrial sources by land plants can be transported *via* the ocean by the atmosphere (particles, dusts issued from evaporation, adsorption on dusts, forest fires) or rivers from erosion of drainage basins. For marine internal sources, if phytoplankton is predominantly the source of organic carbon, other actors, producing or recycling carbon, must be considered such as benthic algae, zooplankton, bacteria, viruses, fungi.

These sources exchange carbon with the two dissolved and particulate organic carbon reservoirs through various processes involving biosynthesis, respiration, excretion, food transfer, selective accumulation. Besides this biological aspect of the oceanic reactor, chemical processes do occur and participate to the creation and transformation of organic material. One can mention in the ocean

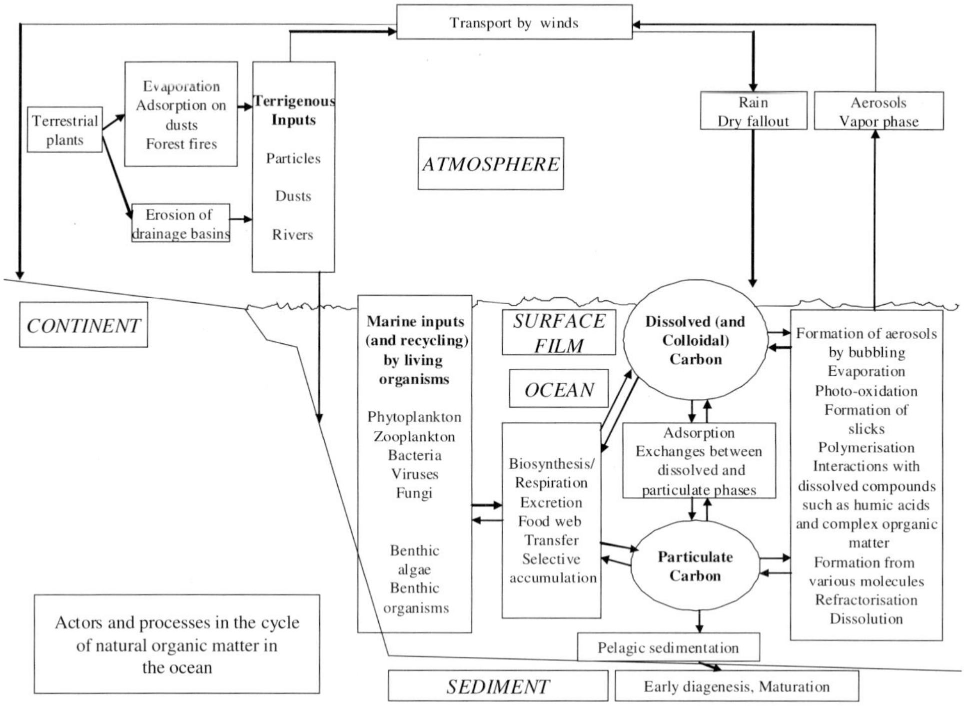

Figure 2. Representation of the ocean reactor, driven by organic carbon sources, both marine and terrigenous and associated biological and chemical processes.

surface: evaporation, formation of aerosols, photo-oxidation and in the deep ocean poly-condensation, refractorisation into high molecular compounds such as fulvic and humic compounds and dissolution. The challenge of the chemist will be there to identify the main sources and processes that control the oceanic machine, using a combination of isotopic and molecular indices.

Based on the photosynthetic process the biomass pool is the main source of energy for the organic carbon oceanic machine, as it constitutes the first step of the various marine food chains and the main actor for the biological pump of CO_2. Our ability to predict the impacts of global warming is limited by a number of key uncertainties, significant among which is the role of biotic feedbacks [9]. The response of biota in the surface ocean is particularly pertinent and still not understood. However, the potential for multiple feedbacks between climate, ocean circulation and mixing, and photosynthetic primary production has been manifestly evident for some time [11,12].

2.2 The CO_2 biological pump; the first step: Photosynthesis

Phytoplankton plays a major role in the oceanic carbon cycle [13]. These floating, microscopic, single-cell plants are the foundation of the marine food webs. Like land plants, phytoplankton fixes carbon through photosynthesis, making it available for higher trophic levels. The pigment equipment in the phytoplankton composed of a combination of chrorophylls and carotenoids is used as an energy source to fuse water molecules and carbon dioxide into carbohydrates.

Photosynthesis can simply be considered as follows: solar energy is absorbed by phytoplankton cells and algae and converted into biological energy stored in the form of organic compounds; the reaction is reversible *via* respiration:

$$H_2O + CO_2 \overset{h\nu}{=} CH_2O + O_2$$

Inorganic nutrients and trace metals are also involved in the reaction that can be summarized as follows:

$$106CO_2 + 122H_2O + 16NO_3^- + HPO_4^{2-} + 18H^+ \underset{\text{trace metals}}{\overset{h\nu}{=}} C106H263O110N16P1 + 138O_2$$

Although the stoechiometry of this last reaction is different for various aquatic systems and algae, it is remarkable that the complex dynamics of the photosynthesis/respiration system can be expressed by simple reactions such as:

$$\Delta C : \Delta N : \Delta P \approx 106 : 16 : 1(\text{Redfieldratio})[14]$$

Additionally three stages can be identified in the photosynthetic process:

1) Absorption of photons of light by photosynthetic pigments (chrorophylls and carotenoids), contained in the chromatophores. The resonating systems stabilize the molecule and provide electrons that are easily excited to orbitals of higher energy when light is absorbed.

2) Part of the energy of these excited electrons is converted to chemical energy through a cyclic series of enzymatic reactions involving Cytochrome I, which leads to the production of high energy adenosine triphosphate (ATP), from adenosine diphosphate (ADP) and orthophosphate (P).

$$ADP + P = ATP \tag{1}$$

The remainder of energy of the electrons is used in a series of enzymatic reactions of riboflavin phosphate and nicotinamide adenosine dinucleotide phosphate (NADPH).

$$4NADP + H_2O + 2ADP + 2P = 4NADPH + O_2 + 2ATP \tag{2}$$

The protons come from water and the excited electrons are from the reduced form of NADP. The hydroxide from the water yields molecular oxygen and gives electrons to chrorophyll *via* the Cytochrome I chain.

3) The CO_2 is assimilated in a cycle series of reactions utilizing the reducing action of NADPH and the phophorylating power of ATP. These reactions, which can take place in the dark, lead to the production of carbohydrate (CH_2O).

$$CO_2 + 4NADPH + ATP = CH_2O + H_2O + 4NADH + P \tag{3}$$

Combining reactions (1), (2), and (3) gives:

$$H_2O + CO_2 = CH_2O + O_2$$

CO_2 finds its way very quickly into compounds other than carbohydrates. Compounds such as lipids and amino acids are synthesized from intermediates in the carbon cycle in addition to being produced from CH_2O.

The major environmental factors that influence phytoplankton growth are light, temperature, turbulence and inorganic nutrients. When favourable conditions are encountered, phytoplankton can undergo rapid population growth usually referred to as "blooms", such as in upwellings. However, most of the time phytoplankton is either limited by light (e.g. in winter at high latitudes) or nutrient limited (e.g. in the subtropical gyres). Because light attenuates exponentially with depth, phytoplankton growth is restricted to the euphotic layer (max 200m in very transparent waters, e.g. the Sargasso Sea). Limiting inorganic nutrients such as nitrogen, phosphorus, silicium and iron are constantly removed from the surface waters by the growing phytoplankton. Dissolved inorganic carbon is also consumed by phytoplankton. It is exchanged at the sea surface and always plentiful in the surface layer, contrary to limiting nutrients.

Most of the phytoplankton is consumed locally by zooplankton, so that the nutrients associated with its biomass are regenerated at the surface and more available for another round of production. The regenerated production is that portion of primary production fuelled by the inorganic nutrients reminalized within the euphotic zone. The export production is the fraction of primary production that finds its way to the deep sea trough the settling of dead cells and detritus, trough zooplankton diel migrations, or by downwelling or mixing. Most of the export production is ultimately assimilated by bacteria, viruses and fungi, which regenerate it into inorganic forms (end-products of respiration and excretion). At steady state and at a large scale, the biotically mediated downward flux of organic matter is balanced by an upward return flux of inorganic nutrients that fuels the new production.

The collective action of this so-called "biologically pump" is to create a sharp vertical gradient of nutrients. This pump plays a central role in the global carbon cycle because it sequesters carbon away from the atmosphere to the deep sea. The structure of biological communities in the upper ocean for example, between diatoms and coccolithophorids can also influence climate change with the calcium carbonate pump. Calcifying phytoplankton species such as coccolithophorids build their carbonate test in the surface ocean together with photosynthesis. Calcium carbonate production releases CO_2 in the surface layer. Thus strengths of the organic carbon pump and calcium carbonate pump largely determine the biologically-mediated ocean-atmosphere CO_2 exchange.

2.3 From the concept of oceanic province to the global observation

First global estimations of the primary production followed the development of the ^{14}C method by Steeman-Nielsen [15] and the famous expedition of the Galathea around the Atlantic, Pacific and Indian Oceans in 1950-1952. A first estimation of the production of biological carbon was proposed: 15-18 GtC/year. The first maps showing the distribution of primary production, e.g. [16-17] were in agreement with the general concept proposed by Sverdrup et al. in 1955 [18]. High primary production areas are located close to continents because of nutrient inputs in surface waters by rivers and streams and by coastal upwellings. The contribution of satellite observations by CZCS, SeaWIFS, as indicators of the ocean productivity is major in the comprehensive studies of open-sea and coastal provinces.

This view has been revisited after the JGOFS programme by Ducklow [19]. He addressed the question of biogeochemical provinces in the ocean. JGOFS embodied the ocean biogeochemistry paradigm, that is, the idea that the ocean is an organized system of physically-driven, biologically controlled chemical cycles which regulate the planetary climate over large spatial and temporal scales.

Most schemes to partition the ocean into a system of bounded regions are based on physical climate and circulation or have been biogeographic, based on the occurrence of distinctive species assemblages. Longhurst [20], pioneered a more encompassing ecological scheme. Four primary domains or biomes are identified in the ocean: polar, where the mixed layer is constrained by a surface brackish layer formed each spring in the marginal zone ($>60°$ latitude), westerlies, when the mixed layer depth is forced largely by local winds and irradiance (ca. 30-60° latitude), trades, where the mixed layer depth is forced by geostrophic adjustment on a basin scale to often-distant wind forcing (ca. 30°N to 30°S latitude), and coastal, where diverse coastal processes (e.g. tidal mixing, estuarine runoff) force mixed layer depth (all latitudes). In the map of Longhurst "Ecological Geography of the Sea", 83 provinces are identified, some of them being intensively visited such as the Arabian Sea, the trades wind domain at the Hawaii Ocean Time series HOT, the Bermuda Atlantic Time Series BATS, and the EUMELI oligotrophic stations, and stations in the Pacific ocean and in the Southern ocean. JGOFS estimates yield a new global total primary production, excluding the coastal domain of ca 45 GtC/year. This estimate should be viewed with reservation since the areal coverage is patchy and the productive coastal zones are excluded.

2.4 The exportation of organic carbon in open and coastal oceans

The net efficiency of the CO_2 pump must be referred not only to the surface production but to consider what reaches the ocean floor which is the net result of production in the surface ocean, subsequent alteration in the deeper water column and net incorporation in the sediment. The export of biogenic particles from the productive upper layer of the ocean removes a small and variable fraction of algal biomass from the euphotic zone, transferring biologically bound carbon and associated elements into the deeper water layers. If small-size particles ($< 60\,\mu m$) are representative of the mass of particles, they are transported by currents over long distances and reach the sediment after a long time (100-1000 years). Large-size particles ($> 60\,\mu m$), that can be collected with moored sediment traps of by direct large-volume *in situ* pumping, are mainly responsible for the vertical flux, with sinking velocities $>100\,m/day$. Several processes contribute to this flux: 1) grazing of phytoplankton by zooplankton, leading to the formation of faecal pellets, that often constitute most of material collected in sediment traps, 2) agglomeration of individual algal cells forming relatively large and rapidly sinking aggregates, which also scavenge other particles and dissolved organic compounds from the water column, 3) carcasses of dead animals. Oceanic regions characterized by seasonal phytoplankton blooms are known for high and episodic sedimentation while oligotrophic oceanic gyres are characterized by low and more constant particle fluxes. One can mention the first compilation of carbon fluxes obtained at different depths by sediment traps made by Suess [21]. He demonstrated that a quasi-exponential relation was obtained between the ratio carbon flux/primary production and depth. The relation was as follows:

$$C(Z) = C_{PP}/0.0238Z + 0.212 \quad (n = 33; r^2 = 0.79; Z > 50m)$$

C(Z) is the carbon flux measured at depth Z in $g/m^2/y$; C_{PP} is the net primary production in $g/m^2/y$; n is the number of data.

Since that pioneer work the number of programmes providing data flux with sediment traps has considerably increased. Berger et al. [17] presented a comprehensive model of the alteration processes during sedimentation from the primary production to the incorporation in underlying sediment. For an open-sea region numbers varied accordingly: 30 $gC/m^2/y$ for primary production, 3 $gC/m^2/y$ for exported production just under the euphotic fertile zone, 0.3 $gC/m^2/y$ for gross deposition on the sediment and 0.01 $gC/m^2/y$ for net incorporation in the sediment. For a model coastal ocean, following numbers were estimated: 120, 30, 8 and 1 $gC/m^2/y$. This means that very small part of what is produced escapes to alteration processes and return to CO_2 during respiration and degradation processes: 30/0.01 and 120/1 $gC/m^2/y$ for open-sea and coastal areas, respectively.

As a more global scale, The JOGFS bank of data presents maps providing organic carbon flux values reaching 2000 m, expressed as percentages of organic carbon exported from surface waters, from deep-sea moored sediment traps or pore-water modelling. Strong differences are observed depending on the ocean province: high values are encountered in the eastern equatorial Atlantic (10-11%) and eastern equatorial Pacific (19%) and middle equatorial Pacific (9-12%), northern Arabian Sea (10-12%), intermediate values in the north Atlantic (5-7%) and tropical Pacific (4-5%) and low values at high latitudes in the Atlantic and Pacific (1-3%).

A recent synthesis has been presented for benthic processes and the burial of carbon [22], including processes of transport and turnover of material in the deep ocean, estimates of carbon deposition and carbon turnover.

The situation for the coastal ocean is less documented. The North Sea is a rather surprising example. In the shallow southern North Sea, production and respiration processes of the heterotrophic food web occur in the mixed layer and the area is a source of CO_2 for the atmosphere. On the contrary in the deep northern North Sea respiration processes mainly occur in the separated subsurface layer, which is subjected to exchange circulation with the North Atlantic Ocean. That area is globally a sink for the atmospheric CO_2 (up to 2.5 $mol/m^2/y$), with marked seasonal variations [23].

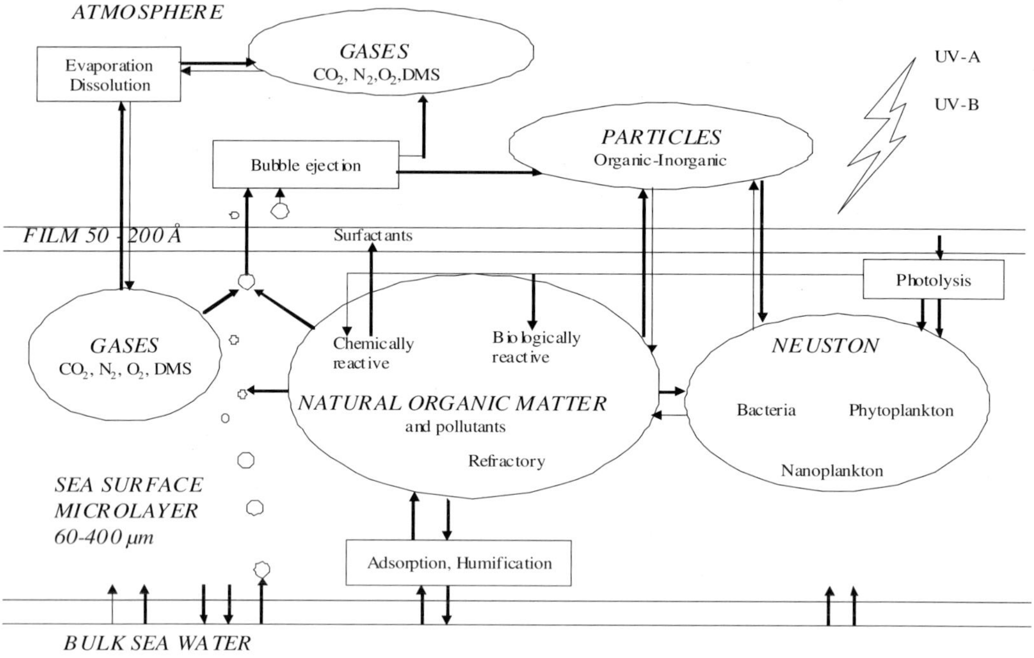

Figure 3. Conceptual model of the sea-surface microlayer chemical reactor, illustrating main processes occurring at the atmosphere-ocean boundary.

3. BIOCHEMICAL PROCESSES AT THE OCEAN-ATMOSPHERE INTERFACE

3.1 The sea-surface microlayer

Most biogeochemical processes take place at different phase discontinuities, among which the largest one is the ocean-atmosphere boundary. The water-air interface, covering 71% of the Earth's surface, has specific physical, chemical and biological properties that govern exchange processes between the ocean and the atmosphere and cause interfacial fractionation of trace elements, leading to the formation of the sea surface microlayer [24] (Fig. 3).

The sea-surface microlayer is often been operationally defined as roughly the top 1 to 1000 micrometers of the ocean surface [25-26]. There has been considerable new research in this area over the past 5-10 years. The microlayer is known to concentrate, to various degrees, many chemical substances both mineral and organic, particularly those that are surface active. Additionally many micro-organisms live and/or find food here. The exchange of gases across the sea surface and the fate and effects of airborne contaminants and particulate inputs into the sea are strongly influenced by the sea surface microlayer.

Organic films and surface tension forces provide physical stability of the sea-surface microlayer [27]. Surface films are classified into coastal and open-sea types. Surface films are less elastic near the Pacific coast of North America than in areas distant from the latter, reflecting the difference in the chemical composition of marine surfactants. A lower carbohydrate and lipid to protein ratio (C/N ratio) in more productive coastal areas is reflected in a less elastic surface film. At the contrary in open sea, low productive waters, an increased C/N ratio results in an increased film elasticity. Marine surfactants are primarily derived from phytoplankton exudates and their degradation products. Production of this material appears to be seasonal and is linked to biological production cycles [28-29]. Autotrophic and heterotrophic neuston (surface dwelling biota) range in size from less than 2 μm (piconeuston, mainly bacteria and virus) to much larger organisms (phytoplankton, zooplankton and macroneuston). Neuston occurs in much greater densities in the top ocean surface than their sub-surface counterparts, the plankton and nekton [30].

3.2 Gas exchange

Various parameterizations of gas exchange with wind speed at the ocean surface are poorly constrained by field measurements using natural and artificial tracers. One of the factors leading to uncertainty for *in situ* estimates of the gas transfer velocity is the presence of organic films at the sea-air interface. Organic films can affect sea-air gas exchange through both static and dynamic mechanisms. The static effect arises from the presence of additional mass transfer resistance due to the physical barrier provided by the film. This effect is not considered to be important at the sea surface whereas more significant is the hydrodynamic effect of a film that arises from the viscoelastic property of a surfactant-influenced interface [31].

Results from laboratory investigations of gas exchange at surfactant-influenced water-air interfaces in various flow systems have been reviewed [31-32]. These results provide evidence of first-order effects of surfactant films in reducing the mass transfer coefficient. Gas transfer is significantly reduced by even slight surface accumulations of organic materials. This effect would be most important for coastal waters where biological productivity is high. However, even at the low surface film pressures typical of the ocean surface, the viscoelastic modulus is estimated to be sufficient to cause gas exchange reductions. Evidence from wind-wave tank studies show strong inverse correlations between gas exchange rate and various measures of organic matter and surfactant concentrations. Thus, significant spatial and temporal variations in gas transfer are expected for different geographical and biological regimes for a given set of physical forcing conditions, e.g. wind stress and fetch. A high degree of correlation exists between gas transfer velocity and mean square wave slope. Thus, mean square wave slope may provide a confident parameterization of gas transfer that integrates the surfactant film effect. Such a parameterization could be applied over large spatial scales using satellite-based microwave sensors.

3.3 Bubbles and aerosols

Bubbles in the upper ocean are involved in many physical processes of geochemical significance (Fig. 3). Bubbles are generated at the sea surface by various mechanisms including the entrainment of air in breaking waves and the biological activity. Bubbles contribute to sea-air gas transfer, the transport of organic material that will form the sea-surface microlayer and the transport of material into the atmosphere, such as sea salt particles.

Bubbles will scavenge material from the surrounding water, thus contributing to the cycling of dissolved and particulate material. The role that bubbles play in generating organically rich particles is also important. Since the dissolved organic pool (DOC) exceeds the particulate organic pool (POC) (Fig. 1) processes that convert DOC to POC may have a significant impact on the POC budget. Aggregation of dissolved and colloidal material by rising bubbles at the sea-air interface is known as surface coagulation and contributes to global scavenging of DOC into POC [33]. This rate of DOC scavenging by bubbles represents a 1.7% per year turnover of the global oceanic POC pool, and each year produces an amount of POC equal to 17% of the POC pool in the uppermost 300 m of ocean [34].

When they burst at the sea surface these bubbles generate a sea salt aerosol enriched with material scavenged from the sea-surface microlayer and below. Gases will be exchanged between a bubble and the surrounding water while it is submerged. Additionally, the breaking waves and surfacing bubble plumes disrupt the surface microlayer, and this may enhance transfer of gases through the sea surface. The sea-surface microlayer serves as a concentration point for trace metals such Cu, Cd, Pd and Zn, and organic contaminants such hydrocarbons and pesticides that have low water solubility or are associated with floating particles. Thus a high level of films contamination plays a key role in the transfer of pollutants between the ocean surface and the atmosphere.

3.4 Photochemistry

Due to photochemical production and atmospheric deposition of highly reactive species at the sea surface, the microlayer could well act as a highly efficient micro-reactor, effectively sequestering and transforming selected materials brought to the interface from the ocean and the atmosphere by physical processes, such as bubbling (Fig. 3). Based on measurements of light absorbing material in the microlayer and employing photochemical yields obtained for bulk waters, photochemical production rates and fluxes can be estimated for the microlayer. The microlayer fluxes are generally small with respect to atmospheric deposition and the water column fluxes. The higher photochemical production rates at the surface should lead to more rapid oxidative turnover of materials at the interface and potentially to reactions and processes not observed in bulk waters. The particular aspect of photodegradation process, i.e. the heterogeneous visible light-induced degradative reactions associated with phytodetritus has been reviewed by Rontani [35]. Lipid photoproducts could constitute good "stress indicators" of the cells. This approach presents considerable interest since it could allow to follow the evolution of phytoplanktonic assemblages within the euphotic layer of the oceans and their responses to a stress such as pollution.

4. PERSPECTIVES

4.1 Deposition of aeolian dust on the ocean by the atmosphere

Aeolian dust deposition over the oceans provides a biogeochemical link between climate change, and terrestrial and marine ecosystems [36]. *In-situ* iron-fertilisation experiments have demonstrated in the equatorial Pacific and Southern Ocean that rising the iron level in the water by a few nanomoles per litre produces a significant increase in phytoplankton biomass. The majority of iron deposition to the ocean occurs in the Northern hemisphere and is principally associated with dust export from the major arid zones such as the Sahara and Taklamakan Deserts. The North Atlantic and North Pacific Oceans receive 48% and 22% of global iron deposition to the oceans, while the Indian Ocean receives 18% and the Mediterranean Sea receives 4%. The South Atlantic and South Pacific Oceans receive only 4% and 2% respectively, while the polar regions in both hemispheres also receive very low iron inputs, with the Arctic receiving only 0.9% and the Antarctic 0.5% [37].

Recently, Gabric et al. [38] reported evidence for a coupling between satellite derived (SeaWIFS) aerosol optical thickness and chlorophyll concentrations in the upper Southern Ocean. The coupling was evident at monthly, weekly and daily timescales. The shorter time scale coupling supports the hypothesis that episodic atmospheric delivery of iron is stimulating phytoplankton growth in the region.

The fact that atmospheric iron fluxes appear to play an important role in ecosystem dynamics in many locations underscores the interwoven nature of the links between climate changes, the biochemical cycles of carbon, nitrogen and sulphur and the potential for the oceans to sequester carbon.

4.2 Production of atmospheric DMS by oceanic phytoplankton

Dimethylsulfide (DMS) is the most abundant form of volatile sulphur in the ocean and is the main source of biogenic reduced sulphur to the global atmosphere. The ocean to atmosphere flux is currently estimated to constitute about 40% of the total atmospheric sulphate burden. During its synthesis and cycling in the upper ocean, DMS is ventilated to the atmosphere, where it is rapidly oxidized to form SO_4^{2-} and methanesulfonate (MSA) aerosols. Sulfate aerosols (of both biogenic and anthropogenic origins) play an important role in the earth's radiation balance both directly through scattering, absorption and reflection of solar and terrestrial radiation, and indirectly, by modifying cloud physical properties. The flux of DMS from the ocean to the atmosphere is an important concern for atmospheric modellers since the net effect of DMS is believed to be a cooling effect for the global climate.

Various phytoplankton species synthesize differing amounts of dimethylsulfonioproprionate (DMSP), the precursor of DMS. The function of DMSP in algal physiology seems to be varied, and it is thought to act as an osmolyte, a cryoprotectant and also relieve oxidative stress in the algal cell. In general coccolithophorids and small flagellates have higher intracellular concentrations of DMSP. Correlations have been found between algal biomass and DMS concentrations for dinoflagellate and coccolithophorid blooms. However DMS production is also suggested to be due to release from grazing by zooplankton.

4.3 Changes in oceanic community composition by climate changes

Contemporary ecological data indicate that phytoplanktonic populations can respond extremely sensitively and quickly to ocean variability [39]. Open ocean regions may be affected by changes in the thermohaline pump, in which temperature induced changes in density leading to mixing *via* overturning of cold dense surface waters. This mechanism is responsible for transport of oxygen and nutrients to the deep ocean and would adversely impact organisms and communities in the deep ocean. CO_2 is also transported to the deep ocean, but a slow down would not affect deep ocean organisms directly because there is no light for CO_2 photosynthesis and the pressures at depth disallow formation of carbonate structures. However, some suggest that decreasing transport to depth would increase the amount of CO_2 at the surface, decrease CO_2 uptake by the ocean from the atmosphere and exacerbate CO_2 buildup in the atmosphere. This is offset by those that believe that global warming will lead to increased stratification of the surface waters, which would trap more autotrophs at the surface, causing an increase in photosynthesis and uptake of CO_2, which in turn would lead to uptake by the ocean of atmospheric CO_2 and a decrease in atmospheric CO_2.

Direct nitrogen fixation by diazotrophs is observed over much of the subtropical and oligotrophic oceans and is likely major input to the marine and global nitrogen cycle. This additional process actively participates to the efficiency of the biological pump of CO_2.

Increasing atmospheric CO_2 concentration is reducing ocean pH and carbonate concentration and thus the level of calcium carbonate saturation. Experimental evidence suggests that if these trends continue, key marine organisms, such as corals and some calcifying plankton, will have difficulty maintaining their external calcium carbonate skeletons [40]. Southern Ocean surface waters would begin to become undersaturated with respect to aragonite, by the year 2050. By 2100, this undersaturation could extend throughout the entire Southern Ocean and into the subarctic Pacific Ocean.

International ocean science of a new kind evolved with JGOFS. It is created by the efforts of individuals who do not sea boundaries, only opportunities. A thriving community of students and post Docs. emerges each year, and happily spreads across international borders seeking excellence. And they often find it in the laboratories of scientists whose work is presented in [2].

References

[1] Riley J.P., 1965 Historical Introduction, Chemical Oceanography, 1, J.P. Riley and G. Skirrow Eds. (Academic Press, London, 1965) pp. 1-41.

[2] Fasham M.J.R. Ed., Ocean Biogeochemistry. The Role of the Ocean Carbon Cycle in Global Change, (Springer, Berlin, 2003) 297 pp.

[3] SCOR, Oceans, Carbon and Climate Change: an introduction to the Joint Global Ocean Flux Study. Scientific Committee on Oceanic Research, Halifax, Canada, (1990) 61 pp.

[4] IGBP, Joint Global Ocean Flux Study: Implementation Plan, IGBP Report N° 23, IGBP Secretariat, Stockholm, Sweden (1992).

[5] Prentice I.C., Farquhar G.D., Fasham M.J.R., Goulden M.L, Heimann M., Jaramillo V.J., Khesghi H.S., Le Quéré C., Scholtes R.J., Wallace D.W.R., Archer D., Ashmore M.R., Aumont O., Baker D., Battle M., Bender M., Bopp L., Bousquet P., Caldeira K., Ciais P., Cox P.M., Cramer W., Dentener F., Enting I.G., Field C.B., Friedlinstein P., Holland E.A., Houghton R.A., Ishida A., Jain A.K., Janssens I.A., Joos F., Kaminski T., Keeling C.D., Keeling R.F., Kicklighter D.W., Kohfeld K.E., Knorr W., Law R., Lenton T., Lindsay K., Maier-Reimer E., Manning A.C., Matear R.J., Mcguire A.D., Melillo J.M., Meyer R., Mund M., Orr J.C., Piper S., Plattner K., Rayner P.J., Sitch S., Slater R., Taguchi S., Tans P.P., Tian H.Q., Weirig M.F., Whorf T., Yool A. The Carbon Cycle and Atmospheric Carbon Dioxide. In: Climate Change 2001, The Scientific Basis, Intergovernmental Panel on Climate Change, (Cambridge University Press, Cambridge 2002).

[6] Watson A.J., Orr J.C. (2003). Carbon dioxide fluxes in the global ocean., Ocean Biogeochemistry. The role of the Ocean Carbon Cycle in Global Change, Fasham M.J.J., Ed. (Springer, Berlin, 2003) pp. 123-143.

[7] Hedges, J.I., *Mar. Chem.*, **39** (1992) 67-93.

[8] Saliot A. Biogéochimie organique marine. *Océanis* Institut Océanographique, Paris (1994) **20**, 197 pp.

[9] IPCC. Climate change 2001 : the Scientific Basis. Contribution of working group I to the third assessment report of the Intergovernmental Panel on Climate Change, Houghton et al., Eds., (Cambridge University Press, Cambridge, New York).

[10] Suzuki Y., Sugimura Y., *Mar. Chem.*, **24** (1988) 105-131.

[11] Falkowski P., Scholes R.J., Boyle E., Canadell J., Canfield D., Elser J., Gruber N., Hibbard K., Hogberg P., Linder S., Mackenzie F.T., Moore B., Pedersen T., Rosenthal Y., Seitzinger S., Smetacek V., Steffen W., *Science*, **290** (2000) 291-296.

[12] Gildor H., Follows M.J., *Israel J. Chem.*, **42** (2002) 15-27.

[13] Falkowski P.G., Laws E.A., Barber R.T., Murray J.M., Phytoplankton and their role in primary, new, and export production. Ocean Biogeochemistry. The role of the Ocean Carbon Cycle in Global Change, Fasham M.J.J., Ed. (Springer, Berlin, 2003) pp. 99-121.

[14] Redfield A.C., Ketchum B.H., Richards F.A., The influence of organisms on the composition of sea-water, The Sea, Hill M.N. Ed (Interscience, New York 1963) pp. 26-77.

[15] Steeman-Nielsen E.J, .*Cons. Int. Explor. Mer*, **18** (1952) 117-140.

[16] Romankevitch E.A., Ed., Geochemistry of Organic Matter in the Ocean. Springer, Berlin, (1984) 334 pp.

[17] Berger W.H., Smetacek V.S., Wefer G., Ocean productivity and paleoproductivity, An overview. Productivity of the ocean: Present and Past, Berger W.H., Smetacek V.S., Wefer, Eds. (John, Wiley, Chichester, 1989) pp. 1-34.

[18] Sverdrup H.U., Johnson M.W., Fleming R.H., The Oceans Their Physics, Chemistry and General Biology. Prentice Hall, Englewood Cliffs, N.J. (1964). 1087 pp.

[19] Ducklow H.W., Biogeochemical provinces: towards a JOGS synthesis, Ocean Biogeochemistry. The role of the Ocean Carbon Cycle in Global Change, Fasham M.J.J., Ed.(Springer, Berlin, 2003) pp. 3-17.

[20] Longhurst A.R. Ed., Ecological Geography of the Sea. Academic Press, San Diego (1998). 398 pp.

[21] Suess E., *Nature*, **288** (1980) 260-263.

[22] Lochte K., Anderson R., Francois R., Jahnke R.A., Shimmield G., Vetrov A., Benthic processes and the burial of carbon. Ocean Biogeochemistry. The role of the Ocean Carbon Cycle in Global Change, Fasham M.J.J., Ed. (Springer, Berlin, 2003) pp. 195-216.

[23] Thomas H., Bozec Y., Elkalay K., De Baar H., *Science*, **304** (2004) 1005-1008.

[24] Liss P.S., Duce R.A. Eds., The Sea Surface and Global Change. Cambridge University Press, Cambridge, (1997) 519 pp.

[25] Zuev B.K., Chudinova V.V., Kovalenko V.V., Yagov V.V., *Geochem. Int.*, **39** (2001) 702-710.
[26] Momzikoff A., Brinis A., Dallot S., Gondry G., Saliot A. and Lebaron P., *Limnol. Oceanogr.: Methods*, **2** (2004) 374-386.
[27] Ćosović B. (2005). Surface-active properties of the sea surface microlayer and consequences for pollution in the Mediterranean Sea, The handbook of Environmental Chemistry The Mediterranean Sea Vol 5, part K 2005, A. Saliot Ed. (Springer, Berlin, 2005) pp. 269-296.
[28] Gašparović B., Kozarac Z., Saliot A., Ćosović B. and Möbius D., *J. Colloid Interface Sci.* **208** (1998) 191-202.
[29] Brinis A., Méjanelle L., Momzikoff A., Gondry G., Fillaux J., Point V. and Saliot A., *Org. Geochem.*, **35** (2004) 1275-1287.
[30] Hardy J.T. (1997) Biological effects of chemicals in the sea-surface microlayer. Liss P.S. and Duce R.A. Eds. The Sea Surface and Global Change. Cambridge University Press, 519 p.
[31] Frew N.M. The role of organic films in air-sea gas exchange. Liss P.S. and Duce R.A. Eds. The Sea Surface and Global Change. Cambridge University Press, pp. 121-172.
[32] Wanninkhof R., *J. Geophys Res.* **97**(C5) (1992) 7373-7382.
[33] Passow U., *Mar. Ecol. Prog. Ser.* **192** (2000) 1- 11.
[34] Monaham E.C., Dam H.G., *J. Geophys. Res.* **106** (2001) 9377-9383.
[35] Rontani J.-F. Importance of visible light-induced photo-oxidation process in the North-western Mediterranean Sea. The handbook of Environmental Chemistry The Mediterranean Sea Vol 5, part K 2005, A. Saliot Ed. (Springer, Berlin, 2005) pp. 296-317.
[36] Jickells T.D., An Z.S., Anderson K.K., Baker A.R., Bergametti G., Brooks N., Cao J.J., Boyd P.W., Duce R.A., Hunter K.A., Kawahata H., Kubilay N., La Roche J., Liss P.S., Mahovald N., Prospero J.M., Ridgwell A.J., Tegen I., Torres R., *Science,* **308** (2005) 67-71.
[37] Gao Y., Kaufman Y.J., Tanre D., Kolber D., Falkowski P.G., *Geophys. Res. Letters*, **28** (2001) 29-32.
[38] Gabric A.J., Cropp R., Ayers G.P., McTainsh G., Braddock R., *Geophys. Res. Letters* **29** (2002), article number 1112.
[39] Boyd P.W., Doney S.C. The impact of climate change and feedback processes on the ocean carbon cycle. Ocean Biogeochemistry. The role of the Ocean Carbon Cycle in Global Change, Fasham M.J.J., Ed.(Springer, Berlin, 2003) pp. 157-193.
[40] Orr J., Fabry V.F., Aumont O., Bopp L., Doney S.C., Feely R.A., Gnanadesikan A., Gruber N., Ishida A., Joos F., Key R.M., Lindsay K., Maier-Reimer E., Matear R., Monfray P., Mouchet A., Najjar R.G., Plattner G.-K., Rodgers K.B., Sabine C., Sarmiento J.L., Schlitzer R., Slater R.D., Totterdell I.J., Weiring M.-F., Yamanaka Y., Yool A. *Nature* **437** (2005) 681-686.

J. Phys. IV France **139** (2006) 211–237
© EDP Sciences, Les Ulis
DOI: 10.1051/jp4:2006139016

Occurrence and air/sea-exchange of novel organic pollutants in the marine environment

R. Ebinghaus[1,2] and Z. Xie[1]

[1] *Institute for Coastal Research, GKSS Research Centre Geesthacht, Max-Planck-Str. 1, 21502 Geesthacht, Germany*
[2] *Laboratoire de Glaciologie et Géophysique de l'Environnement, 54 rue Molière, BP. 96, 38402 Saint-Martin-d'Hères, France*

Abstract. A number of studies have demonstrated that several classes of chemicals act as biologically relevant signalling substances.

Among these chemicals, many, including PCBs, DDT and dioxins, are semi-volatile, persistent, and are capable of long-range atmospheric transport via atmospheric circulation. Some of these compounds, e.g. phthalates and alkylphenols (APs) are still manufactured and consumed worldwide even though there is clear evidence that they are toxic to aquatic organisms and can act as endocrine disruptors.

Concentrations of NP, *t*-OP and NP1EO, DMP, DEP, DBP, BBP, and DEHP have been simultaneously determined in the surface sea water and atmosphere of the North Sea. Atmospheric concentrations of NP and *t*-OP ranged from 7 to 110 pg m^{-3}, which were one to three orders of magnitude below coastal atmospheric concentrations already reported. NP1EO was detected in both vapor and particle phases, which ranged from 4 to 50 pg m^{-3}. The concentrations of the phthalates in the atmosphere ranged from below the method detection limit to 3.4 ng m^{-3}. The concentrations of *t*-OP, NP, and NP1EO in dissolved phase were 13-300, 90-1400, and 17-1660 pg L^{-1}. DBP, BBP, and DEHP were determined in the water phase with concentrations ranging from below the method detection limit to 6.6 ng L^{-1}.

This study indicates that atmospheric deposition of APs and phthalates into the North Sea is an important input pathway. The net fluxes indicate that the air–sea exchange is significant and, consequently the open ocean and polar areas will be an extensive sink for APs and phthalates.

1. INTRODUCTION

In the last two decades, a number of studies have demonstrated that several classes of chemicals exist that can act as biologically relevant signalling substances, capable to influence the control gene expression at the molecular level and interfering with homoeostatic feedback loops at the development and function level [1, 2]. These substance features are known as endocrine disruption potential, a hypothesis which was first described in the early 1990s [3].

Among these chemicals, many, including PCBs, DDT and dioxins, are semi-volatile, persistent, and are capable of long-range atmospheric transport via atmospheric circulation [4-8]. Some of these compounds, e.g. phthalates and alkylphenols (APs) are still manufactured and consumed worldwide even though there is clear evidence that they are toxic to aquatic organisms and can act as endocrine disruptors. Since 1978, phthalates have been detected in the marine environment and in remote regions such as the Arctic, with concentrations comparable to terrestrial environments [9]. Phthalates are a group of chemicals that are primarily used as plasticizers in huge quantities. Since they are not chemically bound they can be liberated from the consumer products into the environment. Alkylphenols are typically not directly released but are rather formed as anaerobic biological breakdown products of widely used non-ionic surfactants, alkylphenol ethoxylates (APEOs) [10]. Concentrations of APs and their parent compounds have been measured worldwide in all compartments of the environment and even in food products for human consumption [11-14].

International conventions aiming at the protection of the marine environment (such as OSPAR) have listed phthalates and APs together with more persistent compound classes, such as brominated flame retardants (BFRs), musk compounds and chlorinated paraffines as emerging organic pollutants of priority concern. Concentrations of nonylphenol (NP), tertiary octylphenol (*t*-OP) and nonylphenol monoethoxylate (NP1EO) and a number of phthalates have been simultaneously determined in the surface seawater and atmosphere of the North Sea. Air-sea exchange of these compounds was estimated using the two-film resistance model based upon relative air-water concentrations [15].

2. SOURCES OF ALKYLPHENOLS AND PHTHALATES AS ENVIRONMENTAL POLLUTANTS

Phthalates are manufactured worldwide on a large scale, being mainly produced for use as plasticizers in resins, polymers and especially as a softener in polyvinylchloride (PVC). Other industrial applications include the manufacturing of cosmetics, insect repellents, insecticide carriers, and propellants [16]. In the early 1980s, world production of phthalates was estimated to be 2 million t yr^{-1} [17]. A market volume of phthalates in European Union (EU) was estimated around 1000,000 t yr^{-1} in 1990s, of which approximately 480,000 t yr^{-1} was for DEHP and 45,000 t yr^{-1} for BBP. The production volume of DBP in the EU was estimated at 49,000 t yr^{-1} in 1994, and 26,000 t yr^{-1} in 1998, showing clear decreasing trends in production. As plasticizers, phthalates are not physically bound to the polymer and can thus migrate out of plastic and leach into the environment. Phthalates can enter the environment via emission from household and industrial products; from wastewater from production and processing activities, including losses during phthalates synthesis, resin and plasticizer compounding, and during the production of adhesives and coatings; from the use and disposal of materials containing phthalates or landfills with refuse and other waste. In the United States, it was estimated that in 1989, 500 t of DEHP were released into the environment through manufacturing facilities. Recent data on releases of phthalates in the EU are not available. However, it was estimated that 3-5% of the market volume of the phthalates are released into the environment.

Akylphenol ethoxylates (APEOs) are non-ionic surfactants with a worldwide production of approximately 700,000 t yr^{-1} and are used in a wide range of applications [18]. In mid of 1990s, the EU had a market volume of 79,000 t yr^{-1} for technical nonylphenol (NP) and nonylphenol ethoxylates (NPEOs). In 1995, Germany had an alkylphenols (APs) consumption of 20,000 t, of which approximately 14,000 t was nonylphenol (NP). The rest of 6,000 t was made up of octyl-, butyl- and other alkylphenols. The main application was in industrial and household detergents. Moreover, APEOs were used in many other industrial applications, e.g. as wetting agents, emulsifiers of pesticides, dispersants, solubilizers, foaming agents and polymer stabilizers [19]. APEOs are discharged in high quantities in sewage or industrial wastewater treatment plants. They can also directly enter the environment in areas without wastewater treatment [20]. It was estimated that NP and NPEOs enter into the air and water at a rate of 850 t yr^{-1}.

2.1 Environmental fate and distribution of phthalates

The release of phthalates directly into the atmosphere is believed to be the most important input pathway into the environment [21]. Following release they will distribute between different environmental compartments according to their physicochemical properties. Dimethylphthalate (DMP) and diethylphthalate (DEP) are more present in ambient air and water as a result of high vapor pressure and high solubility in water. Dibutylphthalate (DBP) is moderately adsorbed to soil [22-24], but forms complexes with water-soluble fulvic acids that might increase its mobilization and reactivity in soil to some degree [25]. A similar effect can be observed with diethylhexylphthalate (DEHP) that has a strong tendency to be adsorbed by soil however, this effect can be less pronounced in the presence of fulvic acids.

Additionally, phthalates as a group may enter the aquatic environment via wastewater treatment plants, rainfall, runoff and atmospheric deposition. The latter has shown to be a significant input pathway for the Great Lakes in Canada [26].

Several degradation pathways have been described for phthalates, e.g. photo degradation in the atmosphere, bio-degradation in water, and anaerobic degradation in sediments and soils [21]. Contribution of hydrolysis to the overall environmental degradation of phthalates is expected to be low, whereas photo oxidation by OH-radicals contributes more to the elimination of phthalates from the atmosphere.

As presented in Table 1, reported half-lives are specified as a range to indicate differences that are expected due to the OH radical concentrations between pristine (3×10^5 radical cm^{-3}) and polluted (3×10^6 radical cm^{-3}) air. Results are indicating that susceptibility to photo degradation of phthalates increases as alkyl chain length is increasing. The photo degradation half-lives presented in Table 1 are calculated with air oxidation program (AOP) [21] developed by Atkinson and recalculated (the brackets) with an updated version of (AOP, AOPWIN 1.89) [27]. Obviously, the recalculated values are at the lower level as compared to the previous calculation. These values may significantly influence the prediction for the persistence and transport of the phthalates in the atmosphere. Concerning the photo degradation half-lives of particle-associated phthalates, Behnke et al. [28] have investigated the photo degradation rate for DEHP adsorbed to various particulate aerosols. They reported a first-order rate constant of 1.4×10^{-11} cm^3 molecule s^{-1} for the reaction of DEHP with hydroxyl radicals when adsorbed as a monolayer on Fe_2O_3 or SiO_2 aerosols. This rate for inert particle absorbed photo degradation corresponds to a half-life of 0.6 d, using the global average hydroxyl radical concentrations of 9.7×10^5 molecule cm^{-3}, which is not much longer than that calculated for the vapor phase. It seems that sorption to atmospheric particles have no significant effect on the overall rate of indirect photo degradation of the phthalates.

Table 1. Half-lives of phthalates for aqueous hydrolysis, microbial degradation and atmospheric photo degradation.

Phthalate	DMP	DEP	DnBP	BBP	DEHP	DOP
Aqueous hydrolysis (years)[e]	3.3	8.8	22	<0.3	2000	107
Biodegradation (aerobic) (days)	–(1.4-3.0)[g]	2.5 (0.39-4.33)	2.9 (0.87-5.78)	3.1 (0.32-5)	14.8 (0.4-30)	–(1.0)
Biodegradation (anaerobic) (days)	–(21.0)[g]	33.6 (-)	14.4 (2.2-19.3)	19.3 (9.1-13.6)	34.7 (1.0-53.3)	–
Atmospheric photo degradation (days)	9.3-93 (14.41)[g]	1.8-118 (2.39)	0.6-6.0 (0.89)	0.5-5.0 (0.75)	0.2-2.0 (0.38)	0.3-3.0 (0.40)

[e] [21][f] [29]
[g] the values in the brackets are recalculated with an updated version of AOP. Peterson, D. and Staples, C.A.2003.

The persistence of phthalates was predicted with EQC level II modeling [30]. Increasing half-lives and tendency could be expected with increasing alkyl chain length, which ranges from 9.9 to 34 days for the phthalates. It seems the phthalates are not as persistent as the well-known POPs, e.g. α-HCH and PCBs. However, based on the estimated overall persistence for emission to air, travel distances ranging from 220 km for DEHP to 1000 km for DEP were predicted, which is beyond or close to the distance from the European continent to the North Atlantic ocean and Arctic circle. In these cold areas, the phthalates will undergo a slow degradation processes as compared to that predicted with temperate conditions. These facts suggest the need for a detailed investigation of the occurrence and turnover of phthalates in the cryosphere.

2.2 Environmental fate and distribution of nonylphenolethoxylates and their metabolites

NPEOs are produced by the based-catalyzed reaction of NP with ethylene oxide (EO). During the production, a mixture of NP isomers with branched hydrocarbon chains is typically used to form the NPEOs. Biodegradation of NPEOs results in a series of transformations that shorten the ethoxylate chain. The proposed aerobic and anaerobic biological degradation mechanism for NPEOs is shown in Figure 1. It was suggested that under aerobic conditions, NPEOs degrades to

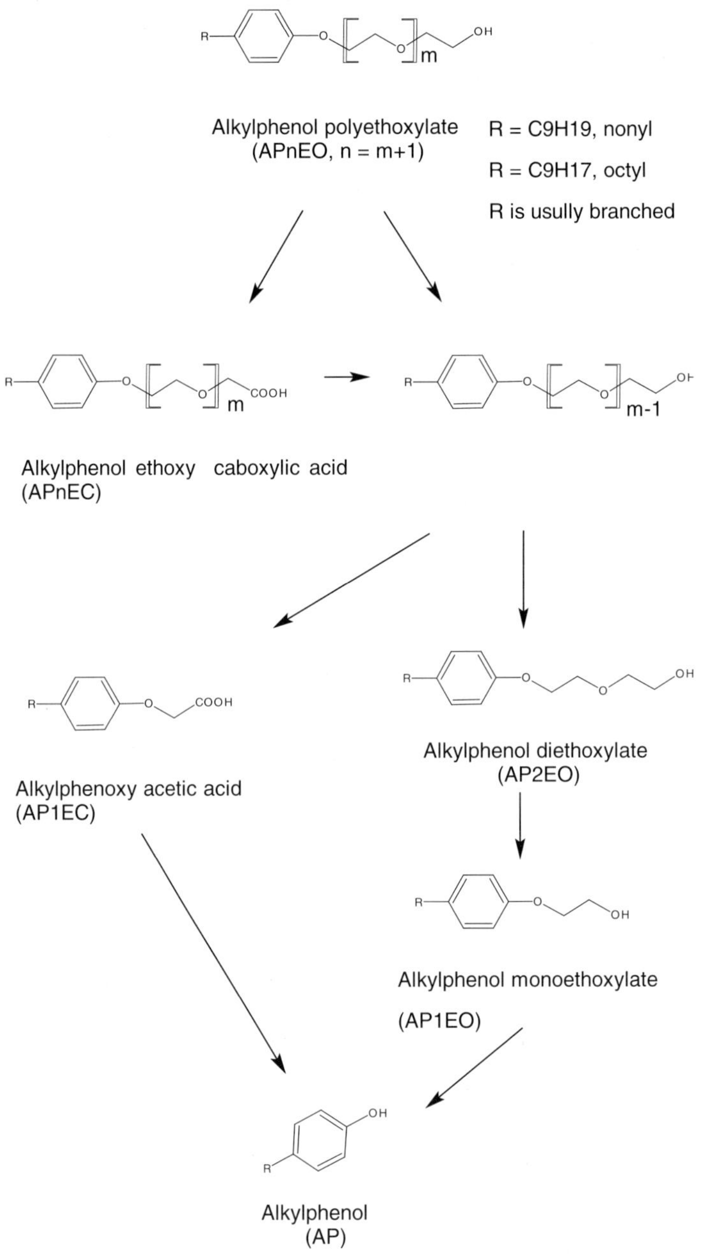

Figure 1. Degradation pathways of Alkylphenol ethoxylates [19].

nonylphenol ethoxylates with short-chained ethoxylates groups or to nonylphenol ethoxycarboxylates with carboxylated ethoxylate and carbon chains, e.g. nonylphenol diethoxylates (NP2EO), nonylphenol monoethoxylate (NP1EO). Complete deethoxylation with formation of NP has been observed under anaerobic conditions [10]. The three most common groups of intermediates reported were summarized as follows [19]: (a) NP, (b) short chain NPEOs having one to four EO unites; (c) a series of ether carboxylates including alkylphenoxy acetic acid and alkylphenoxy ethoxy acetic acid. NP tends to be formed as the final product. Studies have shown that the metabolites of NPEOs, e.g. NP, NP1EO and NP2EO are more hydrophobic, persistent and toxic in the environment [31, 32]. Previous investigations showed that NPEOs metabolites degraded more easily under aerobic, than under anaerobic conditions [33]. The removal rates of NPEOs through sewage treatment plants (STPs) were from 86% to 99% in autumn and from 66% to 99% in winter in Japan, indicating the temperature dependence of degradation of NPEOs [34].

Knowledge on the photo-degradation of APs is very poor. However, it is supposed that photo-degradation dominates the atmospheric fate of NP and NPEOs. A half-life for NP was estimated as 0.3 day [35]. Pelizzetti et al. [36] studied photocatalytic degradation of NPEOs with TiO_2 particulates as photocatalyst. They found that a competitive attack of OH radicals on the ethoxy chain and on the aromatic ring occurs. As a result, complete conversion to CO_2 has been demonstrated. Therefore, photodegradation of NPEOs may quantitatively minimize the accumulation of NP during the sample handling.

This work has been designed to improve our understanding of the distribution pattern and transport mechanisms of APs and phthalates in the coastal and marine environment, with special emphasis on the air-sea exchange to derive flux estimates between the atmosphere and the North Sea/North Atlantic.

3. EXPERIMENTAL

3.1 Reagent preparation

The solvents (methanol, acetone, hexane, dichloromethane, acetonitrile, diethyl ether (Promochem GmbH, Germany) used were residue analysis or HPLC grade, and were distilled prior to use. Milli-Q water (18.2 MΩcm) was generated by a Millipore Ultra-pure water system (Millipore S.A., Molsheim France) and additionally purified with XAD-2 or PAD-2 resins. All glassware was rinsed with Milli-Q water and acetone and then baked at 450°C for at least 8 hours before use.

Analytical standards (*t*-OP, technical NP and NP1EO, dimethyl phthalate (DMP) diethyl phthalate (DEP), di-n-butyl phthalate (DnBP), di-i-butyl phthalate (DiBP), butylbenzylphthalate (BBP), DEHP and dioctyl phthalate (DOP)), internal standards (4-n-NP *d*8 and dibenzylphthalate) and the surrogates (4-n-OP, 4-n-NP, technical NP1EO *d*2 (NP1EO *d*2), DMP *d*4, DEP *d*4, DBP *d*4, DEHP *d*4) were supplied by Dr. Ehrenstorfer (Augsburg, Germany). Stock solutions of each chemical or mixture of chemicals were made by dissolving approximately 5-10 μg of the neat chemicals in liquid, solid or in solution into 10 mL of hexane. The standard solutions used in these experiments were made from appropriate dilutions of these stock solutions. Calibration solutions for preparing GC-MS calibration curves were made by diluting 1-200 μl of the standard solutions in hexane (final volume 200 μL). Stock solutions were prepared every half-year; internal standards and surrogates were prepared for the entire sampling campaign and the measurements.

3.2 PUF/XAD-2 column, PAD-2 column and glass fiber filter (GF/F) preparation

Amberlite XAD-2 resins (particle size: 20-60 mesh) were obtained from Supelco Germany. PAD-2 resins (particle size: 0.3-1.0 mm) were obtained from SERVA Electrophoresis GmbH (Heidelberg, Germany). To prepare the PUF/XAD-2 column, 30 g of XAD-2 resin were packed into a glass column with a glass frit. A piece of polyurethane foam (PUF, 2 cm x 5 cm Ø) was placed on the top to cover the XAD-2 resin. The packed column was cleaned with methanol, acetone and hexane (twice with each solvent) in turn using a modified soxhlet extractor for 72 hours. The residue solvent was removed using purified N_2 (300 mL for 20 min).

To prepare the PAD-2 resin column, 50 g of PAD-2 resin were first rinsed with 500 mL Milli-Q water, and then, the water was replaced with acetone. The PAD-2 resins and acetone were packed into a glass column with a glass frit. The column was filled to about 2/3 with PAD-2 resin. The PAD-2 column was rinsed with 200 mL acetone and then cleaned with acetone and DCM (twice with each solvent) using a modified soxhlet extractor for 72 h. Finally, DCM was replaced by purified milli-Q water (200 mL).

Glass fiber filters (GF/F 8 and GF/F 52) were obtained from Schleicher and Schuell Corporation (Dassel, Germany). GF/F 8 (diameter: 155 mm, pore size: 0.45 μm) was used for atmospheric particles and GF/F 52 (diameter: 142 mm, pore size: 0.7 μm) was used for total suspended matter (TSM) in sea water. Filters were wrapped in a single layer of aluminium foil that was sealed around the filter to create a 'bag'. The filters and the aluminium bag were then baked for 12 h at 450°C in a muffle furnace.

After purification, the PUF/XAD-2 and PAD-2 columns were covered by a pair of pan-like and ball-like caps and sealed by sliding clips. Columns were stored before and after sampling in heat-sealed airtight polypropylene/aluminium/polyethylene bags (PP/AL/PE, Tesseraux, Germany) at 7°C for water samples and at –20°C for air samples, respectively. Cleaned filters were wrapped between aluminium foil in PP/AL/PE bags and used filters were closed in fused test tubing and stored at –20°C.

3.3 Sampling and sample preparation

3.3.1 Water and air sampling

Water sampling was conducted with a modified Kiel In-Situ Pump (KISP) which has been widely applied to the extraction of marine trace organic chemicals [37-39]. Petrick et al [40] described the technical design and principle and tested its performance in the Atlantic Ocean. Although low blanks and extremely low detection limits obtained from KISP samples could satisfy the demands for reliably detecting PCBs and HCHs, the system still presents a blank risk for the determination of trace APs and phthalates as several parts of the KISP are manufactured with or contained PVC material. Therefore, modifications were made to the frame of KISP. All plastic parts were removed and replaced with parts made from stainless steel or glass.

As shown in Figure 2, the in-situ pump includes a filter holder, a PAD-2 column, a pump and a flow meter. The pump and the flow meter were operated on board. The pumping rate can be selected from 0.01-2 L $\min^{-1}$ by adjusting the power supply. The glass fiber filter (GF/F 52) was placed on the glass filter holder. Stainless steel tubing was used to connect the pump to the filter plate. Glass tubing connects the filter plate to the PAD-2 resin column. Sea water samples were taken from beneath the bottom of the ship. In the North Sea, typical water sample volumes were from 20 to 100 L in the area near the coast and from 200 to 400 L in the open sea. In the Atlantic Ocean, up to 1000 L of sea water can be extracted due to the low concentration of total suspended matter (TSM).

Air samples were collected using a high-volume air sampler that was operated at a constant flow rate of 200 L $\min^{-1}$. As shown in Fig 3. (left), the high volume air sampler consists of a high volume pump (ISAP 2000, Schulze Automation & Engineering, Asendorf, Germany), a digital flow meter, a metal filter holder and a PUF/XAD-2 column. The filter holder and the PUF/XAD-2 were linked with a Teflon connector that could protect the glass column while it works under stormy weather. To eliminate the blank risk from the Teflon, the connector was cleaned ultrasonically, three times with acidified water (pH: 2.0) and three times with acetone, respectively. All parts of the filter holder were washed with a washing machine and rinsed with acetone. The pump and the flow meter were set up separately in metal boxes. All electronic plugs were wrapped with waterproof stick film for work outside. GF/F 8 was used to collect atmospheric particles. The filter was changed in the laboratory with tweezers pre-cleaned by thermal treatment. The ship-borne air samples were collected on the upper deck of the research vessel

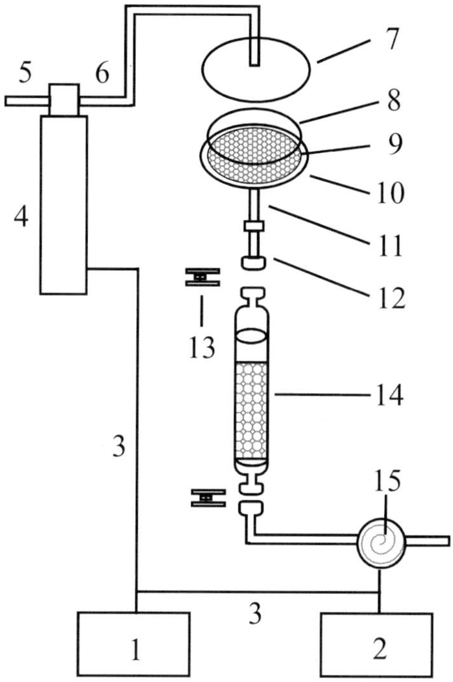

Figure 2. Schematic of the in-situ pump. 1: flow meter controller; 2: flow meter; 3: cable connections; 4: pump; 5: pump inlet; 6: pump outlet; 7: stainless steel deck of filter holder; 8: GF/F 52 filter; 9: glass plate; 10: filter holder; 11: stainless steel tubing; 12: glass connect; 13 adjustable clip; 14: PAD-2 resins column; 15: counter of flow meter.

(see Figure 3, right). Typical air sample volumes were from 400 to 1000 m^3. As reported by Lohmann et al. [41], there is always the potential for contamination by air from ship-board samples. In order to avoid emissions from the ship's funnel, therefore, air sampling was performed on headwind and was halted at station or wind speeds lower than 3 m s^{-1}.

3.3.2 Extraction

The PUF/XAD-2 columns were spiked with the internal standards (50 μL of 200 ng mL^{-1} 4-n-NP d8 50 μL of 1.0 μg mL^{-1} NP1EO d2) and extracted for 16 h using 300 mL of 10% (v/v) diethyl ether in hexane solution with the modified Soxhlet extractor. The PAD-2 columns were extracted for 16 h using 250 mL DCM with the modified Soxhlet extractor after spiking with the internal standards (50 μL of 200 ng mL^{-1} 4-n-NP d8 50 μL of 1.0 μg mL^{-1} NP1EO d2). Both air and water filter samples were spiked with surrogate standards (50 μL of 200 ng mL^{-1} 4-n-NP and 4-n-OP, 50 μL of 1.0 μg mL^{-1} NP1EO d2, 50 μL of 0.5-1.25 μg mL^{-1} deuterated phthalates) and extracted for 16 h using 150 mL of DCM with the Soxhlet extractor. After Soxhlet extraction, the samples were stored in the freezer for rotation evaporation. Several PUF/XAD-2 columns, PAD-2 columns and filters were extracted for a second time in order to check the extraction efficiency.

3.3.3 Evaporation

The inner system of the rotation evaporator was cleaned with 100 mL of acetone prior to and after use. A self-designed adaptor was used to connect the round flask to the evaporator. The special design prevents condensate solvent flow backward into the round flask to eliminate potential contamination from inner tubing of the evaporator. The volume of the extracts were reduced to ~20 mL using rotation evaporator

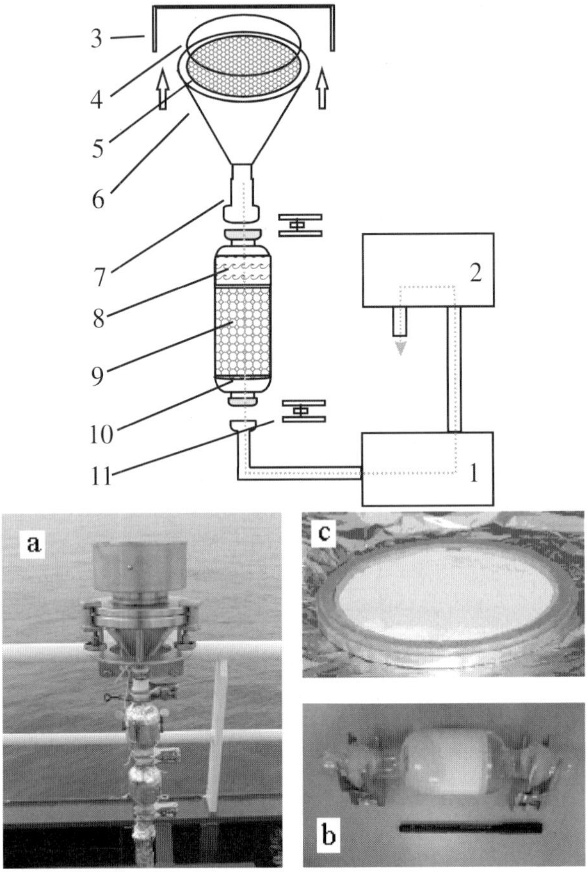

Figure 3. Schematic of the air sampler and operation on board. 1: high volume pump; 2: flow meter; 3: filter shelter; 4: GF/F 8 filter; 5: metal frame for holding up glass filter 6: stainless steel filter holder; 7: teflon connector; 8: PUF sheet; 9: XAD-2 resins; 10: glass frit; 11: adjustable clip; a: air sampler; b: PUF/XAD-2 column; c: filter and particles.

at 30°C under reduced pressure (500-600 mPa for DCM, 220-290 mPa for the mixture of hexane and diethyl ether, 340 for acetone). 20 mL hexane was added to the flask and the solution was continually evaporated to 10-20 mL. The extracts were transferred to another 25 mL pear-bottom flask. The volume of the extracts was further reduced to 1-2 mL before clean-up. In order to remove small residues of water that might be present, the extracts were stored overnight in the freezer at –20°C prior to clean-up.

3.3.4 Silica gel clean-up

All the extracts were purified through a 5% H_2O deactivated silica gel column (2.5 g silica gel packed in a 15 cm × 1 cm i.d. glass column). The silica gel (0.063-0.200 mm, Merck, Darmstadt, Germany) was prepared as follows: extracted using acetone and baking out at 450°C for 12 h to remove organic contamination and deactivation by addition of 5% (w/w) of milli-Q water (purified by PAD-2 resin). After the extracts were transferred into the column, purification was performed by passing 10 mL of hexane through the column in order to remove non-polar compounds. The column was then eluted with 30 mL of hexane and diethyl ether (3:1 v/v) for the APs and phthalates fraction. It was followed with a 25 mL hexane and diethyl ether (1:1 v/v) fraction for NP_2EO. Eluates were reduced in volume in a rotatory evporator and subsequently concentrated in a nitrogen evaporator to 100 μL.

3.3.5 Derivatization

The extracts were derivatized in a glass vial by the addition of N,O-bis(trimethylsilyl)trifluoroacetamide and 1% trimethylcholosilane (TMCS) (BSTFA +1% TMCS) (Part No. 701 490.201, Macherey-Nagel GmbH, Dueren, Germany). 40 μL of 500 ng mL^{-1} surrogate standard mix 5 were spiked as internal standard (if it is not spiked before extraction). The volume was reduced to 100 μL under a gentle stream of nitrogen (99.999%). 100 μL of BSTFA + 1% TMCS was added to the glass vial. The mixture was allowed to react for 1 h at 70°C. After cooling for 5 min, the final sample volume was adjusted to 200 μL using hexane. After derivatization, the extracts were ready for GC-MS without further treatment.

3.4 GC-MS analysis

Quantification of APs and phthalates was performed with an Agilent system consisting of a 6890 N gas chromatograph equipped with an Agilent 7683 series autosampler, a 7683 split-splitless temperature and pressure-programmed injector, and an Agilent 5973 quadrapole mass selective detector (GC-MS). Chemstation Software (2000 version) was used for data processing. The injector was equipped with a deactivate PTV multi-baffle liner. Ions detected were generated by electron impact ionization and monitored in the selective mode (EI-SIM) and total ion scan mode by two injections. A 30 m × 0.25 mm fused silica capillary column (5%-phenyl-95% methylpolysiloxane, HP-5ms) with 0.25 μm film thickness was used for the separation. General conditions for GC-MS analysis are shown in Table 2.

Table 2. GC-MS conditions for the determination of APs and phthalates.

GC-MS	APs	Phthalates
Column	HP-5ms (30 m × 0.25 mm i.d., 0.25 μm film thickness; J&W Scientific, Folsom, CA, USA)	HP-5ms (30 m × 0.25 mm i.d., 0.25 μm film thickness; J&W Scientific, Folsom, CA, USA)
Injection	1 μL	1 μL
Injector temperature program	280°C (pulse splitless mode, 20 psi for 2 min) (Program 1) 80°C (1min), 300°C min^{-1} to 250°C (10 min)^{b} (Program 2)	300°C (pulse splitless mode, 20 psi for 2 min)
Carrier gas	Helium, 1.0 mL min^{-1}	Helium, 1.0 mL min^{-1}
Purge gas	Helium, 250 mL min^{-1}	Helium, 250 mL min^{-1}
Oven temperature program	80°C (1 min), 30°C min^{-1} to 130°C, 3°C min^{-1} to 240°C, 10°C min^{-1} to 300°C, then 300°C (5 min)	80°C (1 min), 30°C min^{-1} to 150°C, 5°C min^{-1} to 300°C (5 min)
Ionization energy	70 eV	70 eV
Interface temperature	280°C	290°C
Ion source temperature	230°C	230°C
Quadrapole	150°C	150°C

3.5 Calibration and quantification

Stock solutions containing all the analytes at accurately defined concentrations were prepared in hexane by dilution in the peak-bottom glass vials. The solvent was removed under a gentle nitrogen stream to 100 μL. These solutions were derivatized as described above. Quantification was carried out using calibration curves based on the peak area of the internal standards 4-n-NP d8 and the surrogate standard mix 5. NP and NP1EO were quantified by each of the isomer peaks. Calibration curves were made with concentrations from 12.5 to 500 ng mL^{-1} for *t*-OP, NP and NP1EO and from 5 to 5000 ng mL^{-1} for the phthalates. The limits of detection (LODs) were set as 3 times the signal to noise ratio. The detection limits of the method (MDLs) were derived from the blanks and quantified as mean field blanks plus

Table 3. Instrumental limit of detection (LOD) and method detection limits obtained in this method.

Compound	LOD (pg)	Sea water (200 L) (pg L^{-1})		Air (500 m^3) (pg m^{-3})	
		Dissolved	TSM	Vapor	Particle
t-OP	0.4	5	5	5	5
NP	3.5	40	5	15	5
NP1EO	3.7	25	10	5	5
DMP	0.8	65	15	5	5
DEP	1.2	75	125	10	10
DiBP	0.3	40	15	5	5
DnBP	0.3	25	30	5	5
BBP	1.8	5	5	2	2
DEHP	1.8	200	150	100	40
DOP	1.4	5	5	2	2

three times the standard deviation (3σ) of field blanks according to the sample volumes (typically, sea water: 200 L, air: 500 m^3). The LODs and MDLs calculated for the analytes are listed in Table 3.

As compared to those reported in the literature, the instrument detection limits for *t*-OP, NP and NP1EO were quite comparable to those obtained with GC-MS [42, 43] GC-MS/MS [44, 45], LC-MS and LC-MS/MS [46]. For phthalates, it was found that GC-MS provided LODs from 0.03 to 0.5 pg, which are 1-3 orders of magnitude lower than those obtained with LC-ESI-MS. The detection limits of the method were found to be comparable between GC-MS and LC-ESI-MS [47]. In this work, coupling GC-MS analysis with large volume sampling, except for DEHP, the detection limits for APs and phthalates could reach a few pg L^{-1} in sea water and a few pg m^{-3} in the atmosphere, which are 1 - 2 orders of magnitude lower than the reported MLDs [42, 46, 48-51].

4. THEORETICAL CONSIDERATIONS ON AIR/SEA GAS EXCHANGE AND WASH-OUT RATIOS

4.1 Washout ratio

The washout ratio (W) is defined as the dimensionless ratio of chemical concentrations in precipitation to that in air. Considering both vapor and particle scavenging mechanism, washout can be expressed as [5]:

$$W = (1 - \phi)\frac{RT}{H} + \phi W_P \tag{1}$$

where ϕ is the fraction of the chemical on the particle, RT/H is the dimensionless Henry's law constant at the ambient temperature, W_P is the particle scavenging coefficient. Based on our experimentally determined ϕ values (see section 5), using Henry's law constants estimated by Cousins and Mackay [52], and a representative Wp value of 20,000 [53], the washout ratio calculated is 9,200 for DBP, 15,000 for BBP, and 15,600 for DEHP at average sampling temperature of 5°C. These values are very comparable to the washout ratios estimated by Staples et al. [21]. Obviously, the washout ratios appear to be temperature dependent, because of both Henry's law constant and vapor pressure are determined by the temperature. In fact, this phenomenon has been observed for the atmospheric removal of DBP and DEHP in Sweden [54].

4.2 Air-sea vapor exchange model

In this study, air-sea vapor exchange fluxes were estimated using the modified version of two-film resistance model [55, 56] which was extensively used for the evaluation of PCBs and PAHs fluxes through air-water interfaces [57-61]. It assumes that the rate of transfer is controlled by the compound's ability to diffuse across two thin stagnant films at the air-water interface, the water film and the air film. The molecular diffusivity of the compound (dependent on the amount of resistance encountered in the liquid and gas films) describes the rate of transfer while the concentration gradient drives the direction of transfer.

The overall flux calculation is defined by

$$F = K_{OL}\left(C_w - \frac{C_a}{H'}\right) \tag{2}$$

where F is the flux (ng m^{-2} day^{-1}), C_w (ng m^{-3}) and C_a (ng m^{-3}) are the dissolved- and vapor-phase concentrations, (C_w-C_a/H') describes the concentration gradient (ng m^{-3}), H' is the dimensionless Henry's law constant, and K_{OL} (m day^{-1}) is mass transfer coefficient comprising resistances to mass transfer in both water (k_w) and air (k_a). Since averages of water temperatures were ranging from 3.8 to 6.7°C on this cruise, the Schmidt number for CO_2 at 5°C (Sc_{CO2} = 1395) was applied for the estimation of k_w [62]. H' was corrected with water temperatures (T, K) and averaged salt concentrations (C_s, 0.5 mol L^{-1}) based on following equations (3-5) [62, 63]

$$\ln H = \ln H_0 + \frac{\Delta H_v}{R}\left(\frac{1}{T_0} - \frac{1}{T}\right) \tag{3}$$

$$H_{corrected} = H \cdot 10^{KsCs} \tag{4}$$

$$H' = \frac{H_{corrected}}{RT} \tag{5}$$

where R is the ideal gas constant (8.314 × 10^{-3} kJ mol^{-1} K^{-1}), H (pa m^3 mol^{-1}) is the Henry's law constant at given T, H_0 is Henry's law constant at T_0 (298.15 K), K_s is the salting constant of 0.3 [62], and $H_{corrected}$ is the corrected Henry's law constant for salinity. ΔHv is the enthalpy of vaporization at 298.15 K, which is supposed to be constant over the ambient temperature range [63]. ΔHv values for phthalates (see Table 2) were estimated from their enthalpies of vaporization at the boiling point [62, 64]. H_0 values of phthalates estimated using the 'three solubility' approach [52] were used for the calculation.

4.2.1 *Uncertainty analysis*

The uncertainty in the F (Equation 2) was evaluated using a propagation of error analysis derived from Shoemaker et al. [65], which has been used in previous studies [57, 59]. The summation of the various random errors in the flux are described by

$$\sigma^2(F) = \left(\frac{\delta F}{\delta K_{OL}}\right)^2 (\sigma K_{OL})^2 + \left(\frac{\delta F}{\delta C_w}\right)^2 (\sigma C_w)^2 + \left(\frac{\delta F}{\delta C_a}\right)^2 (\sigma C_a)^2 + \left(\frac{\delta F}{\delta H}\right)^2 (\sigma H')^2 \tag{6}$$

Total propagated variance σ^2(F) is the linear combination of the weighted contribution of the variances (σ^2) of the mass transfer coefficient, H' and measured concentrations. The error in H' was assumed to be zero because it is a systematic and not random error [57, 59]. The errors of C_w and C_a were assumed to be 15% including the sampling and analytical errors. The uncertainty in K_{OL} was determined by propagating random errors in the air- and water-side transfer velocities, which was summated to be 40% following Wanninkhof et al. [66] and Nelson et al. [57]. The overall propagated error in F is thus, 45% (Table 4). It was shown that most of the uncertainty associated with the fluxes was attributed to K_{OL} (78%), which was a factor of 7 higher than the uncertainties associated with C_w and C_a (11%). As a source of systematic error, H' has a standard error of prediction that is approximately a factor of 3 on the arithmetic value [52], which can affect either K_{OL} or the overall concentration gradient. Therefore,

Table 4. Water and air sampling.

Sample	Date	Station	Volume (L)	Temperature(°C)	Wind speed (m s^{-1})	Salinity (L')
Water						
W1	29/2/2004	3 – 4	38	4.3	9.0	28.5
W2	1/3/2004	4 – 6	174	4.5	8.7	30.4
W3	1/3/2004	6 – 8	82	3.8	10.0	27.8
W4	2/3/2004	10 – 13	175	4.5	10.0	30.1
W5	3/3/2004	16 – 19	254	6.0	6.8	34.1
W6	4/3/2004	23 – 25	410	6.7	14.0	34.9
W7	5/3/2004	22 – 36	381	6.3	11.6	34.6
W8	6/3/2004	29, 30, 36	335	6.2	4.3	34.5
W9	7/3/2004	35 – 37	201	6.0	4.2	34.5
W10	8/3/2004	33 – 34	162	5.3	7.2	32.2
W11	9/3/2004	34 – 39	81	4.8	11.4	31.6
Air			(m^3)			
A1-1*	29/2 – 2/3/2004	1 – 11	494			
A1-2*			549			
A2-1	2/3 – 6/3/2004	11 – 25 – 36	1147			
A2-2			927			
A3-1	6/3 – 9/3/2004	33 – 39	670			
A3-2			589			

accuracy of H' is significant for the estimation of air-sea exchange fluxes, which keeps a need for better understanding of their temperature dependences and improvements for the estimation.

5. AQUEOUS AND ATMOSPHERIC CONCENTRATIONS OF PHTHALATES AND ALKYLPHENOLS AND DERIVED AIR/SEA GAS-EXCHANGE IN THE NORTH SEA

The air and water samples in the North Sea (German Bight) for both phthalates and alkylphenols were collected during the cruise no. 414 with the research vessel *'Gauss'* from February 29th to March 10th in 2004. Integrated water samples were collected at 4.5 m depth during the ship steaming. The air samples were collected on the upper deck of *'Gauss'* about 9 m above sea level. Detailed information on the sampling stations, temperatures, wind speeds, salinities and sample volumes, are given in Figure 4 and Table 4.

5.1 Concentrations of phthalates in sea water

The concentrations of phthalates in dissolved and TSM phases together with air concentrations are given in Table 5. The results showed that DEHP and DBP dominated phthalates' concentrations in the water phase of the North Sea. The concentrations were in the range of 0.45 – 6.6 ng L^{-1} for DBP and 0.52 – 5.3 ng L^{-1} for DEHP. DMP and DEP were detected in more than 80% water samples with concentrations ranging from 0.02 to 4.0 ng L^{-1}. Due to the artefacts of high volume sample collection and their high water solubility, the concentrations of DMP and DEP might be under-estimated. BBP was unexpectedly found in most of the water samples, with concentrations ranging from below MDL to 0.26 ng L^{-1}, although it is released in lower quantities as compared to those of DBP and DEHP. Since DOP concentrations were lower than the method detection limits ($\leq$0.01 ng L^{-1} in water and $\leq$ 0.002 ng m^{-3} in air) in all the samples, this compound is not discussed in this paper.

Fromme et al. [67] investigated phthalates in the surface water of various rivers in Germany (Rhine, Elbe, Ruhr, Mosel, Havel, Spree, Oder), and reported that the concentrations were ranging from 0.33 to 97.8 μg L^{-1} for DEHP and from 0.12 to 8.80 μg L^{-1} for DBP, respectively. They were 2 or 3 orders of magnitude higher than the concentrations determined in the North Sea in our study. Moreover, the

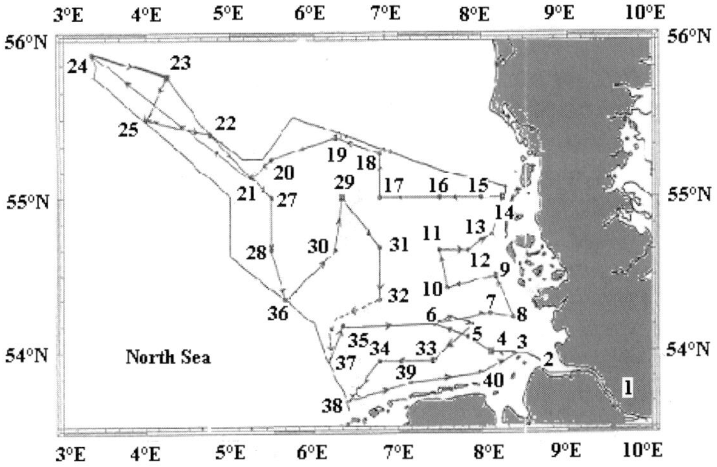

Figure 4. Sampling stations in the North Sea during the cruise 414 with research vessel "*Gauss*", February 29-March 10, 2004.

Table 5. Field blanks (ng) and the concentrations of phthalates in the sea water and in the atmosphere of the North Sea. (The averages of field blanks were subtracted before calculating the concentrations of the samples.)

Water	**Dissolved**					**TSM**				
	DMP	**DEP**	**DBP**	**BBP**	**DEHP**	**DMP**	**DEP**	**DBP**	**BBP**	**DEHP**
Mean blank (ng)	9.7	10.2	3.7	0	43.7	2.8	9.0	2.5	0	30.8
SD (n = 3)	4.1	2.4	0.7	0	8.0	0.4	0.9	0.4	0	9.0
W1(ngL^{-1})	0.68	4.0	6.6	0.26	4.4	0.07	0.49	0.2	0.03	5.8
W2	0.18	0.60	2.0	0.03	1.6	0.05	0.53	0.04	ND	1.0
W3	0.26	0.71	2.2	0.02	NA	NA	NA	NA	NA	NA
W4	0.25	0.71	2.2	0.02	0.93	0.01	0.05	0.01	ND	0.56
W5	0.08	0.17	1.4	0.01	0.88	0.01	4.1	0.01	ND	0.16
W6	ND	0.03	0.94	0.01	5.3	0.01	0.13	0.01	ND	4.0
W7	ND	0.05	0.67	ND	3.4	NA	NA	NA	NA	NA
W8	0.02	0.08	0.52	0.01	0.99	ND	0.02	0.01	ND	0.27
W9	0.05	0.14	0.45	ND	0.58	ND	0	0.01	ND	0.34
W10	0.19	0.51	0.78	0.01	0.52	ND	ND	0.02	0.01	0.64
W11	0.1	0.35	1.4	0.08	3.5	0.01	0.09	0.04	0.02	1.4
Average	0.2	0.67	1.7	0.05	2.2	0.03	0.68	0.04	0.02	1.6
Air			**Vapor**					**Particle**		
Mean blank (ng)	5.1	17.1	4.6	0.3	63.1	4.0	17.6	3.2	0	11.6
SD (n = 3)	0.3	1.0	0.4	0.05	12.4	0.8	5.6	0.6	0	2.0
A1(ng m^{-3})	0.54	3.4	1.1	0.04	0.36	ND	0.18	1.2	0.05	0.95
A 2	0.16	0.64	0.34	0.01	0.30	ND	ND	0.10	0.05	0.97
A 3	0.19	0.75	0.17	0.01	0.22	ND	0.01	0.32	0.06	1.1
Average	0.30	1.6	0.53	0.02	0.29	ND	0.06	0.53	0.05	1.0

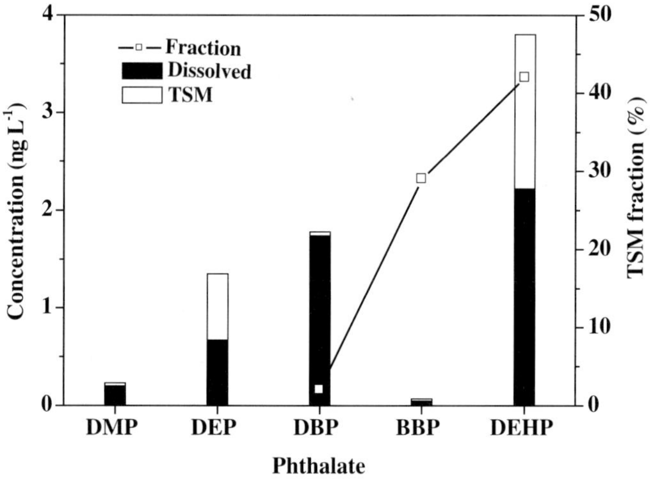

Figure 5. Total concentrations of phthalates in sea water and TSM fractions.

concentrations of DBP and DEHP in the samples collected in the plume of Elbe were one order of magnitude higher than that in the central part of the North Sea. It was suggested that the river-carried contaminations are a significant input source of phthalates into the North Sea.

DBP and DEHP were also detected in most of TSM samples. The concentration ranged from 0.01 to 0.20 ng L^{-1} for DBP, and from 0.16 to 5.8 ng L^{-1} for DEHP. BBP was found in three samples with concentrations ranging from 0.01 to 0.03 ng L^{-1}. As shown in Figure 5, TSM-associated DBP, BBP and DEHP fractions were 2%, 29% and 42%, respectively. DMP and DEP were also found in some TSM samples. Considering the underestimated water concentrations (see above), the TSM fractions were not calculated for DMP and DEP.

TSM associated phthalate fractions were examined previously in surface water samples. Germain and Langlois [68] reported a TSM fraction of 14% for DBP and 53% for DEHP. No other phthalate was detectable in TSM associated fractions. Preston and Al-omran [69, 70] studied the distribution of phthalates in the River Mersey Estuary in UK. 14–34% of DBP was found to be TSM bound (TSM = 1524 mg L^{-1}). Furtmann [71] reported that 15–17% of the low molecular weight phthalates DMP and BBP were TSM bound. Ritsema et al. [72] estimated that 2% of DBP and 67% of DEHP was TSM bound in the Lake Yssel and the Rhine River (Netherlands). As compared to the reported values, our results are in the lower part of the ranges. The differences might be result from the sampling device and separation techniques, especially the TSM concentrations in aquatic phase. In the North Sea, TSM concentrations ranging from 1.36 to 12.69 mg L^{-1} were recorded during another cruise in April 2003. Moreover, the organic matter fraction in the TSM phase is also an important factor for the partitioning of phthalates in the aquatic phase. In addition, the water temperature significantly influences the partition of phthalates between aquatic and TSM phases. Finally, the effect of the salinity should be taken into account for the salting out effect. Nevertheless, these results provided evidence that high molecular phthalates likely partition to suspended matter either in river or the open sea. Particularly, DEHP has a very low solubility in water and high TSM binding affinity, consequently, it was expected to be accumulated in the sediment and to undergo degradation processes only slowly.

5.2 Air concentrations of phthalates

Phthalates concentrations in the atmosphere were shown in Table 5. The average concentrations were 0.30, 1.6, 0.53, 0.02, 0.29 ng m^{-3} for DMP, DEP, DBP, BBP and DEHP in the vapor phase, respectively.

In the particle phase, average concentrations were 0.06, 0.53, 0.05 and 1.0 ng m^{-3} for DEP, DBP, BBP and DEHP. DMP concentration in the particle phase was below the detection limit in all air samples. The low level of particle associated DMP fractions is likely to be the result of its high vapor pressure.

Atlas and Giam [4] reported that DBP and DEHP mean concentrations at Enewetak Atoll in the North Pacific Ocean, which were 0.9 ng m^{-3} for DBP and 1.4 ng m^{-3} for DEHP, can be considered as background data for selected organic pollutants. Giam et al. [9, 73] reported the total vapor and particle concentrations of phthalates in marine atmosphere of the North Atlantic and the Gulf of Mexico. The average concentrations were 1.30 and 1.16 ng m^{-3} for DBP and DEHP in the Gulf of Mexico and 1.0 and 2.9 ng m^{-3} for DBP and DEHP in the North Atlantic. Weschler [74] reported DBP in the Arctic aerosol at Barrow, Alaska, at a concentration of 1 ng m^{-3} indicating that phthalates are ubiquitous in the marine atmosphere. The total concentrations of DBP and DEHP in the atmosphere over the North Sea were determined as 1.0 and 1.3 ng m^{-3}, which agree well to those previously reported values. Similar DEHP concentrations from 0.5 to 5 ng m^{-3} have been detected in the Great Lakes [26] and in the Swedish atmosphere [54]. Moreover, DEHP were detected in relatively high concentrations near contaminated areas with levels of 29 – 132 ng m^{-3} in Antwep, Belgium [75, 76], 300 ng m^{-3} in City of Hamilton and Ontario, Canada [77], and 38-790 ng m^{-3} in Japan [78]. Hoff and Chan [79] reported mean DBP concentrations of 1.9 ± 1.3 ng m^{-3} in the vapor phase and 4.0 ± 2.2 ng m^{-3} in the particle phase along the Niagara River in Ontario, Canada. DBP has also been detected in ambient air in Barcelona, Spain, with a concentration ranging from 3.0 to 17 ng m^{-3} in winter and 1.1-10 ng m^{-3} in summer for coarse (>7.2 μm) and fine particulate (<0.5 μm), respectively [80]. Cautreels et al. [75] reported the concentrations of DBP from 24 to 74 ng m^{-3} in the particulate phase of the air in a residential area of Antwerp, Belgium, and 19 to 36 ng m^{-3} in a rural area in Bolivia. BBP have been also previously determined in the ambient air in Barcelona, Spain, with a concentration ranging 0.25 – 8.0 ng m^{-3}, associated with coarse (>7.2 μm) and fine particulate (<0.5 μm), respectively [80]. DEP was determined in Newark (USA) in the indoor air and outdoor air, with concentrations ranging from 1.60 to 2.03 μg m^{-3}, and from 0.40 to 0.52 μg m^{-3} [81], respectively. As compared to the concentrations of phthalates in urban or contaminated area, the concentrations in the atmosphere over the North Sea are 1-2 orders of magnitude lower. This level of phthalates determined in the North Sea, therefore, appears to represent a regional background concentration.

As shown in Figure 6, the particle-associated fractions were calculated as 2%, 46%, 75% and 78% for DEP, DBP, BBP and DEHP. Comparing to the values of 43% for DEHP, 32% for DBP reported by Giam et al. [73], these values are factor of 1.4 to 1.8 higher. The difference may be a result of the temperature difference of the sampling period or by the velocity of the sampling units. The distribution

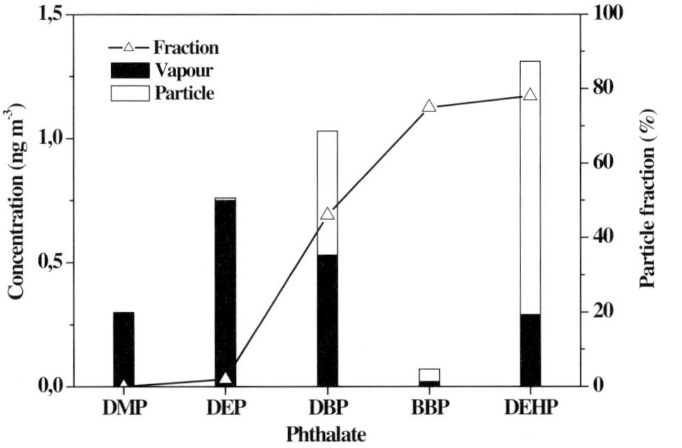

Figure 6. Total concentrations of phthalates in air and particle fractions.

of phthalates in air between vapor and particles have been estimated by Staples et al. [21], based on the Junge model [26]. The estimated fractions of phthalates on the particles were 0.019, 0.039, 1.4, 7.2, and 80%, at ambient temperature. The estimated value of DEHP agrees with that determined in this study although the temperatures were different. However, the estimated values of DEP, DBP and BBP are 1-2 orders of magnitude lower than those determined by Giam et al. [9] and in this work. It suggests that temperature significantly determines the partition of phthalates between the vapor and particle phases, especially for the lighter phthalates. Consequently, both vapor and particle phases are important media for the transport of these phthalates from contamination sources to the coastal margins and the open sea.

5.3 Air/sea gas-exchange fluxes of phthalates

The estimated fluxes of DBP, BBP and DEHP in the North Sea and overall mass transfer coefficients were shown in Table 6. The negative values indicate a net deposition into water; in reverse, the positive values indicate a net volatilization to the atmosphere. Since the water concentrations of DMP and DEP may be underestimated, therefore, their fluxes were not calculated in this work.

Table 6. Air-sea vapor exchange fluxes of DBP, BBP and DEHP in the North Sea. The errors were calculated at 45% level for F and at 40% level for K_{OL}.

Sample	DBP		BBP		DEHP	
	K_{OL} (10^{-3}m day^{-1})	Flux (ng m^{-2}day^{-1})	K_{OL} (10^{-3}m day^{-1})	Flux (ng m^{-2}day^{-1})	K_{OL} (10^{-3}m day^{-1})	Flux (ng m^{-2}day^{-1})
W1	4.4 ± 1.8	−686 ± 309	6.6 ± 2.6	−24 ± 11	86 ± 34	+178 ± 80
W2	4.4 ± 1.8	−685 ± 308	6.6 ± 2.6	−25 ± 11	86 ± 34	−58 ± 26
W3	4.5 ± 1.8	−773 ± 348	6.8 ± 2.7	−28 ± 13	–	–
W4	5.0 ± 2.0	−238 ± 107	7.5 ± 3.0	−7 ± 3	98 ± 39	−95 ± 43
W5	4.3 ± 1.7	−173 ± 78	6.5 ± 2.6	−5 ± 2	85 ± 34	−57 ± 26
W6	8.8 ± 3.5	−327 ± 147	13 ± 5.2	−10 ± 4	180 ± 72	+686 ± 309
W7	7.1 ± 2.8	−278 ± 125	11.0 ± 4.4	−8 ± 4	140 ± 56	+279 ± 125
W8	3.1 ± 1.2	−61 ± 27	4.6 ± 1.8	−4 ± 2	60 ± 24	−7 ± 3
W9	3.0 ± 1.2	−60 ± 27	4.4 ± 1.8	−4 ± 2	57 ± 23	−32 ± 14
W10	4.1 ± 1.6	−91 ± 41	6.2 ± 2.5	−5 ± 2	81 ± 32	−60 ± 27
W11	5.8 ± 2.3	−132 ± 59	8.6 ± 3.4	−7 ± 3	110 ± 44	+245 ± 110
Average	4.9 ± 2.0	−338 ± 152	7.4 ± 3.0	−13 ± 6	97 ± 39	+53 ± 24

The net fluxes of DBP ranged from –60 to –686 ng m^{-2} day^{-1} in the North Sea. It indicates that a net deposition dominates the air–sea vapor exchange process of DBP. The net fluxes of BBP were in a range from –4 to –28 ng m^{-2} day^{-1}, which indicate also a net deposition. The results obtained for DEHP were ranging from –95 to +686 ng m^{-2} day^{-1}, indicating a more complex pattern of the air-sea vapor exchange for DEHP.

As shown in Table 6, the averages of K_{OL} values were 0.0049, 0.0074 and 0.097 m day^{-1} for DBP, BBP and DEHP, respectively. Similar effects of wind speeds on k_w as indicated by Wanninkhof [82] and Bamford et al. [59] can be seen in air-sea exchange fluxes (Table 6). Besides, although water temperatures were only varying from 3.8 to 6.7°C during this cruise, the H' values estimated at higher temperatures were higher by factor of approximately 1.5 than the average value. Both impacts from temperature and wind speed significantly increased K_{OL} values for water samples W6, W7, and by a factor of 1.5 compared to average values. Consequently, relative higher exchange fluxes were estimated for these samples. It was demonstrated that a better understanding of the mass transfer coefficient would rather improve the estimation of the vapor exchange fluxes than a higher accuracy of the concentration [58].

Eisenreich et al. [26] have previously estimated total deposition (dry and wet) of DBP and DEHP to the Great Lake as 3.7-16 tons per year, which indicated that the atmosphere is the major contamination source. Because of the absence of H values and air/water fluxes calculations for phthalates in their report, it was suitable for comparison to only a limited extent. Therefore, comparison was performed with air-water vapor exchange fluxes of PAHs and PCBs in coastal regions. Nelson et al. [57] studied air-sea exchange of PAHs and PCBs in the Chesapeake Bay. Moreover, the vapor exchange fluxes of PCBs across Baltimore Harbor and the North Chesapeake Bay were well studied by Bamford and co-workers [59]. The high ratios of volatilization fluxes to deposition fluxes indicated the dominance of the water–side concentration of PCBs in the overall concentration gradient. As compared to their results, our findings for DBP indicated that the air-side concentrations dominated the overall concentrations gradients, which were comparable to those of lighter PAHs. As compared to PCBs, the difference for overall concentration gradients can be addressed to their H' values. For example, H values of PCBs estimated by Bamford et al. [59] are 1-3 orders of magnitude higher than that of DBP. It indicates that, besides the effect on K_{OL}, temperature determined H values could change the overall concentration gradients with the interface temperature increasing in warm season. If it is supposed that the concentrations of phthalates were less variable through all the seasons, air-sea exchange fluxes and directions might significantly change in warm season. Consequently, decreasing deposition fluxes for DBP could be expected in summer. BBP might show a similar pattern like Chrysene [57], based upon their comparable K_{OL} values and their concentration gradients. As for DEHP, it can be expected that increasing temperature in warm season will increase the potential of volatilization based on the significantly increasing H value.

6. AQUEOUS AND ATMOSPHERIC CONCENTRATIONS OF ALKYLPHENOLS AND DERIVED AIR/SEA GAS-EXCHANGE FLUXES

6.1 Concentrations of alkylphenols in sea water

Total concentrations of NP in sea water and TSM ranged from 90 to 1400 pg L^{-1} and less than MDL (10 pg L^{-1}) to 86 pg L^{-1}, respectively. Water-phase *t*-OP was determined with concentrations ranging from 13 to 300 pg L^{-1}. In the TSM-phase, *t*-OP concentrations were very close to the level of MDL. NP1EO concentrations ranged between 28 and 1660 pg L^{-1} in the dissolved phase, and from below the MDL to 68 pg L^{-1} in the TSM.

Figure 7 shows that high concentrations occurred in the water samples that were collected near River plumes, and an obviously spatial distribution profile with decreasing concentrations from the coast to the open sea. These results agree well with the distribution pattern determined by Heemken et al. [43] and Bester et al. [83]. W1 was taken from station 2 to station 3, which is located within the plume of the Elbe. The highest concentrations of *t*-OP, NP and NP1EO found in this sample were 300, 1400 and 1660 pg L^{-1}, respectively. Compared with other coastal waters and open ocean the concentrations of APs in the North Sea are in a relatively low level. As compared to the concentrations of NP and *t*-OP determined in previous sampling campaigns in the North Sea, present results are 8-40 times lower.

In Figure 8, the dissolved concentrations of *t*-OP, NP and NP1EO are plotted against the salinities of the water samples. The salinity changed from 27.5 – 34.9L' in our sampling area. Typically, the high concentrations were found in the body of water with low salinity, and contrarily, low concentrations were found in the body of water with high salinity. The correlation coefficients between salinity and concentrations of *t*-OP, NP and NP1EO are, 0.70, 0.88, and 0.90, which indicates that changing concentrations could be linked to the origin of the current. Together, with the investigations done by Heemken et al. [43], Bester et al. [83], and Jonkers et al. [85], it shows that the input of River Elbe and Rhine represent a significant source for alkylphenols in the North Sea.

In this study, *t*-OP, NP and NP1EO were also found in TSM in the samples collected near the coast. The concentrations were in a range of 1-11 pg L^{-1} for *t*-OP, 8 – 86 pg L^{-1} for NP and 7 – 68 pg L^{-1} for NP1EO. As compared to the dissolved concentrations, TSM bound fractions were 5 $\pm$ 1% for *t*-OP, 6 $\pm$ 2% for NP and 3 $\pm$ 1% for NP1EO, which are very comparable to their log K_{OW} sequence: NP (4.48)>NP1EO (4.17 - 4.20)>*t*-OP (4.12) [86]. However, these fractions were lower by factor of

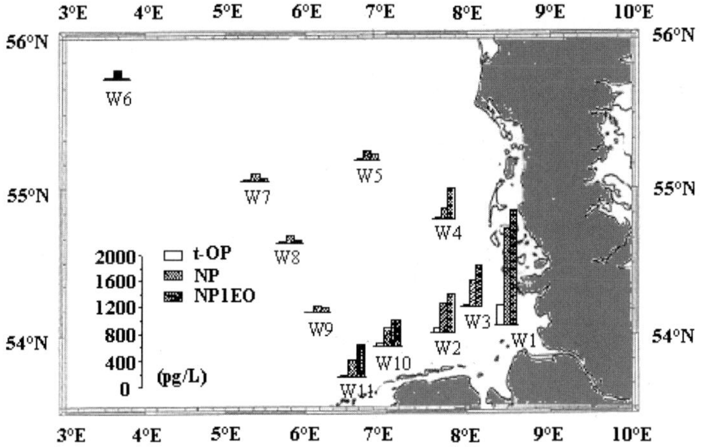

Figure 7. Spatial distributions of *t*-OP, NP and NP1EO in the North Sea. The concentrations are marked on the mean sample locations.

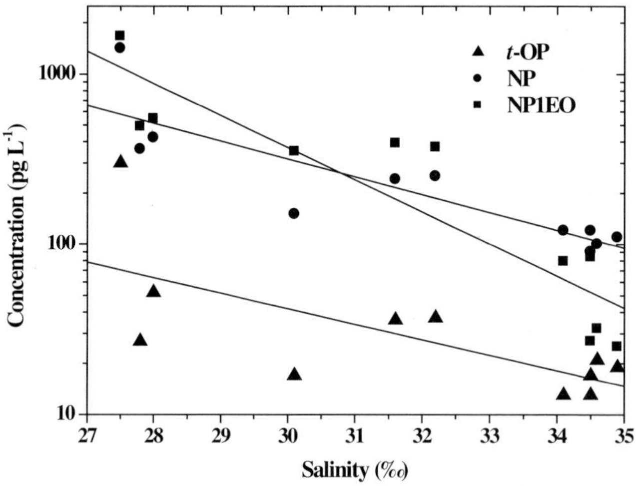

Figure 8. Changing of concentrations of *t*-OP, NP and NP1EO with salinity gradients.

2 – 5 than the reported values found in Tokyo Metropolitan area [87] and in the Lower Hudson River Estuary [88]. The likely reason is the low amount of TSM present in the North Sea as compared to those in rivers and estuaries.

6.2 Air concentrations of alkylphenols

NP, *t*-OP were detected in the vapour phase of the air samples. Relatively high concentrations were present in the air samples A1 and A2, which were collected while the ship was steaming along the Northeast German Coast. The concentrations in the air samples A3-A6 are lower by a factor of 3-5. NP1EO was unexpectedly determined in the particle phase with a concentration range from 14 to 50 pg m^{-3}. Consequently, we re-examined the NP1EO in the vapour phase. Concentrations of 31 and 14 pg m^{-3} were found in A1 and A2, and 4 - 9 pg m^{-3} were found in A3 - A6. As a result of the internal standard of NP1EO D2 was not applied for PUF/XAD-2 extraction, therefore, there is considerably uncertainty concerning the vapour concentrations. NP1EO has been detected in indoor dusts [14] and in the atmosphere surrounding an industrialised area of Prato (Italy) [89]. This was the first time that this particular compound was found in the marine atmosphere. The air samples collected over the sea surface are often influenced by wind pattern. During this cruise, Southwest and Southeast winds were dominant. It meant that air masses were originating from the German and Dutch coasts. Furthermore, when the wind speed is higher than $7\,m\,s^{-1}$, the aerosol formation might be dominated by sea spray and sea salts [90]. Therefore, atmospheric samples were a mixture of land-based, coastal and oceanic air and aerosols. A1 – A4 were collected always under heavy winds and high humidity conditions. This may be an explanation for the high concentrations present in the samples A1 and A2, especially for the occurrence of NP1EO. As the data of aerosol composition and particle concentrations of *t*-OP and NP was not available, the uptake of APs during the aerosol generation over the North Sea is still not clear.

Comparisons of atmospheric APs concentrations in the North Sea described here have been made with previously reported data in the forest area in Southeastern Germany [42] and urban and coastal sites at the Lower Hudson River Estuary [10, 88]. Figure 9a, 9b shows that atmospheric *t*-OP and NP

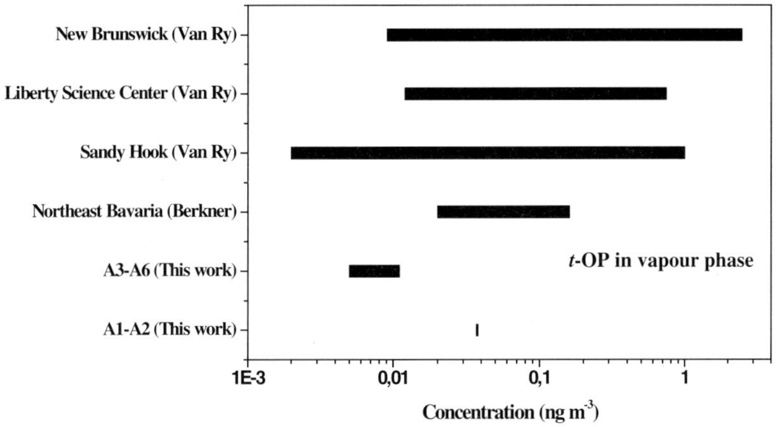

Figure 9a. Range of measured *t*-OP concentrations in the atmosphere over the North Sea and comparison with other reported values.

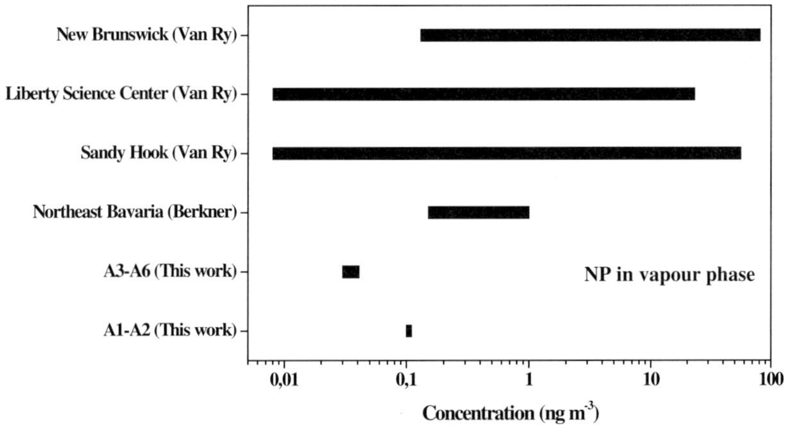

Figure 9b. Range of measured NP concentrations (in total isomers) in the atmosphere over the North Sea and comparison with other reported values.

in the North Sea were in the lower concentration range. Concentration of NP and *t*-OP in air samples A1 and A2 were very comparable to reported terrestrial concentrations in southeast of Germany, and lower than those determined at the Lower Hudson River Estuary. These results provide evidence that, apart from the volatilization from surface waters, the terrestrial use of APs is also a significant source for atmospheric APs [91].

6.3 Air/sea gas-exchange fluxes of alkylphenols

The estimated fluxes of *t*-OP and NP in the North Sea and overall mass transfer coefficients are shown in Table 7. The negative values indicate a net deposition into the water; contrarily, the positive values indicate a net volatilization to the atmosphere. The fluxes ranged from –3 to –30 ng m^{-2} day^{-1} for *t*-OP and –12 to –81 ng m^{-2} day^{-1} for NP, which indicated that net deposition dominate the air–sea vapour exchange process of *t*-OP and NP in the North Sea.

Table 7. Air-sea exchange fluxes of NP and *t*-OP in the North Sea. The errors were calculated at 48% level for F and at 40% level for K_{OL}.

Sample	NP		*t*-OP	
	K_{OL} (10^{-3}m day^{-1})	Flux (ng $m^{-2}day^{-1}$)	K_{OL} (10^{-3}m day^{-1})	Flux (ng m^{-2} day^{-1})
W1	11 ± 4	-62 ± 30	14 ± 6	-23 ± 11
W2	11 ± 4	-71 ± 34	14 ± 6	-26 ± 13
W3	12 ± 5	-81 ± 39	15 ± 6	-30 ± 14
W4	13 ± 5	-30 ± 14	16 ± 6	-6 ± 3
W5	11 ± 4	-22 ± 10	13 ± 5	-4 ± 2
W6	22 ± 9	-41 ± 20	27 ± 11	-8 ± 4
W7	18 ± 7	-35 ± 17	22 ± 9	-7 ± 3
W8	8± 3	-12 ± 6	10 ± 4	-3 ± 2
W9	7 ± 3	-12 ± 6	9 ± 3	-3 ± 2
W10	11 ± 4	-16 ± 8	13 ± 5	-5 ± 2
W11	15 ± 6	-24 ± 12	18 ± 7	-7 ± 4
Average	13 ± 5	-39 ± 19	16 ± 6	-12 ± 6

As shown in Table 7, the averages of K_{OL} values were 0.013 and 0.016 m day^{-1} for *t*-OP and NP, respectively. Both impacts from temperature and wind speed significantly increased K_{OL} values for water samples W6, W7, which were higher by a factor of 2 compared to the average values. Consequently, relative higher exchange fluxes were estimated for these samples. Again it was demonstrated that a better understanding of the mass transfer coefficient would improve the estimation of the vapor exchange fluxes rather than a higher accuracy of the concentrations [58, 92].

Van Ry et al. [88] estimated the air-water exchange fluxes for NP in the lower Hudson River Estuary, which ranged from +25 to +67 μg m^{-2} day^{-1} in the upper bay and from +1.2 to +9.5 μg m^{-2} day^{-1} in the lower bay. As compared to these results, the air-sea exchange fluxes of NP in the North Sea are 2 orders of magnitude lower. The large differences of air-sea exchange fluxes of NP present in these two studies, with the exception of those resulting from the differences in the NP concentrations, could be addressed to the H' values used for the estimations and its temperature dependence [92]. As estimated by Xie et al. [92], the H value of *t*-OP and NP at 298.15 K is 8 times higher than that at 278.15 K. Assuming the concentrations of *t*-OP and NP are less variable in both water and vapour phases of the North Sea through all the seasons, in warm times, the K_{OL} could significantly change, following the increasing H'. Moreover the concentration gradients (Cw-Ca/H') will significantly decrease and might change in directions. Nevertheless, these results indicated that the air-sea exchange significantly interferes with the process of distributions of *t*-OP and NP in the North Sea.

In conclusion, *t*-OP, NP and NP1EO were studied in different compartments of the North Sea using ship-borne samples. Concentrations in sea water were 1-3 orders magnitude lower than those determined in the River Elbe, Rhine and their estuaries. Dilution profiles were present from the coast to the central part of the North Sea. It can be derived that the rivers are main input sources for APs present in the North Sea. Analysis for the compositions of NP and NP1EO isomers showed that NP isomers are relative stable under aerobic conditions, however, selective degradation among NP1EO isomers might be occurring. However, further study of NP and NP1EO isomers in sediment is still needed for a better understanding of their behavior under anaerobic conditions in the marine ecosystem. *t*-OP, NP and NP1EO were studied for first time in the atmosphere over the North Sea. Concentrations were in lower ng m^{-3} range and comparable to those found in the rural area of Southeastern Germany. Air-sea exchange fluxes of *t*-OP and NP showed that atmospheric deposition were dominant in the winter. Re-volatilization might happen in the warm seasons due to the increasing H values with the rise of the temperature. Nevertheless, this study indicated that the North Sea is an important sink for the APs, and might be a potential source for the occurrences of APs in the Oceans and remote area. Therefore, further studies will be conducted to elucidate the spatial and temporal dependences of the distributions of APs in the coastal margins and in the open ocean.

7. SUMMARY AND CONCLUSIONS

High-volume sampling methods with PAD-2 resin columns for seawater and PUF/XAD-2 column for atmospheric samples are powerful and suitable tools for the collection of trace APs and the phthalates in the environment. The field blanks are partly eliminated with a self-designed glass connector for the in-situ pump and active carbon cartridges for the soxhlet extractor and the rotation evaporator. These developments are not only a benefit for reducing the blanks for APs and the phthalates, but also suitable for controlling the blank levels of other organic pollutants. BSTFA + 1% TMCS was selected for the derivatization of t-OP, NP and NP1EO. The instrument detection limits of APs reached picogram in absolute amount. Furthermore, the reagent of BSTFA does not react with the phthalates under the optimized condition, which allows detecting *t*-OP, NP, NP1EO and the phthalates simultaneously. The soxhlet extraction with the modified soxhlet extractor combined with the active carbon cartridge and the distilled solvent is very convenient for the operation and achieves detection of very low contaminations. The recoveries of *t*-OP, NP, NP1EO and the phthalates achieved for the entire procedure are satisfactory. The loss of the phthalates during sampling and laboratory treatments could be fully recovered using the deuterated compounds. Although the large volume sampling and soxhlet extraction procedures are time consuming and labor intensive, they eliminate matrix, feature high enrichments capacity and allow method detections in the pg L^{-1} and pg m^{-3} for sea water and air samples.

Concentrations of NP, *t*-OP and NP1EO, DMP, DEP, DBP, BBP, and DEHP have been simultaneously determined in the surface sea water and atmosphere of the North Sea. The concentrations of *t*-OP, NP, and NP1EO in dissolved phase were 13-300, 90-1400, and 17-1660 pg L^{-1}. In total suspended matters (TSM), the analytes were detectable only in the samples collected close to the coast with concentrations ranging from 1 to 84 pg L^{-1}. DBP, BBP, and DEHP were determined in the water phase with concentrations ranging from below the method detection limit to 6.6 ng L^{-1}. Concentrations in sea water were 1-3 orders magnitude lower than those determined in the River Elbe, Rhine and their estuaries. Dilution profiles were present from the coast to the central part of the North Sea. It indicated that the rivers are main input sources for APs and the phthalates present in the North Sea.

Atmospheric concentrations of NP and *t*-OP ranged from 7 to 110 pg m^{-3}, which were one to three orders of magnitude below coastal atmospheric concentrations reported before. NP1EO was detected in both vapor and particle phases, which ranged from 4 to 50 pg m^{-3}. The concentrations of the phthalates in the atmosphere ranged from below the method detection limit to 3.4 ng m^{-3}.

Air-sea vapor exchanges of *t*-OP and NP, DnBP, BBP and DEHP were estimated using the two-film resistance model based upon relative air-water concentrations. The average of air-sea exchange fluxes was -12 ± 6 ng m^{-2} day^{-1} for *t*-OP, -39 ± 19 ng m^{-2} day^{-1} for NP, -338 ng m^{-2} day^{-1} for DBP and -13 ng m^{-2} day^{-1} for BBP which indicates a net deposition is occurring. The air-sea exchange fluxes of DEHP were ranging from –95 to +686 ng m^{-2} day^{-1}. The average value of +53 ng m^{-2} day^{-1} for DEHP suggested a net volatilisation from the North Sea. These results suggest that the air–sea vapor exchange is an important process that intervenes in the mass balance of alkylphenols and the phthalates in the North Sea.

The concentrations of *t*-OP, NP and NP1EO present over land and the North Sea suggested that apart from the emission of a highly contaminated water body, terrestrial applications of NPEOs and APs are significant sources for their atmospheric occurrence. Consequently, via atmosphere, both APs and the phthalates may undergo a long distance transport and accumulate in the cold region. Air-sea exchange fluxes of *t*-OP and NP and the phthalates showed that atmospheric deposition were dominant in the winter. Re-volatilization might happen in the warm seasons due to the increasing H values with the rise of the temperature. Nevertheless, this study indicated that the North Sea is an important sink for the APs and the phthalates. The net fluxes indicate that the air–sea exchange is significant and, consequently the open ocean and polar areas will be an extensive sink for APs and the phthalates.

As many physicochemical properties governing the environmental fate of organic pollutants are temperature dependent, there is a need to determine the gas-particle partitioning coefficients of APs and phthalates and the Henry's Law Constants of the phthalates over the range of ambient temperatures. Knowledge of the transport mechanisms of APs and phthalates via the atmosphere and the air-sea exchanges might be significantly increased by a re-evaluation with improved data of physicochemical properties.

Acknowledgements

This manuscript was prepared while R. E. was Visiting Professor at Laboratoire de Glaciologie et Géophysique de l'Environnement (LGGE) in Grenoble, France. This stay was made possible by financial support of the Institut Universitaire de France (IUF), which is gratefully acknowledged.

References

[1] McLachlan, J. A. 2001. Environmental signalling: What embryos and evolution teach us about endocrine disrupting chemicals. Endocrine Reviews, 22, 319-341.

[2] Myers, J.P., Guillette, Jr., L.J., Palanza, P., Parmigiani, S., Swan, S. H., vom Saal, F. S., 2003. The emerging science of endocrine disruption. Paper presented at the International Seminar on Nuclear War and Planetary Emergencies. 28th session, Erice, Italy 18-23 August.

[3] Colborn and Clement, 1992.Chemically-induced alterations in sexual and functional development: The wildlife/human connection (Vol. XXI). Princeton: Princeton University Press.

[4] Atlas, E., Giam, C.S., 1981. Global transport of organic ambient concentrations in remote marine atmosphere. Science 211, 163-165.

[5] Bidleman T.F., 1988. Atmospheric processes: Wet and dry deposition of organic compounds are controlled by their vapor-particle partitioning. Environmental Science & Technology 22, 361-367.

[6] Eitzer, B.D. and Hites, R.A., 1989. Atmospheric transport and deposition of polychlorinated dibeno-p-dioxins and dibenzofurans. Environmental Science & Technology 23, 1396-1401.

[7] Bright, D.A., Grundy, S.L., Kenneth J. Reimer, K.J., 1995. Differential Bioaccumulation of Non-ortho-Substituted and Other PCB Congeners in Coastal Arctic Invertebrates and Fish. Environmental Science & Technology 29, 2504-2512.

[8] Kalantzi, O.I., Alcock, R.E., Johnston, P.A., Santillo, D.; Stringer, R.L., Thomas, G.O., Jones, K.C., 2001. The Global Distribution of PCBs and Organochlorine Pesticides in Butter. Environmental Science & Technology 35, 1013-1018.

[9] Giam, C.S., Chan, H.S., Nett, G.S., Atlas, E.L., 1978. Phthalate ester plasticizers: a new class of marine pollutant. Science 199, 419-421.

[10] Giger, W., Brunner, P.H., Schaffiner, C., 1984. 4-Nonylphenol in sewage sludge: accumulation of toxic metabolites from nonionic surfactants, Science 225, 623-625.
[11] Dachs, J., Van Ry Da, Eisenreich S. J., 1999. Occurrence of estrogenic nonylphenols in the urban and coastal atmosphere of the lower Hudson River estuary. Environmental Science & Technology 33, 2676-2679.
[12] Kolpin, D. W., Furlong, E. T., Meyer, M. T., Thurman, E. M., Zaugg, S. D., Barber, L. B. et al. 2002. Pharmaceuticals, Hormones, and Other Organic Wastewater Contaminants in U.S. Streams, 1999-2000: A National Reconnaissance. Environmental Science & Technology 36, 1202-1211.
[13] Guenther, K., Heinke, V., Thiele, B., Kleist, E., Prast, H., Raecker, T., 2002. Endocrine Disrupting Nonylphenols Are Ubiquitous in Food. Environmental Science & Technology 36, 1676-1680.
[14] Rudel, R.A., Camann, D.E., Spengler, J.D., Korn, L.R., Brody, J. G., 2003. Phthalates, Alkylphenols, Pesticides, Polybrominated Diphenyl Ethers, and Other Endocrine-Disrupting Compounds in Indoor Air and Dust. Environmental Science & Technology 37, 4543 -4553.
[15] Liss, P.S. and Slater, P.G., 1974. Flux of gases across the air-sea interface. Nature 233, 327-329.
[16] Peakall, D.B., 1975. Phthalate esters: occurrence and biological effects. Residuces review 54, 1-41.
[17] Schmezer, P., Pool, B.L., Klein, R.G., Komitowski, D., Schmaehl, D., 1988. Various short-term assays and two long-term studies with the plasticizer di(2-ethylhexyl) phthalate in the Syrian hamster. Carcinogenesis 9, 37-43.
[18] Jonkers, N, Knepper, T.P, de Voogt, P., 2001. Aerobic biodegradation studies of nonylphenol ethoxylates in river water using liquid chromatography-electrospray tandem mass spectrometry. Environmental Science & Technology 35, 335-340.
[19] Ying, G.G., Williams, B., Kookana, R., 2002. Environmental fate of alkylphenols and alkylphenol ethoxylates—a review. Environment International, 28, 215-22.
[20] James, M. R., 1999. Review of the persistence of Nonylphenol and Nonylphenol Ethoxylates in aquatic Environments, Water Quality Research Journal of Canada 34, 37-38.
[21] Staples, C.A., Peterson, D.R., Parkerton, T.F., Adams, W.J., 1997. The environmental fate of phthalate esters: a literature review. Chemosphere 35, 667-749.
[22] Howard, P.H., 1989. Handbook of environmental fate and exposure data for organic chemicals. Chelsea, Michigan, Lewis Publishers Inc., 574.
[23] Zurmühl, T., Durner, W., Herrmann, R., 1991. Transport of phthalate esters in undisturbed and unsaturated soil column. J. Contamination Hydrology 8, 111-133.
[24] Environmental Health Criteria 189. Di-n-butyl phthalate. 1997. World Health Organization.
[25] Matsuda, K. and Schnitzer, M., 1971. Reactions between fulvic acid, a soil humic matirial and dialkyl phthalates. Bullitin of Environmental contaminant and Toxicology 6, 200-204.
[26] Eisenreich, S.J., Looney, B.B., David, J.B., 1981. Airborne organic contaminants in the Great Lakes ecosystem. Environmental Science & Technology 15, 30-38.
[27] Staples, C.A., Peterson, D.R., Parkerton, T.F., Adams, W.J., 1997. The environmental fate of phthalate esters: a literature review. Chemosphere 35, 667-749.
[28] Behnke, W., Nolting, F., Zetzsch, C., 1987. An aerosol smog chamber for testing biotic degradation of compounds with low volatility. In: Greenlegh R, Roberts TR (eds) Pestic Sci Biotechnol, Proc Int Congr Pestic Chem Blackwell, Oxford, UK, P 401.
[29] Yuan, S.Y., Liu, C., Liao, C.S., Chang, B.V., 2002. Occurrence and microbial degradation of phthalate esters in Taiwan river sediments. Chemosphere 49, 1295-1299.
[30] Cousins, I., Mackay, D., Parkerton, T., 2003. Physical-chemical properties and evaluative fate modelling of phthalate esters. The Handbook of Environmental Chemistry 3, 57-84.
[31] Soto, A.M., Justicia, H., Wray, J.W., Sonnenschein, C., 1991. p-Nonylphenol: an estrogenic xenobiotic released from modified polystyrene, Environmental Health Perspectives 92, 167–173.

[32] Renner, R., 1997. European bans on surfactant trigger transatlantic debate, Environmental Science & Technology 31, 316A-320A.

[33] Brunner P.H., Capri, A., Marcomini, A., Giger, W., 1988. Occurrence and behaviour of linear alkylbenze-sulphonates, nonylphenol, nonylphenol mono- and di-ethoxylates in sewege and sewage sludge treatment. Water Research 22, 1465-1472.

[34] Nasu, M., Goto, M., Kato, H., Oshima, Y., Tanaka, H., 2001. Study on endocrine disrupting chemicals in wastewater treatment plants. Water Science & Technology 43, 101-108.

[35] CSF/01/12. Chemicals Stakeholder Forum Fifth Meeting. 4-t-noylphenol and 4-t-octylphenol. 11 September 2001.

[36] Pelizzetti, E., Minero, C., Maurino, V., Sciafani, A., Hidaka, H., Serpone, N., 1989. Photocatalytic degradation of nonylphenol ethoxylated surfactants. Environmental Science & Technology 23, 1380-1385.

[37] Wodarg, D., Kömp, P., McLachlan, M.S., 2004. A baseline study of polychlorinated biphenyl and hexachlorobenzene concentrations in the western Baltic Sea and Baltic Proper. Marine Chemistry 87, 23-36.

[38] Bruhn, R., Lakaschus, S., Mclachlan, M.S., 2003. Air/sea gas exchange of PCBs in the southern Baltic Sea. Atmospheric Environment 37, 3445-3454.

[39] Lakaschus, S., Weber, K., Wania, F., Bruhn, R., Schrems, O., 2002. The air-sea equilibrium and time trend of hexachlorocyclohexanes in the Atlantic Ocean between the Arctic and Antarctic. Environmental Science and Technology 36, 138-145.

[40] Petrick, G., Schuz, D.E., Martens, V., Scholz, K., Duinker, J.C., 1996. An in-situ filtration/extraction system for the recovery of trace organics in solutions and on particles tested in deep ocean water. Marine Chemistry 54, 97-105.

[41] Lohmann, R., Jaward, F.M., Durham, L., Barber, J.L., Ockenden, W., Jones, K.C., Bruhn, R., Lakaschus, S., Dachs, J., Booij, K., 2004. Potential contamination of ship-board air samples by diffusive emissions of PCBs and other organic pollutants: implications and solutions. Environmental Science and Technology 38, 3965-3970.

[42] Berkner, S., Streck, G., Herrmann, R., 2004. Development and validation of a method for the determination of trace levels of alkylphenols and bisphenol A in atmospheric samples. Chemosphere 54, 575-584.

[43] Heemken, O.P., Reincke, H., Stachel, B., Theobald, N., 2001. The occurrence of xenoestrogens in the Elbe River and the North Sea. Chemosphere 45, 245-259.

[44] Jeannot, R., Sabik, H., Sauvard, E., Dagnac, T., Dohrendorf, K., 2002. Determination of endocrine-disrupting compounds in environmental samples using gas and liquid chromatography with mass spectrometry. J. Chromatography A 974, 143-159.

[45] Hoai, P.M., Tuunoi, S., Ike, M., Kuratani, Y., Kudou, K., Viet, P.H., Fujita, M., Tanaka, M., 2003. Simultanious determination of degradation products of nonylphenol polyethoxylates and their halogenated derivatives by solid-phase extraction and gas chromatography-tandem mass spectrometry after trimethylsilylation. J. chromatography A 1020, 161-171.

[46] Loyo-Rosales, J., Schmitz-Afonso, I., Rice, C.P., Torrents, A., 2003. Analyis of octyl- and nonylphenol and their ethoxylsates in water and sediments by liquid chromatography/tandem mass spectrometry. Analytical Chemistry 75, 4811-4817.

[47] Lin, Z.P., Ikonomou, M.G., Jing, H., Mackintosh, C., Gobas, F. A. P. C., 2003. Determination of Phthalate Ester Congeners and Mixtures by LC/ESI-MS in Sediments and Biota of an Urbanized Marine Inlet. Environmental Science & Technology 37, 2100-2108.

[48] Teil, M.J., Blanchard, M., Chevreuil, M., 2005. Atmospheric fate of phthalate esters in an urban area (Paris-France) Science of The Total Environment, Vol. 354, 204-215.

[49] Cincinelli, A., Stortini, A. M., Perugini, M., Checchini, L., Lepri, L., 2001. Organic pollutants in sea-surface microlayer and aerosol in the coastal environment of Leghorn—(Tyrrhenian Sea). Marine Chemistry 76, 77-98.

[50] Díaz, A., Ventura, F., Galceran, M. T., 2002. Development of a solid-phase microextraction method for the determination of short-ethoxy-chain nonylphenols and their brominated analogs in raw and treated water. J. Chromatography A 963. 159-167.

[51] Kuch, H.M. and Ballschmitter, K., 2001. Determination of endocrine-disrupting phenolic compounds and estrogens in surface and drinking water by HRGC-(NCI)-MS in the picogram per litre range. Environmental Science & Technology 35 (2001), pp. 3201–3206.

[52] Cousins, I., Mackay, D., 2000. Correlating the physical-chemical properties of phthalate esters using the 'three solubility' approach. Chemosphere 41, 1389-1399.

[53] Ligocki, M.P., Leuenberger, C., Pankow, J.F., 1985b. Trace organic compounds in rain – III. Particle scavenging of neutral organic compounds. Atmospheric environment 19, 1619-1626.

[54] Thurén, A. and Larsson, P., 1990. Phthalates in the Swedish atmosphere. Environmental Science & Technology 24, 554-559.

[55] Achman, D.R., Hornbuckle, K.C., Eisenreich, S.J., 1993. Volatilization of polychlorinated biphenyls from Green Bay, Lake Michigan. Environmental Science & Technology 37, 75-87.

[56] Hornbuckle, K.C., Jeremlason, J.D., Sweet, C.W., Eisenreich, S.J., 1994. Seasonal variation in air-water exchange of polychlorinated biphenyls in Lake Superior. Environmental Science & Technology 28, 1491-1501.

[57] Nelson, E.D., Mcconnell, L.L., Baker, J.E., 1998. Diffusive exchange of gaseous polycyclic aromatic hydrocarbons and polychlorinated biphenyls across the air-water interface of the Chesapeake. Environmental Science & Technology 32, 912-919.

[58] Bamford, H.A., Offenberg, J.H., Larsen, R.K., Ko, F.C., Baker, J.E., 1999. Diffusive exchange of polycyclic aromatic hydrocarbons across the air-water interface of the Patapsco River, and urbanized subestuary of the Chesapeake Bay. Environmental Science & Technology 33, 2138-2144.

[59] Bamford, H.A., Ko, F.C., Baker, J.E., 2002. Seasonal and annual air-water exchange of polychlorinated biphenyls across Baltimore Harbor and the Northern Chesapeake Bay. Environmental Science & Technology 36, 4245-4252.

[60] Gioia, R., Offenberg, J.H., Gigliotti, C.L., Totten, L.A., Du, S., Eiswnreich, S.J., 2005. Atmospheric concentrations and deposition of organochlorine pesticides in the US Mid-Atlantic region. Atmospheric Environment 39, 2309-2322.

[61] Totten, L.A., Gigliotti, C.L., Vanry, D.A., Offenberg, J.H., Nelson, E.D., Dachs, J., Reinfelder, J.R., Eisenreich, S.J., 2004. Atmospheric concentrations and deposition of polychlorinated biphenyls to the Hudson River Estuary. Environmental Science & Technology 38, 2568-2573.

[62] Schwarzenbach, R.P., Gschwend, P.M., Imboden, D.M. Environmental Organic Chemistry. John Wiley & Sons, New York, 1993.

[63] ten Hulscher, T.E.M., Van Der Velde, L.E., Bruggeman, W.A., 1992. Tempearture dependence of Henry's Law Constant for selected chlorobenzene, polychlorinated biphenyls, and polycyclic aromatic hydrocarbons. Environmental Toxicology and Chemistry 11, 1595-1603.

[64] California EPA. 2001. Fact Sheet: Correcting the Henry's Law constant for soil temperature.http://www.dtsc.ca.gov/ScienceTechnology/JE_Models/Henrys-law-factsheet.pdf.

[65] Shoemaker, D.P., Garland, G.W., Steinfeld, J.I., 1974. Propagation of errors, 3rd ed., McGraw-Hill: New York, pp 51-58.

[66] Wanninkhof, R., Ledwell, J., Crusius, J., 1990. In Air-Water Mass Transfer, Wilhelms, S., Gulliver, J., Eds., American Society of Civil Engineers: New York, pp 441-458.

[67] Fromme, H., Küchler, T., Otto, T., Pilz, K., Müller, J., Wenzel, A., 2002. Occurrence of phthalates and bisphenol A and F in the environment. Water Research 36, 1429-1438.

[68] Germain, A., Langlois, C., 1988. Contamination des eaux et des sediments en suspension du fleuve Saint-Laurent par les pesticides organochlores, les biphenyls polychlores et d'autres contaminants organiques prioritaires. Water Pollutant Research Journal of Canada 23, 602-614.

[69] Preston, M.R., Al-Omran, L.A., 1986. Dissolved and particulated phthalate ester in the river Mersey estuary. Marine Pollution Bulletin 17, 548-553.

[70] Preston, M.R., Al-Omran, L.A., 1989. Phthalate ester speciation in estuarine water, suspended particulates and sediment. Environmental Pollution 62, 183-193.

[71] Furtmann, K. 1993. Phthalate in der aquatischen Umwelt. PhD Thesis, University Duisburg.

[72] Ritsema, R., Cofino, W.P., Frintrop, P.C.M., Brinkman, U.A.T., 1989. Trace-level analysis of phthalate esters in surface water and suspended particulate matter by means of capillary gas chromatography with electron capture and mass selective detection. Chemosphere 18, 2161-2175.

[73] Giam, C.S., Atlas, E., Chan, H.S., Neff, G.S., 1980. Phthalate ester plasticizers, PCB, DDT residues in the Gulf of Mexico atmosphere. Atmospheric Environment 14, 65-69.

[74] Weschler, C.J., 1981. Identification of selected organics in the Arctic aerosol. Atmospheric Environment 15, 1365-1369.

[75] Cautreels, W., Van Cauwenberghe, K., Guzman, L.A., 1977. Comparison between the organic fraction of suspended matter at a background and an urban station. Science of the Total Environment 8, 79-88.

[76] Cautreels, W. and Van Cauwenberghe, K., 1978. Experiments on the distribution of organic pollutants between airborne particulate matter and the corresponding gas phase. Atmospheric Environment 12, 1133-1141.

[77] Thomas, G.H., 1973. Quantitative determination and confirmation of identity of trace amounts of dialkylphthalate in environmental samples. Environment Health Perspectives 3, 23-28.

[78] Environmental Agency of Japan, 1989. Chemicals in the Environment. Environmental Agency of Japan, Tokyo, p. 418.

[79] Hoff, R.M. and Chan, K.W., 1987. Measurement of polycyclic aromatic hydrocarbons in air along the Niagara River. Environmental Science & Technology 21, 556-561.

[80] Aceves, M. and Grimalt, J.O., 1993. Large and small particle size screening of organic compounds in urban air. Atmospheric environment 27, 251-263.

[81] Shields, H.C., Weschler, C.J., 1987. Analysis of ambient concentrations of organic vapor with a passive sampler. Journal of the Air Pollution Control Association 37, 1039-1045.

[82] Wanninkhof, R., 1992. Relationship between wind speed and gas exchange over the ocean. J. Geophysical Research 97, 7373-7382.

[83] Bester, K., Theobald, N., Schroeder, H.Fr., 2001. Nonylphenols nonylphenol-ethoxylates, linear alkylbenzensulfonates (LAS) and bis (4-chlorophenyl)-sulfone in the German Bight of the North Sea. Chemosphere 45, 817-826.

[84] Kannan, N., Yamashita, N., Petrick, G. and Duinker, J.C., 1998. Polychlorinated biphenyls and nonylphenols in the Sea of Japan. Environmental Science & Technology 32, 1747–1753.

[85] Jonkers, N., Lanne, R.W.P.M., de Voogt P., 2003. Fate of nonylphenol ethoxylates and their metabolites in two Dutch estuaries: evidence of biodegradation in the filed. Environmental Science & Technology 37, 321-327.

[86] Ahel, M. and Giger, W., 1993. Partitioning of alkylphenols and alkylphenol polyethoxylates between water and organic solvents. Chemosphere 26, 1471-1478.

[87] Isobe, T., Nishiyama, H., Nakashima, A., Takada, H., 2001. Distrabution and behaviour of nonylphenol, octylphenol and nonylphenol monoethoxylate in Tokyo Metropolitan area: Their assciation with aquatic particles and sedimentary distributions. Environmental Science & Technology 35, 1041-1049.

[88] Van Ry D.A., Dachs J., Gigliotti, C. L., Brunciak, P.A., Nelson, E.D., Eisenreich, S.J., 2000. Atmospheric seasonal trends and environmental fate of alkylphenols in the Lower Hudson River estuary. Environmental Science & Technology 34, 2410-7.

[89] Cincinelli, A., Mandorlo, S., Dickhut, R.M., Lepri, L., 2003. Particulate organic compounds in the atmosphere surrounding an industrialised area of Prato (Italy). Atmospheric Environment 37, 3125-3133.

[90] Spiel D.E. and De Leeuw G., 1996. Formation and production of sea spray aerosol. J. of Aerosol Science 27, S65-S66.

[91] Gehring, M., Tennhardt, L., Vogel, D., Weltin, D., Bilitewski, B. 2003. Release of Endocrine Active Compounds Including 4-Nonylphenol and Bisphenol A with Treated Wastewater and Sewage Sludge to the Environment in Germany. Poster Joint UK Government/SETAC Conference on Endocrine Disrupters. York, UK.

[92] Xie, Z., Ebinghaus, R., Temme, C., Caba, A., Ruck, W., 2005. Atmospheric concentrations and air-sea exchanges of phthalates in the North Sea (German Bight). Atmospheric Environment 3209-3219.

J. Phys. IV France **139** (2006) 239–256
© EDP Sciences, Les Ulis
DOI: 10.1051/jp4:2006139017

Inorganic aerosol formation and growth in the Earth's lower and upper atmosphere

R.W. Saunders[1] and J.M.C. Plane[1]

[1] *School of Chemistry, University of Leeds, Leeds LS2 9JT, UK*
e-mail: j.m.c.plane@leeds.ac.uk

Abstract. This chapter describes the photo-chemical production of aerosol particles in two very different regions of the atmosphere: iodine oxide particles in the marine boundary layer (MBL), and meteoric smoke particles that form in the upper mesosphere from the ablation of interplanetary dust. These two systems are surprisingly analogous – the source of the condensable inorganic vapours is external to the atmosphere, being injected into the atmosphere from the ocean or from space – and the particles are formed by homogeneous nucleation. The purpose of the chapter is to describe a laboratory and modelling study to understand at a fundamental level how the nucleation and growth of the particles occurs. Iodine oxide particles were produced from the photo-oxidation of gaseous I_2 with O_3, which is most likely the primary photo-chemical route to produce the bursts of new particles observed in the MBL at seaweed-rich coastal locations. The captured particles were observed to be fractal-like (i.e., with open or non-compact structures), and to be composed of the stable oxide I_2O_5. Meteoric smoke analogues of iron oxide, silicon oxide, and iron silicate composition were similarly formed from the photo-oxidation of iron- and silicon-containing gas-phase precursors in the presence of O_3. Imaging of the iron-containing particles showed them to be extended, fractal aggregates. For each system, models were developed to elucidate the growth kinetics of the particles and to characterise them in terms of standard fractal parameters. I_2O_5 particles were found to have a fractal dimension (D_f) value of 2.5 at long growth times, consistent with a particle-cluster diffusion-limited aggregation (DLA) mechanism, whereas smoke analogues had lower D_f values (1.75) which appear to result from a magnetic aggregation process.

1. INTRODUCTION

The presence of solid or liquid particles (aerosols) in the atmosphere, either directly transported from the surface or formed in situ, plays an important role in the Earth's climate [1-3]. Of increasing interest is the influence of aerosols on the Earth's radiation budget ('radiative forcing' contribution), the potential for activation of cloud formation, and their participation in catalytic gas-aerosol reaction cycles [4-6]. The degree to which aerosols contribute to these processes is dependent upon their fundamental properties, including chemical composition, shape, size and density. In order to understand (and model) these properties, laboratory studies of their formation mechanisms and kinetics, and their physical and chemical properties, provide a critical complement to field observations.

One process that occurs in the troposphere, stratosphere and mesosphere is *secondary aerosol formation* [7], resulting from the photo-oxidation of precursor gas-phase species which leads to the nucleation of aerosol particles. This chapter is concerned with inorganic aerosol-forming chemical systems specific to the troposphere (iodine oxide particles) and the mesosphere (iron oxides and silicate particles – meteoric smoke). For both systems, major questions include:

- What are the mechanistic pathways which lead to particle formation?
- What are the rates of particle formation and growth?
- What structures, shapes and compositions do the particles have?
- How do their properties vary with changing conditions, such as humidity?
- What are the optical properties of the particles?

The remainder of this section provides a tutorial on aerosol physical parameters and growth kinetics.

1.1 Particle sizes and growth kinetics

The most fundamental parameters of particles are their size (radius r/diameter d), surface area ($\propto r^2$), volume ($\propto r^3$), mass ($\propto r^3$ and particle density ρ) and number concentration N (particle cm^{-3}). A number of size (diameter) definitions exist with respect to aerosol particles. Depending on the analytical technique used to measure particle size and the morphology of the particles concerned, care must be taken in discussing the physical properties of size, volume and density. The recent paper by DeCarlo et al. [8] gives a comprehensive explanation of these different definitions and so only a brief summary of relevant terms will follow.

Modern studies of aerosol systems, in both the field and laboratory, invariably employ a combination of *condensation particle counter* or CPC (to measure number concentration, N) with a *differential mobility analyzer* or DMA which allows the particle size distribution to be determined (the size distribution is the variation in N with d). Particle size derived from such techniques is termed the *electrical mobility diameter* (d_m), defined as the diameter of a sphere moving at the same velocity in a constant electric field as the particle concerned. However, because aerosol particles can form in a variety of shapes which can deviate from the idealized spherical case, a more useful definition is that of the *volume equivalent diameter* (d_{ve}), which is the diameter of a spherical particle having the same volume as that under analysis. For 'compact' aggregates, d_{ve} and d_m are approximately equal, but as the particle shape becomes more irregular, $d_m > d_{ve}$. Two further size definitions when considering non-spherical particles are the *collision diameter* (d_{coll}) and the *fractal diameter* (d_{frac}). d_{coll} is defined as the distance between the centroids (centres of mass) of two colliding particles, whereas d_{frac} is the largest diameter of a non-spherical particle [9].

Similarly, when discussing particle volume, precise definitions are required. Values obtained from measured size distributions relate to the *particle volume* (V_p), which encompasses the volume of both solid phase material and the internal void space contained within the structure. This is related to the *material volume* or V_m (the value of the solid phase only) by the relationship

$$\frac{V_m}{V_p} = 1 - \omega \tag{1a}$$

where ω is the volume fraction of void space within the particle structure [8], equivalent to the *porosity* of the particle. It follows that *particle density* (ρ_p), also known as the *effective density* (ρ_{eff}) will vary from the *material bulk density* (ρ_b) as the particle size increases and the void space changes:

$$\frac{\rho_p}{\rho_b} = 1 - \omega \tag{1b}$$

i.e. for spherical, compact particles, $\omega = 0$, $\rho_p = \rho_b$ and $V_p = V_m$.

The transport and collision of electrically neutral and non-magnetic particles in air is governed by random diffusion known as Brownian motion [7, 9]. Modelling the particle growth that results from collisions between particles requires consideration of collisions between all possible permutations of different size particles. The critical parameter required for such calculations is the *Brownian collision kernel* (β^B) which describes the frequency of collisions. For collisions between two spherical particles i and j (size r_i and r_j, respectively) under the *free molecule regime* (particle radius (r) « mean free path of an air molecule (λ_{air})):

$$\beta^B_{i,j} = \pi \left(r_i + r_j\right)^2 \sqrt{\bar{v}_i^2 + \bar{v}_j^2} \tag{2a}$$

where $\bar{v} = \sqrt{\frac{8k_BT}{\pi M}}$ is the thermal velocity of a particle of mass M at temperature T, and k_B is the Boltzmann constant.

For the *continuum regime* ($r > \lambda_{air}$),

$$\beta^B_{i,j} = 4\pi\left(r_i + r_j\right)\left(D_i + D_j\right) \tag{2b}$$

where D is the particle diffusion coefficient.

Finally, for particle growth at sizes corresponding to the *transition* regime ($r \sim \lambda_{air}$), the collision kernel takes the form

$$\beta^B_{i,j} = \frac{4\pi\left(r_i + r_j\right)\left(D_i + D_j\right)}{\frac{r_i+r_j}{r_i+r_j+\left(\delta_i^2+\delta_j^2\right)^{1/2}} + \frac{4\left(D_i+D_j\right)}{\left(\bar{v}_i^2+\bar{v}_j^2\right)^{1/2}\left(r_i+r_j\right)}} \tag{2c}$$

where δ is a parameter with dimensions of length [9-10].

Additional coagulation mechanisms such as *convective diffusion* and *gravitational collection* along with *turbulent* processes can also play roles at larger (micron) sizes. However, at nanometer dimensions, as is the case with the aerosol systems discussed in this paper, Brownian coagulation is the dominant mechanism [9].

In the lower troposphere (e.g. 2 km altitude, $T = 275$ K), $\lambda_{air} \sim 80$ nm, whereas in the mesosphere (e.g. 60 km altitude, $T = 245$ K), $\lambda_{air} \sim 250\,\mu$m i.e. $\sim$3000 times greater. Therefore, particle growth in the upper atmosphere falls in the free molecule and transition regimes, whereas in the MBL, growth can cross into the continuum regime. Consequently, analysis of particle growth processes requires careful consideration of which flow regime is applicable.

1.2 Fractal particles

Equations 2a-c are derived from kinetic theory, and assume that the particles are spherical and compact [7]. Similarly, Mie theory, the main tool for calculating the optical light scattering and absorption properties of aerosol particles, assumes that the particles are spherical and chemically homogeneous [11]. However, a number of processes including combustion [12,13], formation of colloidal suspensions [14], laser ablation of surfaces [15] and aerosol generation in gas flow reactors [7] have been shown to lead to distinctly non-spherical or *fractal* aggregate formation. This is also the case for particle aggregates identified in extraterrestrial environments [16, 17]. Although such aggregates are not truly fractal in the strict definition of the term – 'any shape possessing the properties of self-similarity and scale independence' i.e. at any level of resolution, any and every part maintains the form of the whole [18], they are considered to be *fractal-like* with various properties being related via expressions used to characterize 'true' fractal forms. Consequently, the study and modelling of the growth (variation in particle size, surface area, volume and mass distributions) and other properties (i.e. light scattering) of aerosols formed in such systems must take into account such morphological differences. Methods developed for this purpose therefore require modifications to growth laws by the introduction of parameters required to describe any deviation from the idealized spherical form [7].

From a large body of experimental evidence, it has been established that a fractal-like aggregate is composed of a number of *primary* particles (N_p), given by

$$N_p = A\left(\frac{R_g}{r_0}\right)^{D_f} \tag{3}$$

where R_g is the *radius of gyration*, D_f is the *fractal dimension*, r_0 is the *primary particle radius*, and A is a pre-fractal constant which, for many cases, has a value close or equal to unity. For spherical particles, $D_f = 3$, whereas at the other extreme, a 1-dimensional chain structure has a value of 1. Subsequently, fractal aggregates are typically found to be characterized by $1 < D_f < 3$.

Three distinct processes have been identified to describe the formation of aggregates with specific fractal dimension values. *Diffusion-limited*, *reaction-limited* and *ballistic* mechanisms can all occur via *particle-cluster* or *cluster-cluster* processes. Each specific combination of clustering / aggregation type leads to a reasonably narrow and hence diagnostic range of D_f values i.e. aggregate shape [7].

1.3 Aggregation of magnetic particles

All discussion of aggregation to date has assumed that particles are non-magnetic in nature and, as a result, particle growth is determined by collision and coagulation via Brownian diffusion and all inter-particle/aggregate forces are short range i.e. van der Waals interactions. However, examples of aerosol generation involving magnetic particles indicate that analysis of such systems must account for the long-range magnetic dipole-dipole interactions which can determine aggregate shape and growth [19-21]. Irrespective of the exact magnetic nature of the material of which the particles/aggregates are composed i.e. ferromagnetic (metals such as iron) or paramagnetic (transition metal compounds e.g. iron oxides), aggregate growth can be modelled by implementing a modified *magnetic coagulation kernel* (β^{mag}) in the growth equations. For initially randomly orientated magnetic dipoles, this takes the form [22]

$$\beta_{i,j}^{mag} = \pi \left(\frac{8k_B T}{\pi \bar{m}}\right)^{1/2} \left(\frac{2C}{k_B T}\right)^{1/3} \times \Gamma\left(\frac{2}{3}\right) \tag{4}$$

where $\bar{m} = \frac{m_i \times m_j}{m_i + m_j}$ is the reduced mass of a particle formed through the collision of particles i and j, $C = \frac{2}{3}\frac{(\mu_i \mu_j \mu_0)^2}{k_B T}$ where μ_i and μ_j are the dipole moments of the individual particles prior to collision, μ_0 is the permeability of free space, and $\Gamma\left(\frac{2}{3}\right) = 1.355$.

Armed with such concepts, we now describe the results and analysis from the laboratory generation of iodine oxide particles and meteoric smoke mimics. In addition to compositional analysis of the particles and the determination of their optical properties, we then use a system-specific fractal growth model to account for the observed particle morphology and structure, and to fit the temporal evolution of measured size distribution data.

2. IODINE OXIDE AEROSOL FORMATION

2.1 Atmospheric background

The primary source of atmospheric iodine is the biological activity of marine algae (phytoplankton and macroalgae) in seawater [23]. In coastal areas, macroalgae (seaweed) produce molecular iodine (I_2) and a variety of iodine-containing organic gases such as methyl iodide (CH_3I) and diiodomethane (CH_2I_2), which diffuse across the sea-air interface and into the marine boundary layer (MBL). These gases are subsequently photolysed by sunlight to produce iodine atoms, which are then oxidized in the presence of O_3 and other atmospheric species to form various iodine-containing molecules (HOI, INO_3, I_2O_4 and I_2O_5). These species can either provide condensable vapour to form new particles, or be removed from the gas-phase by being taken up into pre-existing aerosols. Wet precipitation (rain-out) and dry deposition of these particles ultimately returns iodine to the ocean or onto land, thus completing the biogeochemical cycle of iodine [24].

A number of field studies conducted at different coastal sites have established a direct correlation between daylight, low-tide periods, elevated concentrations of gas-phase iodine oxide species in the MBL, and rapid ‘bursts’ of new particles [25-29]. Laboratory work, stimulated by these observations, established that the near UV photolysis of CH_2I_2 [30, 31] and photolysis of I_2 at visible wavelengths [29], followed by oxidation with O_3, rapidly leads to the formation of aerosol particles, most likely of iodine oxide composition. ‘Seaweed chamber’ studies [32, 33] have further proven that the transformation of gas-phase iodine species into aerosol is most readily explained by such a photochemical mechanism. The recent identification of I_2 [34] at appreciably higher levels than those of organo-iodide species in the MBL, and the faster rate of photolysis of this molecule in the atmosphere [35], provide strong evidence that this molecule is the dominant source of iodine atoms leading to new particle formation in seaweed-rich coastal regions [29].

In spite of these recent findings, many areas of uncertainty remain, regarding the precise chemical composition, form and structure, and optical properties of these coastal aerosols. Knowledge of their properties is essential for any evaluation of the likely climatic impacts that they might have. To this end, a series of experiments were undertaken to generate and characterise particles formed by the photo-oxidation of I_2 in an aerosol flow reactor.

2.2 Laboratory study

2.2.1 Particle generation

Iodine oxide particles were produced using an experimental set-up which has been described elsewhere [35, 36]. Briefly, irradiation of a glass reaction vessel through which I_2 and O_3 gases were flowing, was used to form particles which were (i) captured for analysis using a transmission electron microscope (TEM), (ii) counted and sized using a CPC and scanning mobility particle sizer (SMPS) system, and (iii) passed through an optical cell in order to measure their light extinction.

2.2.2 Morphology and composition

Figure 1 shows representative TEM images of particles captured from the aerosol flow reactor and an energy dispersive x-ray (EDX) spectrum of a single particle. At high magnification (Figure 1b), one of these larger structures is revealed to possess a non-spherical or fractal-like appearance and to be composed of a number of smaller, more spherical particles. The growth of particles to larger sizes with fractal morphologies is most likely a result of the initial collision-coalescence of ultra-fine particles (maintaining spherical form and compact structure), followed by particle collision-agglomeration, leading to the formation of non-spherical aggregates [37].

The presence of iodine and oxygen in the particles (Figure 1c) was established using the EDX technique. Quantitative determination of the particle composition by integrating the areas of the elemental peaks gave an average atomic ratio (O/I) of 2.45, consistent with a composition of the stable, solid oxide I_2O_5 (diiodine pentoxide). Electron diffraction showed some degree of crystallinity in these particles, though a detailed comparison with published data for I_2O_5 [38] was not possible.

2.2.3 Gas-phase reaction pathways to aerosol formation

Having identified the aerosol composition as I_2O_5, the next question is whether I_2O_5 forms in the gas phase prior to condensation to the solid phase, or whether an initially mixed iodine oxide solid undergoes internal rearrangement to form I_2O_5. Figure 2 illustrates a possible scenario involving the sequential oxidation of lower gas-phase oxides (I_2O_x, $x = 2$-4) to produce gas-phase I_2O_5. Quantum chemistry calculations were performed to assess the viability of each reaction channel by determining the reaction enthalpy changes. Large, negative enthalpy changes for each oxidation step indicate these to be thermodynamically favourable. In addition, the large electric dipole moment (μ) of the I_2O_5 molecule suggests that once formed, this species is likely to rapidly polymerise to form molecular clusters of sufficient size at which point homogeneous nucleation to the aerosol phase can take place [36].

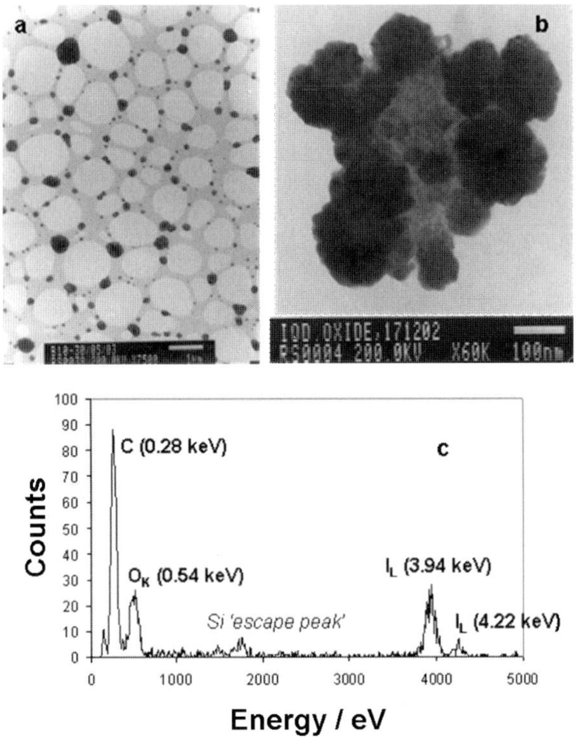

Figure 1. (a and b) Transmission electron microscope images of particles generated and captured from the photo-oxidation of I_2 with O_3 (scale bar in the caption of image (a) = 1 μm) and (c) an x-ray spectrum of a particle, confirming the presence of iodine and oxygen. The carbon peak results from the grid film and the Si peak is a detector artefact. The figure is reproduced with permission from [36].

2.2.4 Particle size distributions

Figure 3a shows measured size distribution data for iodine oxide particles generated at different initial I_2 concentrations as indicated, at a constant irradiance in an excess of O_3 at 295 K and 1 atmosphere. With increasing I_2 concentration (and hence concentration of iodine atoms resulting from photolysis), the total particle concentration increases and the modal diameter remains essentially constant. Figure 3b shows the data from experiments using the same initial I_2 concentration over a range of cell irradiance levels. In this case, both total particle concentration and modal diameter increase with increasing irradiance (I_2 photolysis rate).

2.2.5 Particle growth kinetics

The kinetics of iodine oxide particle growth can now be modelled [39], and tested against the measured time-resolved size distribution data. The model needs to account for the growth of non-spherical particles with densities (ρ_p) lower than the bulk density (ρ_b), and requires the following assumptions:

- All iodine produced from the initial period of I_2 photolysis ends up as I_2O_5 which is assumed to be the nucleating species
- The smallest size of particle treated was set at the molecular diameter of I_2O_5 ($d_1 = 0.60$ nm) i.e. nucleation of single molecules can occur without an associated thermodynamic barrier

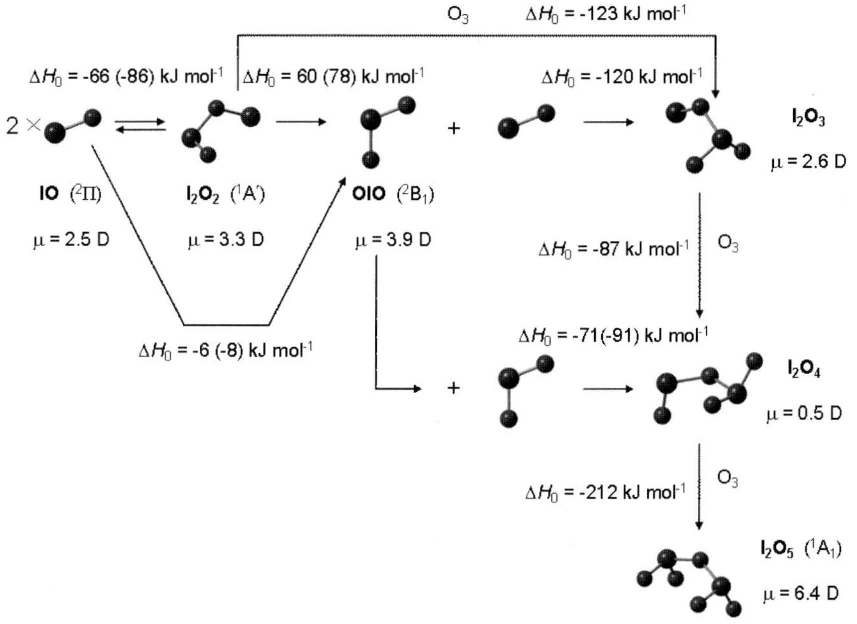

Figure 2. Proposed gas-phase reaction pathways from iodine monoxide (IO), the immediate product of the oxidation of atomic iodine, to the end composition of I_2O_5, which is the iodine oxide stoichiometry in captured particles. Calculated reaction enthalpies (ΔH_0) and molecular dipole moments (μ) are shown. This figure is reproduced with permission from [36].

- Collisions between I_2O_5 particles of diameter (d) less than a 'primary' value (d_0) result in coalescence, where the resulting particle retains spherical shape and a compact structure. Equation 2a is used to calculate collision kernels for these collisions.
- Collisions of particles with a diameter at or larger than d_0 leads to agglomeration, which generates fractal-like particles with a characteristic fractal dimension (D_f) value. Growth at these sizes is determined using equation 2c, with d set at the fractal diameter ($= d_0 \times N_p^{1/Df}$).
- van der Waals and viscous (hydrodynamic) forces between particles were treated using a size-dependent kernel correction factor [9]. An empirically fitted expression was used as there is no published value for the Hamaker constant for I_2O_5. The correction term ($V_{i,j}$) appears as a multiplying factor in equation 2c.
- The density of spherical particles ($d \leq d_0$) is set at the bulk density (ρ_b) of I_2O_5 (4.98 g cm^{-3}), while fractal particles ($d > d_0$) have size-dependent densities $\rho_p < \rho_b$.

Figure 4 shows the model fits to a number of particle size distributions obtained for different growth times as determined from the reactor geometry and total gas flow rate. The fitting procedure uses a weighted least-squares method to optimise the parameters within the model.

The measured distributions show features typical of particle growth through a collision-coagulation process with total number concentration decreasing and modal diameter increasing with increasing growth time. The model provides satisfactory fits to the data sets with $D_f = 2.2$– 2.5, and $d_0 = 6.6$ nm. A D_f value of 2.5 is consistent with the value found for particles formed by the particle-cluster diffusion-limited aggregation (DLA) mechanism [7]. The particle densities will then vary according to the particle size: for example, with $D_f = 2.5$, $d_0 = 6.6$ nm and $\rho_b = 4980$ kgm^{-3}, at $d = 20$ nm, ρ_p is calculated to be $\sim$2860 kgm^{-3} (particle porosity $\omega = 0.43$), whereas for $d = 40$ nm, $\rho_p \sim$2023 kgm^{-3} ($\omega = 0.59$).

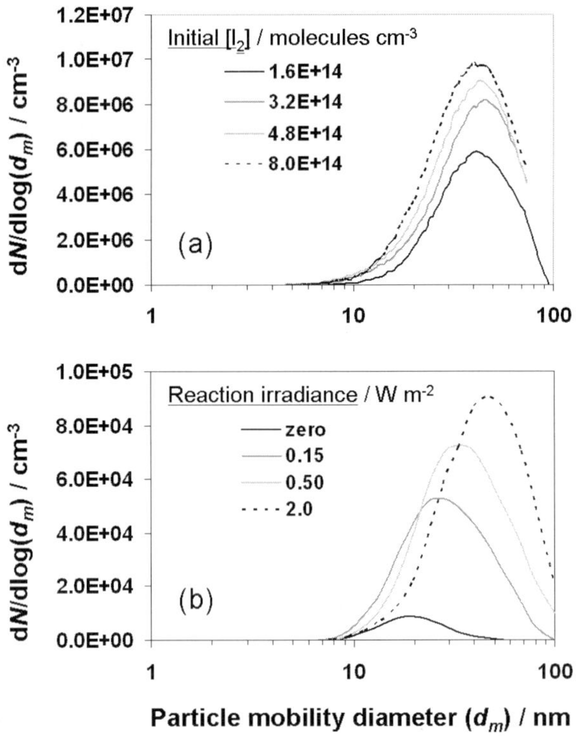

Figure 3. Iodine oxide particle size distributions obtained from experiments of (a) varying initial I_2 concentrations as indicated, at constant photochemical irradiance (10.0 W m^{-2}) and $[O_3] = 2.5 \times 10^{16}$ molecule cm^{-3}, and (b) varying reaction irradiance as indicated, at constant $[I_2] = 3.2 \times 10^{14}$ molecules cm^{-3} and $[O_3] = 2.5 \times 10^{16}$ molecules cm^{-3}. Zero irradiance corresponds to data obtained from the 'dark' or thermal reaction between I_2 and O_3.

2.2.6 Light extinction measurements

The optical properties of the particles were measured by directing the flow of bath gas (N_2) containing the particles through an optical absorption cell with multi-pass optics to measure the small extinction in the near-UV/visible. The optical density (*OD*) spectrum was determined from the unattenuated lamp intensity profile ($I_0\{\lambda\}$) and the intensity with gases / particles present ($I\{\lambda\}$), using the Beer–Lambert Law:

$$OD\{\lambda\} = \ln\left(\frac{I_0\{\lambda\}}{I\{\lambda\}}\right) = \sigma\{\lambda\}cl \tag{5}$$

where $\sigma\{\lambda\}$ is the extinction cross-section (cm^2), c is the concentration (cm^{-3}) of gas or particle species and l is the path-length (cm) over which the extinction is measured. Particle extinction was extracted by numerical fitting and subtraction of the spectra of gas-phase species (I_2 and O_3). The particle extinction coefficient $\varepsilon_p(\lambda) = \sigma_p(\lambda) \times c_p$ shown in Figure 5, was calculated from the measured *OD* and the optical path–length. Particle extinction at wavelengths greater than 500 nm fell below the system detection limit, but increased steadily into the near UV. Due to the poly-disperse nature of the particle distributions (Figures 3 and 4), the calculated *OD* is the sum of the individual contributions from all particle sizes within the distribution. For a more detailed analysis of the optical properties of the particles, calculations

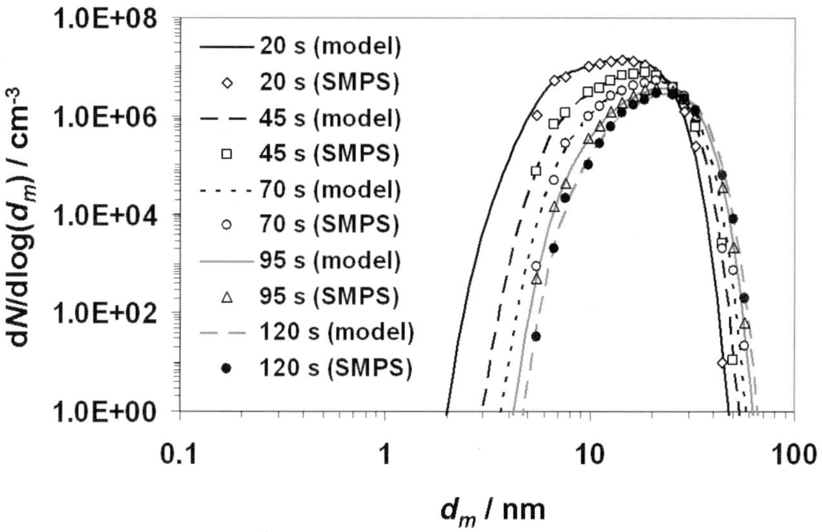

Figure 4. Model best fits (solid/dashed lines) to measured iodine oxide particle size distribution data (discrete points) for a number of growth times as indicated. The fits were obtained with the following optimised model parameter values: $D_f = 2.2$ (20 s), 2.4 (45 s), and 2.5(>45 s) and $d_0 = 6.6$ nm (all times). In all experiments, $[I_2] = 1.0 \times 10^{13}$ molecules cm^{-3}, $[O_3] = 1.0 \times 10^{14}$ molecules cm^{-3} and $J(I_2) = 0.01$ s^{-1}. This figure is reproduced with permission from [39].

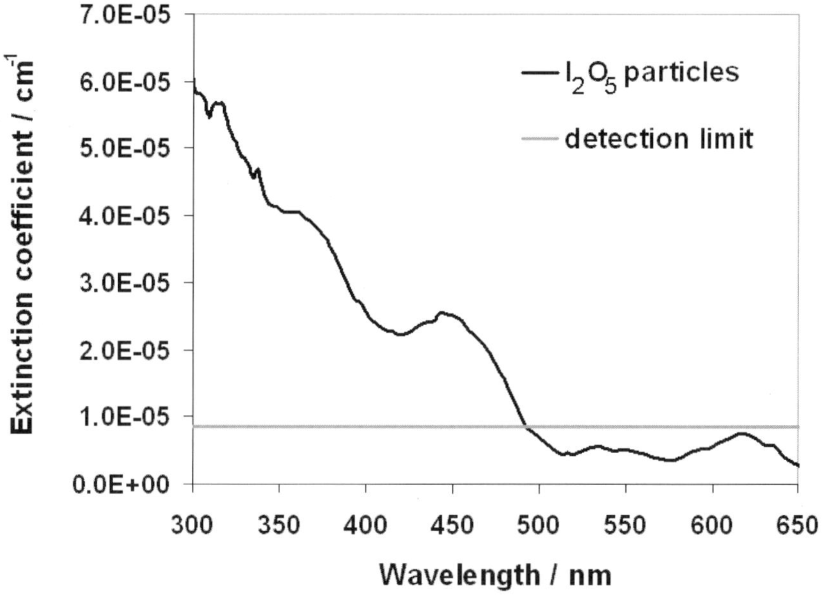

Figure 5. Extinction spectrum obtained from iodine oxide particles generated photochemically from I_2 and O_3.

using methods such as Mie Theory are possible only if reliable refractive index data is available. No such data for I_2O_5 is currently available in the literature.

3. METEORIC SMOKE FORMATION

3.1 Atmospheric background

The continual meteoric bombardment of the Earth's upper atmosphere results in the ablation of various metals (Na, K, Mg & Fe) and silicon [40]. Layers of neutral metal atoms occur globally in the upper mesosphere (80 – 105 km) which, in the presence of oxidising / reducing species such as O_3, O and H, are subject to chemical cycling [41, 42]. Rosinski & Snow [43] first proposed that metallic oxides, hydroxides and silicates would condense to form submicron refractory particles, termed *meteoric smoke*. This concept was developed in detail by Hunten et al. [44], who presented the first modelling study of the coagulation and gravitational sedimentation of these aerosol particles in the upper atmosphere. Subsequent models have been developed in order to further characterise the distribution of this material [45, 46].

Meteoric smoke has been implicated in a number of important atmospheric processes such as the nucleation of noctilucent clouds (NLCs) [47, 48], the condensation of stratospheric sulphate aerosols [49], and the production and destruction of acidic gases such as H_2SO_4 and HNO_3 [50-52]. As a result of the process of *differential ablation* [45, 53], whereby species with lower boiling points are ablated from the parent meteoroid body first, at higher altitudes than those with higher boiling points, it is predicted that Fe and Si are likely to be the predominant meteoric species available for smoke formation, and in nearly equal abundance. Recent observations have provided evidence for the transport of smoke particles through the atmosphere [54, 55] to the terrestrial surface [56]. However, relatively little progress has been made in establishing the details of the processes which lead to particle formation, and the chemical composition of the smoke particles. Once again, laboratory studies can provide significant insights into these questions.

3.2 Laboratory study

3.2.1 Particle generation

As with the iodine oxide particle generation, an aerosol flow reactor was used to photolyse vapours of (i) iron pentacarbonyl ($Fe(CO)_5$), (ii) $Fe(CO)_5$, with H_2O vapour present and (iii) $Fe(CO)_5$ and tetraethyl orthosilicate (TEOS), in order to oxidise Fe/Si atoms with O_3 and form condensable gas-phase species and hence aerosol. Again, the generated particles were captured on TEM grids for analysis, CPC-SMPS instruments were used to measure the particle size distributions and growth kinetics, and an optical absorption cell was employed to determine the particle extinction spectra.

3.2.2 Morphology and composition

Figure 6 shows TEM images of sampled particle aggregates from the three systems. Clearly, these are much larger and more extended than the iodine oxide particles discussed previously. The aggregates are composed of a large number (10^2–10^3) of smaller particles, themselves aggregates composed of yet smaller particles.

Again, EDX analysis was used to identify the particle compositions. In addition, electron energy loss spectroscopy (EELS) was used for further analysis of the iron oxidation state within the particles from system (iii). The photo-oxidation reactions produced particles with elemental ratios consistent with compositions of (i) hematite (α-Fe_2O_3), (ii) goethite (α-FeOOH) and (iii) fayalite (Fe_2SiO_4) – all stable minerals commonly found on the Earth's surface. In all cases, electron diffraction showed that the particles were amorphous in nature. Although non-magnetic materials have been shown to form similar extended structures characterised by D_f values less than 2 [57], the presence of iron within all three types of particle, along with the amorphous material state, suggests that magnetic forces could explain the growth of these fractal aggregates (see section *3.2.5*).

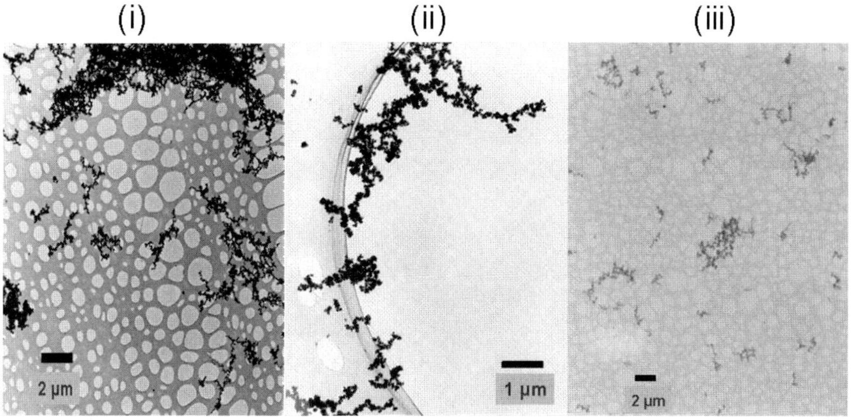

Figure 6. TEM images of fractal particle aggregates of laboratory generated meteoric smoke mimics with compositions identified as (i) Fe_2O_3, (ii) FeOOH and (iii) Fe_2SiO_4.

3.2.3 Gas-phase reaction pathways to aerosol formation

A number of pertinent reactions of neutral Fe atoms have been studied in the laboratory (58-60), and hence some kinetic information is available to deduce likely pathways from gas-phase iron oxide species to the aerosol phase. In contrast, the reaction kinetics of Si and SiO, both likely to be produced during meteoric ablation, are more uncertain although the most likely end-product is the stable mineral form of silica (α-SiO_2). However, in the mixed Fe-Si system, additional possible routes involving gas-phase iron silicate species could lead to the formation of condensable species which result in the nucleation and growth of particles with the compositions identified in section *3.2.2*. Figure 7 outlines possible reaction pathways to the end-compositions of hematite, goethite and fayalite. These involve polymerisation of gas-phase species leading to the formation of initially different solid-phase compositions, which then undergo structural re-arrangement to the forms indentified in the TEM-EDX analysis [22].

3.2.4 Particle size distributions

Figure 8 shows size distribution data for the meteoric smoke mimics. In these systems, in contrast to the iodine oxide particles, distributions were observed to evolve with prolonged cell irradiation time (growth time being constant and equal to the gas residence time in the reactor). Another difference to note is the bi-modal nature of the smoke distributions with a nucleation mode present at smaller sizes and a second accumulation mode, the peak of which was observed to move to larger sizes and decrease in number concentration with increasing irradiation and growth timescales.

3.2.5 Particle growth kinetics

The growth kinetics of the fractal smoke aggregates can be modelled and compared with the measured time-resolved particle size distribution data. As in the iodine oxide model (section 2.2.5), some assumptions are required for the treatment of collisional particle growth in the three Fe-containing particle systems:

- The gas-phase species responsible for nucleation is the same as that identified as the particle composition i.e. Fe_2O_3, FeOOH or Fe_2SiO_4.
- Single molecules can nucleate to the aerosol phase (e.g. for fayalite particles, $d_1 \sim 0.52$ nm).

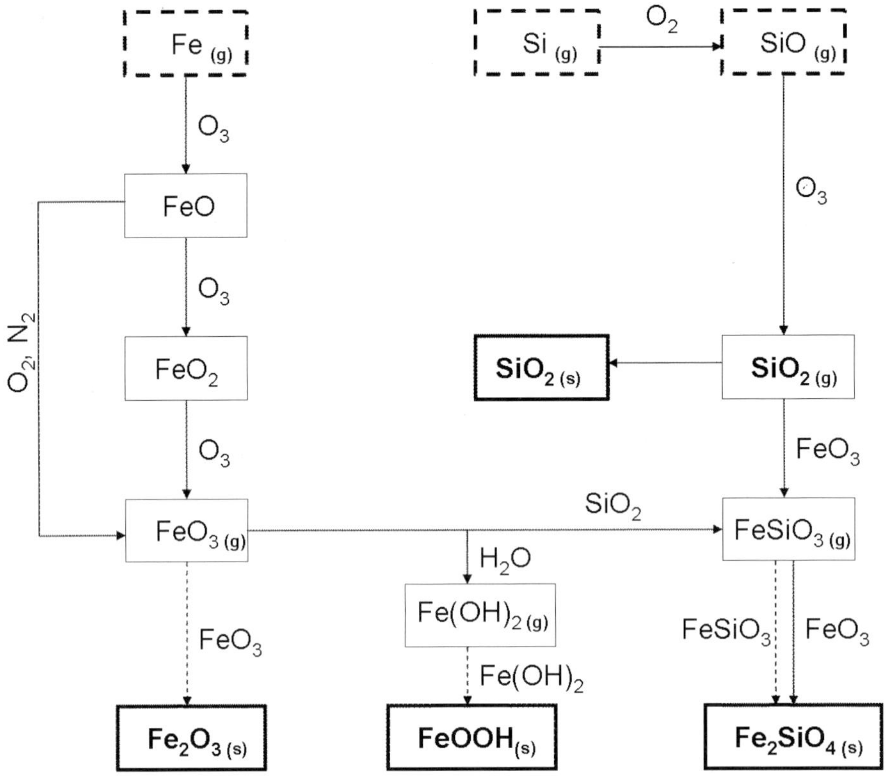

Figure 7. Possible gas-phase reaction pathways leading to the formation of the meteoric smoke mimics with end compositions as identified in the TEM analysis. The proposed polymerisation reactions leading to structural re-arrangement in the solid-phase to form the minerals hematite (Fe_2O_3), goethite (FeOOH) and fayalite (Fe_2SiO_4) are indicated by dashed arrows.

- Collisions between smoke particles with $d < d_0$ result in coalescence, where the resulting larger particles retain their spherical shape and compact structure. Equation 2a is used to calculate collision kernels at these sizes.
- For collisions when $d \geq d_0$ the magnetic collision kernel (equation 4) is used, where the particle dipole moments are calculated using an additional adjustable parameter (z) which reflects the degree of dipole alignment in the particles i.e. a purely crystalline sample of the material of which the respective smoke particles are composed would be characterised by $z = 0$ i.e. no net magnetic dipole moment.
- Particle densities are treated as in the iodine oxide model.

Figure 9 shows the model fit (by least-squares minimization) to the measured data for fayalite particles from system (iii) for a single growth time of 20 s. The model satisfactorily predicts the position and number concentration of both peaks in the bi-modal distribution, with optimised parameters $D_f = 1.75$, $d_0 = 9.2$ nm and $z = 0.24$.

3.2.6 Light extinction measurements

Particle extinction spectra for all three systems show a broad peak in the near UV at $\lambda < 350$ nm, with continually decreasing extinction at visible wavelengths. These features are consistent with those exhibited by poly-disperse distributions of iron-bearing minerals, which result in their distinctive colours. Literature values for the refractive indices of hematite, goethite and fayalite were used

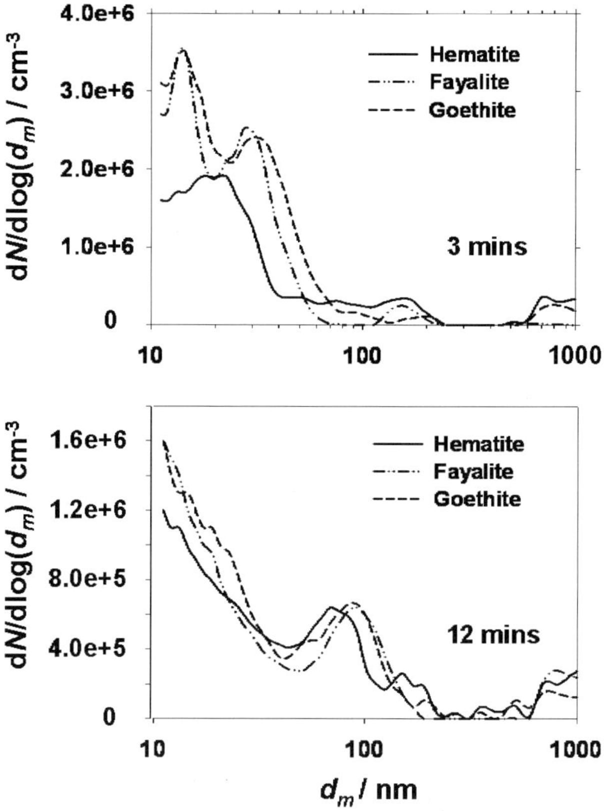

Figure 8. Measured particle size distributions obtained from the photo-oxidation of systems (i)-(iii). Data in the top panel was obtained after a photolysis period of 3 minutes, and that in the lower panel after 12 minutes. Figure adapted from [22].

to corroborate these compositions from the extinction data. A common method of evaluating light extinction by fractal aggregates is to assume that they are composed of a number of monodisperse, spherical 'primary' particles, the individual extinctions of which can be summed to give the total extinction of the aggregate. This approach uses the Rayleigh-Debye-Gans (RDG) approximation which assumes no multiple internal scattering within aggregates. However, for all particle systems, this yielded a calculated total particle extinction of $\sim$100 smaller than measured. Consequently, for the purpose of fitting the measured optical extinction data, we simply treated fractal aggregates as if they were a poly-disperse (log normal) distribution of spherical, compact particles. The distribution parameters of mean diameter ($\bar{d}$) and total particle number concentration (ΣN) were optimised to provide a best fit to the measured data (assuming a distribution spread (σ) value of 1.5, in line with the observed size distribution from the SMPS measurements). Figure 10 shows the measured and fitted extinction data for hematite ($\bar{d} = 70\,\text{nm}$; $\Sigma N = 1 \times 10^8\,\text{cm}^{-3}$) and fayalite ($\bar{d} = 65\,\text{nm}$; $\Sigma N = 2 \times 10^8\,\text{cm}^{-3}$) particles. Good fits were also found for goethite particles.

4. VALIDATION OF THE PROPOSED GROWTH MODELS

Brownian and magnetic coagulation are intrinsically complex processes. The models described here to account for the growth of both iodine oxide and meteoric smoke particles necessarily make some arbitrary initial assumptions, and involve a number of adjustable parameters. Because the experimental

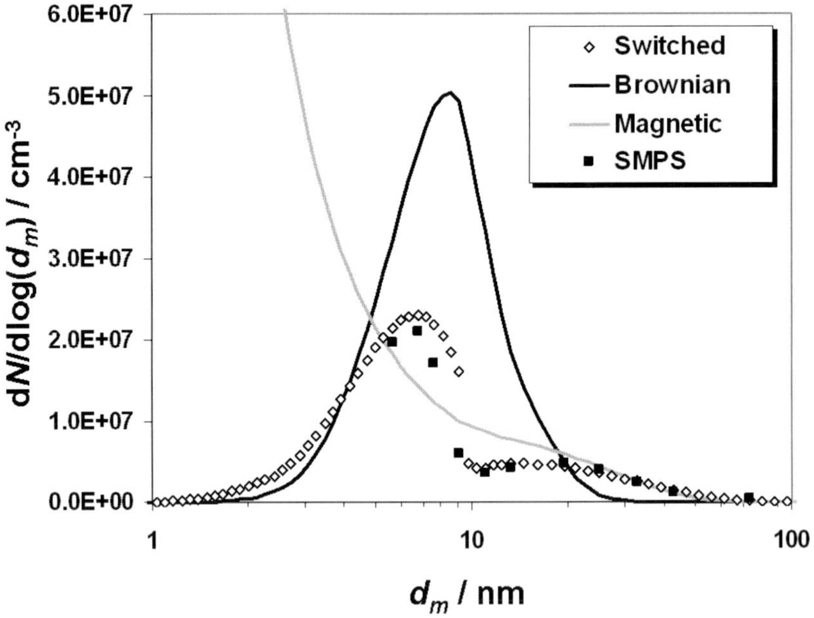

Figure 9. Model best fit (indicated as 'Switched') to SMPS data for a 20 s growth time for fayalite particles formed in system (iii). The fit was obtained by switching the collision kernel from a Brownian form (equation 2a) to that for particles for which interactions are dominated by magnetic forces (equation 4) at $d_0 = 9.2$ nm, with $D_f = 1.75$ and $z = 0.24$. The plot labelled 'Brownian' is that obtained using equation 2c for $d \geq d_0$, and that labelled 'Magnetic' by using equation 4 at all sizes. The figure is adapted from [22].

observable - the particle size distribution as a function of time – is an integration of a large number of elementary steps, the particle growth kinetics cannot be modelled to determine all the parameters unambiguously. Therefore, these models are not unique, definitive descriptions of the growth processes occurring in each system. However, they do provide very satisfactory fits to the measured data, and are based on well-characterised mechanisms and directly identified particle compositions.

There are also other techniques that can be applied to assist in validating the optimised parameters in models of this type. For example, in the case of the meteor smoke particles, inspection of the TEM images of fractal-like aggregates was used to estimate the D_f values [22], which were shown to be in good agreement with the optimised values from the kinetic growth modelling of each system. TEM imaging can also reveal the size of the primary particles from which the larger aggregates are composed, enabling comparison with modelled values. Light scattering is often used to extract structural information on fractal particles [61], while a combination of aerosol sampling and measuring techniques has also been shown to provide a method for accurately determining factors such as particle shape and density [8].

5. ATMOSPHERIC RELEVANCE

Ultimately, the purpose of experimental and modelling studies of the kind described here is to understand at a fundamental level the processes of particle formation and growth, so that these can be modelled under a range of conditions in the atmosphere. Consequently, an important challenge is to relate the results of laboratory studies to aerosol particles formed under atmospheric conditions of temperature, pressure and the presence of species such as water vapour.

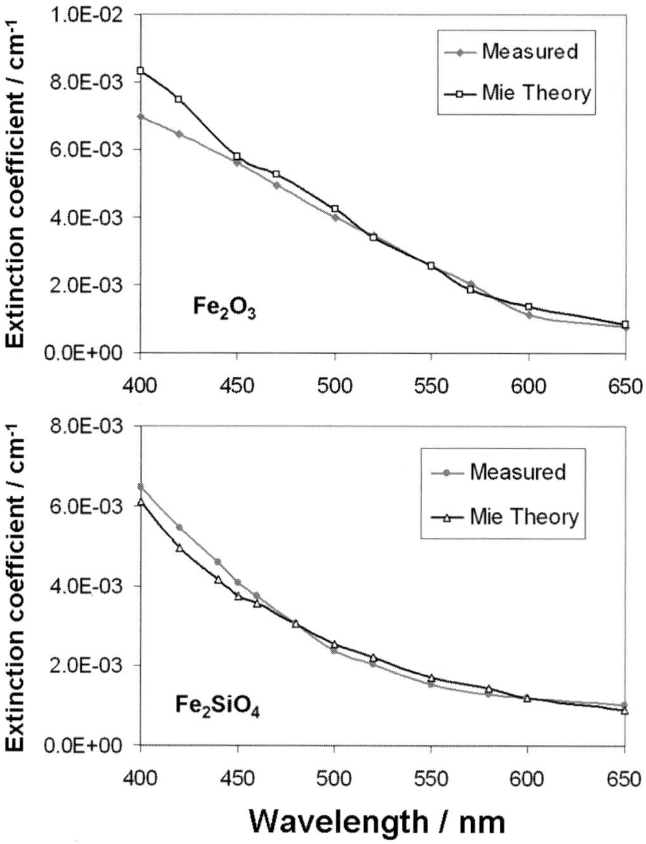

Figure 10. Comparison of measured particle extinction values obtained for hematite (top panel) and fayalite (bottom panel) particles with values determined using Mie Theory (treating particle aggregates as a log-normal distribution of spherical particles – see text). This figure is reproduced with permission from [22].

The experiments described in the earlier sections of this chapter were performed at room temperature (295 K) and atmospheric pressure (1.01×10^5 Pa), in N_2 bath gas. Taking the case of iodine oxide formation in the MBL first, these conditions are reasonably close to those found in the lower troposphere. The major points of difference are (i) the lower atmospheric concentrations of I_2 and O_3 and (ii) the presence of water vapour (humidity) in the MBL. On point (i), previous laboratory studies, conducted at atmospherically realistic levels of CH_2I_2/I_2 and O_3 have shown that iodine oxide formation rapidly leads to the homogeneous nucleation of particles [27, 29-31]. The lower MBL I_2 concentration (by ~ 3 orders of magnitude) would result in much slower reaction timescales for the formation of the higher iodine oxides, since these are formed by reactions between IO and OIO or their self-reactions, and are thus at least second-order in IO (Figure 2). However, previous modelling studies [62] indicate that iodine oxide particles will still form at a fast enough rate to explain the observed coastal particle bursts. Hence, in spite of the lower concentrations of iodine oxides in the MBL, the nucleation and growth processes discussed in section *2* appear to be viable in the atmosphere.

The effect of increased humidity levels on pre-formed particles has been shown to result in size-dependent growth behaviour [29] with larger, lower density particles promoting water uptake within their structures. However, the effect of the presence of water vapour prior to particle nucleation remains to be investigated. The identification of I_2O_5 particle composition could have important atmospheric implications as this oxide is very hygroscopic and therefore such particles could act as efficient cloud condensation nuclei (CCN) – particles which promote the formation of cloud-water droplets. The particles could thus play a role in indirect radiative forcing. In terms of direct radiative forcing, the light extinction measurements indicate that iodine oxide particles in the MBL could potentially make a contribution at wavelengths below ~500 nm, but there is no evidence that the particles observed in coastal regions grow large enough to do this. Finally, the identification of both iodine and sulphur in particles captured at a coastal site [63] indicates the participation of other gas-phase species, for example H_2SO_4, in particle formation and growth.

With regard to the meteoric smoke experiments, the laboratory temperature and pressure conditions are very different from the upper mesosphere (at 80 km, $T \sim 200$ K and $P \sim 1$ Pa). In terms of the particle growth kinetics, such low pressures should have little effect on association reactions of molecules such as FeO_3 and $FeSiO_3$; these should be close to their high-pressure limits even at 1 Pa because of the large calculated binding energies and internal degrees of freedom (vibrations and rotations) involved [22]. Lower temperatures will lead to an increase in the magnetic-coagulation rates due to the small inverse dependence predicted by equation 4. A number of the reaction channels shown in Figure 7 remain to be confirmed through laboratory studies, and gas-phase species (e.g. FeOH) and reaction pathways other than those proposed in Figure 7 may play important roles in meteoric smoke formation in the upper atmosphere. However, it seems likely that polymerisation of such species will lead to the nucleation of smoke particles with similar compositions to those identified in our laboratory generated mimics. Of course, atmospheric smoke particles will also contain smaller quantities of other ablated metals such as sodium and magnesium. These issues will only be resolved through future experiments and the direct sampling of meteoric smoke particles from the atmosphere.

Extended, low density or 'fluffy' smoke aggregates are likely to have different optical properties to those of idealised spherical, compact particles. Also, enhanced surface areas of fractal, porous particles, compared with compact particles, are likely to be available for heterogeneous reactions with species such as H_2O, H_2SO_4 and HNO_3, resulting in a more important role for such reactions in the stratosphere. Finally, the transport and atmospheric lifetime of smoke particles are also likely to be significantly affected due to the reduced aerodynamic drag experienced by such extended shapes and porous structures in comparison with compact, spherical particles. These effects must be taken into account when modelling the coagulation and sedimentation of these particles through the atmosphere.

6. SUMMARY

This chapter discusses the formation of secondary inorganic aerosol in the atmosphere, using two examples from opposite ends of the atmosphere. The photochemical generation in the laboratory of iodine oxide aerosols and analogues for meteoric smoke particles, with appropriate diagnostic tools, provides the fundamental information required to model these systems in the atmosphere. Electron microscope imaging and quantitative elemental analysis provides the shape, structure and composition of the particles. Optical extinction measurements can corroborate the particle composition by comparing with the extinction predicted by Mie theory with published refractive index data. The formation of the condensable (nucleating) species can be inferred through models using available measured rate coefficients and mechanisms based on quantum chemistry calculations. Kinetic measurements of the particle size distribution provide a reasonably rigorous test of the growth mechanism and the coagulation kernels. The final result is a model, starting from the gas-phase precursors, which can be used to predict the rate of formation and growth of a secondary aerosol in the atmosphere.

Acknowledgments

This work was supported by funding from the UK Natural Environment Research Council.

References

[1] Finlayson-Pitts B.J. and Pitts Jr. J.N., Chemistry of the Upper and Lower Atmosphere: Theory, experiments and applications (Academic Press, San Diego, 2000).
[2] Andreae M.O. and Crutzen P.J., *Science* **276** (1997) 1052-1058.
[3] Buseck. P.R. and Pósfai M., *P. Natl. Acad. Sci. USA* **96** (1999) 3372-3379.
[4] Jacobson M.Z., *J. Geophys. Res.* **106** (2001) 1551-1568.
[5] Lohmann U. and Feichter J., *Atmos. Chem. Phys.* **5** (2005) 715-737.
[6] Ravishankara A.R., *Science* **276** (1997) 1058-1065.
[7] Friedlander S.K., Smoke, Dust and Haze: Fundamentals of aerosol dynamics, 2nd edn. (Oxford University Press, New York, 2000).
[8] DeCarlo P.F., Slowik J.G., Worsnop D.R., Davidovits P. and Jimenez J.L., *Aerosol Sci. Technol.* **38** (2004) 1185-1205.
[9] Jacobson M.Z., Fundamentals of Atmospheric Modeling, 2nd edn. (Cambridge University Press, New York, 2005).
[10] Fuchs N.A., Mechanics of Aerosols (MacMillan, New York, 1964).
[11] Bohren C.F. and Huffman D.R., Absorption and Scattering of Light by Small Particles (Wiley-Interscience, New York, 1983).
[12] Samson R.J., Mulholland G.W. and Gentry J.W., *Langmuir* **3** (1987) 272-281.
[13] Xiong C. and Friedlander S.K., *P. Natl. Acad. Sci. USA,* **98** (2001) 11851-11856.
[14] Amal R., Raper J.A. and Waite T.D., *J. Colloid Interf. Sci.* **140** (1990) 158-168.
[15] Ullmann M., Friedlander S.K. and Schmidt-Ott A., *J. Nanopart. Res.* **4** (2002) 499-509.
[16] Cabane M., Rannou P., Chassfiere E. and Israel G., *Planet Space Sci.* **41** (1993) 257-267.
[17] Fogel M.E. and Leung C.M., *Astrophys. J.*, **501** (1998) 175-191.
[18] Mandelbrot B.B., The Fractal Geometry of Nature, (W.H. Freeman, New York, 1982).
[19] Eriksson A.B. and Jonson M., *Phys. Rev. B*, **40** (1989) 884-887.
[20] Niklasson G.A., Torebring A., Larsson C. and Granqvist C.G., *Phys. Rev. Lett.* **60** (1988) 1735-1738.
[21] Helgesen G., Skjeltorp A.T., Mors P.M., Botet R. and Jullien R., *Phys. Rev. Lett.* **61** (1988) 1736-1739.
[22] Saunders R.W. and Plane J.M.C., *J. Atmos. Solar-Terr. Phys.* **68** (2006) 2182-2206.
[23] Carpenter L.J., *Chem. Rev.* **103** (2003) 4953-4962.
[24] Baker A.R., Tunnicliffe C. and Jickells T.D., *J. Geophys. Res.* **106** (2001) 28743-28749.
[25] Carpenter L.J., Sturges W.T., Penkett S.A., Liss P.S., Alicke B., Hebestreit K. and Platt U., *J. Geophys. Res.* **104** (1999) 1679-1689.
[26] Carpenter L.J., Hebestreit K., Platt U. and Liss P. S., *Atmos. Chem. Phys.* **1** (2001) 9-18.
[27] O'Dowd C.D., Jimenez J.L., Bahreini R., Flagan R.C., Seinfeld J.H., Hameri K., Pirjola L., Kulmala M., Jennings S.G. and Hoffmann T., *Nature* **417** (2002) 632-636.
[28] O'Dowd C.D., Hameri K., Mäkelä J.M., Pirjola L., Kulmala M., Jennings S.G., Berresheim H., Hansson H.C., de Leeuw G., Kunz G.J., Allen A.G., Hewitt C.N., Jackson A., Viisanen Y. and Hoffmann T., *J. Geophys., Res.* **107** (2002) doi: 10.1029/2001JD000555.
[29] McFiggans G., Coe H., Burgess R., Allan J., Cubison M., Rami Alfarra M., Saunders R., Saiz-Lopez A., Plane J.M.C., Wevill D., Carpenter L., Rickard A.R. and Monks P.S., *Atmos. Chem. Phys.* **4** (2004) 701-713.
[30] Hoffmann T., O'Dowd C.D. and Seinfeld J.H., *Geophys. Res. Lett.*, **28** (2001) 1949-1952.
[31] Jimenez J.L., Bahreini R., Cocker III D.R., Zhuang H., Varutbangkul V., Flagan R.C., Seinfeld J.H., O'Dowd C.D. and Hoffmann T., *J. Geophys. Res.* **108** (2003) doi: 10.1029/2002JD002452.
[32] Sellegri K., Yoon Y.J., Jennings S.G., O'Dowd C.D., Pirjola L., Cautenet S., Chen H. and Hoffmann T., *Environ. Chem.* **2** (2005) 260-270.
[33] Pirjola L., O'Dowd C. D., Yoon Y. J. and Sellegri K., *Environ. Chem.* **2** (2005) 271-281.

[34] Saiz-Lopez A. and Plane J.M.C., *Geophys. Res. Lett.* **31** (2004) doi:10.1029/2003GL019215.
[35] Saiz-Lopez A., Saunders R.W., Joseph D.M., Ashworth S.H. and Plane J.M.C., *Atmos. Chem. Phys.* **4** (2004) 1443-1450.
[36] Saunders R.W. and Plane J.M.C., *Environ. Chem.* **2** (2005) 299-303.
[37] Rogak S.N. and Flagan R.C., *J. Colloid Interface Sci.* **151** (1992) 203-224.
[38] Selte K. and Kjekshus A., *Acta Chem. Scand.* 24 (**1970)** 1912-1924.
[39] Saunders R.W. and Plane J.M.C., *J. Aerosol Sci.* **37** (2006) 1737-1749.
[40] Plane J.M.C., *Int. Rev. Phys. Chem.* **10** (1991) 55-106.
[41] Plane J.M.C., Cox R.M. and Rollason R.J., *Adv. Space Res.* **24** (1999) 1559-1570.
[42] Plane J.M.C., *Chem. Rev.* **103** (2003) 4963-4984.
[43] Rosinski J. and Snow R.H., *J. Meteorol.* **18** (1961) 736-745.
[44] Hunten D.M., Turco R.P. and Toon O.B., *J. Atmos. Sci.* **37** (1980) 1342-1357.
[45] McNeil W.J., Lai S.T. and Murad E., *J. Geophys. Res.* **103** (1998) 10899-10911.
[46] Kalashnikova O., Horányi M., Thomas G.E. and Toon O.B., *J. Geophys. Res.* **27** (2000) 3293-3296.
[47] Turco R.P., Toon O.B., Whitten R.C., Keesee R.G. and Hollenbach D., *Planet Space Sci.* **30** (1982) 1147-1181.
[48] Rapp M. and Thomas G.E., *J. Atmos. Solar-Terr. Phys.* **68** (2006) 715-744.
[49] Murphy D.M., Thomson D.S. and Mahoney T.M.J., *Science*, **282** (1998) 1664-1669.
[50] Turco R.P., Toon O.B., Hamill P. and Whitten R.C., *J. Geophys. Res.* **86** (1981) 1113-1128.
[51] Mills M.J., Toon O.B. and Thomas G.E., *J. Geophys. Res.* **110** (2005) doi:10.1029/2005JD006242.
[52] Prather M.J. and Rodriguez J.M., *Geophys. Res. Lett.* **15** (1988) 1-4.
[53] McNeil W.J., Murad E. and Plane J.M.C., in Meteors in the Earth's Atmosphere, edited by E. Murad and I.P. Williams, (Cambridge University Press, Cambridge, 2002).
[54] Gerding M., Baumgarten G., Blum U., Thayer J.P., Fricke K.H., Neuber R. and Fiedler J., *Ann. Geophys.*, **21** (2003) 1057-1069.
[55] Curtius J., Weigel R., Vössing H.-J., Wernli H., Werner A., Volk C.-M., Konopka P., Krebsbach M., Schiller C., Roiger A., Schlager H., Dreiling V. and Borrmann S., *Atmos. Chem. Phys.* **5**, (2005) 3053-3069.
[56] Gabrielli P., Barbante C., Plane J.M.C., Varga A., Hong S., Cozzi G., Gaspari V., Planchon F.A.M., Cairns W., Ferrari C., Crutzen P., Cescon P. and Boutron C.F. , *Nature*, **432** (2004) 1011-1014.
[57] Meakin P., *Adv. Colloid Interface Sci.* **28** (1988) 249-331.
[58] Helmer M. and Plane J.M.C., *J. Chem. Soc. Faraday Trans.* **90** (1994) 31-37.
[59] Plane J.M.C., Self D.E., Vondrak T. and Woodcock K.R.I., *Adv. Space Res.* **32** (2003) 699-708.
[60] Self D.E. and Plane J.M.C., *Phys. Chem. Chem. Phys.* **5** (2003) 1407-1418.
[61] Sorensen C.M., *Aerosol Sci. Technol.* **35** (2001) 648-687.
[62] Saiz-Lopez A., Plane J.M.C., McFiggans G., Williams P.I., Ball S.M., Bitter M., Jones R.L., Hongwei C. and Hoffmann T., *Atmos. Chem. Phys.* **6** (2006) 883-895.
[63] Mäkelä J.M., Hoffmann T., Holzke C., Väkevä M., Suni T., Mattila T., Aalto P.P., Tapper U., Kauppinen E.I. and O'Dowd C.D., *J. Geophys. Res.* **107** (2002) doi: 10.1029/2001JD000580.

J. Phys. IV France **139** (2006) 257–267
© EDP Sciences, Les Ulis
DOI: 10.1051/jp4:2006139018

Asian dust events: Environmental significance in Beijing

F. Adams[1] and X. Liu[2]

[1] *Department of Chemistry, University of Antwerp, 2610 Wilrijk, Belgium*
[2] *Chinese Research Academy of Environmental Sciences, Beijing 100012, China*

Abstract. The environmental significance of dust storms (impact on air quality, climatic effects, health effects) is described first in a general terms then in context with the local environment of metropolitan Beijing, China. The regional characteristics of dust storms in China, their sources and transport pathways, the chemical composition and the mixing and interaction with other air pollutants are reviewed.

1. INTRODUCTION

The Tyler Prize for Environmental Achievement was awarded in 2002 to Professor Tungsheng Liu of the Institute of Geology and Geophysics, Chinese Academy of Science, for his pioneering contributions in recognizing the origin of terrestrial sediments and understanding their importance for aspects of global environmental change. Liu's research in paleo-climatology over the last five decades convincingly demonstrated that loess deposits provide a complete and accurate continental record of environmental change over the last 2.5 million years (www.usc.edu/dept/LAS/tylerprize/02tyler). Loess is a geologically recent deposit, usually yellowish or brown in colour and consists of tiny mineral particles brought by wind to the places where they now lie. The fine-grained dust constituting loess deposits is now widely considered one of three reliable sources of past environmental information, the other two being deep-sea sediments and Arctic, Antarctic and glacier ice cores [1].

It took years for researchers to figure out that deposited layers, with a total thickness over two hundred meters, were formed by wind blown dust deposition over extended geological timeframes, just like deep-sea sediments accumulated from suspended river particles but also partly received input from atmospheric deposition. For example, a trail of quartz in the North Atlantic shows a distinct pattern that links it to the western coast of North Africa. Many African soils are quartz-based, but North and West Africa does not have rivers large enough to explain the deep-sea quartz deposits [2].

Arid and semi-arid regions constitute the most sensitive regions to climatic variability. The continued predictions of increasing concentrations in greenhouse gases also include predictions of increased aridification of continental interiors and the likelihood of enhanced dust emission. At present it is not understood how regional and global climate change will affect dust emissions and, in turn, how dust emission will affect the climate. Latest estimates of global dust emissions are approximately 2000-3000 million tons per year.

The massive conveyance of mineral dust is going on nowadays, particularly as Sahara dust storms and Asian dust storms. The influence of Sahara dust storms extends up north to northern Europe, to the American continent to the west, and to India and Nepal to the east. Asian dust storms generally spread over East Asia and sometimes, across the Pacific Ocean, reaching North America.

Dust storms are natural processes that occurred for millions of years. On the other hand human activities such as farming and irrigation caused desertification and thus contributed to their occurrence, frequency and variational trends. One of the best well-documented examples of the impact of farming on desertification is the Dust Bowl that occurred in the American Midwest (http://www.ptsi.net/user/museum/dustbowl.html) in the 1930s. The promise of a nutrient-rich soil and "easy get-quick-rich" farming led to mass immigration into the region. The combination of vast tracts of farmland, detrimental agricultural practices such as part-time farming, and the onset in 1930 of what later turned out to be a severe 10 year drought, resulted in an ecological disaster that drove out a fourth of the population from the region. A dust blown cloud that impacted on Washington D.C. in 1934 was essential to help spur the United States Congress to pass its Soil Conservation Act of 1935. The purpose of the Act was to implement improved farming practices and to put into effect a plethora of other measures that would limit soil erosion by both winds and precipitation [3].

Another example of human induced desertification in the United States is the fate of Lake Owens. Lake Owens, which is located in southern California, had a surface area of approximately 280 km^2 in 1913, the year in which the City of Los Angeles started tapping it as a source of drinking water. By 1926 all that remained was a dry lake bed which has since acted as a diffuse source of dust to the Los Angeles metropolitan area [4]. The lake is the largest single source of PM10 dust (the mass fraction of particles with an aerodynamic diameter less than 10 μm, PM2.5 being those below 2.5 μm) in the United States with 800,000 to 9,000,000 tons a year according to Gill and Gillette [5]. Aerosols sampled from dust storms in the former lake Owens commonly contain significant amounts of arsenic concentrated in the <10 μm size fraction with levels in excess of 400 ng/m^3 according to Reid *et al.*, [6];. http://geochange.er.usgs.gov/sw/impacts/geology/owens/.

Desertification has affected Africa more than any other continent. Over 60% of the continent is composed of deserts or dry lands, and severe droughts are very common. One prominent area in Africa where desertification has been caused by both nature and human activities is Lake Chad (West Africa). The surface area of Lake Chad in 1963 was 25,000 km^2; due to regional drought conditions and irrigation practices its surface area is now reduced to one twentieth of its 1963 size ($\sim$1,350 km^2) [7].

China is another country that has suffered from significant desertification problems. Approximately 27 % of the surface of China (ca 2.6 million km^2) is affected by desertification. Between 1975 and 1987 the average annual desertification rate has been estimated at 2,100 km^2/yr [8]. Human factors such as population growth, deforestation (for building, farming, firewood scavenging...), and overgrazing have been identified as contributing factors to the overall desertification in China [9].

Historic dust events were observed regularly and were documented, for example in the famous historical diary of Tonghe Weng (tutor of Emperor Tongzhi and Emperor Guangxu), a grand total of 349 dusty days were identified in North China during the period 1860-1898 AD, which amounts to an annual average of about 9 days per year [10].

Table 1 presents the annual average number of dusty days in Beijing from 1954 to 1999. A dust storm refers to a severe dust event with high atmospheric particulate concentrations, much deteriorated visibility and strong winds. Floating dust days are those with a high level of dust particles, but with no associated strong winds. Both dust storm and floating dust are caused by regional dust events. Wind-blown dust days (those with degraded visibility conditions) could be either a local or a regional phenomenon. Although the data show considerable variability, there is in general a declining trend in the number of dusty days over the past half century [11].

Table 1. Annual average number of dusty days in Beijing from 1954-1999 [11].

Period	Average number of days over period		
	Dust storm	Floating dust	Wind-blown dust
1954 - 1959	3.8	9.5	56.2
1960 - 1969	5.2	4.2	22.5
1970 - 1979	1.1	6.2	20.6
1980 - 1989	0.9	3.6	14.0
1990 - 1999	0.5	2.3	4.4

2. ENVIRONMENTAL EFFECTS

2.1 Impact on air quality

Asian dust events have an impact on air quality and visibility in the source region as well as over muchlarger down-wind regions hundreds, even several thousands kilometers away. For example, in the year 2000 there were 15 dusty days in Beijing, whereas there were 29 days and 20 days in 2001 and 2002, respectively [11]. Surprisingly, there was no single dusty day in 2003. This matched well with the PM10 concentration data as shown in Figure 1 [11]. Due to the dust events, PM10 concentration levels in the spring, i.e. in March and April, and sometimes also in May were the highest of the year in Beijing. The year 2003 turned out to be an exceptional one; there was no peak value in the spring. The Chinese National Ambient Air Quality Standard for PM10 annual is set at an average of 0.100 mg/m^3. The PM10 concentration in Beijing is not in compliance with this national standard all year round. Obviously the dust events make this situation even worse in the spring.

Mostly Asian dust events affect many different regions over the Asian continent. Sometimes they extend over East Asia, including the Korean peninsula and Japan, and the northwestern Pacific Ocean. Occasionally, they move as far as North America, for instance during an event in April 1998. This particular Asian dust storm started on April 19 in the border area between Mongolia and China. It took 6-7 days to travel east and it arrived on the western coast of the United States, causing high level particulate matter levels there from 26 April to May 1. Figure 2 shows the daily PM2.5 dust concentration at three monitoring sites for which long-term records exist: Mount Rainier, Washington, Crater Lake, Oregon, and Boundary Water, Minnesota. On the basis of the characteristic dust size of 2-3 μm, it can be estimated that the total dust mass concentration was 2–3 times higher than the PM2.5 dust concentration [12]. Another huge sandstorm in April 2001 reduced visibility to practically zero in Beijing making ground transportation nearly impossible. The huge aerosol cloud was picked up by orbiting earth satellites and was tracked into the Midwest of the USA (http://www.lakepowell;net/asiandust2.htm; http://ggweather.com/dust.htm).

At each site, the fine particle dust concentration was quite variable but, in general, below 5 $\mu g/m^3$ throughout the decade 1988-1998. The exception is the sharp dust peak at each of these sites reaching 5–10 $\mu g/m^3$ on April 29, 1998. Hence, the April 1998 Asian dust event was 2-4 times more intense than any other dust incursion to the western United States over the past decade [12].

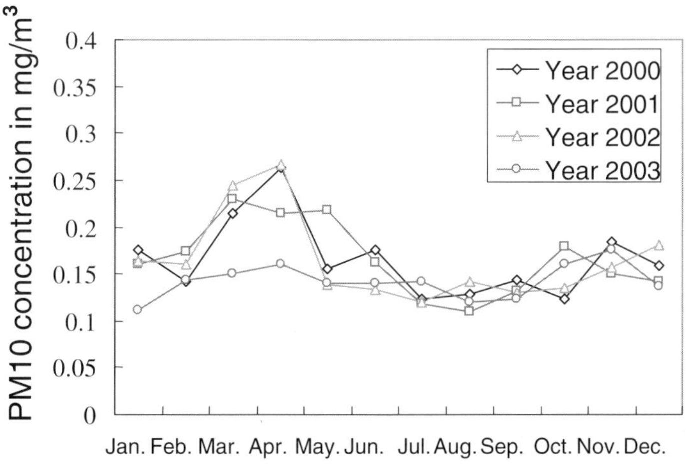

Figure 1. PM10 monthly average data (2000-2003) for Beijing, China [11].

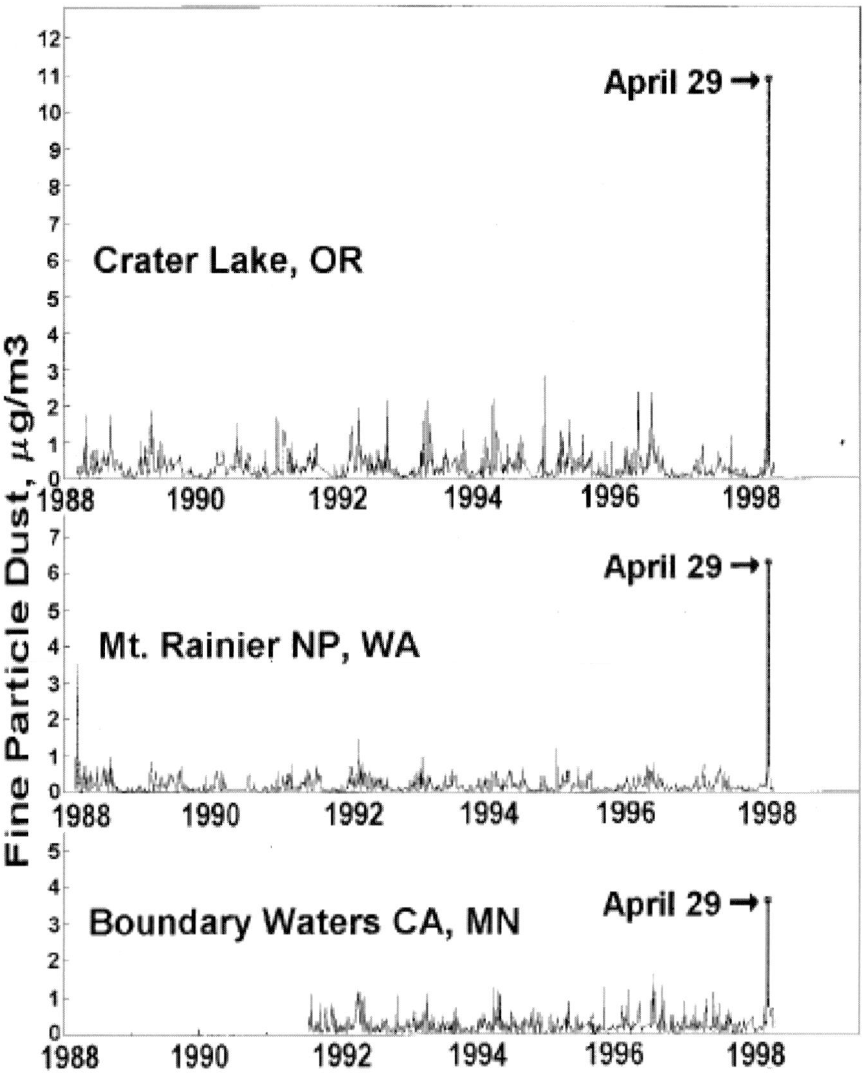

Figure 2. Ten-year trend of PM2.5 dust concentration at three monitoring sites. The fine particle dust mass was reconstructed from the concentration of crustal elements. The simultaneous sharp rise at all three sites on April 29 marks the incoming dust event [12].

2.2 Climatic effects

Aerosol deposition records in ice cores and their variability suggests that they are very sensitive to climate. Mineral dust from soils is known to influence the climate by scattering and absorbing solar radiation. The impact of atmospheric aerosols on visibility and climate depends on the chemical composition and the size distribution of the particles. A detailed knowledge of the aerosol composition is also important for elucidating source identification and apportionment. With 1000-10,000 Tg per year, sea salt is responsible for about 30-75% of all natural aerosols [13, 14] but soil dust with a yearly emission of 1300 Terra g yr^{-1} is practically equivalent at certain intervals [15]. Due to scattering and absorption, the dust component of the aerosol reduces solar radiation from reaching the earth's surface, while simultaneously heating the lower atmosphere by absorption. A detailed discussion of the radiation effects of each of the aerosol components is given by Satheesh and Moorthy [16]. Recent estimates of the radiative forcing of dust shows values ranging between -2 and +0.5 W m^{-2}. Hence, it is not entirely clear whether the overall radiative forcing is positive or negative. Recent work indicates that globally it tends to be negative but that it has a large spatial and temporal variability. A recent detailed study of the influence of size and composition on the cloud nucleating ability of aerosol particles is given by Dusek *et al* [17]. When the temporal variation of chemical effectson cloud condensation nuclei (CCN) activation is neglected, variation in the size distributionalone explains 84 to 96% of the variation in CCN concentrations. The particles ability to act as CCN is largelycontrolled by aerosol size rather than composition.

2.3 Health effects

Dust and other particulate matter that occurs naturally (from volcanic eruptions, forest fires, live vegetation, sea spray...) makes up 90% of airborne aerosols while about 10% of the aerosol in the atmosphere is caused by human activities, mainly as residues from automobile and industrial emissions. Until recently it was this fraction that was considered to be the main concern for human health [18]. Time-series analysis of daily mortality pooled across 20 U.S. cities (1987–1994) showed a 0.5% increase in daily mortality associated with each 10 μg m^{-3} increase in PM10 [19].

In many regions of the world, mineral dust is the dominant constituent of the aerosol, consequently atmospheric dust may constitute a widespread health threat. In addition to the constituents that are associated with wind blown atmospheric dust, soil also can cause human health effects and become responsible for disease. Silicosis is a disease caused by breathing finely grained quartz-type sand, a common component of desert soils. Inhalation of quartz-based soils can cause tissue scarring and lung fibrosis. Symptoms of silicosis include shortness of breath, fever and persisting fatigue; in severe cases, the disease can be fatal. While this disease is usually associated with occupational exposure (mining, cement work, etc.), studies have convincingly shown that individuals residing in silicate rich environments are also at risk of developing the disease [20]. Exposure to quartz soils has also been implicated as a causative agent in lung cancer and auto-immune and non-malignant renal diseases [21]. The World Health Organization reported for China yearly approx. 24,000 deaths due to silicosis (http://www.who.int/inf-fs/en/fact238.html).

The exposure to airborne microbes and biologically derived particulate matter can also cause allergic responses. The National Institute of Allergy and Infectious Diseases of the United States identified atmospheric dust as the most prominent single source of allergic stress worldwide http://www.niaid.nih.gov/pulications/allergens/intro.htm. The WHO estimates that between 100 and 150 million people suffer from asthma with an annual death rate of 180,000 persons http://www.who.int/inf-fs/en/fact206.html. The majority of exposures to allergens occurs outdoors and research has shown that once sensitized to an allergen (e.g., fungal spores), small quantities of the allergen can give rise to acute reactions. Areas such as the Aral Sea (another important and -due to human interference-largely dried up inner sea in central Asia) and the Caribbean, where desert dust activity is common, have some of the highest recorded incidence of asthma on Earth. Barbados, for example, experienced a 17-fold increase in the incidence of asthma between 1973 and 1996 and acute asthma attacks accounted for over 22% of emergency visits at Queen Elisabeth Hospital in 1999 [22]. This observed increase in incidence has coincided with the increased dust flux from the Sahara and the Sahel region to the island [3].

2.4 Ecological effects

Red tides could be taken as an example of many ecological effects with far-reaching consequences caused by the dust storm events. The red tides (also called "sea-saw dust") often found in tropical and subtropical seas are now known to be the result of the growth of a nitrogen fixing cyanobacterium

(*Trichodesmium*) in a iron rich environment such as available in the sea as a result of iron rich dust blown from the African Sahara. Typical red tides arrived on the Gulf of Mexico coast of Florida in the Autumn of 1999 and 2001. A large toxic bloom formed in the fall and drifted for months between Pensacola and St. Petersburg in Florida -a distance of over 400 kilometers- and a correlation between African dust and *Trichodesmium* was observed. The bacteria increased around the time of the summer dust storms in Africa. With the resulting harm to marine life, red tides also hurt Florida's fishing and tourism industries, studies estimating the damage at $25 million per year [18]. More recent reports indicated that iron rich dust from the Gobi desert gave rise to phytoplankton growth in the North Pacific Ocean.

There was also a debate about the origin of the epidemic of foot and mouth disease (FMD) in 2001 in Britain. The British Ministry of Agriculture believes that illegally imported meat is the root cause of FMD introduction. Some scientists proposed another possibility correlating it with a tongue of Sahara dust curling up from the coast of Africa and blowing over the British Isles and parts of Europe from February 13, 2001 and the next days. The FMD outbreak started on February 19, 2001, which is within the average incubation period of 3 - 8 days from the appearance of the Sahara dust. There are seven serotypes of the FMD virus, and the strain that infected Britain is the same serotype as the prevalent strain in northern Africa (Type O). The largest previous outbreak in Britain in 1967 also occurred in February, the time of year when wind patterns are most likely to cause African dust to blow north [18].

3. CHARACTERIZATION OF ASIAN DUST EVENTS

3.1 Regional characteristics of dust storm in China

In what follows we will concentrate on the Beijing metropolitan area and its surroundings in Northern China. The Beijing area is featured by a rapidly expanding population, a high consumption of coal, an increasing number of vehicles and flourishing construction activities. Situated in a semi-amid region in North China and surrounded by mountains in the west and north, the weather conditions usually do not favour dispersion and transport of air pollutants during the entire year.

Regional characteristics of dust storms in northern China were characterized using a rotated empirical orthogonal function (REOF), based on the annual days of dust storms from 1954 to 1998 (Figure 3) [23]. The results show that five leading modes of dust storms exist in the following areas: the Taklamakan Desert (Tarim Basin) over the Xinjiang region (far northwestern China), the eastern part of Inner Mongolia (North China), the Tsaidam Basin, the Tibetan Plateau, and the upper reaches of the Yellow River (Gobi Desert). These areas are associated with an arid climate and frequent winds. For the first mode (the Tarim Basin), most dust storms appear in the 1980s, while dust storms became less frequent in the 1990s. The second mode (North China) shows the highest frequency of dust storms in the mid-1960s but the frequency decreased afterward. The third mode indicates a decreasing trend of annual dust storms after the mid-1960s but with a high inter-annual variability. The fourth mode (Tibet) also shows a decreasing trend but with a low inter-annual variability while the fifth mode (Gobi desert) displays a high frequency of dust storms in the 1970s followed by a decreasing trend since then. For the five modes of dust storm distributions, four of the epicenters are located in the central Asian desert regions. The annual dust storms of a selected station in each mode region are shown to compare the coefficient time series of these modes.

The negative correlation between the winter temperature before the events and dust storm frequency is identified for most stations. There is no consistency in the correlation between the dust storm frequency and the annual rainfall as well as the prior winter rainfall at these stations. The activity of dust storms in northern China is, on the other hand, directly linked to the cyclone activity, especially for the inter-decadal variability.

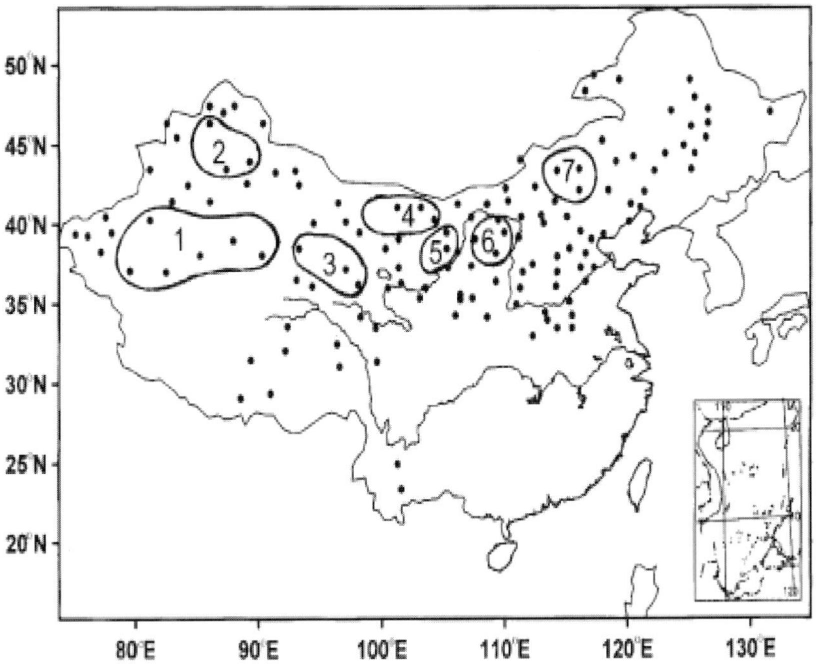

Figure 3. Locations of 135 stations in which dust storms were observed. The numerals mark major deserts: (1) Taklamakan desert in the southern Xinjiang region, (2) northern Xinjiang region, (3) Tsaidam Basin, (4, 5) Gobi Desert in the west of the upper Yellow River valley, (6) Ordos Desert in the middle Yellow River valley and (7) northern area of China [23].

3.2 Transport pathways and sources

According the Wang *et al.*, [24], three major pathways exist (Figure 4). Of the three major pathways, cluster B is the most important one because it has the highest mean PM10 concentration and the highest number of polluted trajectories. By the same reasoning, Clusters D, E and F were considered less important as pathways for pollutants. The air masses associated with cluster A passed through the northern part of Xinjiang Province, then through the southwestern desert and their desert-margin regions of Mongolia, moving into China along the northeast part of the Badian-Juran Desert, then to the Loess Plateau, finally turning northeast to Beijing. The air masses associated with cluster B were mainly originating in Kazakhstan and passed through desert and semi-desert regions in southwestern Mongolia, and then through the middle of Inner Mongolia and onwards to Beijing. The air masses associated with cluster C were mainly from high latitude areas, and passed through eastern Mongolia and the northeastern sandy soil region north of Beijing before reaching the Beijing metropolitan area.

Four major potential sources were identified: (1) the border areas between Kazakhstan and China, (2) the desert and semi-desert regions in western Mongolia, (3) the northern high dust desert and (4) the Loess Plateau of China. The pathways and sources for pollution aerosol are also associated with Asian dust emissions. All this suggests a significant contribution of Asian dust to PM10 pollution in the Beijing area.

3.3 Chemical characterization of mineral dusts

Knowledge of both particle size distribution and size resolved composition is necessary for an adequate characterization of a given aerosol. Size distribution can be measured by cascade impactor sampling

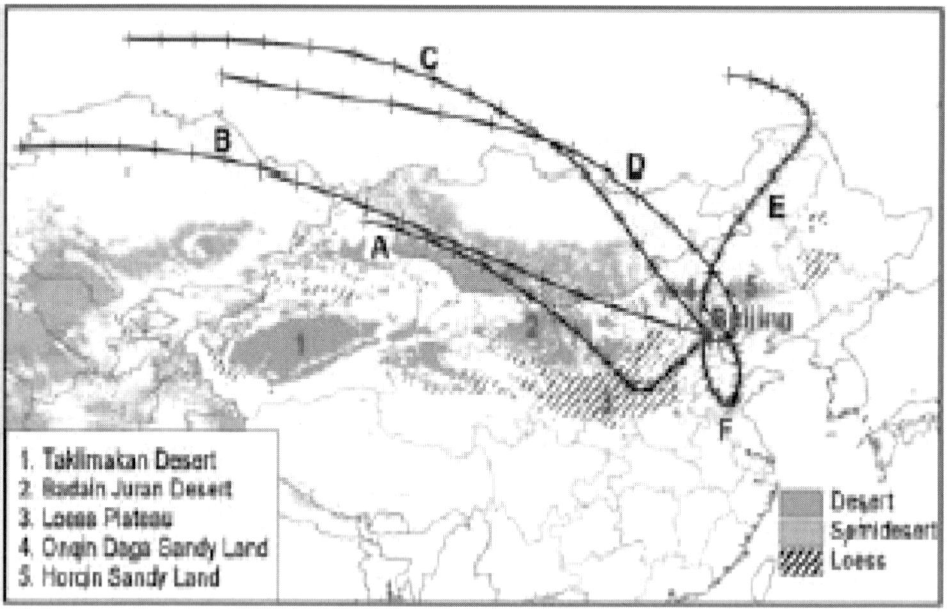

Figure 4. Cluster-mean back trajectories of Beijing during the spring of 2001 to 2003 [24].

devices with sampling according to size classes, or else, with cut-off devices to sample particles up to a certain aerodynamic dimension (e.g. PM10 and PM2.5). The measurement of chemical composition of dust event aerosol may serve as chemical fingerprints for a variety of applications. For instance, it can be used to trace back source regions, differentiate contributions of regional dusts from that of local emissions for an urban area, interpret environmental records such as loess, deep-sea deposition, and dust in ice core over geological times. Bulk analysis on the Chinese dust has been documented by Yang *et al.*, [25], and Noguchi and Hara., [26], while single particle analysis has been reported by Ma *et al.*, [27], Liu *et al.,* [28] and Shi *et al.*, [29]. Regional contributions were differentiated from local emissions by Han *et al.*, [30].

The application of single particle analysis with methods such as scanning electron microscopy-energy dispersive X-ray analysis (SEM-EDX) provides complementary and more detailed information than that issuing from the application of bulk methods of elemental analysis. In recent years SEM-EDX has been used to study aerosol particles collected over the North Sea [31, 32], the Atlantic Ocean [33, 34] and the Antarctic [35] Urban areas around the world were also investigated by this single particle analysis method such as e.g. in Phoenix, Arizona, USA [36], Antwerp, Belgium [37], Seoul and other cities in Korea [38].

Hierarchical cluster analysis can be used for classification of the particles in groups or classes according to their composition or morphology. Mineral dust particle classes are dominant in coarse fractions for both Asian dust and non-dust samples [28]. The dust seems to be mainly alumino-silicates, quartz, calcite, and calcium silicate. However, in non-dust samples there are additional classes such as Ca-S-Si (the major elements detected), likely Ca-Si particles partially reacted with sulfuric acid. Noteworthy a CaS class is present. This class, which has been observed in many previous studies [37], is considered to be $CaSO_4$ (e.g., gypsum). Sulfur is not necessarily always a major element in the various sulfur containing particles. The acidification reactions between sulfur species and the particles may take place on the particle surface, so that the sulfur becomes enriched on the surface.

Similar mineral dust particles were found abundantly in fine fractions, particularly in fine dust samples. There are also several sulfur-containing particle classes, in low numbers in dust samples, but

Table 2. Major particle classes identified using SEM for coarse and fine particulate samples [39].

Coarse fraction		Clay	Quartz	Calcite	Ca,Si	Ca-S-Si	Ca-S	K-rich	S-rich
Dust	Apr.7	79.4	10.4	1.8	3.6	0	0	0	0
Non-dust	average	61.9	9.3	3.9	3.4	3.2	7.2	0.3	0.2
Fine fraction	Clay	Quartz	Calcite	Ca, K-S	K-rich	S-rich	S-K	Si-S-K	
Dust	Apr.7	57.4	4.8	6.6	0	0.4	0.6	0	0
Non-dust	average	18.3	5.0	0.8	8.2	2.2	11.0	23.2	13.4

in high numbers in non-dust fine samples. It is the sulfur-containing particle classes that make the difference.

Table 2 shows a number of results obtained by SEM/EDX and illustrates the different particle types that can be detected in the aerosol of Changdao Island, a remote sampling location north of the province of Shandong in eastern China during a dust episode in April 2002 [39].

Scanning electron microscopy-energy dispersive X-ray analysis was used for single particle analysis of 2500 particles in five atmospheric aerosol samples collected with a cascade impactor in the spring and summer of 2000 in Beijing [28]. Mineral dust appeared to be the dominant particle class during an Asian dust episode, while in other samples mineral dust and S-containing particles constitute the major particle classes. In pollution episodes during the summer a large abundance of S-containing particles featured the atmospheric aerosol. Chemical and size distribution characteristics are discussed for Ca-S, K-S, and Ca-K-S particle classes. Formation of Ca-K-S and other S-containing particle classes with high abundance was closely related to meteorological conditions such as relative humidity and cloud coverage. Particles, with an elongated crystalline morphology were detected which appear to be formed through aqueous phase oxidation, such as occurring in cloud processes.

Mineral dust, inorganic secondary aerosol mainly as sulfate and nitrate, and organic aerosol are the three major components of airborne particulates in Beijing, of which mineral dust accounts for 32-67% of total suspended particles (TSP), 10-70% of fine particles (PM2.5), and as high as 74% and 90% of TSP and PM2.5, respectively, during dust storm episodes [30]. A major problem is distinguishing the local emissions from the long range transported component. The ratio of Mg/Al in the aerosol is a feasible element tracer to distinguish between the sources inside and outside Beijing. The sources from outside Beijing contribute 62% (range 38-86%) of the total mineral aerosols in TSP, 69% (range 52-90%) in PM10, and 76% (59-93%) in PM2.5 in the spring, and 69% (52-83%), 79% (52-93%), and 45% (7-79%) in TSP, PM10, and PM2.5, respectively, in winter, while only ~20% in summer and autumn. The sources from outside Beijing contributed as high as 97% during dust storm and were the dominant source of airborne particulates in Beijing. The contributions from outside Beijing in spring and winter are higher than those in summer, indicating clearly that it was related to the various meteorological factors and episodes such as Asian dust events [30].

3.4 Mixing and interaction between mineral dust and other air pollutants

Compared with the dust clouds from Africa, Asian dust shows higher concentrations of human caused air pollutants such as sulfates. For that reason, the Asian aerosols may overall be more problematic in interpretation than those in other locations. At present, China as a whole is experiencing a crisis of airborne dust. The dust storms from the Inner Mongolia region of China start from severely eroded soils, caused by intense grazing in that region. As the dust clouds from these eroded lands pass over Beijing and other large cities, they gather additional industrial pollutants. The resulting dust storms can reduce visibility enough to close airports [18].

Anthropogenic sulfate aerosols are transported ahead of mineral particles during a typical Asian dust event. A humid air mass containing a high concentration of sulfate is often followed by a dry air mass containing a high concentration of calcium. Interactions between aerosol components and anthropogenic gaseous species can give insight into the formation process of sulfate and nitrate in aerosols over the

East China Sea. Nitrate showed a good correlation with calcium except in a highly polluted air mass, indicating that gaseous nitric acid is adsorbed on the surface of coarse particles. There is no correlation between nitrate and ammonium, which also supports this contention. In contrast, sulfate showed an excellent linear correlation with ammonium, indicating quick neutralization by ammonia of free sulfuric acid in the gas phase in a polluted air mass. Exceptions were observed when the calcium content was greater than 4 mg m^{-3}, suggesting that SO_2 gas was adsorbed on the surface of yellow sand dust particles and then oxidized on the surface of the dust particles [40].

During dust storm episodic periods, the concentrations of polyaromatic hydrocarbons (PAHs) and fatty acids (FAs) in the aerosols were much higher than that in the non-dust days according to Hou *et al.*, [41]. In this study the total concentrations of PAHs and FAs reached the highest peak of 111 and 47 ng m^{-3} before the dust storm occurred and reduced to 66 and 8.5 ng m^{-3}, respectively, during a dust storm episode, possibly because the accumulated pollutants in the aerosols were cleared out before the pure dust appeared. This variation is in agreement with the mechanism of the concentration change of those pollution components in the super dust storm in 2002 in Beijing described in an earlier study by Guo *et al.*, [42]. These results suggest that in dust storm episodes the PAHs and FAs accumulate in the aerosol because the pollutants from the long range transport from the source areas or from the pathways of the dust storm overlap/mix with locally emitted pollutants, which would increase the total concentrations of these pollutants in the aerosols collected in the dust storm days [42].

4. CONCLUSION

As a rapidly growing metropolis Beijing, China suffers from various local air pollution sources and has difficulties to comply with local emission standards. The situation is aggravated by frequent dust episodes from the arid regions in central Asia. Mixing and interaction between the dust from sand storms with locally generated air pollution can only be studied by refined characterization techniques involving cascade impactor sampling and the application of trace elemental and microscopic analytical techniques.

References

[1] S.L. Yan, F. Ding, Z.L. Ding, Geochim. Cosmochim Acta, 70(7) 1695 (2006).
[2] J.M. Prospero, Proc. Natl. Acad. Sci., 96, 3396 (1999).
[3] D.W. Griffin, C.A. Kellogg, E.A. Shinn, Global Change & Human Health, 2 (1) 20 (2001).
[4] M.C. Reheis, J. Geophysical Research (Atmospheres), 102 (25), 999 (1997).
[5] T.E. Gill, D.A. Gillette, Geological Society of America, Abstracts with Programs, 23, no. 5, 462 (1991).
[6] J.S. Reid, R.G. Flocchini, T.A. Cahill, R.S. Ruth, D.P. Salgado, Atmos. Environ., 28, 1699 (1994).
[7] M.T. Coe, J.A. Foley, J. Geophysical Research (Atmospheres), 106, 3349 (2001).
[8] Z.D. Zhen, W. Tao, Desertification Control Bulletin, 22, 27 (1993).
[9] M. Fullen, D. Mitchell, Geographical Magazine, 63, 26 (1993).
[10] J. Fei, J. Zhou, Q.Y. Zhang, H.Z. Chen, Atmos. Environ., 39, 3943 (2005).
[11] Anon, in: Internal project report, Energy resources and environmental analyses for Municipal Beijing (2004).
[12] R.B. Husar, D.M. Tratt , B.A. Schichtel, S.R. Falke, F. Li, D. Jaffe, S. Gasso, T. Gill, N.S. Laulainen, F. Lu, M.C. Reheis, Y. Chun, D. Westphal, B.N. Holben, C. Gueymard, I. McKendry, N. Kuring, G.C. Feldman, C. McClain, R.J. Frouin, J. Merrill, D. DuBois, F. Vignola, T. Murayama, S. Nickovic, W.E. Wilson, K. Sassen, N. Sugimoto, W.C. Malm, J. Geophysical Research, Atmospheres, 106, D16, 18, 317, (2001).
[13] B. Winter, P. Chylek, Tellus B, 49 (1) 72 (1997).
[14] D.C. Blanchard, A.H. Woodcock, Annals of the New York Academy of Sciences, 338, 330 (1980).

[15] M.O. Andreae, in: Climatic effects of changing atmospheric aerosol levels, edited by A. Henderson-Sellers, World Survey of Climatology, vol. 16, Future Climates of the world, Elsevier, New York, p. 341 (1995).
[16] S.K. Satheesh, K.K. Moorthy, Atmos. Environ., 39, 2089 (2005).
[17] U. Dusek, G.P. Frank, L. Hidebrandt, J. Curtius, J. Schneider, S. Walter, D. Chand, F. Drewnick, S. Hings, D. Jung, S. Borrmann, M.D. Andreae, Science, 312 1375 (2006).
[18] D.A. Taylor, Environmental Health Perspectives, 110 (2) A80 (2002).
[19] J.M. Samet, F. Dominici, F.C. Curriero, I. Coursac, S.L. Zegler, New Engl. J. Med., 343, 1742 (2000).
[20] R. Patial, Journal of the Association of Physicians of India, 47, 503 (1999).
[21] W.T. Sanderson, K. Steenland, J.A. Deddens, American J. Industrial Medicine, 38, 389 (2000).
[22] M.E. Howitt, R. Naibu, T.C. Roach, Journal of Respiratory and Critical Care Medicine, 157, A 624 (1998).
[23] W.H. Qian, X. Tang, L.S. Quan, Atmos. Environ., 38, 4895 (2004).
[24] Y.Q. Wang, X.Y. Zhang, R. Arimoto, J.J. Cao, Z.X. Shen, Geophys. Res. Letters, 31, L14110 (2004).
[25] F.M. Yang, B.M. Ye, K.B. He, Y.L. Ma, S.H. Cadle, T. Chan, P.A. Mulawa, Science of the Total Environment, 343, 221 (2005).
[26] I. Noguchi, H. Hara, Atmos. Environ., 38, 6969 (2004).
[27] C.J. Ma, M. Kasahara, R. Holler, T. Kamiya, Atmos. Environ., 35 2707 (2001).
[28] X.D. Liu, J. Zhu, P. Van Espen, F. Adams, R. Xiao, S.P. Dong, Y.W. Li, Atmos. Environ., 39:36. 6909 (2005).
[29] Z.B. Shi, L.Y. Shao, T.P. Jones, S.L. Lu, J. Geophys. Res., 110 D01303 (2005).
[30] L.H. Han, G.S. Zhuang, Y.L. Sun, Z.F. Wang, Science in China Ser. B, Chemistry, 48, 3 253 (2005).
[31] H. Van Malderen, S. Hoornaert, R. Van Grieken, Environ. Sci. Technol., 30, 489 (1996).
[32] S. Hoornaert, H. Van Malderen, R. Van Grieken, Environ. Sci. Technol., 30(5), 1515 (1996).
[33] M. Posfai, J.R. Anderson, P.R., Buseck, H. Sievering, J. Geophys. Res., 100, 23063 (1995).
[34] J.R. Anderson, P.R. Buseck, T.L. Patterson, R. Arimoto, Atmos. Environ., 30, 319 (1996).
[35] P. Artaxo, M.L.C. Rabello, W. Maenhaut, W, R. Van Grieken, Tellus, 44B, 318 (1992).
[36] K.A. Katrinak, J.R. Anderson, P.R. Buseck, Environ. Sci. Technol., 29, 321 (1995).
[37] W.A. Van Borm, F.C. Adams, Atmos. Environ., 23(5), 1139 (1989).
[38] C.U. Ro, H. Kim, K.Y. Oh, S.K. Yea, C.B. Lee, M. Jang, R. Van Grieken, Environ. Sci. Technol., 36, 4770 (2002).
[39] X.D. Liu, F. Adams, unpublished results.
[40] S. Hatakeyama, A. Takami, F. Sakamaki, H. Mukai, N. Sugimoto, A. Shimizu, H. Bandow, J. Geophys. Res., 109, D13304, doi:10.1029/2003JD004271 (2004).
[41] X.M. Hou, G.S. Zhuang, Y.L. Sun, Z.S. An, Atmos. Environ., 40 1205 (2006).
[42] J.H. Guo, K.A. Rahn, G.S. Zhuang, Atmos. Environ., 38 855 (2004).

J. Phys. IV France **139** (2006) 269–294
© EDP Sciences, Les Ulis
DOI: 10.1051/jp4:2006139019

Elemental speciation analysis, from environmental to biochemical challenge

P. Jitaru[1,2] and C. Barbante[2,3]

[1] *University "Al. I. Cuza" of Iasi, Faculty of Chemistry, Department of Inorganic and Analytical Chemistry, 11 Carol I Blvd., 700506 Iasi, Romania*
[2] *Institute for the Dynamics of Environmental Processes (CNR), Dorsoduro 2137, 30123 Venice, Italy*
[3] *University of Venice Ca' Foscari, Department of Environmental Sciences, Dorsoduro 2137, 30123 Venice, Italy*

Abstract. Information regarding the distribution of metallic/metalloid chemical species in biological compartments is required for understanding their biochemical impact on living organisms. To obtain such information implies the use of a dedicated measurement approach, namely *speciation analysis*. The current trend in (elemental) speciation analysis regards bioinorganic applications. New analytical methodologies are therefore necessary for identification, detection and characterization of metal(loids) complexed or incorporated into biomolecules. The established element-speciation approaches developed for the determination of low molecular mass metal(loid) species (e.g. organometallic compounds) in environmental, food, toxicological and health sciences are presently being adapted for the determination of high molecular mass metal-species, generally related to biological processes. This is one of the newest approaches in terms of element speciation and is called *metallomics*; this concept refers to the totality of metal species in a cell and covers the inorganic element content and the ensemble of its complexes with biomolecules, particularly proteins, participating in the organisms' response to beneficial or harmful conditions. Compared to conventional elemental speciation analysis, the approach applied to bioinorganic analysis is challenging, particularly given the difficulties in identification/characterization of the organic (e.g. protein) content of such species. In addition, quantification is not feasible with the conventional approaches, which led to the exploitation of the unique feature of (post-column) online isotope dilution-mass spectrometry for species quantification in metallomics.

1. GENERAL CONSIDERATIONS ON ELEMENTAL SPECIATION ANALYSIS

It has become largely accepted in the last decades that the information in terms of the total content of a metallic/metalloid or non-metallic element is not generally adequate for understanding its actual impact on the environment and human health. Consequently, information regarding the distribution of individual chemical species is increasingly being required for understanding the global potential hazards, both in terms of environmental and biochemical impact. To obtain such information is more difficult and it commonly implies the use of a dedicated measurement approach called *speciation analysis*. In accordance to the IUPAC recommendations [1], the 'chemical species' is a *specific form of an element defined as to isotopic composition, electronic or oxidation state, and/or complex or molecular structure*. Based on the same IUPAC guidelines, *speciation analysis* represents *the analytical activity of identifying and/or measuring the quantities of one or more individual chemical species in a sample*.

A schematic representation of the difference between the conventional trace analysis and the speciation approach is shown in Fig. 1. In classical (trace) analysis, the total content of an element of interest is measured, whereas when carrying out speciation analysis the individual elemental chemical species (e.g. A_1, A_2, A_3, M) are identified and quantified.

It is worth mentioning that speciation analysis is of utmost interest and is especially challenging when applied to determination of traces and ultra-traces (generally below pg g^{-1} level). In such cases,

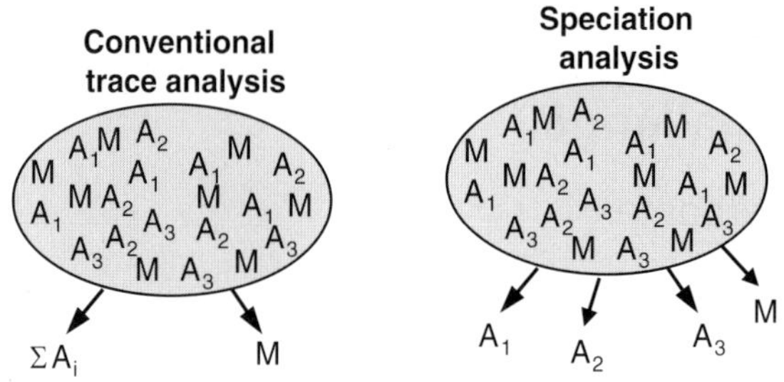

Figure 1. Schematic comparison between conventional trace and speciation analysis.

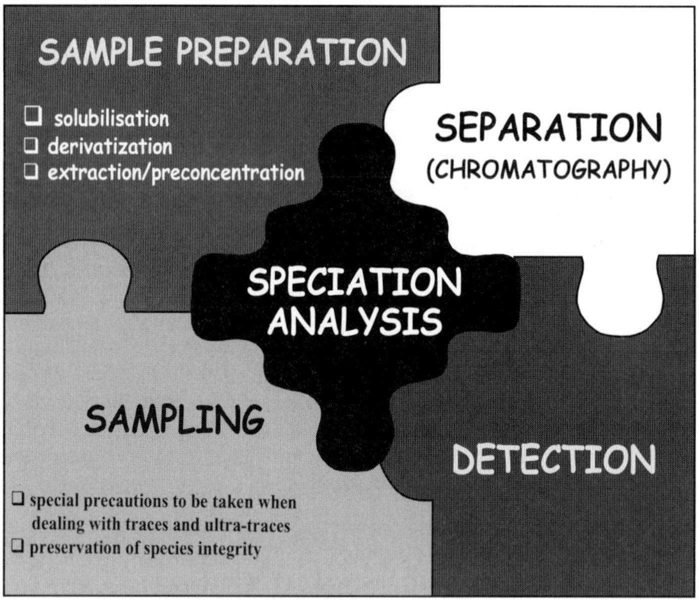

Figure 2. Main steps of trace speciation analysis.

special precautions should be taken in comparison with conventional trace analysis, which are mainly related to preservation of the species integrity.

Fig. 2 illustrates the main steps of speciation analysis, with emphasis on the experimental part. From a practical point of view, trace elemental analysis and in general any chemical analysis begins with the sampling step (the process of taking the sample from the larger quantity of material to be investigated). In most cases, in order to make the analytes amenable to analysis, a sample preparation step is also involved. Further, separation of analytes using chromatography, sequential extraction or physical separation is required in most cases for trace speciation analysis.

The final step of trace speciation analysis is the sequential detection of the species, which is generally carried out using atomic or molecular selective spectrometric techniques. The combination of a chromatographic technique with such a detector (atomic or molecular selective) is generally regarded as a *hyphenated technique* (*tandem* or *coupled*). Nowadays, elemental speciation analysis is carried out almost exclusively using hyphenated techniques, and the progress in the speciation field is undoubtedly related to the improvements of such techniques in all aspects.

Each step of the elemental speciation process shown in Fig. 2, with emphasize on the analytical challenges when applied to trace and ultra-trace levels will be described briefly in the next sections. More details regarding elemental speciation analysis and the hyphenated techniques employed for such applications can be found in a several books, which cover various aspects in terms of environmental, food and human health [2-13].

In the second part of this work, some examples in terms of challenging issues regarding the application of elemental speciation to environmental studies will be addressed. Finally, elemental speciation analysis applied to biochemical samples (*bio-inorganic speciation analysis*) will be briefly overviewed, with emphasis on the main differences and particularities compared to environmental speciation analysis.

It appears that speciation analysis is scarcely carried out on a routine basis *when applied to trace and ultra-trace levels*. This statement, at least for the case of mercury, is supported by a recent inter-comparison exercise (World-wide Intercomparison Exercise for the Determination of Trace Elements and Methylmercury in Marine Sediment IAEA-433) [14], for which participated 103 laboratories from 47 countries but only 11 provided results for methylmercury (CH_3Hg^+).

2. SAMPLING

A *sample* generally represents a portion selected from the bulk of investigated material. It is straightforward that sampling should be carried out in such a fashion that the material analyzed is highly representative of the bulk. Also, the handling procedure, sample size and sample preservation depends on the detail and specificity of the analytical technique intended to be used [15].

When dealing with trace and ultra-trace analysis, the most important precautions to be taken are the avoidance of sample contamination and the preservation of species integrity. Samples may be preserved by the addition of chemicals, but care should be taken that the chosen method of preservation does not interfere with the subsequent laboratory examination. In general, for storage over short periods (up to 24 h), cooling to 4°C may be sufficient; the samples, which cannot be analyzed within a day, should be stabilized or preserved. The most common way to do so, e.g. for total trace metals determination, consists of the addition of HNO_3 to the sample to achieve a pH <2 [16]. However, for speciation studies particular precautions should be taken in order to preserve the species in the original sample. For instance, a special case is the stabilization of mercury species solution at ultra-trace levels particularly during storage over long periods of time, as some mercury species are photosensitive. Other species such as those of chromium can undergo redox reactions in solution in the presence of iron necessitating alkaline extraction conditions. So far, there is no general agreement in terms of the most appropriate approach for this problem. The main factors affecting the stability of mercury species (inorganic and methylmercury) during sample storage are described in detail elsewhere [17].

For sample storage over long periods (i.e. more than one month), freezing to -20°C is recommended. Particular samples such as polar snow/ice should be transported and stored frozen, generally at -20°C, until their analysis. Moreover, given the extremely low level of the target analytes in such samples, addition of preservation/stabilization agents is generally avoided, in order to prevent contamination. Finally, it should be mentioned that when dealing with trace and ultra-trace speciation analysis, it is mandatory that decontamination and sample preparation is carried out in a ultra-clean laboratory (eventually a cold/clean bench for handling snow/ice samples), and reagents of the highest purity available should be used.

In comparison with the other (following) steps of the chemical analysis depicted in Fig. 2, it is clear that, once the sample is contaminated, the bias introduced in this very first step is propagated over the entire analytical procedure since in most cases the sampling process cannot be easily repeated, as for instance, the sample preparation, or separation can be.

One of the current challenges to the analytical chemist is the investigation of pollution of remote areas such as the poles. In other words, in looking for real-life applications of the very sensitive and accurate analytical approaches nowadays available, the analytical chemist has shifted the ‘interest’ from the analysis of environmental compartments in the ‘vicinity’ of the laboratory, to remote regions, such as high altitudes and polar areas, including ice cores from the poles. Such an approach requires special measures to be taken, especially in terms of sampling, sample decontamination and species preservation. This is a consequence of the very low content of the target analytes usually encountered in such samples (the polar ice is considered the ‘purest water on the Earth’). It is beyond the purpose of this work to present a comprehensive description of the sampling and sample treatment of snow/ice samples, but such information can be found in a recent book dealing with the environmental contamination in Antarctica [18].

3. SAMPLE PREPARATION

Obviously, the analyst’s dream is the use of a direct non-destructive method, which allows the examination of the material in its original state. However, for trace speciation analysis such methods are rare [19]. Hence, in laboratory practice, it is necessary to consider the preliminary treatment of the sample, which is further examined by a direct method, generally a spectrometric one.

Sample preparation is considered the most difficult step of a chemical analysis and it is also believed that it contributes to the highest proportion of the bias (over 60%) in the total measurement error, as shown in Fig. 3 (analytical procedure for volatile organic compounds exemplified here) [20].

One of the first steps in terms of sample preparation consists of the particle size reduction (homogenization) of the solid samples (when required) followed by the analytes’ solubilization. In some cases, for speciation analysis the analytes should be derivatized before their detection. This means that they are converted into derivative species with improved analytical properties, such as greater volatility, etc. Also, for trace speciation analysis it is highly advisable to extract and preconcentrate the trace constituents from the sample matrix.

In terms of speciation analysis, the sample preparation should provide the derivatized and extracted/preconcentrated species in a medium compatible with the chromatographic separation approach to be used in the next step, but it is also mandatory that in this process the integrity and equilibration of the species be preserved.

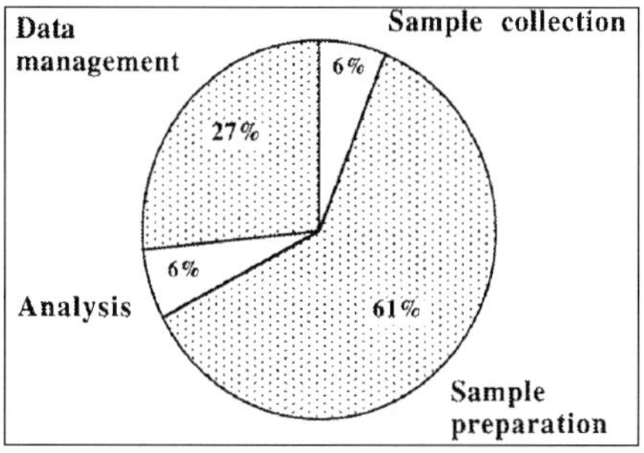

Figure 3. Relative error of different steps for analytical procedure for organic compounds (from ref. [20], with permission).

3.1 Analytes' solubilization

The conventional mode of extraction of various species from (solid) environmental/biological matrices uses either aqueous mixtures of acidic or basic reagents, and water-immiscible (organic) solvents, depending on the nature of the target analyte.

For the analytes that do not participate in geological/mineralogical processes (thus are not integrated into the soil/sediments but only adsorbed onto their surface) or are not intimately bound to the structure of a biological matrix, solubilization can be carried out by leaching with water-miscible solvents such as methanol or acids such as acetic, hydrochloric, sulfuric and nitric acids or bases, etc. The advantage of such procedures is the absence of organic solvents and hence the compatibility with derivatization (if needed) in the aqueous phase and also extraction/preconcentration with techniques 'free' of organic solvent. However, in most cases, leaching should be assisted by either vigorous agitation or sonication, usually requiring prolonged time for homogenization of the sample with the leaching agent. It is common that shaking/sonication of the sediment mixed with the leaching agent is carried out over a period of a few hours [21]. In this context, a significant improvement in terms of analysis time is achieved by using microwave-assisted extraction (MAE). This is nowadays the state-of-the art method for solubilization/analytes leaching of a wide range analytes from various matrices. Detailed information regarding MAE and its application to solubilization of both environmental and biological samples can be found elsewhere [22].

When the analytes are integrated into the matrix, as is generally the case of biological samples, the complete solubilization of the matrix is necessary. In such a case the species preservation issue becomes more important, due to the stronger solubilization procedure employed. MAE can also be used in combination with various solvents to solubilize completely a matrix in order to release the target analytes. For instance, alkaline hydrolysis with tetramethylammonium hydroxide (TMAH) is a common approach for the solubilization of biological tissues for analysis of various species, including organometallic compounds.

3.2 Derivatization

In many cases, trace species cannot be analyzed directly in their native state, either because of the impossibility of extraction/preconcentration, separation (chromatographically) or detection. The common example is the determination of metallic (inorganic) species by UV-VIS molecular absorption spectrometry. Because many metallic ions exhibit either a weak or no absorbance in the UV-VIS spectral domain, they are involved in a chemical reaction that changes their absorptive properties. For this, the metallic cations are reacted with an organic reagent, leading to a (coordinative) complex showing enhanced absorptivity, so that it can be detected by molecular absorption spectrometry (UV-VIS). Such a process of converting analytes into derivatives with enhanced analytical properties is generally called *derivatization.*

In other cases, some species are present in an ionized (polar) state and/or have low volatility. As a consequence, when their separation by gas chromatography (GC) is necessary, they should be converted to volatile and/or thermal stabile derivatives. For this purpose a reaction with a donor of a functional organic group, such as ethyl, propyl etc, which enhances the analytes' volatility, is commonly employed. It should be emphasized that in the case of speciation analysis, the main requirement of a derivatization reaction is the preservation of the species integrity.

3.3 Extraction/preconcentration

Extraction/preconcentration is required in trace elemental speciation analysis in order to: *(i)* isolate/separate the analytes from the matrix, thus alleviating the matrix effects and *(ii)* preconcentrate the analytes in the final analyzed solution, for increased sensitivity and achievement of lower limits of detection.

3.3.1 *Preconcentration capability of an extraction technique*

The preconcentration capability of an extraction technique can be expressed by means of a preconcentration factor (PF) as following:

$PF = \frac{C_{final}}{C_{initial}}$ ($C_{final} \geq C_{initial}$, thus $PF \geq 1$). If the concentration is expressed in terms of mass per volume units (e.g. $\mu g/mL$), PF can be written as:

$$PF = \frac{\frac{m_{final}}{V_{final}}}{\frac{m_{initial}}{V_{initial}}} = \frac{m_{final}}{m_{initial}} \times \frac{V_{initial}}{V_{final}}$$

Further, assuming that the extraction takes place with maximum yield, then $m_{final} = m_{initial} \Leftrightarrow m_{final}/m_{initial} = 1$, and the PF becomes: $PF = V_{initial}/V_{final}$. Although this result is somehow straightforward, it has important implications when dealing with trace and ultra-trace analysis. Given that, handling high (initial) volumes of samples is not feasible in the context of trace chemical analysis, it is clear that, in order to achieve a high preconcentration factor, as small as possible final volumes in which the analytes are extracted should be obtained. In this context, taking into account that volumes of order of tens of milliliters are needed for trace and ultra-trace speciation analysis, final volumes at the μL level are necessary to achieve a preconcentration factor of 4-5 orders of magnitude, as required in practice. This overcomes, for instance, the use of 10 L of sample (e.g. natural water) to be preconcentrated to 0.1-1.0 mL (in order to gain a 10^4-10^5preconcentration factor), which is practically very difficult to achieve.

Historically, the most common way of preconcentrating solutions is with solvent evaporation to the minimum volume still indispensable for analysis. However, this approach has many disadvantages, especially when employed for trace and ultra-trace analysis. More advanced extraction/preconcentration techniques are used nowadays and their number is continuously increasing. A criterion for their classification is based on the involvement of organic solvents for the extraction/preconcentration, namely techniques employing organic solvents and 'free of organic solvents', respectively.

3.3.2 *Extraction/preconcentration techniques employing organic solvents*

Traditional extraction/preconcentration in elemental speciation analysis uses liquid-liquid extraction (LLE), which is based on the partioning of the analyte between the aqueous phase and a suitable organic solvent. Although it is not very selective, this method is still used nowadays for speciation studies, as long as the selectivity of the global procedure is accomplished by species separation combined with element-selective detection.

Besides their toxicity, the use of organic solvents leads also to an overall poor sensitivity of the analytical method, as explained further. Using conventional LLE, at least 100 μL of solvent is commonly needed to achieve the extraction of the target analytes from a volume of solution of order of tens of milliliters. Although a preconcentration factor of ca 2 orders of magnitude seems to be obtained (e.g., if $V_i = 10$ mL and $V_f = 100\ \mu L$, $PF = 10^2$), this preconcentration achievement is lost because of the very small volume (of the extracted analytes) injected into a separation device such as a gas (GC) or a liquid (LC) chromatograph. For instance, given that 1 μL is commonly injected in GC, a loss in sensitivity of 10^2 occurs if the analytes are extracted as mentioned previously. In addition, serious suppression of analytes' intensity as well as instrumental drift is observed when using, for instance, an ICP-MS detector because of the carbon deposition on the sampler and skimmer cones (leading to clogging of the orifices). To avoid this process, addition of oxygen before the analytes reach the detector is mandatory, but such an approach may lead to other inconveniences such as spectral interferences. Thus, an organic solvent matrix is generally more critical in terms of introducing side effects for trace speciation analysis.

A recent development of conventional LLE is *liquid-phase microextraction* (LPME). In this technique, a micro-drop of solvent is suspended from the tip of a conventional micro-syringe after its immersion in the sample solution in which it is immiscible, or suspended in the headspace (above the sample). Although LPME has not gained importance for speciation analysis, it is briefly mentioned here as a proof of the challenge in obtaining high preconcentration of the analytes before their quantification, by transferring them into tiny volumes of solution (ideally < μL).

3.3.3 Extraction/preconcentration techniques 'free' of organic solvents

To reach the state-of-the art of speciation analysis, emphasis has been placed recently on developing sample preparation techniques that are 'free' of organic solvent. The most common approaches in this respect are *static* and *dynamic headspace* (*purge-and-trap injection*) and *solid-phase microextraction*, which will be described briefly in the next paragraphs.

3.3.3.1 Static headspace

Static headspace is one of the simplest 'organic solvent-free' sample preparation techniques. The liquid sample is placed in a vial (usually thermo-stated) during a predetermined time until the equilibrium between the analytes in the liquid and gaseous phase is achieved. Then an aliquot of gaseous analyte(s) is sampled from the headspace (HS) using a (gas-tight) syringe and then usually introduced into a gas chromatograph (GC). Static headspace is a very advantageous method for speciation studies primarily due to the very simple instrumentation. In addition, as the volatile analytes are sampled from the headspace, the carry-over and formation of artifacts are diminished. On the other hand, special attention must be paid to sample preparation in order to maximize the concentration of volatile derivatized compounds in the HS. For the non-volatile analytes, a derivatization step should be considered.

The main drawback of static headspace arises from the limited volume/amount of analyte that is sampled from the HS and consequently analyzed, which drastically limits the method sensitivity. A review on the applications of headspace analysis in modern gas chromatography was recently published by Snow *et al.* [23].

3.3.3.2 Dynamic headspace. Purge-and-trap injection

Dynamic headspace is basically a development of the static headspace so that complete (exhaustive) removal of the volatile analyte from the HS is accomplished. In principle, the volatile (derivatized) analytes are stripped out from the solution by a flow of inert gas and then pre-concentrated by trapping them on various media (by adsorption or cryogenically). Such an approach involving on one hand, analytes's purging and on the other hand their trapping is also called *purge-and-trap injection* (PTI). Unlike static headspace, for volatile analytes/derivatives, PTI is unsurpassed in sensitivity due to its exhaustive extraction/preconcentration capability [23–25]. The major drawback is the limitation to analysis of relatively volatile compounds with boiling points below 200°C. Also, this technique requires a fully dedicated and complex instrumentation, as described in detail elsewhere [26].

3.3.3.3 Solid-phase microextraction

Solid-phase microextraction (SPME) is one of the most efficient 'organic solvent free' techniques for sampling, extraction and analytes' preconcentration. SPME was first reported by Arthur and Pawliszyn [27] and its theoretical and practical aspects are now well documented [28-34].

In SPME the extraction/preconcentration is carried out by means of a tiny amount of extracting phase (usually selective towards the target analytes) that is deposited on a small (1 cm) fused silica rod. A schematic representation of the SPME device, commonly called *fiber* is shown in Fig. 4a (10 cm in total length). The sorptive material is sometimes called a *stationary phase*. The fiber is mounted in a device similar to a syringe, namely a *fiber holder* (Fig. 4b), of *ca* 15 cm in length. During storage and manipulation, the fiber is retracted into the needle of the device, while during the extraction and desorption the fiber is exposed. The main purpose of the needle is to protect the fiber during piercing the septum of the sampling vial and GC injector port.

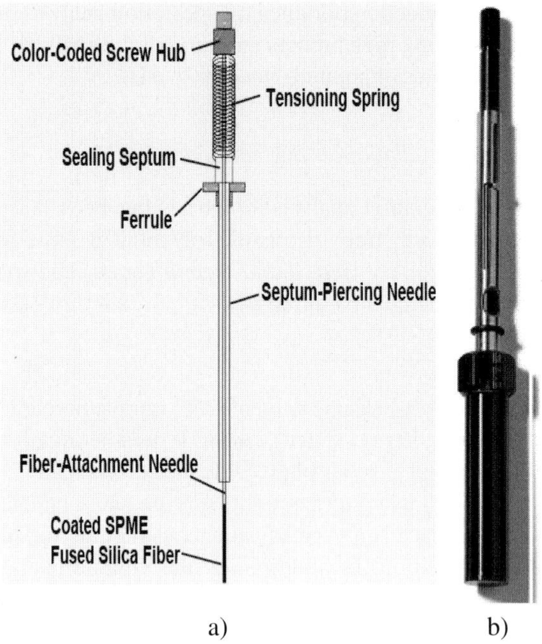

Figure 4. Design of the SPME fiber (a) and the manual holder (b).

SPME successfully combines the advantages of both static and dynamic headspace (the common 'solvent free' extraction/preconcentration techniques).

The major improvement is the very efficient preconcentration capability while using for this purpose a selective sorptive material deposited on a miniaturized fused silica rod, thus eliminating the need of intricate and expensive instrumentation (as the case of PTI).

3.3.3.4 Preconcentration capability of SPME

The coating volume (V_c) of a SPME fiber is calculated as $V_c = \pi L(f^2 + af)$, where L = fiber coating length; f = the coating thickness; a = diameter of the supporting rod. For instance, taken as an example a fiber with $f = 75\,\mu m$, $a = 0.011$ cm and $L = 1$ cm, the volume of the coating phase and hence of the preconcentrated analytes is $V_c = V_{final} = 0.44\,\mu L$. Using an initial 10 mL volume of solution and assuming an exhaustive extraction (the amount of analyte remained non-extracted at equilibrium is negligible), the preconcentration factor is:

$$PF = \frac{V_{initial}}{V_{initial}} = \frac{10\,\text{ml} = 10^4\,\mu L}{0.44\,\mu L} \approx 10^4$$

A preconcentrated factor of 4 orders of magnitude demonstrates the usefulness of SPME in trace speciation analysis, where, besides analytes extraction from the matrix (thus eliminating the matrix effects) the analytes are highly preconcentrated before detection hence contributing to considerable improvement in terms of sensitivity and limits of detection.

4. SEPARATION

After sample preparation, speciation analysis requires that the species of the same or different elements be separated as cleanly as possible before introduction to the atomic or molecular selective detector. Chromatography is the most widely applied separation method for elemental speciation analysis. The separation is achieved by distributing the components of a mixture between two phases, a stationary phase and a mobile phase. Those components held preferentially in the stationary phase are retained longer in the system than those that are distributed selectively in the mobile phase. As a consequence, solutes are eluted from the system as local concentrations in the mobile phase in the order of their increasing distribution coefficients with respect to the stationary phase; *ipso facto* a separation is achieved [35]. Hence, chromatography can separate in a single step process a mixture into its individual components and simultaneously provide a quantitative estimate of each constituent.

Although alternative separation techniques such as *supercritical fluid chromatography* and *capillary electrophoresis* have also been used for speciation studies in the last years, gas (GC) and liquid chromatography (LC) are still the most widely used for speciation analysis. Hence, only these techniques will be referred to in this work. More attention will be devoted to GC given its somehow simple theoretical aspects; on the other hand, the applications of liquid chromatography (LC) and the retention mechanism to be chosen is very much related to both the analytes and the chemical nature of the matrix, a comprehensive description of this technique is beyond the scope of this work. Instead, the main advantages and disadvantages in comparison with GC will be addressed. More details both in terms of GC and LC can be found in a recent published book on chromatography [36].

4.1 Gas chromatography

A GC separation is carried out by introducing the sample (eventually after derivatization and/or extraction/preconcentration) into a fused silica column using a gaseous transport medium (mobile phase). Separation can be carried out using either packed or capillary columns but developments over the last decades led to almost exclusive use of capillary GC (CGC) for speciation studies. The high resolution of CGC is essential for the speciation analysis of complex samples [37]. Besides high resolution and sensitivity, a major advantage of CGC in hyphenated techniques/speciation analysis is the quantitative transfer of the separated analytes into the detector, which has a major benefic impact on the detection limits but also on the method robustness.

Multicapillary versus capillary gas chromatography

Besides efficient separation, the chromatographic run time should be kept as short as possible in order to achieve high sample throughput. In many cases, complex mixtures are resolved using temperature-programmed separation and such an approach is intrinsically disadvantageous because of the relatively long chromatographic time. Fast and efficient chromatography is thus desirable for speciation studies, which inherently is a multi-step procedure, as discussed in a previous section. In practice there are several approaches to achieve fast separation by GC and a detailed description of the parameters to be optimized for this purpose can be found in two recent reviews [38, 39]. A different approach introduced several years ago to achieve fast and efficient GC separation was multicapillary (MC) gas chromatography (MCGC).

A MCGC column consists of a bundle of a very large number ($\sim$900-1000) of very thin capillaries ($\sim$40 μm internal diameter) and it was successfully applied to speciation analysis of organometallic compounds at trace and ultra-trace levels [40-45]. The advantages of such a column compared to conventional CGC are described briefly below.

It is straightforward that a decrease in the separation time by CGC can be achieved simply by shortening the column (generally CGC columns of 25-30 m are used). However, separation efficiency drops significantly when doing so unless this is compensated by a significant decrease of the column inner diameter (id). This leads, on the other hand, to reduced column capacity and allowable sample load. The most feasible way to overcome these limitations is to increase the cross-section of the carrier gas flow-rate by assembling a large number ($\sim$900) of very thin capillaries with very small id ($\sim$40 μm). In this way, decreasing the column length ensures a shorter chromatographic run time, whereas the separation efficiency is retained by decreasing the individual column id. Simultaneously,

Figure 5. Design and internal structure of the MCGC column (from ref. [46], with permission).

the increase in the number of individual capillaries compensates the reduced sample load. The result is the multicapillary GC column, whose design and internal structure can be seen in Fig. 5. [46].

4.2 Liquid chromatography

The most important advantage of LC separation is that it eliminates the derivatization step, as the separation of non-volatile compounds is possible. Besides the elimination of a potential source of uncertainty in the final result, the elimination of the derivatization step may decrease the analysis time and the risk of contamination.

The main drawback of LC separation is the significantly lower resolution compared to GC. In addition, the introduction of a liquid phase into an atomic spectrometric detector, especially inductively coupled plasma-mass spectrometry (ICP-MS) requires nebulization with an efficiency generally below 10%. This may lead to loss in sensitivity of about one order of magnitude in comparison with GC. Due to the presence of a liquid mobile phase the physical interferences related to sample transportation as well as side effects caused by the introduction of organic eluents into the detector (e.g. plasma fluctuations and/or quenching etc) are considerable more elevated than with GC. One way to remove the liquid LC eluent consists of post-column derivatization, generally using hydride generation [47-49] or ethylation [50] with consequent separation of the gaseous analytes from the bulk of the mobile phase using gas-liquid separators. This approach introduces an additional step in the analysis, hence canceling the main advantage offered by LC compared to GC.

5. HYPHENATION

To be amenable for detection, the chromatographically separated analytes should be transported from the chromatograph to the detector and hence a dedicated connection of the separation device and the spectrometer is necessary. The part of a system that ensures the connection (hyphenates) between the two instruments is called the *transfer line* (TL). In a hyphenated system a TL should correspond to the following characteristics, such as: quantitative transportation of the analytes to the spectrometer; preservation of the species integrity; minimum temporal broadening of the chromatographic zone; adaptation of the magnitude of the flow (liquid or vapor) from the separation device to the physical environment of the spectrometer [51]; is simple, robust, flexible and easy to couple/decouple from both the separation and detection device.

In case of LC separation the TL is simple, usually consisting of Teflon of polypropylene tubing, given that the analytes in the liquid mobile phase do not condense during their transportation to the

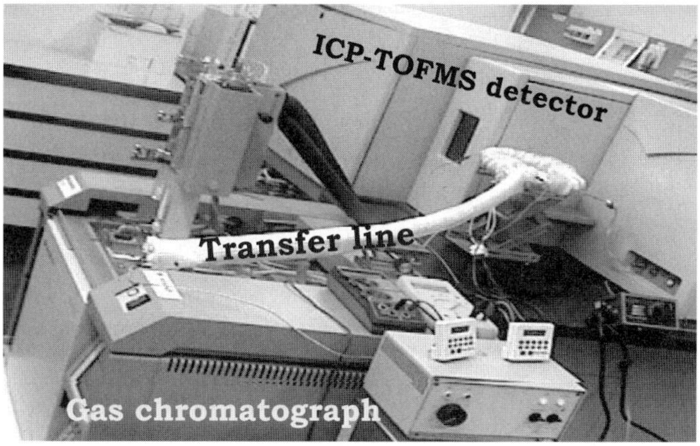

Figure 6. Picture of coupling the GC with ICP-TOFMS instrument [26].

detector. On the other hand, the hyphenation of a GC instrument to a detection device is considerably more intricate. This is the consequence of the gaseous nature of the analytes, which generally requires heating the TL to avoid the presence of cold spots and thus preventing condensation of the less volatile analytes on the walls of the TL. In addition, given that in most cases atomic spectrometers are designed for analysis of solutions, a modification of the original design is necessary to allow the introduction of gases from the outlet of the GC column.

Due to the increased interest in ICP-MS for detection (described more in detail in section 6) in elemental speciation analysis, most papers published over the last years focused on GC-ICP-MS coupling and various designs of TLs were described [52–58]. Because in most cases speciation analysis has not become yet a routine activity, the coupling is generally made 'in house' and the set-up of the TL is dependent on the design of both the GC and the ICP-MS instrument. A picture of such 'in house' hyphenation of a GC and an ICP-time-of-flight-MS developed by the group of Adams *et al.* [52] is shown in Fig. 6 [26].

The core part of such a TL consists of a deactivated fused silica capillary. This silica tubing is housed in a stainless steel tube, which allows, on one hand, protection of the fused silica capillary against breakage and, on the other hand, the actual hyphenation of the GC with the ICP torch of the mass spectrometer. Electrical wire or tape is used for heating up the TL and a thermocouple is attached to the metal tubing (housing the fused silica capillary) for monitoring its temperature.

An important aspect is also the connection of the TL with the plasma torch. This can be achieved by means of an *in-house* made adapter consisting of a PTFE cylinder with aluminum caps on both sides (see Fig. 7) [26, 52].

The metal tube of the TL is passed through the adapter inside the central channel of the torch, as shown in Fig. 7. The fused silica capillary, through which the analytes are transported to the ICP-MS is passed from side to side of the TL metal tubing and inserted into the plasma torch by means of the adapter.

In most cases coupling a GC instrument with an ICP-MS is performed by removing the spray chamber and the nebulizer of the spectrometer, thus the optimization of the mass spectrometer with standard solutions becomes impossible. In such circumstances the optimization with Xe, as a constituent of the carrier gas is a feasible approach [59, 60]. Some authors also developed interfaces allowing either nebulization of solutions and GC separation concomitantly, which eliminate the major inconvenience regarding the decoupling the TL when the mass spectrometer is used in routine mode for analysis of solutions [53].

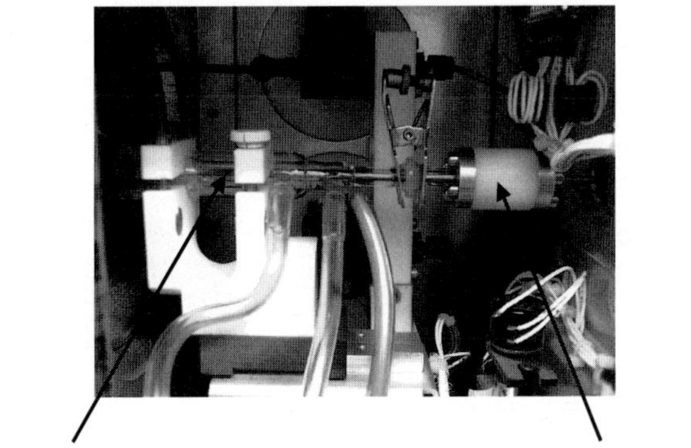

Figure 7. Connection of the TL with the ICP torch [26].

6. DETECTION

Because the analytes are separated before detection, there is no need for molecule-selective detectors. However, conventional GC detectors, such as electron capture, thermal conductivity, photoionisation or infrared cannot be used, however, for speciation studies, as they are unselective. Selective measurements are based on the use of atomic spectrometric detectors, such as atomic absorption spectrometry (AAS), atomic emission spectrometry (AES) with excitation by inductively coupled plasma (ICP-AES) or microwave-induced plasma (MIP-AES), atomic fluorescence (AFS), commonly in combination with cold vapor generation (CV-AFS), and mass spectrometry (MS), mostly using inductively coupled plasma (ICP) as an ionization source (ICP-MS). Among these detectors, the ICP-MS has gained the greatest popularity for elemental speciation analysis, and this technique will be briefly described below.

6.1 Inductively coupled plasma-mass spectrometry

ICP-MS was developed by the groups of Gray and Houk [61] in the early 1980s. Earlier the ICP was already well established as a powerful excitation source in ICP-AES. The use of ICP-MS for speciation studies shows several advantages in comparison with ICP-AES mainly due to the improved detection limits, the multi-element capability and the possibility to obtain isotopic information, including isotopic ratios. However, it appears that among the ICP-MS detectors, time-of-flight (TOF) instruments (ICP-TOFMS) are considered the most suitable for the analysis of transient signals, in particular fast chromatographic signals. As a consequence, although ICP-TOFMS is one of the latest ICP-MS technologies launched on the analytical instrumentation market, it has received considerable attention for speciation studies [62-64] due its particular advantages for measuring transient signals, as described in the next section. More informaton regarding the theory of ICP-MS and particularly ICP-(TOF)MS can be found elsewhere[65-67].

6.2 ICP-TOFMS *versus* sequential mass analyzers (quadrupole or magnetic sector field)

The combination of MS with a chromatographic separation involves 2 axes of information with significantly distinct time frames [68]. As illustrated in Fig. 8, for each chromatographic point recorded, an entire mass spectrum must be acquired for a full deconvolution of the separation process. With the current developments in GC separation, especially using MCGC providing peaks with full width at half maximum (FWHM) down to ~0.1 s, the requirement of a very fast detection is evident. Although ICP-MS based on quadrupole mass analyzer (ICP-QMS) is well accepted for routine analyses, some

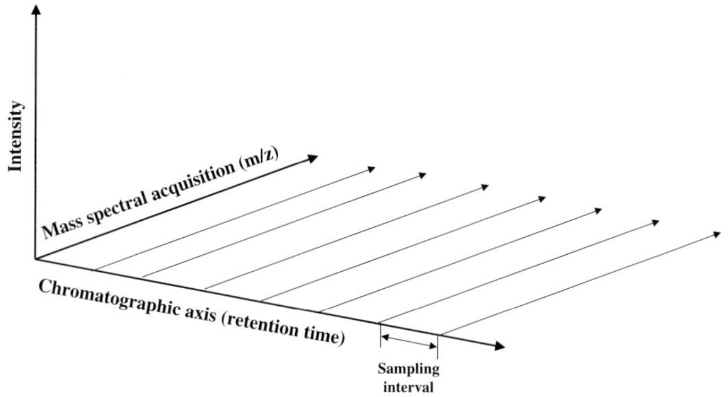

Figure 8. Three-dimensional nature of GC-MS data. (adaptation from ref. [68])

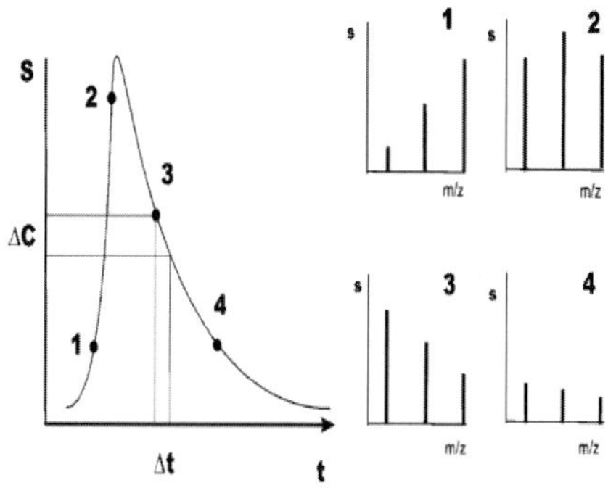

Figure 9. Representation of the spectral skew principle (from ref. [70], with permission): transient profile (left) depicting signal (S) as a function of time (t) undergoes a change in concentration (Δc) during the time (Δt) required to complete one scanned spectral acquisition. The result (right) is a skewing of apparent intensity of three equally abundant m/z dependent upon the point (1–4) at which they were observed during the transient.

intrinsic limitations are still remaining when applied to monitoring fast chromatographic signals, as described below.

In ICP-QMS (the most representative of scanning mass analyzers) only a single m/z value can be measured at any time. This leads to a compromise between the mass range that can be scanned in a given period of time and the limit of detection or precision [69]. Such limitation is critical when fast transient signals should be monitored, in other words, when the analyte's ions are present only for a very brief period of time in the MS. In addition, the total time of a measurement using ICP-QMS is directly proportional to the number of isotopes determined within one run. As a result of these (sequential) features of ICP-QMS, during detection of a fast chromatographic signal, small changes in time during the course of a spectral scan can lead to significant changes in the recorded intensity and consequently in measured concentration, as illustrated in Fig. 9. This effect is known as *spectral skew* and more details about it can be found elsewhere [70].

As shown in Fig. 9, for a given overall sensitivity (e.g. total peak area) the narrower the peak, the larger is the difference in the intensity during a scan window event. With other words, the ideal detector must be able of sampling the narrowest peak achievable without significant peak distortion. This is fully accomplished by ICP-TOFMS, where all ions are simultaneously extracted from the plasma and hence multielemental and multi-isotope analysis of transient signals is in principle achieved without spectral

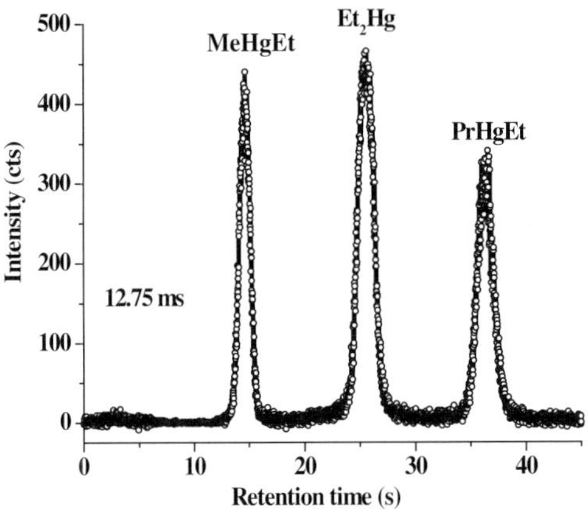

Figure 10. Chromatogram obtained for three ethylated Hg species using 12.75 ms integration time (LECO *Renaissance* ICP-TOFMS) [26].

skew. In addition, unlike ICP-QMS, that can produce at best 10 complete mass spectra per second [71], ICP-TOFMS can generates more than 20,000 complete mass spectra every second.

Due to the combination of simultaneous character and the very high data acquisition rate, multi-elemental quasi-simultaneous analysis of fast transient signals can be carried out by ICP-TOFMS without sacrificing precision or sensitivity regardless of how many ions are monitored or how short the transient signal is. Nevertheless, with the data acquisition hardware nowadays available it is difficult to collect the high amount of data provided by the ICP-TOFMS. As a consequence, in practice, data collection is carried out by the summing of spectra over a given period of time. For instance, using a LECO Renaissance ICP-TOFMS instrument, the minimum integration time achievable is 12.75 ms; this corresponds to a maximum sampling rate of 78 Hz ($1.0/0.01275 \cong 78$) and each recorded point corresponds to 256 integrated mass spectra ($20{,}000/78 \cong 256$).

A peak profile obtained for three ethylated mercury species (methylethylmercury, MeHgEt; diethylmercury, Et_2Hg and propylethylmercury, PrHgEt) using detection by ICP-TOFMS at the minimum available integration time available (12.75 ms) by an ICP-TOFMS (LECO Renaissance) is shown in Fig. 10 [26]. For a peak width of $\sim$1 s at the base peak definition is excellent. In the particular case of a chromatographic peak with a width of $\sim$0.1 s at the base (somehow the 'ideal' case in chromatography), 8 points still can be recorded, which provide an adequate measurement of the peak profile.

7. CHALLENGES OF ENVIRONMENTAL SPECIATION ANALYSIS

As mentioned earlier, the *chemical species* defined by IUPAC [1] are those regarding: *(i) isotopic composition; (ii) electronic and oxidation states; (iii) inorganic compounds and complexes; (iv) organic complexes; (v) organometallic compounds and (vi) macromolecular compounds and complexes.* Among the categories of species listed above, in the last 2 decades the organometallic compounds have been extensively studied in terms of their presence in the environment and the impact on the humans. Most common metals/metalloids such as Hg, Sn, Pb and As are able to form organometallic species. Among these, most attention has been paid to the organometallic species of Hg, Sn and Pb, as they have been released in significant amounts in the environment *via* their industrial/commercial use. It is well known that tetraalkyllead compounds (TAL) were used worldwide as anti-knocking agents in the leaded gasoline. Given its enormous consumption, this was in the past by far the most important source of lead in the atmosphere [52, 72]. However, once the use of the leaded fuel was banned in most countries, the release of organolead compounds in the environment decreased significantly, which also led to a lower interest in their (speciation) monitoring.

Similar considerations can be formulated for organotin compounds. They were widely used as biocides and for wood preservation (triorganotins) and as industrial stabilizors and catalyzers (e.g. mono- and di- organotins) for various products and processes. Most significant from an environmental viewpoint was the use of tributyltin (TBT) as an anti-fouling paint for the boats. This led to 'transportation' of TBT over long distances and contributed to a significant impact on the environment. As a result, at least compared to organolead speciation, the development of analytical methodologies for butyltin determination is still receiving consideration nowadays.

Among the heavy metals, and particularly in comparison with tin and lead, mercury is one of the most studied environmental pollutants. This is mainly the consequence of the existence of considerable natural inputs of mercury to the environment [73], but also of its high mobility, via the biomethylation of inorganic mercury to methylmercury (CH_3Hg^+) and dimethylmercury ($(CH_3)_2Hg$) (organometallic species), which seems to lead to an endless cycle of mercury in the environment. In addition, mercury has received great consideration also due to its extreme toxicity, particularly in the methylated form and the fact that it can be transported thousands of kilometers through the atmosphere, which makes it a global pollutant. Details regarding the toxicity, sources and biogeochemical cycle of mercury were given elsewhere [73].

For speciation analysis of mercury in water-based environmental compartments, particularly ice/snow, CH_3Hg^+ and inorganic mercury (Hg^{2+}) are the species of interest. Other chemical species such as $(CH_3)_2Hg$ and elemental mercury (Hg^0) are very problematical for analysis by common speciation methods. Although the latter species may be present in the aqueous based environmental compartments at the moment of sampling, they volatilize easily during sampling, sample transport and storage because of their very low water solubility [74].

Given the previous considerations, from an environmental point of view, the focus is still placed on speciation analysis of mercury (CH_3Hg^+ and Hg^{2+}). An analytical challenge seems to be the analysis of mercury species in snow and ice from remote environments, such as high mountain altitudes and polar areas. The assessment of individual mercury species in such environments is important for understanding the biogeochemical cycle of mercury, as described briefly in the next sections.

7.1 Speciation analysis of mercury in snow/ice archives from remote areas

7.1.1 Assessment of regional mercury pollution by means of high altitude snow/ice archives

Little attention has been paid to mercury depositions in the successive snow/ice layers from high altitude (e.g. Alpine) glaciers. Such sites are characterized by low water percolation from one layer to the other during the summer months and hence are useful archives for assessment of pollution by heavy metals. As they are close to populated and industrialized areas they have considerable potential to provide insights into historical change in atmospheric heavy metals on a regional scale in Europe [72]. Although there are several studies focused on alkyllead [52, 72] and metallic contaminants such as Li, B, Al, Ti, V, Cr, Mn, Fe, Co, Cu, Zn, Cd, Bi, Sn, Mo, Ag, Sb, Ba, Au, Rh, Pd, Pt, Pb and U [72, 75-87], reports on the determination of mercury in such areas are very scarce [88]. Apart from the challenge in analytical process, the main problem associated with mercury determinations at ultra-trace levels (below pg g^{-1} level) in snow/ice and particularly speciation analysis is sample contamination [89] and lack of stability of mercury species solutions during long-term storage [90–93]. Studies focused on cleaning of the ice cores as well as a description of the clean lab facilities required for determination of mercury in ice and snow samples can be found elsewhere [18, 89, 94, 95].

A preliminary effort to speciate mercury in Alpine snow/ice over the last century was reported by Jitaru *et al.* [26, 42]. These data provide the first continuous record of changes in mercury species concentration (CH_3Hg^+ and Hg^{2+}) in Europe during the past century. The concentration of both

CH_3Hg^+ and Hg^{2+} revealed a maximum at the year 1965, and this profile is similar with that obtained for organolead species analyzed previously in the same laboratory in ice snow/samples [75]. Whereas the high level of organolead compounds measured around the year 1965 can be explained by the opening of the Mont Blanc road tunnel in July 1965 (thus leading to a considerably higher consumption of leaded gasoline and consequently higher emission of organolead species), the origin of mercury species, in particular CH_3Hg^+, in this investigated site is not completely understood. To explain the presence of CH_3Hg^+ in this area and its rough correlation with the organolead profile in the same samples [75], it was hypothesized [26] that transalkylation processes from organolead species to Hg^{2+} may have occurred in the snow/ice. Such a transalkylation process from organolead compounds to Hg^{2+} was reported by Hempel *et al.* [96], in a study focused on transethylation occurring in soil contaminated with alkyllead compounds and Hg^{2+}. It appears that organolead compounds and particularly trimethyllead (TML) can be a source of methyl for mercury transmethylation. However, the mechanism of such a process is not clear yet. In the case of soil/sediments it could be at least partially microbial but the kinetics in snow/ice may be significantly lower than in natural soil conditions. However, the old age of the snow/ice samples might be sufficient to lead to transalkylation. More studies on transalkylation of Hg^{2+} in snow/ice are necessary in order to assess the impact of such processes on the fate of Hg^{2+} in the environment.

It is interesting that Lindberg *et al.* [97] measured a so-called 'bioavailable' Hg^{2+} species in Arctic snow. Bioavailable Hg^{2+} is the fraction of Hg^{2+} that is available to genetically modified bacteria. The study showed that the fraction of bioavailable Hg^{2+} was 8% in fresh snow and increased to 55% in a slushy snow. The assessment of such bioavailable Hg^{2+} may have important implications for understanding the fate of Hg^{2+} in snow/ice. It appears that the bioavailable Hg^{2+} is more prone to participate in methylation processes caused by the microorganisms in the snow/ice [97]. This observation may help to understand the elevated levels of CH_3Hg^+ in certain snow/ice sites but further investigations must be conducted to prove such an assumption.

7.1.2 Assessment of the mercury fate in Polar areas

In the global change discussion it is also important to understand the biogeochemical cycle of mercury and in particular its transport and deposition in the polar environment [98]. Toxic mercury species tend to accumulate in the polar areas hence the Arctic and Antarctic ice sheets are valuable archives for assessing ancient mercury deposition [99].

A rather large number of studies [100-124] have focused in the last years on the determination of various metallic pollutants (Al, Sc, Sr, V, Cr, Mn, Fe, Co, Cu, Ag, Ba, Cd, Zn, Pb, Bi, Ir, Pt, Pd, Rh and U) in snow and/or ice from Antarctica and Greenland but reliable information concerning the mercury levels in polar snow and ice and especially the speciation of mercury is still lacking. Apart from the difficulties mentioned earlier (contamination, stability during storage etc.) this is mainly because of difficulties in the analytical process. The concentration of Hg^{2+} reported earlier in Antarctic ice [125] and snow [126] is generally below 2.5 pg g^{-1} and 0.5 pg g^{-1}, respectively and the fraction of CH_3Hg^+ in polar snow [43, 44] is generally below 0.1 pg g^{-1}. Highly sensitive analytical methods are thus required for the quantification of mercury species in Polar Regions.

It was rather recently discovered that at the beginning of the polar spring, thus with the onset of the solar activity, a significant and sudden loss in atmospheric Hg^0 occurs. This unusual phenomenon is called 'Mercury Depletion Event' (MDE) and it was first noted in the Arctic [127] and later also in the Antarctic [128]. It appears that during a MDE, atmospheric Hg^0 and ozone concentrations considerably drop together with an increase of concentration of reactive gaseous mercury (RGM) and also total particulate mercury (TPM) in the atmosphere. The snow pack seems to be an environmental sink of mercury, which accumulates in the snow during the spring and could then be a source of mercury to the aquatic reservoirs during snowmelt. Because the snowmelt occurs during a rather short period of time,

MDE can be the cause of a large pulse of mercury to the environment. More detail on MDEs can be found elsewhere [129].

Studies regarding the fate of mercury species in the seasonal snow pack in Greenland, before and after the sunrise in 2002 were conducted by the group of Ferrari *et al.* [43]. For snow sampled just after polar sunrise, Hg^{2+} concentrations increased with depth from extremely low levels (close to method detection limit, MDL) for snow on the surface to $\sim$ 4 pg g^{-1} for bottom snow. For snow sampled after a snowmelt event, the Hg^{2+} concentration variation with depth following the same pattern as observed shortly after polar sunrise, but with lower concentrations measured in the deeper layers. For CH_3Hg^+, the concentrations after sunrise were 5 times lower than observed before melt event except for the upper layer of the snow pack, where CH_3Hg^+ concentrations were fairly similar.

With the results of mercury speciation in the snow and the data on the interstitial Hg^0, Dommergue *et al.* [130] estimated that 1 m^3 of snow (density $\sim$0.25 g cm^{-3}) contained $\sim$80 pg of interstitial Hg^0, $\sim$(8-50) μg of CH_3Hg^+ and $\sim$(150-1550) μg of Hg^{2+}. Therefore, mercury in the snow pack consisted mainly of Hg(II) ($\sim$98%) with a contribution of CH_3Hg^+ of $\sim$2% while interstitial Hg^0 represented less than $\sim$0.01%. A detailed description of the possible physical and chemical phenomena occurring in the snow pack based on the present speciation data and the measurement of Hg^0 in the interstitial air of the snow pack is presented elsewhere [130].

One of the key questions in terms of MDEs is related to the fate of mercury species after sunrise, in other words, it is not clear whether mercury is reemitted to the atmosphere or incorporated into the melt water. Studying the fate of mercury species (CH_3Hg^+ and Hg^{2+}) in the sub-Arctic snow-pack after the polar sunrise Dommergue *et al.* [130] demonstrated that the first day of snowmelt coincides with a removal of Hg^{2+} from surface snow. It appears that the beginning of annual snowmelt initiates a massive incursion of mercury in the water system, leading to a potential contamination of Arctic and sub-Arctic ecosystems. Therefore, a better understanding of the role of the snow pack on the mercury cycle in the Polar Regions is of particular interest for understanding its environmental impact on both short time and longer time scales, such as in the case of global warming, because these regions could be strongly affected by the toxic mercury species.

The examples shown above demonstrated that new analytical methodologies for fast and highly accurate measurement of environmental species at ultra-trace levels (<pg g^{-1}) must be developed and validated. This can be achieved by putting effort into the development of all the steps of the trace speciation analysis (sampling, sample preparation, separation and detection) as discussed in the previous paragraphs.

8. CHALLENGE OF ELEMENTAL SPECIATION ANALYSIS IN BIOMOLECULES

The current trend in speciation analysis is the orientation towards bioinorganic applications, particularly elemental speciation analysis of metal/metalloid biomolecules, with emphasis on protein-heteroatom compounds. The established element-speciation approaches developed for the determination of low molecular mass metal(loid) species (e.g. organometallic compounds) in environmental, food, toxicological and health sciences are being adapted to determination of high molecular mass metal-species (e.g. metalloproteins), generally related to biological processes [131]. The evolution of the elemental speciation concept is shown schematically in Fig. 11.

One of the newest approaches in terms of elemental speciation analysis in biological systems is *metallomics,* which was first introduced by Haraguchi [132]. It has been demonstrated that the function of many proteins, referred to as *metalloproteins*, critically depends on their interaction with a metal, usually a transition one, such as e.g. Cu, Fe, Zn or Mo [133]. In this context, the chemistry of a cell needs to be characterized not only by its protein content but also by the distribution of the metals/metalloids among the different cell compartments and the species contained within them [134]. The *metallomics* concept refers to determination of the individual species of trace metallic/metalloids with endogenous or bio-induced biomolecules such as organic acids, proteins, sugars or DNA fragments [133]. The

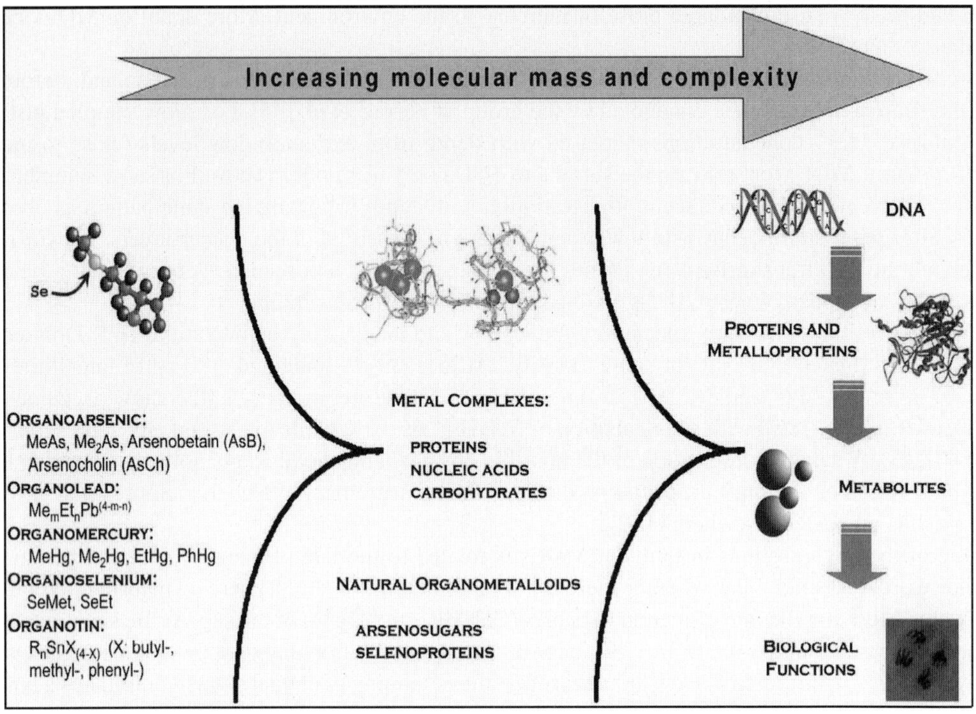

Figure 11. Evolution of metals speciation concept (from ref. [131], with permission).

metallomics concept has been introduced as an analogy to *proteomics*, defined as the study of the proteome of an organism (the entire protein complement of a given genome, that is, the entirety of the proteins that are expressed by the genome) [133]. Although it is relatively a new scientific field, several excellent reviews on proteomics/metallomics have been published so far [133, 135-145], including several monographs [146-148].

The development of analytical approaches to study the bioavailability and the human health impact of metals, i.e. Hg and non-metals such as Se on living organisms offers a continuing challenge for analytical chemistry, eco- and clinical toxicology and related disciplines. To elucidate the mechanisms that control the uptake of such species by organisms exposed to metals/non-metals as well as metal transport, metabolism and detoxification processes, reliable information is needed about trace-element speciation. New analytical methodologies for detection, identification and characterization of metal(loids) and/or non-metals complexed with or incorporated into biomolecules are therefore required.

The analytical methodologies employed for metallomics are based on the use of hyphenated techniques, namely a combination of a chromatographic (or electrophoresis) separation and an atomic or molecular selective detector. In contrast to conventional trace elemental analysis, where optical detectors, such as (cold vapor) AAS, ICP-AES or AFS are still widely used, the shift of the speciation approach towards species with increased molecular mass and complexity demands almost exclusively the use of mass spectrometric techniques for detection; hence, it appears that the *photonic* analytical tool is being substituted by the *ionic paradigm* [131].

A brief overview of the main particularities of each step mentioned previously for speciation analysis of metal/metalloid-high molecular mass species (e.g. with protein content), including some current issues of interest, will be addressed in the next sections.

8.1 Sample preparation in metallomics

In metallomics a rather laborious sample preparation step is commonly required, given the complexity of the biological samples, either of vegetable, animal or human origin. The procedure is strongly dependent on the physical form of the analyte and the matrix and is also linked to the species expected in the sample. Therefore, whereas general guidelines can be given for chromatographic separation and the detection steps, it is rather difficult to summarize within the purpose of this work the issues related to sample preparation in metallomics. The challenge is to extract the species of interest with the highest recovery, and to separate them from the bulk matrix without changing the original species form [136]. In addition, the myriad of metalloprotein species present in biological systems represents a challenge to the analyst, especially in terms of sample preparation. Also, the type of the biological sample plays an important role in the choice of sample preparation; for instance, there can be very different approaches in terms of sample preparation of selenium species in plant or mammalian tissue, or body fluids.

The extraction of metal-proteins species from biological matrices is commonly carried out by homogenization, centrifugation, precipitation, solvent extraction, LLE, solid-phase extraction (SPE), (micro)-dialysis and ultrafiltration. Another interesting approach is the enzymatic digestion of the metal- or heteroatom-protein macromolecules followed by quantification on the basis of the resulting individual amino acids [136, 149].

Using a sample preparation technique or a combination of several sample-preparation techniques is sometimes insufficient to selectively extract the target analyte from a biological matrix [150]. As a result, in such cases the complete sample preparation is actually achieved by means of employing a further step, namely a (liquid) chromatographic separation. More details regarding bioinorganic speciation analysis by hyphenated techniques, with emphasis also on sample preparation of various metal species in different types of biological materials can be found in several recent reviews [142,144,149-151].

8.2 Species separation

The separation technique is the key in the speciation analysis of bioinorganic molecules. Given the low-volatility of the species of interest in metallomics research, selectivity as well as high sensitivity (allowing the speciation of ultra-trace element concentrations) is commonly achieved by a powerful separation technique such as high performance liquid chromatography (HPLC). Gas chromatography in bioinorganic speciation analysis has a limited area of application but can be of choice in some cases, such as, e.g. analysis of sulfur- and selenium-containing amino acids (after derivatization), or of volatile sulfur and selenium species [133].

Given the complexity of the bioinorganic species in a natural tissue, it is common that at least two (liquid based) chromatographic approaches are used, thus providing a so-called *multi-dimensional hyphenated technique*. Generally, as a first step, size-exclusion chromatography (SEC) is used before a finer separation by reversed-phase (RP), anion exchange (AE) or affinity (AC) liquid chromatography is carried out. More details regarding the principle and the applications of these chromatographic techniques can be found elsewhere [36].

8.3 Detection

The main peculiarity of bioinorganic speciation analysis compared to conventional (low molecular mass compounds) studies is that, besides the requisite in terms of identification of the inorganic moiety, in the case of bio-molecules, the characterization of the organic fragment (e.g. protein content) is required. Although the advantages offered by ICP-MS to speciation analysis are crucial for species quantification at trace and ultra-trace levels, its elemental selectivity 'destroys' the chemical structure of the original species. Hence, the limits of ICP-MS for bioinorganic studies are its inability to elucidate the structure or identify the metallo-biomolecules for which retention/migration time standards are not available.

Therefore, it is necessary to employ in parallel a molecular (or moiety)-selective detector to establish the identity of the eluted species.

Recent reports have indicated exciting potential opportunities offered by electrospray ionization-mass spectrometry (ESI-MS) for the determination of molecular weight and the structural characterization of molecules at trace levels in fairly complex matrices [152–154]. ESI is an ionization technique in MS and it is especially useful in producing ions from macromolecules because it overcomes the tendency of these molecules to fragment when ionized [155]. This technique provides information that can unambiguously identify biochemical species [154].

A competitive technique to ESI-MS is Matrix-Assisted Laser Desorption/Ionization-Mass Spectrometry (MALDI-MS). This is a soft ionization technique used in mass spectrometry, allowing the ionization of biomolecules (biopolymers such as proteins, peptides and sugars), which tend to be more fragile and quickly lose structure when ionized by conventional ionization methods. It is most similar in character to electrospray ionization both in relative softness and ions produced [156]. However, this technique is less vulnerable to matrix effects than ESI-MS, and produces mostly singly charged ions with little fragmentation, which makes it an attractive technique to detect species that would have remained undetected with ESI-MS [135]. As a consequence, it appears that ICP-MS, ESI-MS and MALDI-MS are truly complementary techniques for the acquisition of information on the metal and metalloid-protein species in biological matrices [135].

8.4 Species quantification in metallomics

The challenges in metallomics are the quantification of the target analytes as well the validation of newly developed analytical methods. This is so because, in contrast to the classical speciation analysis, where standards for most of the anthropogenic pollutants are available, the majority of species of interest in bioinorganic trace analysis have not yet been isolated in sufficient purity to be used as (external) standards. In this context, it appears that the unique feasible approach of species quantification in metallomics is the use of on-line isotope dilution-mass spectrometry (IDMS) using a species-unspecific spike, as first described by Heumann and Rottman [157]. This approach is applied when the structure of the compound is unknown, or difficult to determine, and standards are not available. Here the (chromatographic) separation is carried out first and the eluted species are spiked post column with an isotopically enriched material (known as the 'spike') analogue of the analyte, which in this particular case is commonly in an inorganic state. An enriched material has the relative abundances different from the natural values, which means that an isotope is 'artificially' enriched compared with its natural state. For instance, in the case of Cu (natural abundances $^{63}Cu = 68.17\%$ and $^{65}Cu = 30.83\%$), ^{65}Cu can be enriched up to a value above 99%, with the difference up to 100% being of course attributed to ^{63}Cu. Provided that the enriched isotope is present in an equilibrated and equivalent state to the natural isotope, such an enriched isotope can perform the role of the ideal internal standard [158]. Details regarding the concentration calculation using species-unspecific IDMS and the particularities related to the on-line addition of the spike can be found elsewhere [159, 160].

IDMS is considered one of the most powerful and accurate methods for determining chemical amounts, on the basis of only the measured isotope ratios of the elements (when applied in a conventional way, IDMS is a primary method of analysis). It has several advantages over conventional calibration methodologies. Partial loss of the analyte after equilibration of the spike (enriched isotopic species) and the sample will not influence the accuracy of the determination. Furthermore, even though the complete separation of the species is desired, SIDMS permits incomplete separation. Thus, the low resolution of the chromatographic separation will not influence the deconvolution of the species concentration. Physical and chemical interferences have a smaller influence on the determination as they have similar effects on each isotope of the same element (given that the isotopic ratios are measured). It is also worth noting that when a species-specific spike is added to the sample (e.g. CH_3Hg^+ species synthesized using an enriched Hg^{2+}precursor), speciated IDMS is the only method capable of assessing species interconversions (in this instance, methylation or demethylation reactions) and to correct mathematically for them. This approach is unfortunately impossible to apply at the current state of development of the metallomics research.

Acknowledgements

Petru Jitaru is very grateful to Prof. Freddy Adams from the University of Antwerp, Belgium for his great collaboration during the Ph.D. research and guidance in the field of speciation analysis. The authors thank Dr. Warren Cairns (Institute for the Dynamics of Environmental Processes, CNR, Venice, Italy) for the useful scientific discussions. This work is a contribution to the Marie-Curie Intra-European project (MEIF-CT-2006-024156/ELSA-BIM) funded by the European Commission.

References

[1] Templeton D.M., Ariese F., Cornelis R., Danielsson L.-G., Muntau H., Van Leeuwen H.P. and Lobinski R., *Trends Anal. Chem.* **72** (2000) 1453-1470.

[2] Cornelis R., Caruso J., Crews H. and Heumann K.G. (Eds.), Handbook of Elemental Speciation, II:Species in the Environment, Food, Medicine and Occupational Health (Wiley, UK, 2005).

[3] Cornelis R., Caruso J., Crews H. and Heumann K.G. (Eds.), Handbook of Elemental Speciation: Techniques and Methodology (Wiley, UK, 2003).

[4] Szpunar J. and Lobinski R., Hyphenated Techniques in Speciation Analysis (Royal Society of Chemistry, UK, 2004).

[5] Weltje L., Bioavailability of Lanthanides to Freshwater Organisms: Speciation, Accumulation & Toxicity (Delft Univ Press, The Netherlands, 2002).

[6] Ebdon L., Pitts L., Cornelis R., Crews H., Donard O.F.X. and Quevauviller Ph. (Eds.), Trace Element Speciation for Environment, Food and Health (Royal Society of Chemistry, UK, 2001).

[7] Ure A.M. and Davidson C.M. (Eds.), Chemical Speciation in the Environment (Blackwell Publishing, USA, 2001).

[8] Buffle J. and Horvai G. (Eds.), In Situ Monitoring of Aquatic Systems: Chemical Analysis and Speciation (John Wiley & Sons, UK, 2000).

[9] Caruso J.A., Sutton K.L. and Ackley K.L. (Eds.), Elemental Speciation (Elsevier, 2000).

[10] Donald T. Reed, Sue B. Clark and Rao Linfeng (Eds.), Actinide Speciation in High Ionic Strength Media. Experimental and Modeling Approaches to Predicting Actinide Speciation and Migration in the Subsurface (Kluwer, 1999).

[11] Quevauviller Ph., Method Performance Studies for Speciation Analysis (Royal Society of Chemistry, UK, 1998).

[12] Caroli S. (Ed.), Element Speciation in Bioinorganic Chemistry (John Wiley & Sons, 1996).

[13] Tessier A. and Turner D.R. (Eds.), Metal Speciation and Bioavailability in Aquatic Systems (John Wiley & Sons, 1995).

[14] International Atomic Energy Agency, Report No. IAEA/AL/147-IAEA/MEL/75, Monaco, July 2004: *World-wide Intercomparison Exercise for the Determination of Trace Elements and Methylmercury in Marine Sediment IAEA-433:* http://www.iaea.org/monaco/files/IAEA433.pdf.

[15] Markert B. (Ed.), Environmental Sampling for Trace Analysis (VCV, Weinheim, Germany, 1994).

[16] http://www.ec.gc.ca/nopp/docs/rpt/2MM5/en/C6.cfm

[17] Li-Ping Y. and Xiu-Ping Y., *Trends Anal. Chem.* **22(4)** (2003) 245-253.

[18] Barbante C., Turetta C., Capodaglio G., Cescon P., Hong S., Candelone J.P., Van de Velde K. and Boutron C.F., Trace element determination in polar snow and ice. An overview of the analytical process and application in environmental and paleoclimatic studies in 'Environmental

Contamination in Antarctica: A Challenge to Analytical Chemistry', edited by Caroli S., Cescon P. and Walton D.W.H. (Elsevier Science Ltd., Oxford, UK, 2001) pp. 193-195.
[19] Adams F.C., *J. Anal. At. Spectrom.* **19(9)** (2004) 1090-1097.
[20] Letellier M. and Budzinski H., *Analusis* **27** (1999) 259-271.
[21] Szpunar J., Schmitt V.O., Donard O.F.X. and Lobinski R., *Trends Anal. Chem.* **15(4)** (1996) 181-187.
[22] Iyer S.S., Development and Optimization of Integrated Microwave-Enhanced Extraction as a Sample Preparation Technique: Environmental, Clinical and Green Chemistry Applications, Ph.D. Thesis, Bayer School of Natural and Environmental Sciences of Duquesne University, 2005.
[23] Snow N.H. and Slack G.C., *Trends Anal. Chem.* **21** (2002) 608-617.
[24] Abeel S.M., Vickers A.K. and Decker D., *J. Chromatogr. Sci.* **32** (1994) 328-338.
[25] Silgoner I., Rosenberg E. and Grasserbauer M., *J. Chromatogr. A* **768** (1997) 259-270.
[26] Jitaru P., Ultra-Trace Speciation Analysis of Mercury in the Environment, Ph.D. Thesis, 2004, Promotor Prof. F. Adams, University of Antwerp, Belgium.
[27] Arthur C.L. and Pawliszyn J., *Anal. Chem.* **62** (1990) 2145-2148.
[28] Pawliszyn J., *Anal. Chem.* **75** (2003) 2543-2558.
[29] Pawliszyn J., Solid Phase Microextraction: Theory and Practice (Wiley, New York, USA, 1997).
[30] Alpendurala M. de Fatima, *J. Chromatogr. A* **889** (2000) 3-14.
[31] Prosen H. and Zupancic-Kralj L., *Trends Anal. Chem.* **18** (1999) 272-281.
[32] Zeng E.Y. and Noblet J.A., *Environ. Sci. Technol.* **36** (2002) 3385-3392.
[33] Zhang Z. and Pawliszyn J., *Anal. Chem.* **65** (1993) 1843-1852.
[34] Gorecki T. and Pawliszyn J., *Analyst* **122** (1997) 1079-1086.
[35] http://www.chromatography-online.org/topics/efficiency.html.
[36] Poole C.F., The Essence of Chromatography (Elsevier, Amsterdam, 2003).
[37] Lobinski R., *Analusis* **22** (1994), 37-48.
[38] Mastovska K. and Lehotay S.J., *J. Chromatogr. A* **1000** (2003) 153-180.
[39] Matisova E. and Domotorova M., *J. Chromatogr. A* **1000** (2003) 199-221.
[40] Jitaru P., Goenaga Infante H. and Adams F.C., *J. Anal. At. Spectrom.* **19** (2004) 867-875.
[41] Lobinski R., Sidelnikov V., Patrushev Y., Rodriguez I. and Wasik A., *Trends Anal. Chem.* **7** (1999) 449-460.
[42] Jitaru P., Goenaga Infante H., Ferrari C.P., Dommergue A., Boutron C.F. and Adams F.C., *J. Phys IV* **107** (2003) 683-686.
[43] Ferrari C.P., Dommergue A., Boutron C.F., Jitaru P. and Adams F., *Geophys. Res. Lett.* **31** (2004) L03401.
[44] Dommergue A., Ferrari C.P., Gauchard P-A., Boutron C.F., Poissant L., Pilote M., Jitaru P. and Adams F., *Geophys. Res. Lett.* **30(12)** (2003) 1621.
[45] Jitaru P., Goenaga Infante H. and Adams F.C., *Anal. Chim. Acta* **489** (2003) 45-57.
[46] Alltech Associates, Inc., *on-line catalogue*, 2001.
[47] Yang H.J. and Jiang S.J., *J. Anal. At. Spectrom.* **10** (1995) 963-968.
[48] Harrington C.F. and Catterick T., *J. Anal. At. Spectrom.* **12** (1997) 1053-1056.
[49] Howard A.G., *J. Anal. At. Spectrom.* **12** (1997) 267-272.
[50] Blais J.S. and Marshall W.D., *J. Anal. At. Spectrom.* **4** (1989) 641-645.
[51] Snell J., Mercury species determination in organic solution by plasma spectrometry, Ph.D. thesis, 2001, Umea University, Sweden.
[52] Heisterkamp M., Towards faster and more reliable methods for the speciation analysis of organolead compounds in environmental samples, Ph.D. thesis, 2000, Promotor Prof. F. Adams, University of Antwerp (UIA), Belgium.
[53] Peters G.R. and Beauchemin D., *Anal. Chem.* **65** (1993) 97-103.

[54] Pritzl G., Stuer-Lauridsen F., Carlsen L., Jensen A.K. and Thorsen T.K., *Intern. J. Environ. Anal. Chem.* **62** (1996) 147-159.
[55] De Smaele T., Verrept P., Moens L. and Dams R., *Spectrochim. Acta Part B* **50** (1995) 1409-1416.
[56] De Smaele T., Moens L., Dams R. and Sandra P., *Fresenius J. Anal. Chem.* **355** (1996) 778-782.
[57] Montes Bayon M., Gutierrez Camblor M., Garcia Alonso J.I. and Sanz-Medel A., *J. Anal. At. Spectrom.* **14** (1999) 1317-1322.
[58] Barnes J.H., Schilling G.D., Sperline R.P., Denton M.B., Young E.T., Barinaga C.J., Koppenaal D.W. and Hieftje G.M., *J. Anal. At. Spectrom.* **19** (2004) 751-756.
[59] Moens L., De Smaele T., Dams R., Van Den Broeck P. and Sandra P., *Anal. Chem.* **69** (1997) 1604-1611.
[60] Ritsema R., De Smaele T., Moens L., De Jong A.S. and Donard O.F.X., *Environ. Pollut.* **99** (1998) 271-277.
[61] Houk R.S., Fassel V.A. and Flesch G.D., *Anal. Chem.* **52** (1980) 2283-2289.
[62] Leach A.M., Heisterkamp M., Adams F.C. and Hieftje G.M., *J. Anal. At. Spectrom.* **15** (2000) 151-155.
[63] Goenaga Infante H., Heisterkamp M., Benkhedda K., Van Campenhout K., Blust R. and Adams F.C., *Spectra Analyse* **220** (2001) 23-31.
[64] Bings N.H., Costa-Fernandez J.M., Guzowski J.P., Leach A.M. and Hieftje G.M., *Spectrochim. Acta. Part B* **55** (2000) 767-778.
[65] Guilhaus M., *Spectrochim. Acta Part B* **55** (2000) 1511-1525.
[66] Ray S.J. and Hieftje G.M., *J. Anal. At. Spectrom.* **16** (2001) 1206-1216.
[67] Cotter R.J. (Ed.), Time-of-Flight Mass Spectrometry (American Chemical Society, Washington DC, USA, 1994).
[68] Holland J.F., Allison J., Watson J.T. and Enke C.G., Achieving the Maximum Characterizing Power for Chromatographic Detection by Mass Spectrometry, in 'Time of-Flight Mass Spectrometry', edited by Cotter R.J. (American Chemical Society, Washington DC, USA, 1994).
[69] Mahoney P.P., Ray S.J. and Hieftje G.M., *Appl. Spectrosc.* **51** (1997) 17A-28A.
[70] Ray S.J., Andrade F., Gamez G., McClenathan D., Rogers D., Schilling G., Wetzel W. and Hieftje G.M., *J. Chromatogr. A* **1050** (2004) 3-34.
[71] Thomas R., A Beginner's Guide to ICP-MS. Part VIII-Mass Analyzers: Time-of-Flight Technology (www.spectroscopyonline.com).
[72] Heisterkamp M. and Adams F. C., *J. Anal. At. Spectrom.* **14** (1999) 1307-1311.
[73] Jitaru P. and Adams F., *J. Phys. IV France* **121** (2004) 185-193.
[74] Puk R. and Weber J.H., *Appl. Organomet. Chem.* **8** (1994) 293-302.
[75] Heisterkamp M., Van de Velde K., Ferrari C., Boutron C.F. and Adams F.C., *Environ. Sci. Technol.* **33** (1999) 4416-4421.
[76] Barbante C., Cozzi G., Capodaglio G., Van de Velde K., Ferrari C., Veysseyre A., Boutron C.F., Scarpoini G. and Cescon P., *Anal. Chem.* **71** (1999) 4125-4133.
[77] Rosman K.J.R., Ly C., Van de Velde K. and Boutron C.F., *Earth Planet. Sci. Lett.* **176** (2000) 413-424.
[78] Barbante C., Cozzi G., Capodaglio G., Van de Velde K., Ferrari C., Boutron C. and Cescon P., *J. Anal. At. Spectrom.* **14** (1999) 1433-1438.
[79] Schwikowski M., *Environ. Sci. Technol.* **38 (4)** (2004) 957-964.
[80] Van de Velde K., Boutron C., Ferrari C., Bellomi T., Barbante C., Rudnev S. and Bolshov M., *Earth Planet. Sci. Lett.* **164** (1998) 521-533.
[81] Barbante C., Van de Velde K., Cozzi G., Capodaglio G., Cescon P., Planchon F., Hong S., Ferrari C. and Boutron C., *Environ. Sci. Technol.* **35** (2001) 4026-4030.

[82] Vallelonga P., Van de Velde K., Candelone J.P., Ly C., Rosman K.J.R., Boutron C.F., Morgan V.I. and Mackey D.J., *Anal. Chim. Acta* **453** (2002) 1-12.
[83] Van de Velde K., Ferrari C., Barbante C., Moret I., Bellomi T., Hong S. and Boutron C., *Environ. Sci. Technol.* **33** (1999) 3495-3501.
[84] Barbante C. *et al., Environ. Sci. Technol.* **38** (2004) 4085-4090.
[85] Veysseyre A., Moutard K., Ferrari C., Van de Velde K., Barbante C., Cozi G., Capodaglio G. and Boutron C., *Atmos. Environ.* **35** (2001) 415-425.
[86] Van de Velde K., Barbante C., Cozzi G., Moret I., Belommi T., Ferrari C. and Boutron C., *Atmos. Environ.* **34** (2000) 3117-3127.
[87] Barbante C., Boutron C., Moreau A.L., Ferrari C., Van de Velde K., Cozzi G., Turreta C. and P. Cescon, *J. Environ. Monit.* **4** (2002) 960-966.
[88] Ferrari C.P., Dommergue A., Veysseyre A., Planchon F. and Boutron C.F., *Sci. Total Environ.*, **287** (2002) 61-69.
[89] Ferrari C.P., Moreau A.L. and Boutron C.F., *Fresenius J. Anal. Chem.* **366** (2000) 433-437.
[90] Ping L. and Yan X.P., *Trends Anal. Chem.* **22** (2003) 245-253.
[91] Fadini P.S. and Jardin W.E., *Analyst* **125** (2000) 549-551.
[92] Lansens P., Meuleman C. and Baeyens W., *Anal. Chim. Acta* **229** (1990) 281-285.
[93] Leermakers M., Lansens P. and Baeyens W., *Fresenius J. Anal. Chem.* **336** (1990) 655-662.
[94] Boutron C., *Fresenius J. Anal. Chem.* **337** (1990) 482-491.
[95] Candelone J.P., Hong S. and Boutron C.F., *Anal. Chim. Acta* **229** (1994) 9-16.
[96] Hempel M., Kuballa J. and Jantzen E., *Fresenius J. Anal. Chem.* **366** (2000) 470-475.
[97] Lindberg S.E., Brooks S., Lin C.J., Scott K.J., Landis M.S., Stevens R.K., Goodsite M. and Richter A., *Environ. Sci. Technol.* **36** (2002) 1245-1256.
[98] Morel F.M.M., Kraepiel A.M.L. and Amyot M., *Annu. Rev. Ecol. Syst.* **29** (1998) 543-566.
[99] EPICA community members, *Nature* **429** (2004) 623-628.
[100] Hong S., Kim Y., Boutron C. F., Ferrari C.P., Petit J.R., Barbante C., Rosman K. and Lipenkov V.Y., *Geophys. Res. Lett.* **30(22)** (2003) 2138, doi:1029/2003GL018411.
[101] Boutron C.F. and Paterson C.C., *Nature* **323** (1986) 222-225.
[102] Wolff *et al., Nature* **440(7083)** (2006) 491-496.
[103] Planchon F.A.M., Boutron C.F., Barbante C., Wolff E.W., Cozzi G., Gaspari V., Ferrari C.P. and Cescon P., *Anal. Chim. Acta.* **450** (2001) 193-205.
[104] Gabrielli P. *et al., Anal. Chem.* **78 (6)** (2006) 1883-1889.
[105] Hong S., Candelone J. P., Patterson C.C. and Boutron C.F., *Science* **265** (1994), 1841-1843.
[106] Ferrari C.P., Hong S., Van de Velde K., Boutron C.F., Rudniev S.N., Bolshov M., Chisholm W. and Rosman K.J.R., *Atmos. Environ.* **34** (2000) 941-948.
[107] Ferrari *et al., Atm. Environ.* **9 (39)** (2005) 7633-7645.
[108] Bolshov M.A., Boutron C.F. and Zybin A.V., *Anal. Chem.* **61(15)** (1989) 1758-1762.
[109] Barbante C., Veysseyre A., Ferrari C., Van de Velde K., Morel C., Capodaglio G., Cescon P., Scarpini G. and Boutron C.F., *Environ. Sci. Technol.* **35** (2001) 835-839.
[110] Gabrielli P., Varga A., Barbante C., Boutron C., Cozzi G., Gaspari V., Planchon F., Cairns W., Hong S., Ferrari C. and Capodaglio G., *J. Anal. At. Spectrom.* **19** (2004) 831-837.
[111] Planchon F.A.M., Boutron C.F., Barbante C., Cozzi G., Gaspari V., Wolff E.R., Ferrari C.P. and Cescon P., *Earth Planet. Sci. Lett.* **200** (2002) 207-222.
[112] Barbante C., Boutron C., Morel C., Ferrari C., Jaffrezo J.L., Cozzi G., Gaspari V. and Cescon P., *J. Environ. Monit.* **5** (2003) 328-335.
[113] Boutron C.F., Gorlach U., Candelone J.P., Bolshov M.A. and Delmas R.J., *Nature* **353** (1991) 153-156.
[114] Ferrari *et al., Mem. Nat. Inst. Polar Res.* **59** (2006) 168-184.
[115] Rosman K.J., Chisholm W., Boutron C.F., Candelone J.P. and Gorlach U., *Nature* **362** (1993) 333-334.

[116] Planchon F.A.M., Boutron C.F., Barbante C., Cozzi G., Gaspari V., Wolff E.W., Ferrari C.P. and Cescon P., *Sci. Total. Environ.* **300** (2002) 129-142.
[117] Gaspari V. *et al.*, *Geoph. Res. Lett.* **33(3)** 2006 L03704.
[118] Burton G.R., Morgan V.I., Boutron C.F. and Rosman K.J.R., *Anal. Chim. Acta* **469** (2002) 225-233.
[119] Planchon F.A.M., Van de Velde K., Rosman K.J.R, Wolf E.W., Ferrari C.P. and Boutron C.F., *Geochim. Cosmochim. Acta* **67(4)** (2003) 693-708.
[120] Gabrielli P., Barbante C. and Boutron C.F., *Terra Antartica Rep.* **8** (2003) 87-89.
[121] Chisholm W., Rosman K.J.R., Boutron C.F., Candelone J.P. and Hong S., *Anal. Chim. Acta* **311** (1995) 141-151.
[122] Cozzi G. *et al.*, *Terra Antartica Rep.* **8** (2003) 29-32.
[123] Gabrielli P. *et al.*, *Nature* **432** (2004) 1011-1014.
[124] S. Hong *et al.*, *J. Environ. Monit.* **7** (2005) 1326-1331.
[125] Vandal G.M., Fitzgerald W.F., Boutron C.F. and Candelone J.-P., *Nature* **362** (1993) 621-623.
[126] Capelli R., Mingianti V., Chiarini C. and De Pellegrini R., *Int. J. Environ. Anal. Chem.* **7** (1998) 289-296.
[127] Schroeder W.H., Anlauf K.G., Barrie L.A., Lu J.Y., Steffen A., Schneeberger D.R. and Berg T., *Nature* **394** (1998) 331-332.
[128] Ebinghaus R., Kock H.H., Temme C., Einax J.W., Löwe A.G., Richter A., Burrows J.P. and Schroeder W., *Environ. Sci. Technol.* **36** (2002) 1238-1244.
[129] Poissant L., Dommergue A. and Ferrari C.P., *J. Phys. IV* **12** (2002) Pr10-143.
[130] Dommergue A., Dynamique du mercure dans les neiges de hautes et moyennes latitude: etudes in situ et en conditions simulées des mécanismes de reactivité et d'echanges, Ph.D. thesis, 2003, Université Joseph Fourier, Grenoble, France.
[131] Gomez-Ariza J.L. *et al.*, *Anal. Chim. Acta* **524** (2004) 15-22.
[132] Haraguchi H., *J. Anal. At. Spectrom.* **19** (2004) 5-14.
[133] Szpunar J., *Analyst* **130** (2005) 442-465.
[134] Szpunar J., *Anal. Bioanal. Chem.* **378** (2004) 54-56.
[135] Lobinski R., Schaumloffel D. and Szpunar J., *Mass Spectrom. Rev.* **25** (2006) 255-289.
[136] Polatajko A., Jakubowski N. and Szpunar J., *J. Anal. At. Spectrom.* **21** (2006) 639-654.
[137] Goenaga Infante H., Hearn R. and Catterick T., *Anal. Bioanal. Chem.* **382** (2005) 957-967.
[138] Simone Garcia J., De Magalhaes C. Schmidt and Zezzi Arruda M.A., *Talanta* **69** (2006) 1-15.
[139] Lopez-Bareal J. and Luis Gomez-Ariza J., *Proteomics* **6** (2006) S51-S62.
[140] Bettmer J., *Anal. Bioanal. Chem.* **383** (2005) 370-371.
[141] Jakubowski N., Lobinski R. and Moens L., *J. Anal. At. Spectrom.* **19** (2004) 1-4.
[142] Szpunar J. and Lobinski R., *Pure Appl. Chem.* **71(5)** (1999) 899-918.
[143] Schomburga L., Schweizera U. and Kohrle J., *Cell. Mol. Life Sci.* **61** (2004) 1988-1995.
[144] Szpunar J., *Analyst* **125** (2000) 963-988.
[145] Lobinski R. and Potin-Gautier M., *Analusis* **26(6)** (1998) M21-M24.
[146] Messerschmidt A., Bode W. and Cygler M. (Eds.), Handbook of Metalloproteins, Vol. 3. (John Wiley and Sons, Inc., Hoboken, New York, USA, 2004).
[147] Bertini I., Sigel A. and Sigel H., Handbook on Metalloproteins, (Marcel Dekker, New York, USA, 2001).
[148] Ballou D.P. (Ed.), Essays in Biochemistry. Metalloproteins (Princeton University Press, Princeton, New York, USA, 1999).
[149] Dumont E., Vanhaecke F. and Cornelis R., *Anal. Bioanal. Chem.* **385** (2006) 1304-1323.
[150] Visser N.F.C., Lingeman H. and Irth H., *Anal. Bioanal. Chem.* **382** (2005) 535-558.
[151] Wrobel K. and Caruso J.A., *Anal. Bioanal. Chem.* **381** (2005) 317-331.
[152] Chassaigne H., Vacchina V. and Lobinski R., *Trends Anal. Chem.* **19(5)** (2000) 300-313.
[153] McSheehy S. and Mester Z., *Trends Anal. Chem.* **22(4)** (2003) 210-224.

[154] McSheehy S. and Mester Z., *Trends Anal. Chem.* **22(5)** (2003) 311-326.
[155] http://en.wikipedia.org/wiki/Electrospray_ionization.
[156] http://en.wikipedia.org/wiki/Matrix-assisted_laser_desorption_ionization.
[157] Rottman L. *et al.*, *Fresenius J. Anal. Chem.* **350** (1994) 221-227.
[158] Catterick T., Fairman B. and Harrington C., *J. Anal. At. Spectrom.* **13** (1998) 1009-1013.
[159] Sariego Muniz C., Marchante Gayon J.M., Garcia Alonso J.I. and Sanz-Medel, A. *J. Anal. At. Spectrom.* **16** (2001) 587-592.
[160] Goenaga Infante H., Van Campenhout K., Schaumloffel D., Blust R. and F.C. Adams, *Analyst* **128** (2003) 651-657.

J. Phys. IV France **139** (2006) 295–320
© EDP Sciences, Les Ulis
DOI: 10.1051/jp4:2006139020

Use of aircraft to probe the troposphere

H. Coe[1]

[1] *Centre for Atmospheric Science, School of Earth, Atmospheric and Environmental Sciences, University of Manchester, Manchester, UK*

Abstract. This paper seeks to give an overview of atmospheric research aircraft, offer the reader examples from current research platforms and provide key reference material for further study. The value of aircraft studies to atmospheric measurement science will be discussed in comparison with other measurement research platforms and the capabilities and deficiencies of the different platforms will be considered. The paper will then summarise the key aspects of instrument design that are critical for aircraft operations. Sample inlets are integral to successful sampling from aircraft platforms and these will be discussed in detail, focussing on "sticky" gases, radicals and aerosol particles and clouds. The remainder of the paper will focus on the planning of field missions and offer some examples of major results from aircraft experiments.

1. THE CAPABILITIES OF RESEARCH PLATFORMS AND THE NEED FOR AIRCRAFT

The first question one should ask is why should we use aircraft platforms to study the atmosphere? The answer might seem an obvious one; aircraft can sample parts of the atmosphere that are not accessible by other types of measurement platform. However, aircraft are costly to maintain and to operate and so the value of the data set to atmospheric science has to be assessed very carefully.

Before the advantages of airborne measurement platforms are discussed it is briefly worth presenting an overview of the advantages of other research platforms.

1.1 Ground based stations

Ground based sites are commonly used. They can provide long term records of trace gas and aerosol measurement and also offer locations for deploying a very large number of instruments in a detailed intensive study. Important examples of long term research sites are the Mauna Loa Observatory (MLO) on the island of Honolulu in the mid Pacific Ocean, and the Jungfraujoch high altitude research station in the Swiss Alps. Both these stations are at altitude and therefore can sample air that has not been recently influenced by the Earth's surface and both stations are well removed from local sources of pollution. The biggest contribution the Mauna Loa Observatory has made to atmospheric science is possibly the provision of the longest continuous CO_2 record on Earth. It has been operated since the mid 1950s and has provided a unique contribution to climate science. The first detectable CO_2 increases due to anthropogenic change were observed in 1959 at the MLO and also at the South Pole [1] and since that time continuous increases in annual averaged CO_2 concentration have been observed [2]. The record has also allowed changes in rates of northern hemispheric respiration and photosynthesis to be inferred [3].

The MLO is now part of the Global Atmospheric Watch program of the World Meteorological Organisation (WMO). Several other sites that are part of this network are important long term measurement locations that have contributed effectively to atmospheric science over the last decades. The Jungfraujoch High Altitude Research Station, samples continental air above Europe but, at over 3500 m, is often above the surface boundary layer and so is representative of the lower free troposphere above Europe. It has been used to study lower free tropospheric trace gases, aerosols and clouds [4]. The Mace Head Atmospheric Research Station (MHARS), also a GAW station, is on the western coast of Ireland and hence is an ideal location for observing marine boundary layer processes such as new particle formation from coastal macroalgae [5] and sea spray generation [6]. Surface sites can also be useful for studying cloud processes as the experiments are easier to mount, sample collection suffers far fewer artefacts than on board aircraft and multisite experiments can probe cloud processing and aerosol-cloud interactions in a lagrangian sense by following air parcels over a hillside on which an orographic cloud has formed. Examples of this are the Great Dun Fell series of experiments to investigate sulphur chemistry in clouds [7], and the Aerosol Characterisation Experiment-2, on the island of Tenerife which investigate the influence of aged European pollution on cloud microphysics [8].

The disadvantages of ground based sampling sites are that they are single point sampling locations which can, unless great care is taken, suffer from local effects. These, as is the case at Mace Head, may themselves be important, but care is required to separate short and long range effects. Inevitably the sites are close to the surface and hence are affected by surface behaviour, such as dry and wet deposition, and local emissions. Local wind phenomena can be very important, sea breeze effects separate the local site from the regional background [9], and thermal winds, such as katabatic and anabatic flows, can influence the sampling at mountaintop locations [10].

1.2 Ships as atmospheric platforms

Ships offer the benefit of being mobile platforms and so can sample the atmosphere over the remote oceans. They can transit over large distances, although they are slow moving and usually offer large amounts of power and space for instrumentation. However, they are limited to the marine surface and sampling can be affected by both the motion of the ship and the airflow around such a large structure on what would otherwise be a uniform surface. Examples of some important measurements using shipborne measurements include the Marine Atmosphere Oxidation Capacity Experiment (MARATHON), which took place aboard the German research vessel, the RV Polarstern, in 2001. One of the novel measurements during this cruise were the first measurements of the latitudinal variation in hydroxyl radical concentration from Iceland to the tip of South America. The hydroxyl radical controls the rate at which many species in the atmosphere are oxidised and its cycle is important for understanding ozone formation, the nitrogen and sulphur cycles in the atmosphere and the rate of oxidation of methane and non methane hydrocarbons. However, as it only has a lifetime of one or two seconds in situ measurements are not spatially representative. The Polarstern work showed the latitudinal variation of OH [11].

More recently the RV Ron Brown, a US research vessel, has made measurements of a range of volatile organic carbon compounds and aerosol organic loadings in a cruise off the east coast of the USA during the New England Air Quality Study [12]. These measurements covered the whole of the US east coast and were able to trace the outflow from the US to the NE Atlantic Ocean. The VOC ratios were used to determine air mass age and source footprint and were used to show that the bulk of the oxygenated VOC was secondary in nature after two days and had led to significant production of organic aerosol whose rate of formation cannot be explained by current models of secondary organic aerosol.

1.3 Balloon platforms

Unlike ground and shipborne observations, balloon borne observations are not limited to the surface and readily provide information on the vertical structure of the atmosphere. They are also relatively cheap (compared to aircraft and satellites at least) and in some cases offer the possibility of global measurement. However, there are operational restrictions with these platforms as well. They have a restricted payload that is often quite small, there is only limited power and it is not easy to control the direction of travel of the balloons. Balloons are particularly important for studying the lower stratosphere and upper atmosphere. This is a region where very few aircraft can operate and satellite and ground based remote sensing instrumentation has some difficulty probing. There is a global network of meteorological balloons that are launched daily, and a similar, smaller network of ozone sondes. Many research teams operate small tether balloons to probe the boundary layer structure. The French space agency CNES (Centre National d'Etudes Spatiales) maintain a fleet of balloons that are capable of probing the lower stratosphere. Some of these are dedicated to high payload (100 kg) flights of a few hours above a single locations, whereas other, such as the Montgolfiere Infra-Red (NIR) balloons are capable of maintaining altitudes from which the lower stratosphere can be sampled for many days, circumnavigating the globe. The aluminium Mylar material in the top hemisphere of the balloon traps upwelling infrared and maintains its buoyancy at a level of between 17 and 22 km at night. In the daytime visible downwelling radiation adds to the buoyancy and heights of 28 km are reached [13]. These balloons have been crucial in unraveling the complexities of the Arctic ozone hole [14] and have provided detailed information on the tropical tropopause layer [15].

1.4 Satellite platforms

Satellites have the advantages of providing continuous long term observations, either above a single surface location (geostationary), or offering global coverage. The development of advanced sensors is significantly improving their capability with instruments such as SCIAMACHY, a scanning imaging absorption spectrometer that has viewing angles in the limb and nadir and accesses most visible wavelengths. The instrument can measure a range of trace gases including ozone, sulphur dioxide, BrO, ClO, NO_2, and NO_3. It has been used to determine enhancements in NO_2 above China [16] and global BrO concentrations from space [17]. Though this is a very powerful new probe it, like other instruments (e.g. GOME and MIPAS [18]) has only limited vertical resolution in the troposphere. Tropospheric burdens of CO are now regularly derived from instruments such as MOPITT and have been used to estimate hemispheric variability [19] and also the magnitude of regional biomass burning sources [20].

It has been possible to derive some aerosol and cloud properties from satellite platforms for some years, notably cloud cover from the several generations of Meteosat and aerosol products such as dust, biomass burning aerosols and aerosol optical depth from instruments such as MODIS [21], ATSR and MISR [22]. Each of these instruments has its own strengths and weaknesses such as spatial resolution, overpass time, spectral range and accuracy. New aerosol and cloud satellite instruments are coming online at the present time. These include a Cloud Profiling Radar on board Cloudsat, a space borne LIDAR on board Calipso, and a polarimeter (POLDER) on board PARASOL. These instruments are to fly in close sequence on a train of satellites known as the 'A' train. The close coordination of the satellite overpasses means that they will be providing closely coupled measurements over the coming years.

The disadvantages of satellite platforms are that they are often not able to retrieve all the desired parameters and so cannot probe key processes effectively. The data is for the most part derived and the algorithms used to derive the products often are subject to poorly determined systematic biases. The products often need ground truthing with in situ observations. Lastly, the vertical information provided by satellites is often limited; this is most commonly the case in the troposphere, where horizontal and vertical gradients are often large.

1.5 The need for aircraft platforms

Whilst the combination of ground, ship, balloon and satellite platforms provides a wealth of information, aircraft offer a unique platform from which to conduct atmospheric science. Aircraft cannot provide long term data sets, these come from ground based networks or satellite instruments, nor are they suited to providing detailed information on the marine boundary layer. With one or two notable exceptions, aircraft cannot probe the stratosphere above 20 km. However what aircraft offer is the ability to mount a significant number of instruments together and obtain both remotely sensed and in situ information in three dimensions. In many cases they offer very high levels of spatial resolution, often being able to probe processes at the scales governed by turbulence. They are therefore ideal platforms to investigate detailed atmospheric processes that are taking place throughout the troposphere and lower stratosphere.

As aircraft can rapidly sample large air masses and surface regions in one flight they can test satellite products and provide data to verify large scale transport models in a way that cannot be done easily from other platforms. Some examples of this will be given at the end of this paper.

2. AIRCRAFT CAPABILITIES

There are many different aircraft platforms and their scientific roles vary widely and depend very heavily on the aircraft's capabilities. The European Fleet for Airborne Research (EUFAR – http://www.esf.org/eufar/) is an Integrated Infrastructure Initiative of the 6th Framework Program of the European Commission. It brings together 24 leading European institutions and companies involved in airborne research, operating 24 instrumented aircraft and aims to:

Co-ordinate the network for exchanging knowledge, sharing developments, and building the unified structure that is required for improving access to the infrastructures;
Provide users with transnational access to the infrastructures;
Extend transnational access to national funding sources;
Promote airborne research in the academic community;
Develop research activities in airborne instrumentation.

This resource provides a wide range of information on the different aircraft platforms in the European Fleet and their capabilities and strengths. This paper will present some examples of different types of aircraft within the fleet and what their strengths and weaknesses are.

Aircraft platforms range in size and capability from small aircraft that have limited range and payload and typically have slow airspeeds, through to large multi-engine long range aircraft that can carry many instruments and can probe all altitudes in the troposphere. In between there are medium sized aircraft that offer the opportunity for more complex payloads, but retain some of the flexibility of the smaller aircraft. High level jet aircraft can probe the tropopause region and often have a long duration, but at the expense of some payload and also the lower flight speeds sometimes necessary to investigate smaller scaled processes. High altitude aircraft are one of the few platforms that can probe the lower stratosphere and as a result offer unique capability.

An example of a small light aircraft is the University of Manchester Cessna 182J, a single engine propeller driven aircraft. It has a scientific payload of 130 kg, a ceiling of 4000 m and a cruising speed of 55 ms^{-1}. It is possible to accommodate one scientific crew member in addition to the pilot. The aircraft activity focuses mainly on radiation transfer, aerosols and trace gas pollution studies, and the measurement of turbulence fluxes. Though the low payload of the aircraft means it cannot carry multiple aircraft packages it offers easy access and the payloads can easily be swapped. Its low cruising speed means that it is ideally suited to measuring boundary layer fluxes, convective onset and pollution plumes.

The Natural Environment Research Council in the UK maintains a Dornier Do-228 twin turbo-prop aircraft which has to date mostly been used for surface remote sensing. It has a practical range of around 1000 km and can reach a ceiling of around 7500 m with a normal scientific payload of approximately 500 kg. It carries up to 3 scientific crew. This means it is capable of accommodating several large multispectral radiometers for surface colour imaging and mapping. These include the Compact Airborne Spectrographic Imager (CASI), the Airborne Thematic Mapper (ATM) and Airborne Laser Terrain Mapper (ALTM). These are commonly used for surface characterization of vegetation and coastal waters. The platform is flexible and allows for in situ sampling of trace gases and aerosols in addition to remote sensing of the Earth's surface. Most recently the aircraft provided such measurements to describe the low level inputs to tropical convective storms in the region of Darwin during the SCOUT-O3-ACTIVE experiment in December 2005.

The largest aircraft in the European fleet is the Facility for Airborne Atmospheric Measurements (FAAM), a BAE 146-301 four engine jet aircraft, and is operated by NERC and the UK Met Office. It has a range of 1800 nautical miles, a ceiling of 11000 m, a cruising speed of 796 km/h and a

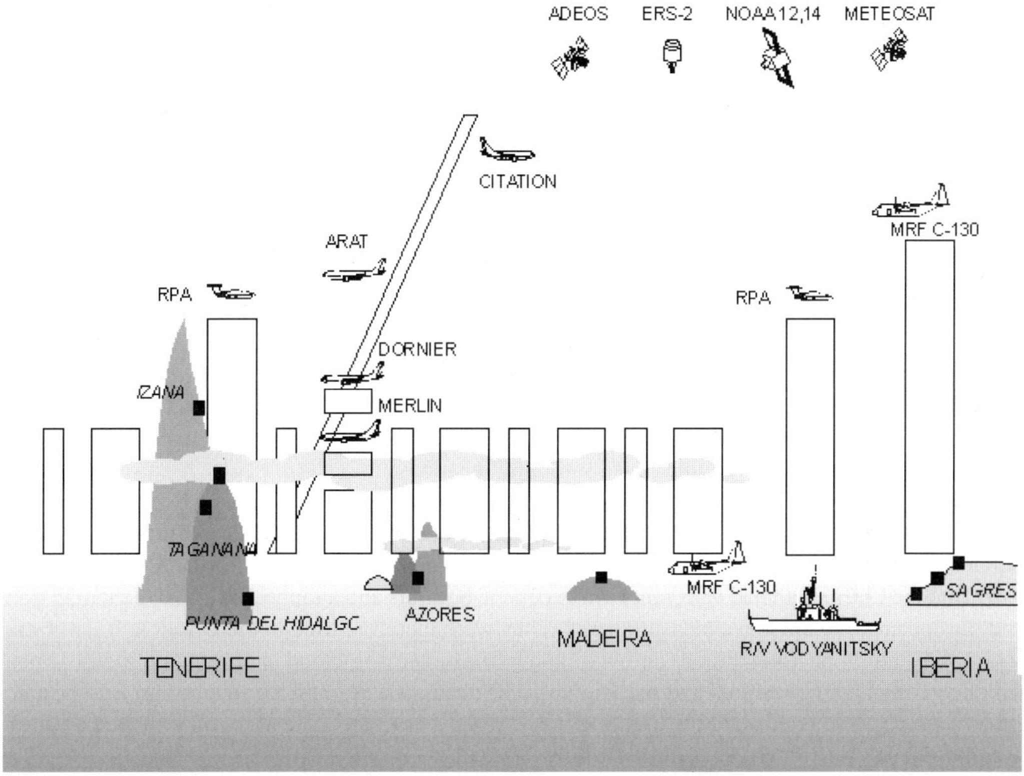

Figure 1. Schematic of field operations during the ACE-2 experiment from [23].

maximum duration of a little over 5 hours. A scientific payload of up to 4600 kg can be carried and the aircraft can accommodate up to 18 scientists on board in addition to the 3 crew members. Hence, the aircraft can house a large range of equipment in up to 24 racks on board and so can offer a multi-measurement platform for conducting detailed process studies focusing on trace gases, aerosol, clouds, radiative properties of the atmosphere and atmospheric dynamics.

These aircraft are all quite different and complementary. They offer different payload options, fly at very different science speeds and hence can cover large spatial scales or provide a slow platform for assessing detailed small scale processes.

The operating characteristics define the capabilities of the aircraft. An example of the way the use of the aircraft can be maximised is provided by the second Aerosol Characterisation Experiment (ACE-2), [23]. Some key results based on this experiment will be illustrated in the final section of this paper but the logistical operation serves to highlight the importance of matching the aircraft characteristics to the science needs. The experiment aimed to investigate the properties, processes and effects of different aerosol types in the sub tropical north eastern Atlantic Ocean. In this region the Azores high pressure advects aerosol from either the clean marine boundary or from the European continent towards the Canary Islands on the north east trade winds, as the marine boundary develops the region to the south becomes heavily impacted by extensive stratocumulus clouds. Overlying the low level outflow, easterly flow often carried Saharan dust into the region. ACE-2 used multiple aircraft platforms, together with ground and shipborne observations and satellites to characterise the region. Figure 1 shows a schematic of the field operations. The experiment was split into several subprojects, each of which had their own discrete set of objectives.

A CLEARCOLUMN experiment investigated the extent to which it was possible to quantitatively describe the radiative impact of aerosol in the atmospheric column above the region when no cloud was present based on a series of so called closures linking the basic physical and chemical properties with the in situ radiative properties at various levels in the column and integrating this information to assess its impact on the radiation transfer through the column. The latter was also measured so that an assessment of uncertainties in predictive capability could be made at every level in the scaling process.

A CLOUDYCOLUMN experiment sought to perform a similar set of closure studies when cloud was present in the column. This also involved the measurement of cloud microphysical properties and knowledge of the cloud dynamics. A LAGRANGIAN study was also conducted. This involved releasing air parcel following balloons with tracking transmitters and GPS on board and revisiting the same air mass as it was advected from the European continent towards the Canary Islands. This was the first time air parcels were followed over distances of a thousand kilometres or more. A study of free tropospheric aerosol, FREETROPE, focused on the dust outflow and its impact on the underlying cloud. Two other studies: a long term measurement campaign and a ground based hill cap cloud experiment [8] were conducted, these studies used ground based sampling locations.

Six aircraft were used in the ACE-2 study: the CIRPAS RPA Pelican; the Cessna Citation-II; a Merlin IV operated by Meteo France, Toulouse; a Dornier 228, operated by the DLR, Germany; and the UK Met Research Flight C-130 Hercules aircraft. All the aircraft were based on the island of Tenerife. The Pelican and the C-130 both had endurances in excess of 10 hours and a range that allowed them to transit to the European coast and make measurements. They were therefore used in the LAGRANGIAN study. The C-130 is a high payload aircraft and was able to make aerosol, cloud and radiation measurements. The Pelican has a mainly aerosol and radiation fit. The clear air was to the north, close to Europe and so these two aircraft also conducted the CLEARCOLUMN experiments. The Merlin-IV carried an aerosol and cloud physics instrument fit and the Dornier, a smaller aircraft, carried a more limited aerosol and radiation fit. These two aircraft conducted the bulk of the CLOUDYCOLUMN measurements along with the C-130 which could also measure cloud properties. The eight hour duration of the MERLIN IV also meant it could transit to the operations area of the LAGRANGIAN when the air parcels came closer to Tenerife and relieve the C130 and Pelican. The Citation II was mainly deployed in the FREETROPE study as it was capable of attaining altitudes of between 12 and 13 km. The Merlin-IV and the C-130 were capable of maintaining altitudes of 8 and 10 km respectively and all the other aircraft were restricted to low level operations. However, the lack of endurance of the Citation II restricted its activity to the Tenerife area and meant it could not participate in the LAGRANGIAN operations.

This example illustrates the need for having a range of different aircraft in the fleet of research platforms at any one time. Their strengths are complementary and they can be coupled together to deliver a large integrated science programme that is not possible in any other way.

3. MAKING MEASUREMENTS ON BOARD AIRCRAFT PLATFORMS

3.1 Airworthiness and certification

Any instrumentation or inlet system has to confirm to strict safety standards that are set in place by the aviation authorities in different nation states where the aircraft is registered. The stringency of these standards is dependent on the registration class of the aircraft. A small aircraft, operated under a private pilot's licence will need to conform to a far more limited set of criteria than an aircraft such as the FAAM BAE-146, which, as a converted 100 seat passenger aircraft, still must conform to rules governing passenger aircraft. The lead-in times are significant and can be costly. Changing an inlet system on a civil registered aircraft affects the aerodynamic drag and hence the aircraft performance. It also impacts on the structural integrity of the airframe and as a result any changes must be carefully designed and documented to ensure that they achieve the desired level of airworthiness that satisfy the regulatory authority. Instrumentation also must be assessed to ensure that it meets the correct weight, and centre of gravity constraints, and does not affect the electrical or safety systems on the aircraft, including telecommunications. Instruments requiring flammable or noxious gases, or utilise high powered lasers or generate significant heat or noise will be subject to more exacting assessment. This process increases both the time required to get an instrument installed into an aircraft and ensure it can be operated effectively and the cost required to do so. The costs of approving instruments are not insignificant on large aircraft. These constraints mean that the instrumentation on board large platforms evolves slowly and new integration is a significant capital investment. Often smaller aircraft can act as flexible test platforms for new instruments before their integration into larger measurement platforms

3.2 Instrumentation

This paper cannot hope to cover instrumentation for all areas of atmospheric science that have ever been mounted on board aircraft. However, there are several key pieces of instrumentation that are essential to successful aircraft operations. Clearly position, pressure and altitude information is crucial to almost all atmospheric aircraft operations. In addition temperature and humidity measurements are also very important. Position, altitude and windspeed information may come from the operational avionics on board the aircraft, as is the case with the DLR Falcon. However, in many cases the cost of approving modifications to integral parts of the aircraft systems is prohibitively large and many aircraft are fitted with separate research sensors for observational work. The accuracy with which measurements of these basic parameters is variable depending on the role of the aircraft. Aircraft conducting land or sea surface imaging require accurate position information; other platforms focusing on measurements of fluxes and turbulence require extremely accurate aircraft attitude measurements as well as position to correct the high frequency wind speed measurements which can only be made relative to the air frame itself.

Temperature, and in particular, humidity measurements are often some of the most difficult make, yet these are crucial measurements of the state of the atmosphere. A temperature sensor on an aircraft is typically a platinum wire protected in some form of housing mounted outside the boundary layer skin of the aircraft and as close to the nose as possible. The temperature probe has a non-negligible thermal inertia. This combined with the thermal response of the housing means that recovery times of around a second are typical for an instantaneous step change in temperature. The measured temperature is often called total temperature as it includes dynamic heating of the air as it is decelerated in the probe housing and an anti-icing heating that is applied to the housing. These need to be corrected for using the measured air speed and pressure. Systematic errors include sensor wetting and environmental deterioration, the latter requiring regular calibration and sensor replacement. Lawson and Coopper [24] discuss the performance of various airborne temperature probes but systematic uncertainties of around 1 °C are not uncommon.

Rapid and accurate measurement of humidity or water vapour over the expected range of atmospheric values is extremely difficult as the dynamic range covers three orders of magnitude, and water can exist in the atmosphere in all three phases. Several possible techniques exist, each of which has its own strengths and weaknesses. Many aircraft use a combination of several methods to calculate a value for humidity. Chilled mirror hygrometers measure the dew point temperature by measuring the loss in reflectivity of a mirror while its temperature is carefully controlled by a peltier cooler. The difficulty with these devices is that they have a slow time response of up to 30 seconds and hence cannot follow rapid changes in temperature likely to be faced during profile ascent and descent. However, they do offer very accurate measurements of the dew point in the mid and lower troposphere. Above these altitudes, the water vapour mixing ratio falls and the absolute abundance of water is very low, leaving the sensor very insensitive and liable to oscillate.

Capacitive sensors are rapid response devices but are only accurate to 2 % relative humidity and are only useful for relative high humidities. They can also suffer from changes in performance as a result of gases and aerosols in the atmosphere, notably the high levels of sea salt aerosol in the marine boundary layer. Measurements based on water absorption of light emitted from the Lyman-a line of hydrogen provide a fast response measurement of water, though their response is to total water as the absorption could come from liquid or solid water as well as the vapour. These measurements, though rapid, suffer drift and so are commonly coupled to devices such as chilled mirrors to provide a robust and rapid measurement of water vapour. Other measurements use VUV dissociation of water molecules and monitor the fluorescent intensity of the decay of the excited OH product, which is proportional to the water vapour mixing ratio. Sensors such as this and modified Lyman-a methods can probe low water vapour mixing ratios close to the tropopause, but can suffer from artefacts that arise from significant water exchange between the atmosphere and surfaces in and around the housing leading to hysteresis behaviour. Furthermore as temperature measurements are not certain to better than 1 °C, calculation of the dew point from mixing ratio and vice-versa at these altitudes is very uncertain. Lastly cloud particles can lead to an erroneous measurement of water vapour.

New techniques based on open path tunable diode laser technology appear to offer the possibility to overcome these deficiencies. An open path TDL system has been operated on board the NASA ER-2 aircraft and flown in the lower stratosphere for a number of years [25] though successfully operating such an instrument is very challenging. TDL type instruments also offer the possibility for one sensor to cover a very wide range of humidities as the same instrument can be used to tune to a weak line at high humidities and a strong one at low values. To date a single instrument that can reliably measure humidity rapidly and accurately over the wide range of values observed in the atmosphere that is free from artefacts on board aircraft platforms remains elusive.

There are many instruments in existence that sample a wide range of dynamical, physical, chemical and radiometric quantities from airborne platforms. The objective of this paper is not to outline all of these in detail but the sections below on sampling and the last examples of airborne experiments both offer some examples of state of the measurement of the atmosphere.

3.3 Sampling

Sampling is a common difficulty with most in situ measurements and this is greatly exacerbated on an aircraft that is travelling at high speed relative the air around it. For gases such as CO_2 that are non sticky, unreactive and not affected by thermal gradients, sampling is straightforward. However, often this is not the case, aerosol and cloud particles have significant momentum and can impact on inlets and the airframe itself; sticky gases may be lost in inlet systems, changes in temperature may affect the phase partitioning of water and other chemicals. Furthermore, the air molecules adjacent to aircraft skin suffer drag and potentially surface loss. The drag slows down the air close to the surface and this propagates outward as the air travels down the aircraft forming a boundary layer that increases from a few mm close to the nose to several centimetres further down the fuselage. Sampling must be conducted outside this zone where surface loss effects, heating and mixing can all affect the sample. These problems offer major design challenges when attempting to measure key parameters from aircraft. There are two basic design options, one is to attempt to counteract the changes induced by the sampling system and essentially engineer out the effects, the second is to attempt to sample as close to the atmosphere as possible, in essence a kind of local remote sensing. Some examples of the both of these approaches are given below.

3.3.1 Trace gases

The oxidizing capability of the troposphere is governed to a large by the hydroxyl radical, OH, which is formed predominately by photolysis of ozone and subsequent reaction of the excited oxygen atom with water vapour. Close to pollution or at altitude other mechanisms may become important. Once formed, OH is converted to the hydroperoxy radical, HO_2, by reaction with CO and methane with a lifetime of around a second. HO_2 rapidly reforms OH by reaction with ozone, leading to net ozone loss, or with NO to form NO_2 if it is present at more than a few tens of parts per trillion by volume. As the OH-HO_2 conversion undergoes several complete cycles before final loss the latter route leads to ozone formation by NO_2. To understand the budget of ozone in the troposphere and the lifetimes of key species such as methane and nitrogen oxides, knowledge of the hydroxyl radical cycle is crucial.

It is therefore important to understand OH chemistry throughout the troposphere and direct measurements are required in order to test chemical mechanisms. However, measuring OH is challenging, even in the laboratory. There are a few techniques available: long path differential optical absorption spectroscopy; chemical ionisation mass spectrometry; and laser induced fluorescence. Of these only the latter two have been deployed on aircraft. Both these systems are point measurements and entrain ambient air into a high vacuum to make the measurement and as a result have to combat possible inlet artefacts. Eisele and Tanner [26] have developed a chemical ionisation mass spectrometer that measures H_2SO_4 by chemically reacting it with nitrate ions, which are formed in a separate chamber and electrically accelerated into the sample flow. Concentrations are determined by measuring the relative ion strengths in the mass spectrometer and knowing the reaction rates and residence times. OH is determined by reacting it with isotopically doped SO_2 at the inlet tip, which rapidly forms isotopically doped H_2SO_4 in the presence of oxygen and water vapour.

The Laser Induced Fluorescence method has been pioneered by Brune and co-workers. It has been deployed on the NASA DC-8 aircraft during several studies, including TRACE-P. The measurement is made by excited OH molecules at 308 nm using a pulsed laser and then observing their fluorescence decay at the same wavelength. At ambient pressure the fluorescence is heavily quenched and as the fluorescence is observed at the same wavelength as the excitation it is not possible to make an accurate measurement. Hence ambient air is entrained into a vacuum through a nozzle. The gas expansion reduces the pressure very rapidly and cools the gas, leading to extended lifetimes of the excited state which can be then measured accurately. The instrument and its associated inlet are discussed by Brune et al. [27] and the inlet system is shown in figure 2.

The inlet is based on a modified jet engine housing that is mounted to underside of the fuselage of the NASA DC-8 aircraft. The aim is to slow the airflow from 240 ms^{-1} to between 8 and 40 ms^{-1} using an aerodynamic nacelle from which it is sub-sampled, minimising losses. Unlike the CIMS instrument the LIF can measure HO_2 by preconversion to OH using excess NO. Two detection cells are employed so that both radicals can be sampled simultaneously. This obviously means that the aircraft needs to carry noxious gases. The instrument, known as ATHOS, has been used, along with other instruments, to test photochemical models of radical budgets. Figure 3 shows an example of one such experiment from measurements made during TRACE-P. The data shown are HO_2 measurements from ATHOS and peroxy radical measurements from a CIMS instrument. There is good agreement between the ATHOS HO_2 measurements and the models over a wide range of concentrations.

The Eisele inlet system is considerably easier to install than the Brune inlet system yet seems to offer artefact free sampling for sticky gases. A modified version has now been used as an inlet for another LIF instrument now installed of the UK FAAM aircraft. It has also been used as a inlet system for sampling sticky gases such as nitric acid and ammonia. Total oxidised nitrogen, so-called NO_y, is difficult to sample as components such as HNO_3 are readily lost to the walls and highly volatile compounds such as peroxy acetyl nitrate (PAN) thermally decompose very readily. Previous studies have focused on the materials necessary for artefact free sampling [28], the results of which have been used in a tested airborne inlet design [29] and have been used in a wide range of field studies [30]. The NO_y measurement methods themselves are based on passing the air through a heated gold convertor held at 200 °C. The technique was initially evaluated by Fahey et al. [31] for use on the ER-2 in the lower stratosphere and has since which been utilised on several platforms. Most recently, two such aircraft systems have been compared by wing tip to wing tip flying [32]. One instrument was operated on a Learjet A35 during the SPURT campaign by ETH Zurich and was configured so that its convertor was mounted outside the aircraft. The other instrument, operated by FZ-Juelich was part of the MOZAIC project to measure

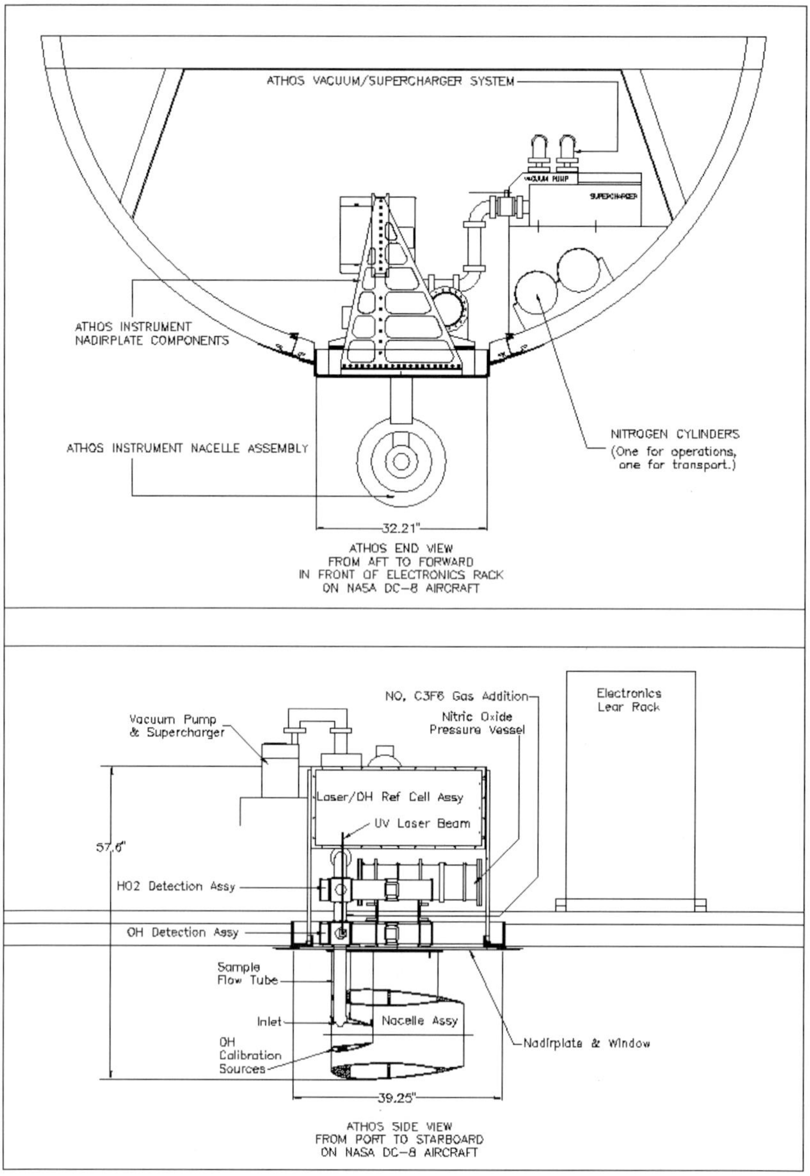

Figure 2. Inlet system used for the OH and HO_2 LIF measurement on the NASA DC-8 aircraft [27].

ozone, water vapour, carbon monoxide and nitrogen oxides aboard an Airbus A340 in-service aircraft. It was not possible to mount the convertor outside the skin of a commercial aircraft and so an inlet line was used after sampling through a Rosemount inlet. During the comparison the concentrations of NO_y ranged from 0.3 to 3 ppb and the instrument agreed to within 6%, which puts an upper limit on the losses of HNO_3 experienced by the Rosemount inlet on the Airbus A340. This comparison demonstrates the complexity of comparing similar instruments on different platforms. Intercomparison flights are necessary involving formation flying over a range of operating altitudes. This is often difficult as the airspeeds and performance characteristics of the aircraft may not be well matched.

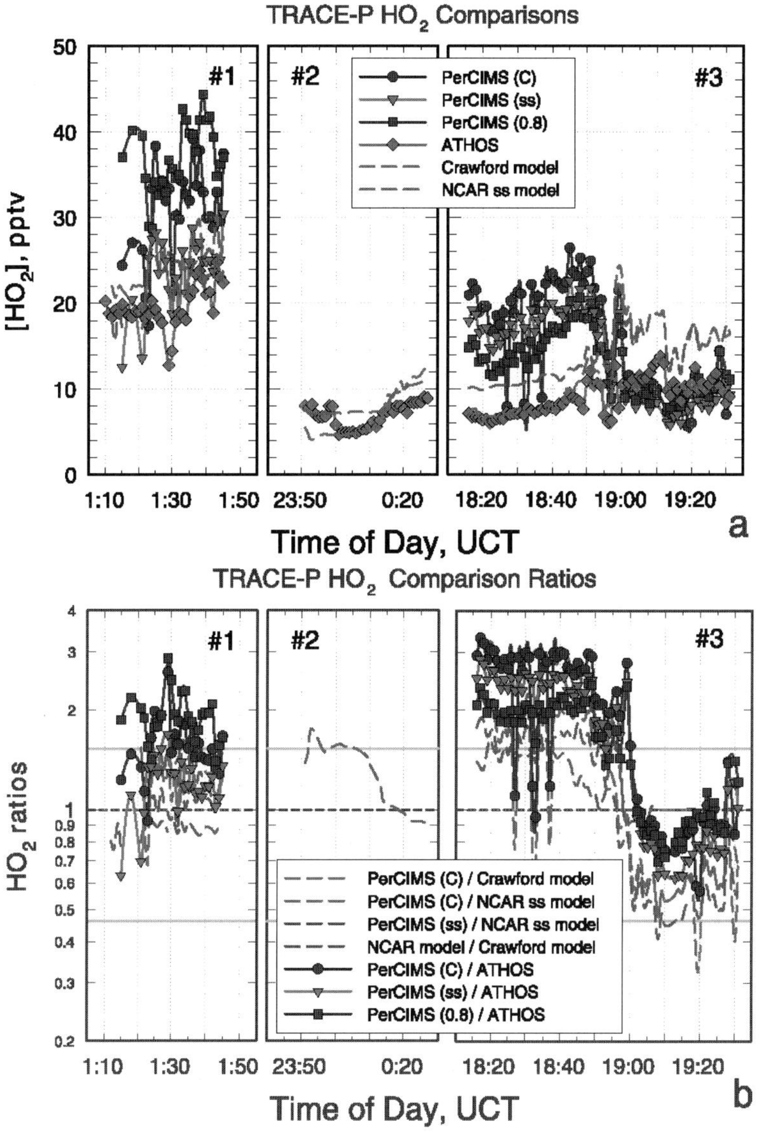

Figure 3. Comparison of radical measurements taken on board the DC-8 aircraft during TRACE-P.

3.3.2 Aerosols

Aerosol particles are important in the atmosphere. They scatter light and so affect the radiative balance of the Earth; they act as sites for cloud droplet formation and so affect the cloud particle number and hence its albedo; they are implicated in the adverse effects on human health; and they are efficient vectors for the long range transport of pollution and the biogeochemical cycling of nutrients. However, many aspects of the aerosol lifecycle in the atmosphere are poorly resolved. They are emitted directly into the atmosphere from human and natural sources and these budgets are not well known, and they can form from condensation of gases in the air. Once formed, they grow by condensation and coagulation, and may be significantly processed by cloud. During this cycle their properties change and this affects both their behaviour and impacts. It is necessary to measure these parameters and so improve our capability

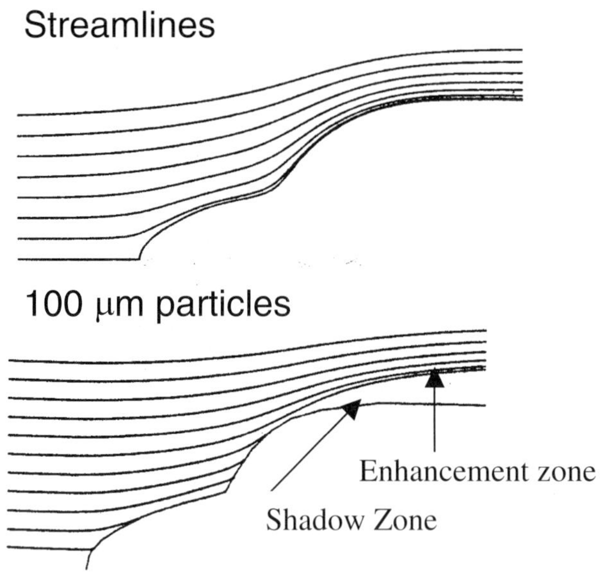

Figure 4. Calculated 10 μm particle tranjectories around the NCAR Elactra aircraft, showing shadow and enhancement zones [39].

to predict behaviour. Direct sampling from airborne platforms clearly has an important role to play here as remote techniques cannot measure all the required physico-chemical properties necessary to study these processes. However, sampling aerosol is also extremely challenging, if not more so than many trace gases. A recent review of aircraft inlets for aerosol sampling is provided by Wendisch et al., [33] and further information is available in Wilson and Seebaugh [34]. This gives a detailed look at the state of the art and offers some recommendations for future research. A summary of the main findings will be discussed here.

Larger particles, greater than 1 micron diameter, have sufficient momentum that their trajectories depart from curved streamlines. This manifests itself in several different ways. The streamlines are perturbed greatly around an aircraft and hence shadow and enhancement zones arise for particles of a given size, where particles have either been removed or injected into air flowing over the aircraft as a result of divergence between the particle path and the streamlines [35, 36, 37, 38]. Figure 4 shows the calculated shadow and enhancement zones for the NCAR Electra aircraft as an example [39].

The flow velocity of the air inside the inlet should be matched to the free stream flow immediately around the inlet tip area. This is known as isokinetic sampling. If this is not achieved then the flow will either diverge or converge at the inlet. This does not matter for gases and small particles that follow streamlines, except for added heating or cooling effects and the potential for initiating turbulence.

However for larger particles, typically above a micron in diameter, their momentum causes them to either be over or under sampled, depending on the relative flow speeds in and outside of the inlet. Figure 5 shows idealised examples of such behaviour.

Small aerosol particles diffuse quickly so long sample lines should not used, especially as the small particles are often charged and so the losses can be significant. Barron and Willeke [40] summarise the diffusion loss rates that would be expected for aerosol in a laminar flow along a sample line of known cross section at a given flow rate as a function of their size. If the flow is turbulent such calculations are not readily available and the losses are much harder to characterise. Once above a few tens of microns diameter particles do not diffuse rapidly and as long as they are less than a micron in size such particles follow streamlines to a great degree. However, larger particles readily inertially impact or are

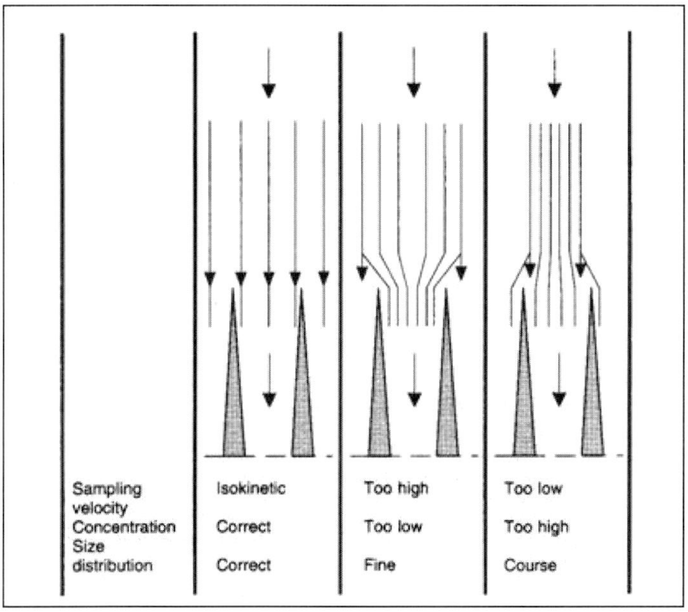

Figure 5. Schematic demonstrating isokinetic and non-isokinetic sampling and the impact on the sampled aerosol distribution compared to that present in the free stream.

lost gravitationally. These losses can also be estimated for straight sample lines but will be enhanced by bends in tubing. Hence curves in inlet pipes should have a large radius of curvature and tube lengths should be minimised. It is especially important to prevent turbulence in sample lines to minimise these effects. Dynamic heating of the air entering the inlet resulting from deceleration may have a role in partitioning material adsorbed onto the particles such as water, ammonium nitrate and the more volatile organic fraction in the gas phase. It is extremely difficult to avoid such potential artefacts.

The aerosol inlet on the Falcon aircraft operated by DLR (Deutsches Zentrum für Luft und Raumfahrt) uses a forward- and a backward-facing configuration. The particle laden air enters the rounded-lip inlet and is decelerated in the diffuser by a factor of 7.1. A large radius, 90° bend leads the aerosol towards the aircraft fuselage. The Falcon aerosol particle inlet is used for near-isokinetic sampling of super- and sub-micrometer aerosol particles with the forward-facing component, whereas only sub-micrometer, interstitial aerosol particles are collected with the non-isokinetic backward-facing part of the inlet system, which samples directly from non-decelerated air flow and so only passes the smaller aerosol particles that follow the streamlines. The diameter at which 50 % of the particle is passed by the backward-facing inlet, i.e., D_{50}, was estimated as a function of altitude using combined measurements with a PMS (Particle Measuring Systems, Boulder, CO, USA) PCASP (Passive Cavity Aerosol Spectrometer Probe), which measures the number of particles as a function of their size in the 300 nm to 10 μm size range and a TSI (Thermo-Systems Inc., St. Paul, MN, USA) CPC (Condensation Particle Counter) which measures the total particle number concentrations above 10 nm. Details of this technique have been described by Fiebig [41]. The CPC is switched between the forward-facing, near-isokinetic inlet, measuring approximately all particles by number, and the backward-facing inlet, measuring only particles smaller than D_{50}.

As already discussed larger particles are more difficult to sample. There are two basic approaches, the first involves engineering mechanisms to transfer the particles efficiently the second to reduce sample artifacts and directly measure in the free stream, examples of both types of probe are given below.

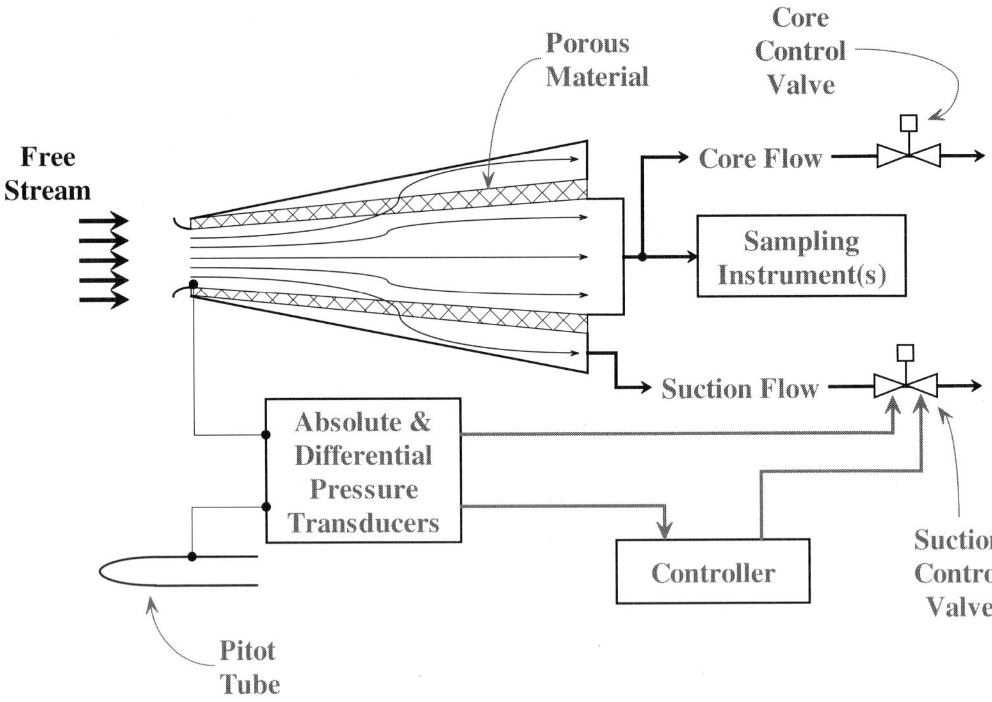

Figure 6. A schematic of the sample and flow control of the Denver University Low Turbulence Inlet.

Typical diffuser type inlets, such as the one described above slow the sample air by increasing the inlet cross section in a short initial section using a gently angled internal cone. This can lead to the inlet being sensitive to the angle of attack of the probe to the local flow field. This can be ameliorated to some extent by placing a shroud around the front tip of the inlet to straighten the flow in the region of the tip and hence make the inlet less sensitive to the local wind field [39]. A second problem with the decelerating cone is that a boundary layer forms at the inlet tip, leading to a strong shear and turbulence. The thickness of this turbulent layer can grow with distance along the inlet and often induce full turbulence down the sample line. This leads to high losses of supermicron particles. To sample these several approaches are possible. Huebert et al., [42] have used a diffuser mounted outside the NCAR C130 that has an internal removable cone with a back plate. This collects all aerosol particles entering the inlet, which are subsequently washed off and analysed. This offers a good measure of the total particle mass but only limited analyses can be performed and no size segregation is possible. The Denver University Low Turbulence Inlet (DULTI) (Wilson and coworkers [43]) uses a porous stainless steel diffuser cone through which boundary layer air is removed and is pumped away. A schematic of the DULTI inlet is shown in figure 6. This allows the sample flow to be retarded to the desired speed whilst remaining laminar. However it is a complicated inlet with very high flow rates and hence powerful pumps are required. It is necessary to balance the sample and extracted flows as a function of pressure to maintain laminarity, which is measured by a hot wire probe. Typically 500 lpm is extracted through the porous inlet and 100 lpm is sampled. The flows mean that large particle concentrations will be enhanced in the sample line as they do not follow the streamlines through the diffuser and furthermore these enhancements increase with the size of the particle. Wilson et al. have used Computational Fluid Dynamics (CFD) modeling to calculate these enhancements and demonstrated that these are consistent with mass and number concentration measurements made as a function of size.

The LTI has been tested using two FSSP-300 instruments (see below) one mounted in the free airstream and one sampled from the LTI. The results were inconclusive as the LTI FSSP had to be modified to sample on the LTI and there were discrepancies between the two FSSPs when sampled in different wing locations.

Other specialised inlets have also been developed. The family of Counter-flow Virtual Impactor inlets (CVI) are such an example. These inlets seek to discriminate large particles from small ones by implementing a porous diffuser in a reverse sense to the DULTI. Particle free air is pushed through a porous steel section in the inlet nozzle at a flow rate that is a little larger than the sample flow. This means that a small fraction of the flow is pushed outwards and a stagnation point is set up close to the inlet tip. As a result only large particles have sufficient momentum to overcome the retarding force and enter the sample stream. The flows are set such that particles of a few microns diameter will be sampled and can be varied with some tolerance [44]. The CVI was originally developed to discriminate cloud particles from smaller aerosol and to measure the residual particles within the cloud elements after the water has been evaporated by the entrained counterflow [45]. Such inlets have undergone substantial development and characterisation to improve the sharpness of the cut-off size. Laboratory and Modelling work by Noone et al., 1988 [46], Anderson et al., 1993[47], and Lin and Heintzenberg, 1995[48] has shown improvements in the sharpness of the cut off diameter. Experimentally determined measurement of the sampled cloud droplet residual number appears to agree with the droplet number concentration to within measurement error [49], though modelling results show that losses are predicted to be higher [50].

The alternative method for large particle sampling is to place the sensor in the free stream. An instrument that has used this approach to measure cloud droplets and has been a standard instrument in its field is the Forward Scattering Spectrometer Probe (FSSP). First probes were flown in the late 1970s and were originally designed by Knollenberg [51]. The probe measures the cloud droplet size distribution for sizes between 1 and 50 microns in three different ranges. The probe is housed in a pod, typical mounted on the wing of the aircraft. It samples the light scattered in the forward direction by cloud droplets passing through the focal point of a 632 nm He-Ne laser. The focal point is positioned in front of the probe and is in the free stream of droplets entering the instrument. The scattering of light by cloud droplets follows Mie theory [52] and so there is a strong relationship between scattered intensity and particle size. Unfortunately, in the size range of cloud droplets, 5 to 20 μm the relationship shows oscillatory behaviour (figure 7).

Since that time the instrument has been used in many experiments. The probe can miss some droplets at high liquid water contents due to the instrument dead time and also due to coincidence [53]. Brenguier et al. [54] introduced the Fast FSSP, an improved version of the instrument, by increasing the size range, number of channels and speed of electronics to reduce the dead time correction. Cloud droplets are spherical and therefore follow Mie scattering predictions. However ice crystals are irregular in shape and this makes retrieval of ice crystal number distributions from FSSPs far from straightforward, especially in mixed phase clouds where ice crystals are often larger in size. Complex radiation calculations have been undertaken for symmetrical oblate spheroids using so-called T-Matrix methods to describe modifications to the Mie functions and hence the scattering signals measured by the FSSP [55]. These have been extended to hexagonal column shapes [56] but the complexity of real ice crystals cannot be replicated and the variety in crystals shapes means that these inversions can only be partially successful.

To assess ice particle shapes imaging probes have previously been used based on diode array technology [57]. However electro-optical limitations have meant that image quality is relatively poor and smaller developing ice is difficult to detect. More recently, combinations of these instruments have been combined in a single probe, for example the Cloud, Aerosol and Precipitation Spectrometer – CAPS [58]. More recently, CCD technology has led to the development of a new generation of probes that also sample in the free stream and image ice crystals with much improved resolution, for example the Cloud Particle Imager, CPI [59], figure 8. The CPI clearly has excellent imaging capability and offers a major advancement in ice physics research. However, establishing the sample volume

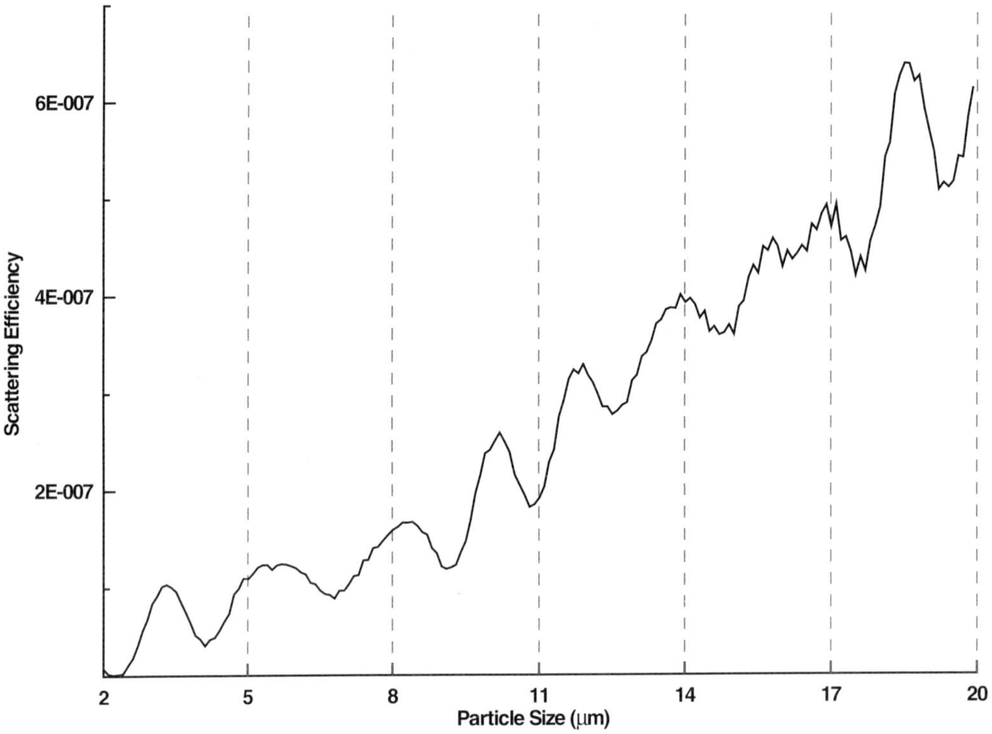

Figure 7. Relationship between particle size and scattering intensity for liquid droplets at a wavelength of 632nm.

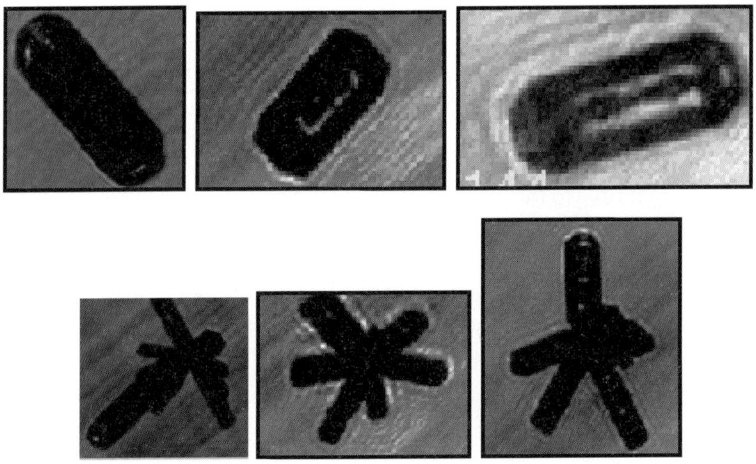

Figure 8. Examples of images from the Cloud Particle Imager [60].

and accounting for image sizing still requires considerable calibration and a combination of sample probes [60].

It is extremely challenging and expensive to characterise such inlets, Wendisch et al [33] discuss both experimental and modelling approaches. For example Twohy [39] discusses both wind tunnel and model testing of a CVI inlet system to establish the benefits of an additional shroud. Chen et al., [61] have also used CFD calculations to test the dependence of angle of attack on their CVI inlet. An aerosol inlet was installed on a commercial Airbus A-340 during the CARIBIC experiment and sampled air into a range of instruments in a container in the hold. The inlet was modelled using CFD over the range of attack angles and airspeeds of the aircraft as the commercial jet flies at higher speeds than are normal for atmospheric measurements and hence there is greater importance on correct inlet design [62]. The CFD modeling allows the leading edges flows and shrouds to be redesigned early in the development process and before large capital is expended interfacing the inlet with the aircraft.

4. EXAMPLES OF AIRBORNE SCIENCE

This section offers some examples of aircraft based research projects to highlight both the way aircraft can be used as an important platform in a range of research areas in atmospheric science and also to offer some interesting scientific results.

Aerosol particles scatter and absorb radiation in the atmosphere and as a result are climatically import. One of the key aerosol types results from the burning of biomass. Savannah fires are widespread in Africa towards the end of the dry season and large areas are affected by significant quantities of biomass burning aerosol which contains significant quantities of absorbing carbon, or soot. The aerosol can both scatter radiation back to space and so cool the surface or absorb the radiation and hence warm the lower layers. Its net effect is not only dictated by its own properties but by the properties of other parts of the column and the surface characteristics. Aircraft can play a key role in this type of science as they can physically and chemically characterise the aerosol in situ and measure its radiative properties over a region. They can therefore probe processes that affect the radiative impact of the aerosol, assess how the aerosol changes in the column affect the radiative budget and validate ground and satellite measurements of the aerosol impact on the column. The South AFrican Aerosol Regional science Initiative (SAFARI) involved a significant number of aircraft studying biomass burning aerosol. The Met Office C-130 aircraft was based in Windhoek, Namibia during the experiment and conducted a wide range of missions [63].

An example of a typical flight plan is shown in figure 9. The aircraft flew to an area above the Etosha AERONET site. The AEROsol robotic NETwork (AERONET) programme is a global system of ground based sun photometers which derive an aerosol optical depth (AOD) in the vertical column by conducting an amlacantor scan. Details of AERONET and the inversion process can be found at (http://aeronet.gsfc.nasa.gov/). During the amlacantor operation the photometer points directly at the Sun and measures the radiative flux. It then performs a series of measurements of the radiative flux whilst precessing around the vertical axis at a fixed angle that is equal to the solar zenith angle. This provides a series of flux measurements which include a range of scattering angles from which the AOD, the integrated loss of radiation in the vertical column due to its extinction by aerosol, can be determined from an inversion based on radiative transfer calculations.

The aircraft flew a stacked profile descent to establish the physical and chemical properties of the aerosol in a series of layers above the AERONET site and to characterise the profile. This was followed by two straight and level runs (SLRs) at low level directly towards and away from the Sun. This served two purposes. Firstly, it provided a low level characterisation of the biomass burning aerosol and secondly it was used to assess the radiative flux through the column and estimate the surface contributions. An obscurer is mounted to the aft of the broad band radiometers so that in down Sun legs only the scattering component of the radiation is measured. This allows a discrimination of the direct radiation in the presence of aerosols to be made. When coupled with model predictions of the radiative flux in the absence of aerosols the AOD along the flight track can be obtained and compared with regional satellite observations if they are conducted close in time and space to a satellite overpass.

The last manoeuvre involved putting the aircraft into a steep banking turn for four complete revolutions at low level under the majority of the aerosol layer. This is not simply a stunt, by banking the aircraft at an angle equal to the solar zenith angle and flying in an orbit the upward pointing broad band radiometers on the aircraft replicate the almacantor operation of the AERONET sites.

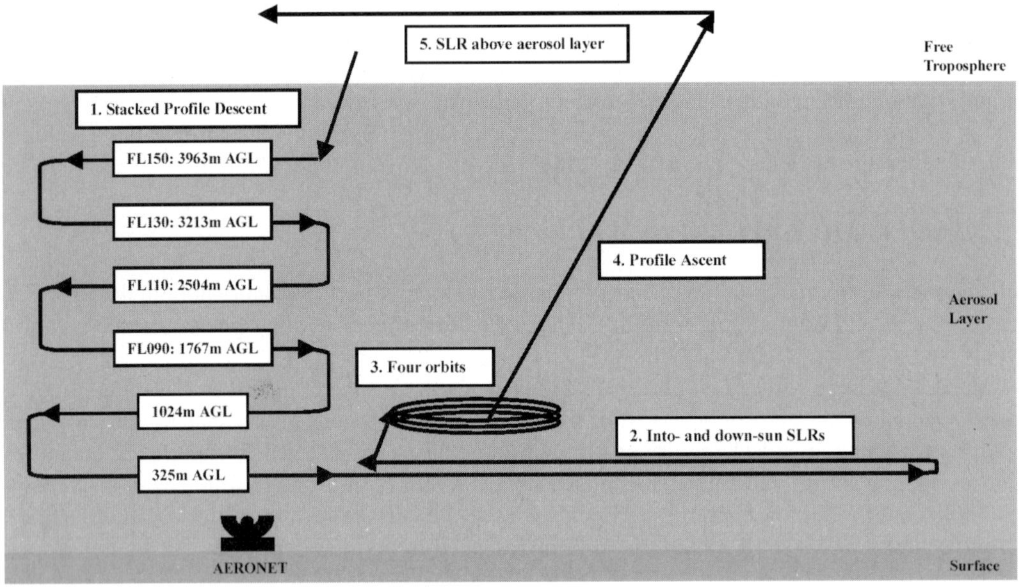

Figure 9. A typical flight plan of the C-130 during the SAFARI experiment [63]. Such maneouvres are typical of many flights focusing on the radiative properties of aerosols.

Solar zenith angles of 40 degrees or more are required to obtain the necessary range of scattering angles. The aircraft then ascended through the layer to obtain a profile and then conducted a final high level straight and level run to establish the radiative flux above the layer in the up and down welling directions.

These data have been used to test global model predictions. Global model studies of biomass burning in the region had previously shown that much of the biomass burning aerosol advects westwards over the Atlantic Ocean. As the model considered the aerosol in isolation the partially absorbing properties of the aerosol did not have a significant impact as the underlying ocean surface was highly absorbing anyway. As a result global models predicted the aerosol was predominately scattering and led to a cooling with a radiative forcing of -3 W m^{-2} or more over much of the region. However, the actual observations taken during SAFARI showed that low level strato cumulus cloud was prevalent over the region throughout the study and that the aerosol advecting from the continent overlaid this layer (figure 10a). The absorbing effects of aerosol were observed to be much more significant above such a reflective layer as a cloud and this leads to a calculated warming. Kiel and Haywood [64] then used climatological occurrences of cloud to predict the radiative forcing which was shown to be largely warming and of a few W m^{-2}(figure10b). This demonstrates the importance of in situ measurement in studying such effects.

Multi-aircraft studies are ideally suited to the investigation of processes occurring on hour to day timescales either through describing regional airmasses for model constraint or for following single parcels in a lagrangian framework. One of the best examples of this latter method was during the ACE-2 study. As previously described the ACE-2 study was based in the north eastern Atlantic Ocean [23]. Part of that study included a lagrangian experiment. Meteorological forecasts were used to predict when European air was likely to be advected southwards towards the Canary Islands and air parcel following balloons were used to tag the air mass. A full description of the experimental details of these lagrangian studies can be found in Johnson et al., [65]. An example of the marine boundary layer development is shown in figure 11 [66].

The above level of detail cannot be gained reliably from operational models but is necessary to interpret the aerosol and cloud transformations during the experiment. This analysis was gained from aircraft profiles obtained through the lagrangian period.

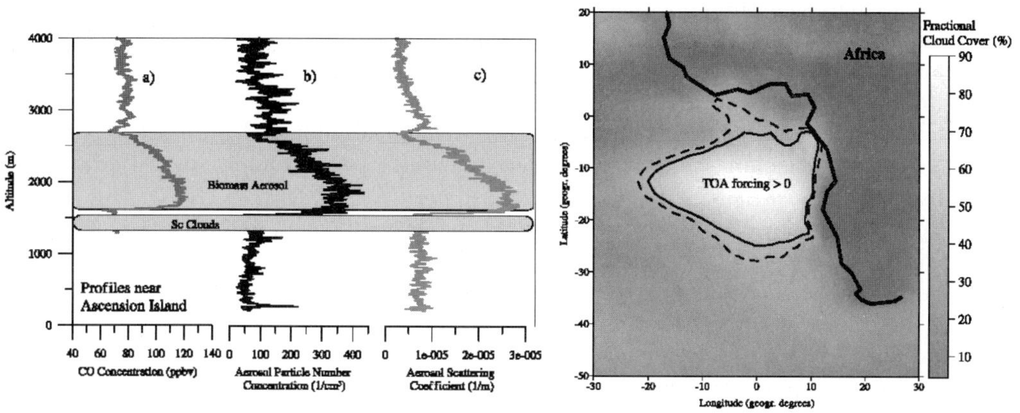

Figure 10. The left hand panel shows vertical profiles of CO, aerosol particle number concentration and scattering coefficient taken through a biomass burning layer overlying a stratocumulus cloud off the Namibian coast. The right hand panel shows the climatological occurrence of cloud in the region and identifies the area under the influence of biomass burning aerosols gives rise to a few W m^{-2} of warming, rather than a cooling as predicted by global models.

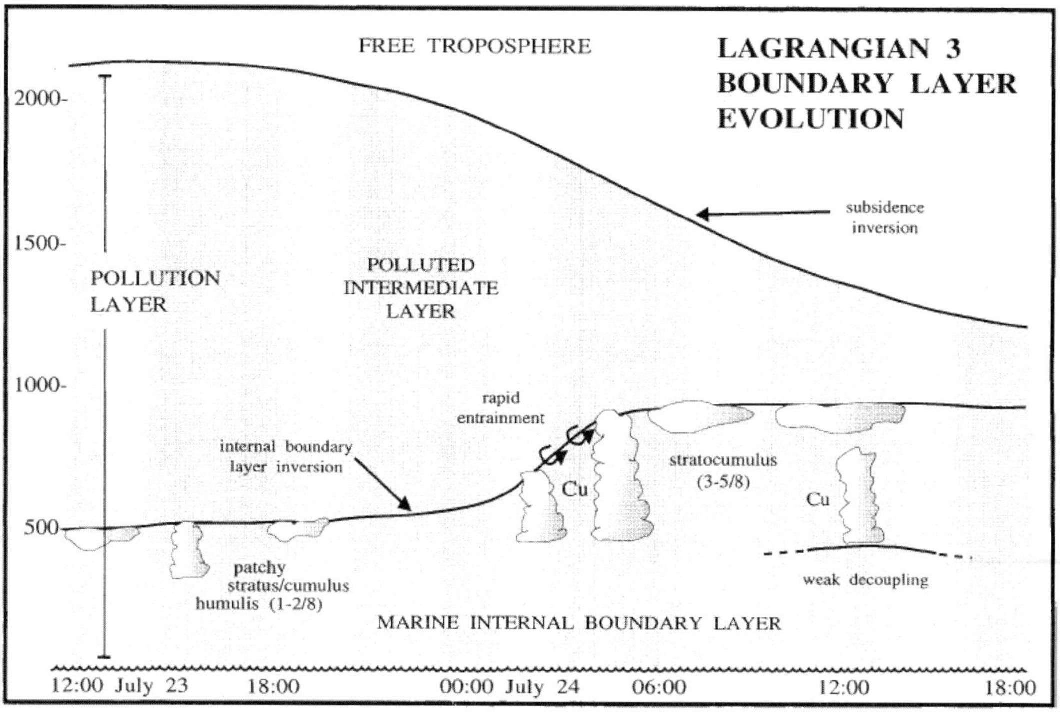

Figure 11. Schematic of the development of the marine boundary layer in the eastern northern mid latitude Atlantic Ocean during the ACE-2 study. The schematic is essentially a north to south transect. Close to the left of the figure the marine boundary layer is smaller in the colder northern waters and the relative dry air produces little cloud. As the air moves southward and becomes moister the boundary layer deepens and fills with cloud. The cloud layer decouples from the surface.

Wood et al. [66] show from this data set that although the cloud filled polluted boundary layer processes aerosol much faster than the dry, overlying polluted layer, the entrainment and exchange between the two layers is sufficient to keep the particulate properties in the two layers similar through the experiment. This extends the lifetime of pollution in the marine boundary layer and delays transition from polluted to marine conditions by several days. Given that pollution controls the cloud droplet number this has important implications for the effect of marine clouds on climate over wide areas of the northern hemisphere.

The lagrangian approach was also used during the International Consortium for Atmospheric Research on Transport and Transformation (ICARTT) experiment during 2004. An overview of the programme is described in detail by Fehsenfeld et al. [67], who also discusses the wing tip to wing tip flights that were carried out between multiple aircraft to intercompare the sensors. This is expensive to carry out and very time consuming but necessary if data from multiple platforms are to be used in a common data set. Part of this work included a lagrangian study of polluted air leaving the eastern seaboard and advecting across the Atlantic Ocean. Four aircraft were involved in this study: the NASA DC-8; the NOAA P-3; the UK BAE-146 FAAM aircraft; and the DLR Falcon. The DC-8 and P-3 were based at Pease airport in New England, the FAAM aircraft was based in the Azores and the Falcon was based at Criel in northern France. Forward trajectories were calculated from forecast wind fields from the main areas of pollution in the north east USA and across the Atlantic Ocean. When air masses carried pollution across the operating regions of more than one of these aircraft the science teams in the three locations were notified and a coordinated mission planning strategy was implemented. This was conducted through a combination of intercontinental conference calls and web forecast products. Once the P3 or DC-8 had sampled the forecasted air parcels, forward trajectories were calculated on forecast wind fields for each point on the aircraft trajectory. This was repeated with every new analysed wind field to tune the forecasted trajectories. These forecasts were used to guide either or both of the BAE146 or Falcon aircraft to sample the same air.

After the experiment, several methods of establishing a lagrangian match were developed by Methven et al. [68]. The first of these involved calculating forward and backward trajectories from each of the aircraft for every point on each of the flight tracks conducted using the analysed wind fields. Each point on each trajectory was tested to see if it coincided with another aircraft track within a certain tolerance. If both forward and back trajectories coincided then the pair of points was identified as a match. This criterium was strengthened by including an assessment of the differences between conserved thermodynamic variables, equivalent potential temperature and water vapour mixing ratio. A similar constraint was also provided by ratios of long lived volatile organic carbon species. Methven et al. used a probability density function analysis to show that the combination of all these methods improves the success of the lagrangian matches over any single technique. These criteria have so far been used by Arnold et al. [69], who have used the lagrangian matches to quantify regional hydroxyl radical concentrations and assess the air mass dilution rates along the parcel trajectories. This was the first time these methods were attempted and have shown that future lagrangian studies using such forecasting tools offer a powerful way of conducting long range transport studies in a lagrangian framework.

Understanding the microphysics of cirrus is of critical importance for assessing the Earth's radiative balance. Cirrus clouds cover substantial areas of the globe at any one time and affect the radiative transfer through the atmospheric column in a number of ways. Visible radiation is reflected by the clouds back to space and hence the cirrus has a cooling effect on climate. In addition, water absorbs upwelling infra-red radiation very efficiently. Low level clouds simply reradiate the absorbed energy at a blackbody temperature that is similar to the Earth's surface and so do not affect the infra red budget. However, cirrus clouds, especially in the tropics, are often at altitudes above 10 km and exist at temperatures as low as 200 to 230 K and hence the absorption of infrared warms the layer and leads to a significant re-radiation in the downward direction that would not be there in the absence of these clouds. At the present time the microphysical processes that govern these clouds behaviour is very poorly known yet must be understood if their effect on climate is to be modelled effectively and feedbacks from future changes in water vapour predicted correctly.

The microphysical processes themselves occur on small scales and cannot be followed in sufficient detail using satellite data alone. Aircraft are required to probe the clouds themselves and assess the properties of the ice crystals and the dynamics in the cloud so that basic cloud processes can be interpreted and linked to the radiative properties of the clouds. The Egret Microphysics with Extended Radiation And LiDar (EMERALD) experiment sought to conduct such an experiment [70]. The experiment was based in Flinders, near Adelaide, Australia and was focused on studying frontal cirrus developing in advance of frontal systems. Frontal cirrus occurs in very thin layers and extends over considerable horizontal distances. Areas of precipitation and development in updrafts and waves occur, making probing the clouds with aircraft challenging. The approach used was to fly two aircraft in tandem in a box pattern. The lower aircraft was the Airborne Research Australia (ARA) Super King Air, which was instrumented with a vertical pointing LIDAR. The LIDAR was able to provide information on the cloud altitude, depth and extent and so guide the upper aircraft into position to directly sample the cloud. This entailed mid air direct communication between the aircraft and direct linkage between the scientists on board the two aircraft but meant that many of the problems of targeting cirrus clouds could be overcome. Figure 12 shows the structure of the cloud as measured by the LIDAR and the superimposed flight track of the upper aircraft being guided by this signal in flight.

The upper aircraft was a Grob 520T also operated by ARA. It sampled cloud particles using an FSSP-100 and a Cloud Particle Imager, water vapour was measured using a chilled mirror hygrometer, specially designed for high altitude work by DLR and turbulence was measured using a "Best Atmospheric Turbulence" (BAT) probe designed by NOAA/ATDL, Flinders University, Australia and ARA. Details of these instruments are available in Gallagher et al., [60], who also describe the ice microphysics and turbulent structure of these clouds in some detail. The CPI provided high resolution images of ice particles and allowed new insight into cirrus microphysics. The experiments showed that a significant feature of the clouds was the occasional columns of active convective columns.

Large variations in the cirrus dynamics and significant variation in the ice crystal habit from cloud top to cloud base and within the evaporating fall streaks of the precipitating ice were observed. Examples of the ice crystal habits can be seen in figure 12. The findings are consistent with a region of high supersaturation in the upper part of the cloud acting as a region for rapid ice generation zone and a location for rapid growth. In this region the CPI showed that the ice was dominated by small irregular or spheroidally shaped particles, some of which can be seen to be initially forming rosettes. As these grow into larger particles, they fall and form larger rosettes or columns which are observed in the middle and lower parts of the cloud. Aircraft coordination such as this, though complex, shows that missions of this type really maximise the aircrafts' capabilities to explore detailed mechanisms in complex atmospheric structures.

5. SUMMARY

This paper discussed the advantages and disadvantages of a range of the most commonly used platforms in atmospheric science. The key role of aircraft is highlighted and the range of aircraft capabilities was demonstrated. The complimentarity of research aircraft was also shown several times in the final section, where a range of aircraft programmes was introduced and the benefits of combined aircraft missions highlighted. The cost of certifying aircraft, inlets and instruments was emphasised and a detailed discussion of instruments and inlets was provided. The difficulties of making even the most basic of measurements accurately on an aircraft were stressed, using examples of temperature and water vapour, which showed that providing accurate and rapid water vapour measurements reliably over the wide ranges of concentrations experienced in the atmosphere is not yet routinely possible. Even for basic measurements inlet systems need to be carefully designed and validated. Examples of more complex inlets for instruments that measure more complex parameters were discussed, examples

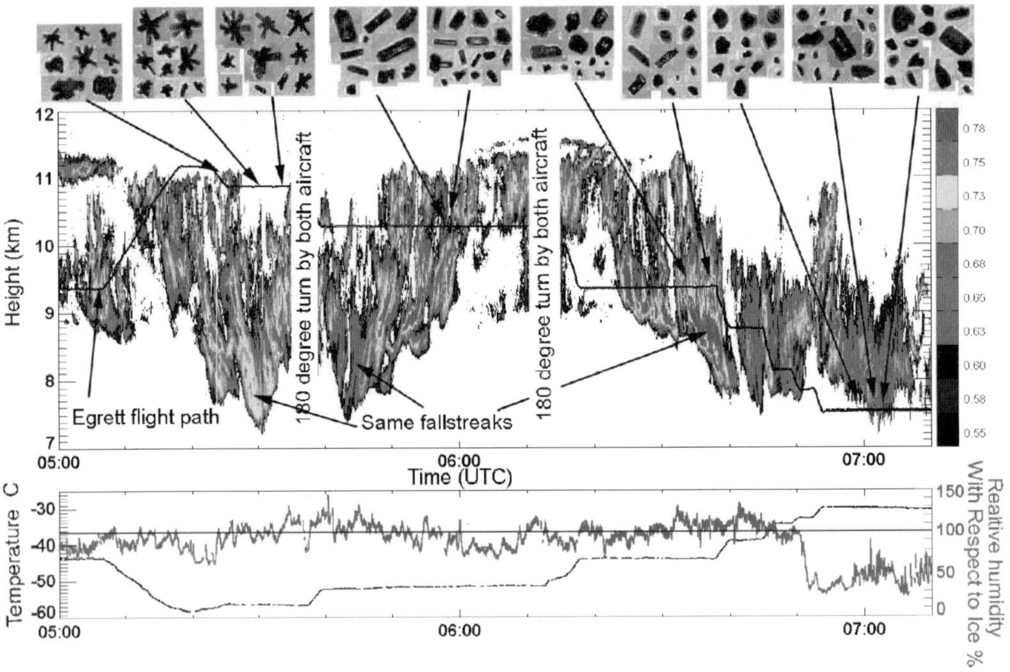

Figure 12. The main panel shows the LIDAR backscatter from cirrus clouds during the EMRALD project as measured from the ARA Super King Air aircraft. Flight maneouvres are annotated on the panel which also shows (black line) the flight path of the Grob Egret aircraft making simulataneous in situ measurements. The Egrett was being guided in flight by the King Air LIDAR information in real time. The bottom panel shows the measured temperature and relative humidity as measured by the Egret aircraft. This can be compared with images from the CPI (top panels) which shows the range of ice crystal habits observed during the flight.

included sticky gases, radicals and aerosol particles. The need for well though out inlet designs was clear and the necessity for computational fluid dynamics modelling and wind tunnel testing was demonstrated. Instruments that sample cloud particles in the free stream were used as examples of the need for minimising the inlet system. Large particle inlets were shown as examples of where the aim was to remove the influence of the inlets on the sample using engineering approaches. What remains clear is that there is no perfect way of treating particles and clouds before sampling but a detailed knowledge of behaviour of the probe is necessary to understand the measurements.

Several examples of complex measurement campaigns were provided, however, the planning requirements necessary to complete these experiments effectively, were not discussed. The planning for such experiments often is well underway before the money for conducting the experiment becomes available. Scientists must first define their scientific objectives, and then carefully plan how they will achieve them by choosing carefully their measuring platforms and their relative roles. If aircraft are being used, their capabilities must be carefully thought through and the fit designed to be appropriate for the work. There are then a series of operational questions that need to be addressed that involve the aircraft management team and the aircrew. These include the location for the field operations and the identification of the operating airfield. This is often not straightforward. The runway length and load capacity need to be considered, together with the crash category of the airport. In addition, to these basic requirements, details such as refuelling capability, airport opening hours, hangarage, and a location for the ground based coordination of the operations may all be very important constraints for the suitability of an airfield. Often there are a large number of people associated with such experiments. These need to be accommodated, transported to and from the airfield and given clearance to the airfield to fly.

The pilots of the aircraft must be able to carry out the scientific plans for flight operation safely. Restricted airspace may limit the area in which operations can take place and flight clearance from the national aviation authorities is often necessary. Airways and military operations must be avoided and other operational restrictions need to be identified and adhered to. For example, when sampling low level clouds it may not be possible to fly underneath them as their cloud base may be below the minimum safe altitude. In addition, if the clouds are at sub-zero temperatures then it may not be possible to sample them at all as freezing cloud water on the aircraft, or riming, must be avoided. The usual avoidance is to descend to warmer levels to melt the ice but this will not be possible if the cloud base is too low and hence the cloud cannot be sampled at all under such circumstances. The scientists and pilots will agree the general flight plans in advance of the start of the experiments but the specific operations on a given day will be carefully worked through by a ground planning team on a daily basis. The team will use a range of meteorological and chemical forecast models and satellite products to produce a plan the day before a given flight. These will be submitted to the pilots who will ensure there are no operational difficulties and amend if necessary. The plan will be submitted to air traffic control ahead of the aircraft take off.

The complexity of aircraft operations and technical and experimental challenges with making the measurements is high, however, aircraft offer unique sampling platforms and as such make a valuable contribution to atmospheric science and are likely to do so for the foreseeable future.

Acknowledgment

The author is grateful to Rachel Burgess for typesetting and editing the final versions of this paper.

References

[1] Keeling C.D., *Tellus*, **12** (1960) 200-203.

[2] Keeling C.D., T.P. Whorf, M. Wahlen and J. Vanderplicht, *Nature* **375** (1995) 666-670.

[3] Keeling C.D., J.F.S. Chin and T.P. Whorf, *Nature* **382** (1996) 146-49.

[4] Henning S., E. Weingartner, M. Schwikowski, H.W. Gäggeler, R. Gehrig, K.-P. Hinz, A. Trimborn, B. Spengler and U. Baltensperger, *J. Geophys. Res.*, **108** (2003).

[5] O'Dowd, C.D., M.C. Facchini, F. Cavalli, D. Ceburnis, M. Mircea, S. Decesari, S. Fuzzi, Y.J. Yoon and J.P. Putaud, *Nature*, **431**, (2004), 7009, 676-680.

[6] O'Dowd, C.D., M.H. Smith, I.E. Consterdine and J.A. Lowe, *Atmos. Environ.*, **31**, (1997), 73-80.

[7] Choularton T.W., R.N. Colvile, K.N. Bower, M.W. Gallagher, M. Wells, K.M. Beswick, B.G. Arends, J.J. Mols, G.P.A. Kos, S. Fuzzi, J.A. Lind, G. Orsi, M.C. Facchini, P. Laj, R. Gieray, P. Wieser, T. Engelhardt, A. Berner, C. Kruisz, D. Moller, K. Acker, W. Wieprecht, J. Luttke, K. Levsen, M. Bizjak, H.C. Hansson, S.I. Cederfelt, G. Frank, B. Mentes, B. Martinsson, D. Orsini, B. Svenningsson, E. Swietlicki, A. Wiedensohler, K.J. Noone, S. Pahl, P. Winkler, E. Seyffer, G. Helas, W. Jaeschke, H.W. Georgii, W. Wobrock, M. Preiss, R. Maser, D. Schell, G. Dollard, B. Jones, T. Davies, D.L. Sedlak, M.M. David, M. Wendisch, J.N. Cape, K.J. Hargreaves, M.A. Sutton, R.L. Storeton-West, D. Fowler, A. Hallberg, R.M. Harrison and J.D. Peak, *Atmos. Environ.*, **31**, (1997), 2393-2405.

[8] Bower K.N., T.W. Choularton, M.W. Gallagher, K.M. Beswick, M.J. Flynn, A.G. Allen, B.M. Davison, J.D. James, L. Robertson, R.M Harrison, C.N. Hewitt, J.N. Cape, G.G. McFadyen, C. Milford, M.A. Sutton, B.G. Martinsson, G. Frank, E. Swietlicki, J. Zhou, O.H. Berg, B. Mentes, G. Papaspiropoulos, H.C. Hansson, C. Leck, M. Kulmala, P. Aalto, M. Vakeva, A. Berner, M. Bizjak, S. Fuzzi, P. Laj, M.C. Facchini, G. Orsi, L. Ricci, M. Nielsen , B.J. Allan, H. Coe, G. McFiggans, J.M.C. Plane, J.L. Collett, K.F. Moore and D.E. Sherman, *Tellus* **52(2)** (2000) 750-778.

[9] Norton E.G., G. Vaughan, J. Methven, H. Coe, B. Brooks, M. Gallagher, and I. Longley *Atmos. Chem. Phys.*, **6** (2006) 433-445.
[10] de Foy, B., A. Clappier, L.T. Molina and M.J. Molina, Atmos. Chem. Phys., **6**, (2006), 2321-2335.
[11] Platt U., J. Rudolph, T. Brauers and G.W. Harris, *J. Atmos. Chem.* **15** (1992) 203-214.
[12] de Gouw J.A., A.M. Middlebrook, C. Warneke, P.D. Goldan and W.C. Kuster, *J. Geophys.Res. – Atmos.* **D16** (2005) Art. No. D16305.
[13] Pommereau J.P., F. Dalaudier, J. Barat, J.L. Bertaux, F. Goutail and A. Hauchecorne, *Adv. Space Res.* (1985) 27-30.
[14] Larsen, N., B.M. Knudsen, S.H. Svendsen, T. Deshler, J.M. Rosen, R. Kivi, C. Weisser, J. Schreiner, K. Mauerberger, F. Cairo, J. Ovarlez, H. Oelhaf and R. Spang, *Atmos. Chem. Phys.*, **4**, (2004), 2001-2013.
[15] Borchi F., J.P. Pommereau, A. Garnier and M. Pinharanda, *Atmos Chem. Phys.*, **5**, (2005) 1381-1397.
[16] Richter A., J.P. Burrows, H. Nüß, C. Granier and U. Niemeier, *Nature* **437** (2005) 129-132.
[17] Sinnhuber B.-M., A Rozanov, N. Sheode, O.T. Afe, A. Richter, M. Sinnhuber, F. Wittrock, J.P. Burrows, *Geophys. Res. Lett.* (2005) doi:10.1029/2005GL023839.
[18] Bracher A., H. Bovensmann, K. Bramstedt, J.P. Burrows, T. von Clarmann, K.-U. Eichmann, H. Fischer, B. Funke, S. Gil-López, N. Glatthor, U. Grabowski, M. Höpfner, M. Kaufmann, S. Kellmann, M. Kiefer, M.E. Koukouli, A. Linden, M. López-Puertas, G. Mengistu Tsidu, M. Milz, S. Noel, G. Rohen, A. Rozanov, V.V. Rozanov, C. von Savigny, M. Sinnhuber, J. Skupin, T. Steck, G.P. Stiller, D.-Y. Wang, M. Weber and M.W. Wuttke, *Adv. Space Res.*, **36** (2005) 855-867.
[19] Edwards, D.P., L.K. Emmons, D.A. Hauglustaine, A. Chu, J.C. Gille, Y.J. Kaufman, G. Pétron, L.N. Yurganov, L. Giglio, M.N. Deeter, V. Yudin, D.C. Ziskin, J. Warner, J.-F. Lamarque, G.L. Francis, S.P. Ho, D. Mao, J. Chan and J.R. Drummond, *J. Geophys. Res.*, 109, (2004) D24202.
[20] Pfister, G., P.G. Hess, L.K. Emmons, J.-F. Lamarque, C. Wiedinmyer, D.P. Edwards, G. Pétron. J.C. Gille and G.W. Sachse, *Geophys. Res. Lett.*, **32**, (2005), L11809.
[21] Kaufman Y.J., I. Koren, L.A. Remer, D. Tanre', P. Ginoux, and S. Fan, *J. Geophys. Res. – Atmos.* **110** (2005) doi:10.1029/2003JD004436.
[22] Kahn R., B. Gaitley, J. Martonchik, D. Diner, K. Crean and B. Holben *J. Geophys. Res. – Atmos.* **110** (2005) doi:jd004706R.
[23] Raes F., T.S. Bates, F. McGovern and M. van Liedekerke, *Tellus* **52B**, (2000) 111-125.
[24] Lawson R.P. and W.A. Coopper *J. Atmos. Oceanic Technol.* **7** (1990) 480-494.
[25] May R.D. *J. Geophys. Res-Atmos.* **103** (1998) *98JS01678.*
[26] Eisele, F.L. and D.J. Tanner, *J. Geophys. Res*, **96**, (1991), 9295-9308.
[27] Brune W.H., I.C. Faloona, D. Tan, A.J. Weinheimer, T. Campos, B.A. Ridley, S.A. Vay, J.E. Collins, G.W. Sachse, L. Jaeglé and D.J. Jacob, *Geophys. Res. Lett.*, **25**, (1998) 1701-1704.
[28] Neuman J.A., L.G. Huey, T.B. Ryerson and D.W. Fahey, *Environ. Sci. Technol.* **33** (1999) 1133.
[29] Ryerson T.B., L.G. Huey, K. Knapp J.A. Neuman, D.D. Parrish, D.T. Seuper and F.C. Fehsenfeld, *J. Geophys. Res.*, **104** (1999) 5483.
[30] Neuman J.A., L.G. Huey, R.W. Dissly, F.C. Fehsenfeld, F. Flocke, J.C. Holecek, J.S. Holloway, G.H.R. Jakoubek, D.K. Nicks Jr., D.D. Parrish, T.B. Ryerson, D.T. Sueper and A.J. Weinheimer, *J. Geophys. Res.*, **107(D20)** (2002) 4436.
[31] Fahey D.W., C.S. Eubank, G. Hübler, and F.C. Fehsenfeld, *J. Atmos. Chem.*, **3**, (1985) 435– 468.
[32] Patz H.-W., A. Volz-Thomas, M.I. Hegglin, D. Brunner, H. Fischer, and U. Schmidt *Atmos. Chem. Phys. Discuss.* **6** (2006) 649-671.
[33] Wendisch, M., H. Coe, D. Baumgardner, J.-L. Brenguier, V. Dreiling, M. Fiebig, P. Formenti, M. Herrmann, M. Kramer, Z. Levin, R. Maser, E. Mathieu, P. Nacass, K. Noone, S. Osborne, J. Schneider, L. Schutz, A. Schwarzenbock, F. Stratmann and J.C. Wilson, *Bulletin Am. Meteorol. Soc.*, (2004), 89-91.

[34] Wilson, J.C. and W.R. Seebaugh, Measurement of Aerosol from Aircraft in Aerosol Measurement: Principles, Techniques, and Applications, ed. P.A. Baron and K. Willeke, (Wiley-Interscience, 2001).
[35] King W.D., *J. Atmos. Oceanic Technol.*, **1** (1984) 5–13.
[36] King, W.D., D.E. Turvey, D. Williams and D.J. Llewellyn, *J. Atmos. Oceanic Technol.*, **1** (1984) 14-21.
[37] King, W.D., *J. Atmos. Oceanic Technol.*, **2** (1985) 539-547.
[38] King, W.D., *J. Atmos. Oceanic Technol.*, **3** (1986) 433-440.
[39] Twohy C.H., *Aerosol Sci. and Tech.*, **29**, (1998) 261-280.
[40] Baron P. and K. Willeke in Aerosol Measurement - Principles, Techniques, and Applications edited by Baron P. and K. Willeke, 2nd Edition (John Wiley & Sons, 2001).
[41] Fiebig M., in Das troposphärische Aerosol in mittleren Breiten – Mikrophysik, Optik und Klimaantrieb am Beispiel der Feldstudie LACE 98. PhD thesis, (Ludwig-Maximilians Universität München, ISSN 1434-8454, 2001).
[42] Huebert, B.J., S.G. Howell, D. Covert, T. Bertram, A. Clarke, J.R. Anderson, B.G. Lafleur, W.R. Seebaugh, J.C. Wilson, D. Gesler, B. Blomquist and J. Fox, Aerosol Sci. Technol., **38**, (2004), 803-826.
[43] Wilson J.C., B.G. Lafleur, H. Hilbert, W.R. Seebaugh, J. Fox, D.W. Gesler, C.A. Brock, B.J. Huebert and J. Mullen, *Aerosol Sci Technol.*, **38**, (2004), 790-802.
[44] Schwarzenböck, A. and J. Heintzenberg, *J. Aerosol Sci.*, **31** (2000) 477-489.
[45] Ogren J.A., J. Heintzenberg and R.J. Charlson, *Geophys. Res. Lett.*, **12** (1985) 121-124.
[46] Noone, K.J., R.J. Charlson, D.S. Covert, J.A. Ogren and J. Heintzenberg, *Aerosol. Sci. Technol.*, **8** (1988) 235-244.
[47] Anderson, T.L., R.J. Charlson and D.S. Covert, *Aerosol. Sci. Technol.*, **19** (1993) 317-329.
[48] Lin, H. and J. Heintzenberg, *J. Aerosol Sci.*, **26** (1995) 903-914.
[49] Glantz P., K.J. Noone and S.R. Osborne, *J. Atmos. Oceanic Technol.*, **20** (2003)133-142.
[50] Laucks M.L. and C.H. Twohy, *Aerosol Sci. and Tech.*, **28** (1998) 40-61.
[51] Knollenberg, R.G. In Clouds: Their Formation, Optical Properties and Effects. Edited by P.V. Hobbs and A. Deepak, (Academic, San Diego, 1981) pp. 15-19.
[52] C.F. Bohren, D.R. Huffmann: Absorption and scattering of light by small particles. New York, Wiley-Interscience, 1983.
[53] Baumgardner D., J.W. Strapp and J.E. Dye, *J. Atmos. Ocean. Tech.* **2** (1985) 626-632.
[54] Brenguier, J.L., T. Bourrianne, A.D. Coelho, J. Isbert, R. Peytavi, D. Trevarin and P. Weschler, *J. Atmos Oceanic Technol.*, **15**, (1998), 1077-1090.
[55] Borrmann, S., B. Luo and M. Mishchenko, *J. Aerosol Sci.* **31**, (2000) 789-799.
[56] Havemann S. and A.J. Baran, *J. Quant. Spectroscopy and Radiative Transfer* **89** (2004) 87-96.
[57] Knollenberg R.G., *J. Appl. Meteor.*, **9** (1970) 86-103.
[58] Baumgardner D., H. Jonsson, W. Dawson, D. O'Connor and R. Newton. *Atmos. Res.*, **59-60** (2001) 251-264.
[59] Lawson R.P., A.J. Heymsfield, S.M. Aulenbach and T.L. Jensen. *Geophys. Res, Lett.*, **25** (1998) 1331-1334.
[60] Gallagher M.W., P.J. Connolly, J. Whiteway, D. Figueras-Nieto, M. Flynn, T.W. Choularton, K.N. Bower, C. Cook, R. Busen and J. Hacker, *Quart. J. Royal Meteorol. Soc.* **607** (2005) 1143-1169.
[61] Chen J., W.C. Conant, T.A. Rissman, R.C. Flagan and J.H. Seinfeld, *Aerosol Sci. Technol.* **39** (2005) 485-491.
[62] Hermann M., F. Stratmann, M. Wilck and A. Wiedensohler *J. Atmos. Oceanic Technol.* **18** (2001) 7-19.
[63] Haywood, J., P. Francis, O. Dubovik, M. Glew and B. Holben, *J. Geophys. Res*, **108**, (2003), 13, 8471.

[64] Keil A. and J.M. Haywood, *J. Geophys. Res.*, **108**, (2003) Art. No. 8467.

[65] Johnson D.W., S. Osborne, R. Wood, K. Suhre, R. Johnson, S. Businger, P.K. Quinn, A. Wiedensohler, P.A. Durkee, L.M. Russell, M.O. Andreae, C. O'Dowd, K. Noone, B. Bandy, J. Rudolph and S. Rapsomanikis, *Tellus* **52B**, (2000) 290-320.

[66] Wood R, D. Johnson, S. Osborne, M.O. Andreae, B. Bandy, T.S. Bates, C. O'Dowd, K. Noone, P.K. Quinn, J. Rudolph and K. Suhre *Tellus* **52B** (2000) 401-422.

[67] Fehsenfeld, F.C., G. Ancellet, T. Bates, A. Goldstein, M. Hardesty, R. Honrath, K. Law, A. Lewis, R. Leaitch, S. McKeen, J. Meagher, D.D. Parrish, A. Pszenny, Russell, H. Schlager, J. Seinfeld, M. Trainer and R. Talbot, International Consortium for Atmospheric Research on Transport and Transformation (ICARTT): North America to Europe: Overview of the 2004 summer field study, *J. Geophys. Res.*, submitted, 2006.

[68] Methven, J., S.R. Arnold, A. Stohl, M.J. Evans, A.C. Lewis, M. Avery, H. Schlager, K. Law, C. Reeves, D. Blake and E. Atlas Establishing Lagrangian connections between observations within air masses crossing the Atlantic during the ICARTT experiment. *J. Geophys. Res.*, submitted, 2006.

[69] Arnold, S.R., M.J. Evans, J. Methven, A. Stohl, A.C. Lewis, M.P. Chipperfield, N. Watson, J. Hopkins, R.M. Purvis, J.D. Lee, J.B. McQuaid, E.L. Atlas, D.R. Blake, B. Rappengluck. Qantification of mean OH concentrations and air mass dilution rates from the ICARTT Lgrangian experiment, *J. Geophys. Res.*, submitted, 2006.

[70] Whiteway, J.A., C. Cook, T. Choularton, M. Gallagher, K. Bower, D. Figueras-Nieto & M. Flynn in Proceedings of EGS ILCS (lidar conference) Meeting, Nice, April 2002.

J. Phys. IV France **139** (2006) 321–336
© EDP Sciences, Les Ulis
DOI: 10.1051/jp4:2006139021

Civil aircraft in global atmospheric chemistry research and monitoring

C.A.M. Brenninkmeijer[1]

[1] *Max-Planck-Institute for Chemistry, Atmospheric Chemistry Division, POB 3060, 55020 Mainz, Germany*
e-mail: carlb@mpch-mainz.mpg.de

Abstract. A system as complex and extensive as the earth's atmosphere with its chemistry involving gases and particles requires many observations for understanding its workings and following its changes. A logical yet unusual, but potentially extremely powerful way is to use civil aircraft in regular service for making measurements and thus lending science a helping hand. There are 3 such aircraft systems in operation and development which are briefly discussed with emphasis on a container based project, named CARIBIC. It is shown which regions of the atmosphere can be probed, what different air masses are encountered, and what one can measure using a container inside an aircraft as a compact automated laboratory. This is illustrated by 3 examples.

1. INTRODUCTION

Even when we merely start to contemplate the problem of weather forecasting, and that of simulating future climate, we quickly become well aware of dealing with truly colossal challenges. The earth's atmosphere is a highly complex, variable, partly chaotic system, and on top of that it constitutes a large, extensive system from pole to pole, and up to tens of kilometre in altitude indeed. The way to deal with weather forecasting is to have a mathematical model that incorporates radiation, evaporation, transport of various kinds, cloud formation etc. etc. One such example is the family of models developed and used by the European Centre for Medium-Range Weather Forecasts [1]. Their models assimilate millions of meteorological observations from around the world. The result is really spectacular, even in remote parts of the globe, where measurements are scarce, temperatures are surprisingly accurately predicted. When we take the temperatures recorded from the aircraft and compare these with the model, agreement is generally much better than 1 degree. Data from hundreds of observational sites and platforms are needed to "keep the model on track". Without these observations the model would "run away". It is clear that the importance of weather forecast for humankind is enormous, and that appropriate funding is supplied to run a very high resolution computer model.

ECMWF write on their homepage "The weather is governed by physical laws". We may now ask ourselves "but what about atmospheric chemistry". Since about 1970 the "chemical workings" of the atmosphere have been unravelled in great detail (science may seem somewhat chaotic at times, yet it is actually extremely efficient). We talk about hundreds of chemical reactions in the gas phase (from extremely fast to slow), the formation of particles that influence radiation and cloud formation, the input of thousands of millions of tons of pollutants (carbon monoxide, methane, nitrogen oxides etc) from a bewildering number of complex emission sources, natural as well as man made. All this complex chemistry inside this extremely complex chemical reactor -namely our atmosphere - and driven to a high degree by uv sunlight has to be incorporated into a sort of weather model to be able to follow and also to understand what is going on, even for trace gases with concentrations as low as parts per trillion. The major incentive to do "atmospheric chemistry" was and is a scientific one. But we know very well that "The climate is governed by physical and by chemical laws". This statement highlights the link chemistry→climate. Thus not only the chemically rather boring molecule CO_2, but also the chemically active greenhouse gases like for instance N_2O and CH_4 need our attention.

When we were to take one of those spatially highly resolved weather models and were to insert all the chemical reaction schemes that need small time steps (there is some pretty fast chemistry under the sun out there), we would get an unwieldy system of such tremendous complexity that no existing computer system could deal with it, not to speak scientists that could program and understand what the system for heaven sake is doing. The cost is extreme, and perhaps it is beyond mankind's present capabilities to realize this. Considering that the chemistry is a "second order factor", one has to device ways of solving this complexity problem without having humongous resources available. Thus less complex, less spatially resolved atmospheric models are used that transport the air and that let the chemistry do its job. Great progress is made in this type of modelling, and the most complex models generate their own "weather" reasonably well, and incorporate all types of chemical modules.

What do we want to achieve with these models? The answer is simply to obtain a sufficiently detailed synthetic picture of the chemical composition of the atmosphere everywhere. But how do we know that this picture is the correct one? Indeed, we do need observations of chemical composition. We can for instance assume that wetlands emanate x million tons of CH_4, but can the model predict the distribution of CH_4 well? Can the model tell us how long a methane molecule on the average survives in the atmosphere? Estimates run from 8 to 10 years. Here we see the emergence of 2 fundamental problems that force us to measure methane in the air. First, we have to know whether chemistry and transport in the model are correct. Second we have to check whether our assumptions about the emissions are correct. CH_4 has been increasing throughout the industrial revolution, but recently its growth waned over a period of years. Why? The corollary is that we have to observe. We have to measure CH_4 and of course other gases all around the globe. How do we go about this daunting but interesting task?

For CH_4, CO, CO_2, N_2O, H_2 and hydrocarbons the simplest recipe is to fill flasks with air at remote site (islands, coastal sites, and mountain tops) and sent these to a laboratory for careful analysis. One would not fill those flasks in a city or forest for instance, because most of the information thus obtained has only local value. Having to cover the entire globe by filling flasks, say monthly, makes you a bit choosy where to sample. Another issue is that some gases, particularly ozone, cannot be stored as it decomposes in captivity, and it is therefore measured in real time. Here there are two options, namely "in situ" and "by remote sensing". When the air passes through an O_3 analyser, it is in situ. When we use the absorption of sunlight by atmospheric O_3, it is remote sensing (Brewer-Dobson). The largest consistent flask network is operated by the United States National Oceanic and Atmospheric Administration's ESRL network [2]. There are several networks all together embraced by the World Meteorological Organization's Global Atmosphere Watch system [3]. GAW is developing new strategies for monitoring using the IGACO theme. This way satellite data and aircraft data will form an integral part of a better optimized global observing system.

Apart from this systematic information gathered in a "monitoring mode", there is a vast amount of data from expeditions. Using all possible observations methods and platforms (land based, ship, aircraft, balloons, and satellites) intensive periods of measurement give highly detailed information over limited regions over a limited period of time. Here the goal is often to better understand the processes themselves. For instance unravelling in detail the role of convective uplift of pollution from the surface into the free troposphere is a vast playground for atmospheric chemists. Such expeditions are expensive.

A research aircraft can cost for each flight hour €10,000.-. To circumnavigate the globe once along a great circle would cost close to €400,000.-. Obviously for monitoring alternative methods are needed. Of great value is remote sensing from satellites in lower orbits around the globe. This is particularly useful for observing trace gases in the stratosphere. Much what we know about the springtime ozone hole formation is based on satellite data. Using satellite measurements tropospheric composition is more difficult, but good progress is made. One obtains this way many "profiles" of a certain trace gas in the troposphere. Needless to say that only those gases that are radiatively active can be gauged this way.

What else do we have? In this article we discuss the use of passenger aircraft for making measurements. Such is an obvious idea, but how well can it be done? The reader will be given a concise view of progress in this interesting application, by discussing the pros and cons, and briefly discussing the current projects of this type. CARIBIC will be described not only for specialists in the field, but also for a broader readership. Finally we show some examples of CARIBIC results.

2. USING PASSENGER AIRCRAFT, SCIENCE AS FREQUENT FLYER

Properly speaking one should say "civil aircraft" then freight aircraft can be used as well, whilst military aircraft are excluded. The flight routes of civil aircraft do not cover neatly the entire globe. Here there are two aspects. One is that flights take place in so called corridors, and one could argue that the air traversed is contaminated by the exhaust gases of other aircraft. Such may be the case on routes that are frequented by many aircraft, like transatlantic routes and some routes in Asia. The other one is that parts of the globe are missing out. While both arguments are correct, and these conditions cannot be circumvented, figure 1 shows an example of a flight from Frankfurt to São Paulo. The back trajectories - that is the movements of the air parcels up to 5 days prior to interception by the aircraft – show a vivid picture of what an aircraft actually "sees". This example shows that a) contamination by other aircraft is not obvious while the air sweeps most of the time through the corridor of flight and that b) the "footprint" or the "catchment area" is very substantial. Furthermore, we emphasize that although civil aircraft use "flight corridors", these appear to be rather wide. As an example we show in figure 2 some pathways of CARIBIC flights. The spread in latitude and longitude obviously is only a real advantage when flight frequencies are high. The conclusion so far is that there is plenty of scope for

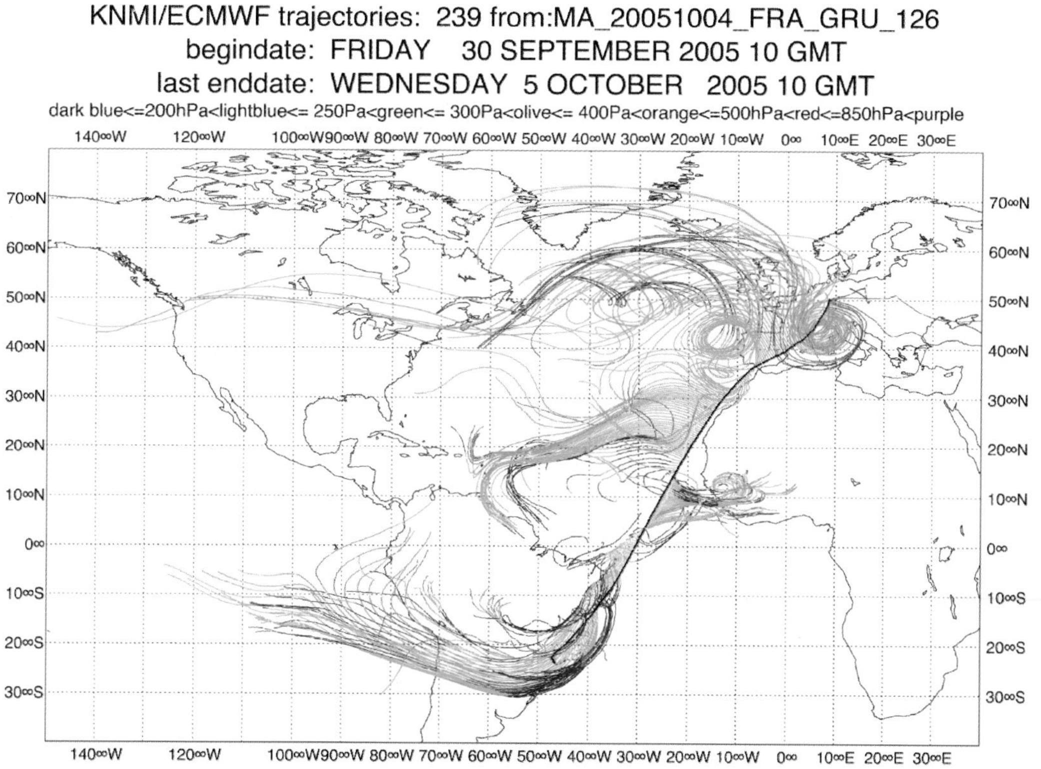

Figure 1. Back trajectories for air intercepted by the CARIBIC aircraft from São Paulo to Frankfurt. Dark blue means that the air is from altitudes of 200 hPA or even higher. Red indicates air that is close to the earth's surface. In this way polluted air can be encountered (e.g. from 5 °N, over Africa). Courtesy Peter van Velthoven, KNMI, de Bilt, the Netherlands.

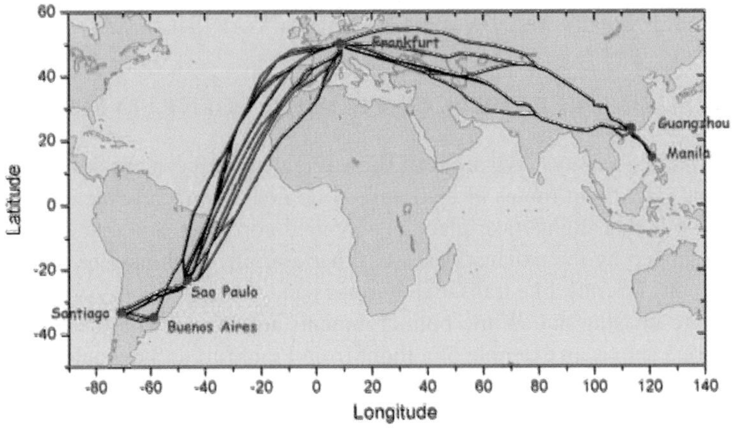

Figure 2. Examples of flight trajectories for CARIBIC illustrating some spread in actual pathways.

valuable observations. Someone may point to the limited altitude range of civil aircraft. Indeed, they cruise at 10 to 12 km altitude. Here is air resistance is sufficiently low to allow optimal efficiency. The question is what the vertical differences are, and what exactly can be "seen" from this altitude range. It happens, perhaps merely by coincidence, that for mid-latitudes this altitude coincides with the UTLS, the Upper Troposphere/Lower Stratosphere. Figure 3 shows a vertical cross section of the PV field, and the aircraft's trajectory. We see that the flight altitude is not constant. With the progressing of

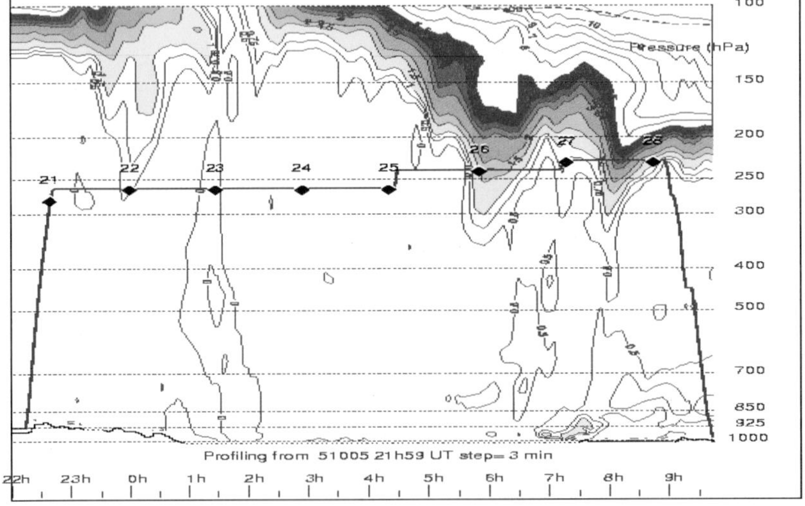

Figure 3. Vertical cross section of the PV field along the flight path (pink), showing increasing altitude, and the locations where air samples were taken. For parts of the flight returning to Europe, the aircraft flies in the stratosphere. Courtesy Peter van Velthoven, KNMI, de Bilt, the Netherlands.

consumption of fuel, the aircraft can fly higher more efficiently. Moreover, it is clear that one samples in the lowermost stratosphere, in the upper troposphere and that in the tropics - where the tropopause height is well over the cruising altitude - free tropospheric air is sampled. When we also revisit figure 1 it is clear that a wide range of air masses is encountered. Furthermore, as figure 1 shows even surface air uplifted by convection can be intercepted. For boundary layer air it is fair to say that this air is of course chemically transformed, mixed or altered even when convection is fast.

The conclusion from this is that although we cannot control the altitude, useful regions of the atmosphere in a vertical sense are sampled. It is noted that when an aircraft returns from a remote destination, its flight path is different in altitude due to this fuel load effect alone. When the time spent at the remote destination is short, one obtains some information about vertical differences. Note in this respect that long range aircraft are not useful for obtaining vertical profiles. If one wishes vertical profiles, one has to use several such aircraft, or use short range aircraft that hop around Europe for instance. But even in that case, the profiles may reflect strongly urbanized regions only.

In the preceding 2 paragraphs we have shown by means of data from ECMWF what air masses are typically encountered by civil aircraft. It is clear that the sampling of the atmosphere when using passenger aircraft is biased. On the other hand we see that regions of the atmosphere can be frequented that are otherwise hardly accessible. Now, a similar type of argumentation can be brought to bear when discussing the question "what can we measure and what can we not measure". There simply are atmospheric constituents that cannot be measured using civil aircraft. These are basically radicals, the highly reactive molecule fragments formed in the photochemical oxidation chains. Such short lived species not only have harrowingly low concentrations needing sophisticated detectors, but are often that reactive, that a large dedicated air inlet system is required. As matter of fact, when scientists approach airline officials and propose to mount a 1 meter long, 0.3 meter wide tube underneath the aircraft, they shake their heads. Some modesty is required in this sense. To recapitulate the flow of thoughts so far, one can state that one cannot fly everywhere and that one cannot measure everything but that significant novel information can be obtained.

3. SUCCESSFUL AIRCRAFT PROJECTS, 3 DIFFERENT APPROACHES

It is highly interesting that there are 3 civil aircraft projects (two in Europe and one in Japan) and that all 3 are different. This situation has emerged through the various strategies that optimize the use of civil aircraft. It shows that the rather unusual type of platform does in fact offer some flexibility. The oldest project is with Japanese Airlines (JAL). Using several Boeing 747 aircraft, air samples were collected using a set of flasks. The use of air sampling that has proven so effective in the NOAA-ESRL network also functioned well in the JAL project. After each flight (mostly to Australia) the sample bottles were analysed. The JAL data complement the NOAA-CMDL dataset for CO_2 and CH_4 neatly. The flight frequency amounted to once-to-twice each month. The second oldest project is MOZAIC initiated in 1993, which entails the measurement of ozone and water vapour using several Airbus A 340-300 aircraft. Here the emphasis is entirely on in situ measurement. Ozone is analysed using a standard uv absorption based detector system. Water vapour is measured using a small sensor that changes its electrical capacity based on the relative humidity. Up to 5 aircraft have at one stage been operational in MOZAIC, leading to tens of thousands of measurement flights. The youngest project is CARIBIC (www.caribic-atmospheric.com). Here again another philosophy is used. Air sampling and in situ analyses are combined and as much equipment as possible is used. This is realized by deploying an entire airfreight container filled with equipment. CARIBIC provides very detailed information, including aerosol counts. All 3 projects have been in operation for 5 years or longer and their current status and development is reviewed in the following sections of this article.

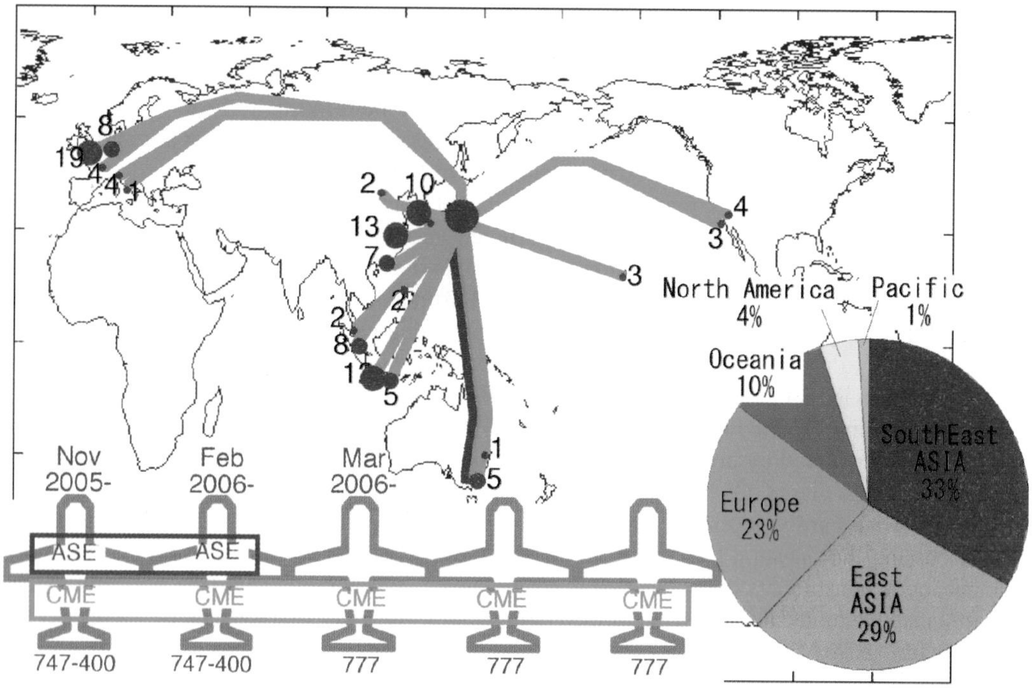

Figure 4. Overview of the scientific project supported by the JAL foundation. "ASE" stands for air sampling equipment and "CME" for continuous measurement equipment for CO_2. Starting from Tokyo, important parts of the globe are covered, supplementing the Atlantic regions covered from Europe.

3.1 The JAL project

The JAL project [4] has been upgraded recently and is now based on 5 Boeing 747 and 777 aircraft that monitor CO_2 in situ, whilst 2 of these can also collect air samples. The latter aircraft are intended to be used for flights to Australia mainly. Figure 4 shows a summary of the JAL system. The CO_2 data are valuable for verifying simulations of models that specifically address the atmospheric cycle of CO_2. Quantifying the distribution of CO_2, and establishing the fluxes from and into the complex biosphere is extremely difficult. The JAL data, with their excellent precision and accuracy will help to check model results. Other observational information is obtained using remote sensing from satellite platforms. Such type of data provides a wide coverage but is inherently less precise and less accurate and needs verification using the actual measurements. In the end, a combination of all methods is needed, but regular accurate aircraft data are a powerful commodity in this all. When considering the cost compared to satellites, one wonders at times why there are not more similar projects.

3.2 The MOZAIC project

The MOZAIC project is based on up to 5 aircraft measuring ozone and water vapour. The challenge is not only the certification of scientific equipment for use in passenger aircraft, which is a complex procedure of documentation and tests. Also the automated operation and the logistics of sensor renewal, calibration and data retrieval whilst dealing with several airline companies (Lufthansa, Air France, Sabena, Austrian) is a major effort. MOZAIC has provided ozone climatologies and given valuable information about water vapour. One discovery that stands out is that parts of the troposphere are supersaturated, a feature not captured by ECMWF analyses. MOZAIC has been extended by including 2 important components, namely NOy (one aircraft) and CO (several aircraft). This extension has increased the scientific value of MOZAIC even further. Namely, despite its importance the information about NOy in the upper troposphere and lower stratosphere is scant. Concerning CO and its relation with ozone, its measurement on a large scale is extremely useful for characterizing plumes, chemical transformation, large scale transport, and validating satellite based observations (MOPITT, Aura, Sciamachy). However, one obstacle that MOZAIC had to face was the weight and size of the equipment that is accommodated below the cockpit in the avionic bay of the aircraft. Basically 100 kg corresponds to one passenger (including luggage), and carrying the MOZAIC package for the ~18/24 hours the aircraft is airborne incurs costs that have to be met somehow. Airlines are basically prepared to help to advance atmospheric science, but with space and efficiency at stake, a conflict emerged. The strategy of the MOZAIC consortium therefore has been to work towards a new, smaller, lightweight version of the equipment package. Also, the lack of aerosol information should be addressed. A new project supported by the EC, named IAGOS [6], is on its way to develop a new generation of measurement package to be certified for A340 aircraft. It is clear that when the airlines are only slightly burdened by scientific activities, the chances for success are higher. Nonetheless, the challenge is to get a useful set of instruments, of a quality that is comparable to standard aviation equipment and is certified. These costs of certification are generally high. Another important aspect of IAGOS is the real time transmission of measurement data. This means that like some aircraft transmit physical data (Amdar [7]) that are used in weather forecast models, IAGOS is to transmit data on ozone, water vapour, carbon monoxide and certain aerosol data.

3.3 The CARIBIC project

The chief distinction wrt to the previous two systems is that the CARIBIC equipment is not part of the aircraft. The equipment is integrated in an airfreight container that is loaded into the aircraft each month for long distance flights. Upon return at the airport of departure the container is disconnected, unloaded and transported to the laboratory. Here data are retrieved, analysers are calibrated, air samples and also aerosol samples are removed for extensive laboratory analyses. Thus basically CARIBIC constitutes a flying observatory, comparable to land based observatories in the GAW network.

CARIBIC (www.caribic-atmospheric.com) was in operation from 1997-2002 based on a Boeing 767. This period proved extremely productive and its success formed the basis for a totally new CARIBIC system implemented on the new series long range A340-600 that airline companies started using in 2004. Thus a new container, new equipment, and a new inlet system were developed for this new aircraft type. Supported by Lufthansa the new system can have an operating horizon of well over 10 years. This is even long compared to the lifetime of satellite equipment. The consortium is based on 11 institutions in Europe, and we refer the reader to the website for more information.

To give an idea about the new container system, the total weight of this compact laboratory is now 1500 kg. The consortium operating CARIBIC pays the airline (Lufthansa Cargo) for the freight cost, thus creating a sound relationship, namely "the user pays". Besides the removable container which to a high degree is "freight" and not aviation equipment, there are permanent provisions in the aircraft, namely the air inlet system, and the transfer tubes between it and the container. Furthermore there are electrical supply and flight data provisions. In the next chapter the CARIBIC system is described.

4. THE CARIBIC SYSTEM

The philosophy underlying the CARIBIC concept is to measure as many as possible relevant gases and aerosol properties simultaneously, and this all without operator assistance for several days at a stretch, even when the aircraft makes stops-overs. To realize the concept, a consortium of now 10

Figure 5. The Lufthansa Airbus A340-600 aircraft equipped for CARIBIC. Shown is the location of the inlet system. The container is above the inlet, 2 meter further to the back. Courtesy Lufthansa AG.

institutions work together [8], each of them supplying equipment (or analytical services) and being expert in their respective application in CARIBIC. All partners were asked to make their equipment comply with strict regulations, and to reduce size and weight as much as possible. As explained before, certain atmospheric species - mainly radicals - cannot be measured from passenger aircraft because of special inlet requirements and or excessive dimensions of equipment. The next consideration is that this whole system should fly as much as possible. Experience has shown that flights can be conducted in a monthly rhythm. It may be possible to increase this by 50%. Also important is if one or more aircraft are to be equipped with an inlet system and tubing and controls.

Concerning where to implement the CARIBIC system inside the aircraft the general consideration is that when the aircraft lands at the airport of destination the measurement container should be left untouched inside the cargo bay connected to the aircraft, ready for the return journey. Because the cargo door of the forward cargo bay of the A340-600 (figure 5) is at the front of the aircraft, the container position has to be towards the wing box. Thus the CARIBIC container is loaded first and rolled to the very end of the long cargo bay. Because the air tubing that connects inlet system with container must be kept as short as possible, the inlet system is about 34 meter from the nose of the aircraft. Ideally the inlet should be close to the nose of the aircraft, but by making it sufficiently tall, air can still be sampled from outside the boundary layer of the hull. Along most of the hull of the aircraft the thickness of the boundary layer (loosely defined as the sheath of air that is in contact with the skin) grows from almost zero to tens of cm. The CARIBIC inlets are 27 to 35 cm away from the hull. We will next discuss this inlet system, the provisions inside the aircraft, and finally the container itself.

4.1 Inlet system

A good inlet system is essential, which brings some degree of complexity (figure 6). Moreover its aerodynamic properties at cruising speed (~ 250 m/s) have been extensively modelled. The long aerosol probe (45 cm long tube) has the function to slow down the air so that particles can be sampled via a concentric forward pointing pick-up tube. The leading shroud (12 cm) makes the system less sensitive concerning particle collection efficiency changes by "guiding" the airflow into the tube, even when the probe does not follow the stream lines perfectly. The smallest probe has two inlets. One at the leading

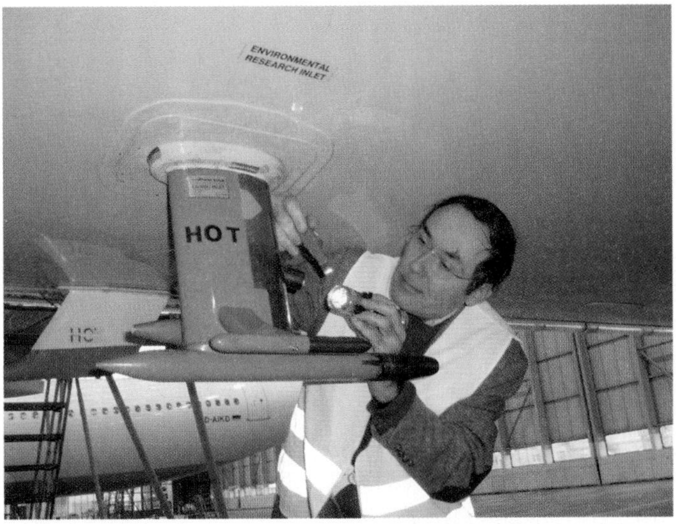

Figure 6. The inlet system being inspected by the author. The lowermost tube is the aerosol tube with the leading shroud. Just above that is the water probe. The aperture in its side is for gaseous water measurement. Not clearly visible at the back is the trace gas probe. The top left, against the aircraft's skin is an exhaust. The grey triangle covers the DOAS windows. Pointing forward (right) slanted downwards is the video camera.

tip collects air inclusive of droplets and ice crystals. The inlet hole positioned at the side only "sees" the gaseous water vapour. Thus clouds can be detected with this system. The medium size probe is for trace gas sampling. The tubing leading to this probe is lined with PFA, and inert polymer. The tips and tubing are heated wherever necessary. A special feature of the inlet is that it contains 3 small telescopes forming a MAX-DOAS system with the spectrographs placed in the container. Finally, there is video camera that allows retrospective inspection of the inlets to verify that no freezing effects occurred (which would alter aerosol collection characteristics). The camera also provides pictures of clouds, contrails etc. Next to the inlet system we need provisions inside the aircraft, which we describe very briefly. Power of up to 7 kw is supplied to the container whereas positional and physical data measured by the aircraft (ARINC) are fed into the computer controlling the container equipment. Furthermore the 3 optical cables and a range of air sampling tubing run between the inlet and container. The container is connected to the aircraft via an interface comprising plugs, quickfit connectors, and standard connectors.

4.2 Container

The measurement container, the compact automated laboratory is central to CARIBIC (figure 7). It weighs 1500 kg, is $\sim$3 meter wide and contains the equipment as listed in Table 1. Besides this equipment there are 2 electrical power supplies, cooling fans, smoke detectors, a central computer, air distribution lines, and a rack with compressed air and gas cylinders for the analysers. It is beyond the scope of this article to describe all equipment. Essential to know is that all equipment is controlled by the master computer and that each apparatus has its own computer for control and data storage. Once the aircraft has taken off, equipment is sequentially initialized and operates when pressure levels have drop and the chance of encountering strongly polluted air small. All parts of the container, all equipment has to be extensively documented and comply with strict aviation rules and tests. Thus structural strength, absence of electromagnetic interference, no combustible materials etc. all is important. Next we discuss the equipment and its detection properties.

Figure 7. The measurement container inside the aircraft facing flight direction with both doors opened. The total width is 3.1 meter.

Table 1. Listing of trace compounds and respective method of analysis in CARIBIC.

	Trace Constituent	Equipment/Analyser
1	O_3fast	Chemiluminescence on an organic dye
2	O_3precise	UV absorption
3	CO	VUV fluorescence
4	H_2O gaseous	Diode laser photo acoustic detector
5	H_2O total	Laser photo acoustic and chilled mirror detectors
6	NO	Chemiluminescence with O_3
7	NOy	Chemiluminescence after conversion to NO
8	Hg	Enrichment and atomic fluorescence
9	CO_2	Non-Dispersive Infrared Absorption (NDIR)
10	O_2 ultra high precision	Electrochemical Cells
11	Methanol, acetone, acetaldehyde	Proton transfer mass spectrometer (PTR-MS)
12	Aerosol particles with diameter >4 nm	Condensation particle counter (CPC)
13	Aerosol particles with diameter >12 nm	Condensation particle counter (CPC)
14	Aerosol particles with diameter >18 nm	Condensation particle counter (CPC)
15	Aerosol size distribution 150 - 5000 nm	Optical particle counter (OPC)
16	Aerosol elemental composition	Impactor collection, analysis by PIXE
17	Particle morphology	Impactor collection, microscopic analyses
18	Samples for VOC	Enrichment and analysis by GC-MS
19	Hydrocarbons, halocarbons, GHGs	Whole air sampler and analysis by GC
20	BrO, HCHO, OClO, O_4	Differential optical absorption spectroscopy, DOAS
21	Cirrus (under certain conditions)	Camera

4.3 What is measured, sensed and collected in flight

Essential are the aircraft data including temperature, speed, coordinates, pressure altitude which all are supplied via the ARINC databus. The second most important information is the one of the mixing ratio of ozone. Ozone is undoubtedly the most widely measured trace gas on earth. Therefore CARIBIC has 2 systems on board. One small apparatus measures ozone through its light emitting reaction with an organic dye. The light is captured by a photomultiplier tube. This system is extremely fast, but needs ongoing calibration in flight. This is achieved by a custom built uv absorption spectrometer. The absorption of uv light by ozone is measured in optical cells and recorded. This is done in a way that gives near absolute values of the abundance of ozone.

Next to ozone, water vapour measurements are of high priority. Water vapour is a tracer, and it is a greenhouse gas. Air at altitude can be supersaturated in water vapour, which is of importance for cloud formation for instance. Water vapour is difficult to measure because water adheres easily to surfaces, and its abundance varies over order of magnitudes, reaching low ppm levels in the stratosphere. On top of that, we want to know not only water vapour, but also the total amount of water which includes vapour and ice crystals. Therefore we have the two separate ports in the inlet system. All sample tubing for water is heated and consist of electropolished stainless steel. One apparatus is based on a gold disc that is cooled and kept at exactly that temperature at which condensation takes place. This is an absolute measure for the dew/frost point of the air sampled. A second system consists of 2 photo acoustic channels. Here infrared light from a laser is shone in short bursts into a cell. Absorption of this light by water molecules leads to heating when these shed the excess energy in collisional processes. This leads to heating, sudden thermal expansion which is captured by small microphones as a "popping sound", of which the amplitude is proportional to the water vapour concentration.

A third trace gas of great importance is NO. Nitrogen oxides are the catalysts that recycle radicals and are emitted from lightning, soils, and combustion processes. Their importance can hardly be overestimated. NO is conveniently detected by its light emission when reacting with an excess of ozone produced from oxygen. This system is very fast, and when exhaust plumes from other aircraft happen to be intercepted at the speed of 250 m/s, clear signals are resolved in time. We note that at night these measurements give no information as all NO has reacted then with ozone to NO_2. Always present is however "NOy". NOy denotes $NO + NO_2 + HNO_3 + HONO$ plus other products. By passing the air over gold with a small amount of hydrogen gas, NOy is converted to NO, which is simultaneously detected with another NO analyser.

The fourth essential trace gas is carbon monoxide, which is a major pollutant and in situ produced gas that burdens a large fraction of the troposphere's OH radical based self cleansing capacity. CO is measured in an optical cell where the molecules are excited by uv light. The resulting fluorescence is captured by a photomultiplier. Like with most detectors/analysers in CARIBIC, calibration takes repeatedly place in flight. This instrument is fast and extremely linear.

Carbon dioxide, the main greenhouse gas is measured as well. Although its direct significance for the chemistry of the atmosphere (CARIBIC's main mission) is nil, it is of such significance and acts as a tracer in pollution plumes that in situ measurements are invaluable. The standard technique based on the absorption of infrared radiation is employed. The challenge is the careful calibration in flight for which several cylinders with compressed calibrated air is performed regularly and frequently.

With the increase of carbon dioxide in the atmosphere, the oxygen content drops accordingly. Here we are dealing with very small changes in the second most abundant atmospheric gas. The measurement of the extremely small changes in oxygen of the order of parts per million is a major challenge solved by using a series of electrochemical cells that show changes of 20 mV over a change in oxygen of 20 % total. Therefore this equipment needs to resolve microvolt variations. The successful operation of the cells requires frequent calibration and extreme temperature and pressure stabilization. The measurement of oxygen by CARIBIC is a true experiment.

Returning to the trace gas level, the most complex pieces of equipment on board is undoubtedly a proton transfer reaction mass spectrometer (PTRMS). Here certain trace gases that have an affinity for H_3O^+ are protonized ($R + H_3O^+ \rightarrow RH^+ + H_2O$), which allows 2 manipulations. One is the separation of R from the air matrix. Basically most of the air can be removed and the ions retained. The second is the separation of RH^+ from other ions and detection of the RH^+ ion with a mass spectrometer. The most interesting gas that can be assayed by PTRMS is perhaps acetone. Acetone is formed in the reactions that degrade a host of organic trace gases. In the upper troposphere acetone plays a role in supplying OH, and thus accelerating the chemistry in the dry and cold part of the troposphere.

Thinking about what one should measure the idea to incorporate a mercury analyser was tabled. The toxic volatile metal is central in many studies dealing with toxicity and food chains. For atmospheric chemists it is not so exciting. However, CARIBIC offers regular surveys of mercury that are unparalleled. The metal is detected by its enrichment through absorption on gold. By heating the gold, the collected mercury is released in a burst, and detected, using fluorescence induced by a mercury discharge lamp.

Not all trace gases can be measured in situ, in real time. Therefore CARIBIC incorporates a system for collecting air. This system uses ultra clean pumps that compress air into glass flasks that are present in two boxes of 14 each. Glass is ideal because its liquid-like surface is small and its chemical affinity is low. Thus the chance that gases can be trapped and kept stored until the extensive laboratory analyses is highest. We measure accurately the greenhouse gases CO_2, CH_4, N_2O, SF_6, and also H_2, followed by non-methane hydrocarbons (e.g. ethane) and halocarbons. Well over 60 gases are assayed by several laboratories.

Certain trace gases in the group volatile organic carbon do however stick to many types of surfaces. Therefore air is also passed through small tubes kept at lower temperatures filled with an absorbing material. Back in the laboratory, the gases are desorbed and measured.

So far we have discussed the analyses of gases. Particles are well known to be of great importance. Therefore CARIBIC has equipment for aerosol analyses. Three analysers detect the number of small particles by making them visible in laser light through the condensation of butanol vapour onto their surfaces. Thus a small flow of air is saturated, and subsequently cooled leading to a "fog". By using different temperatures of saturation, detection tresholds from 5, 12 and 18 nanometer can be achieved. These smallest particles have a short lifetime in the atmosphere and are excellent tracers for mixing of polluted surface air into colder upper tropospheric air. Larger particles that do interact with light obviously are of direct significance for the radiative balance of the atmosphere (they can have a cooling effect at the surface). At the same time such particles can be detected directly by their back-scattering of laser light using an optical particle counter.

If this all were not enough, an ultra pure system on board collects aerosol samples by impaction on thin ultra clean foils. A total of 16 foils, each of which integrating sampling over several hours, is analysed in the laboratory for detecting the various elements, particularly sulphur. Other foils are used to collect particles that are analysed using microscopic techniques, such as atomic force microscopy.

There are satellites carrying optical equipment that peers into the troposphere for obtaining spectrographic information from which the abundance of certain trace gases can be retrieved. CARIBIC has a miniature optical system incorporated in its inlet system. Three small telescopes integrated in a 10 cm small block look sideways above the horizon, below it, and also one points downwards. The light signals are carried by optical cables directly into 3 spectrographs inside the container. This system, called MAX-DOAS (Multi Axis Differential Optical Absorption Spectroscopy) is a unique experiment designed to detect O_4, BrO and SO_2 (when concentrations are high).

Finally, a video camera in the pylon of the inlet system allows in flight inspection of the aerosol shroud to ascertain that no rhyme or ice forms around the leading edge. The camera also gives excellent visual information about clouds and contrails.

5. EXAMPLES OF CARIBIC RESULTS

For the scientific output of CARIBIC we refer to the list of publications on the website where the results of the previous version of CARIBIC using a smaller container and simpler inlet mounted on a Boeing 767 dominate, until the present system is in full deployment. New is certainly the capability to not only obtain accurate data for H_2O, but also the see the presence of clouds. This then is exemplified using figure 8, showing the detection of a field of cirrus clouds over China at ~11.4 km altitude around 30°N, during a flight from Frankfurt to Guangzhou July 5/6 2006. The lower border of the grey band indicates 100% saturation versus ice and the upper border 100% saturation versus liquid water. The ambient temperatures were around -43°C. It can be seen that the green curve (water vapour only) just reaches the saturation line versus ice at 05:02-05:20 and at 05.23-05.42. Note that at about 14 km/min cruising speed, these are larger features. Outside these 2 periods, the total water vapour (blue) and gaseous water vapour (green) agree extremely closely. However during these two periods we see higher total water values, and conclude that the aircraft intercepted ice clouds.

One can calculate how much ice is present in the clouds, and clearly we see pronounced structures, with a maximum of 250 ppmv (~0.13 g/kg air) on the right hand scale. This calculation must take into account that the forward pointing inlet of the water probe "sees" more ice particles than actually are present. Due to the effect of the hull of the aircraft, and the inertia of the particles, there is a particle enhancement, a certain bias, in the amount of solid water intercepted. For this effect, correction calculations are used. A complete calibration of the inlet in this respect has to be carried out. What we can see in the figure is the extreme fine detail of the cloud structure. Thus, despite the rather high speed, details are resolved thanks to the photo-acoustic channels. The red curve is the signal from the chilled mirror frost point hygrometer. The advantage is that absolute values are recorded; the disadvantage is

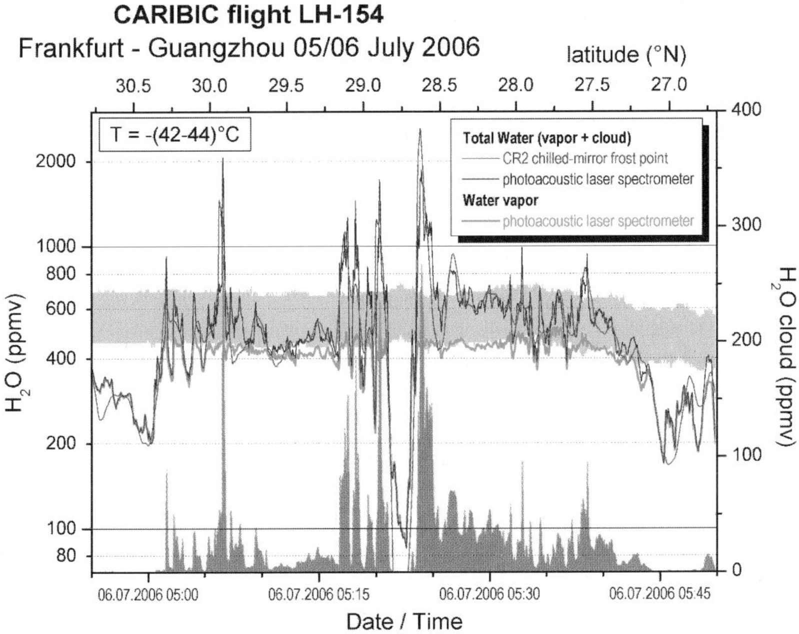

Figure 8. The total and gaseous water content of air at cruising altitude during the interception of cirrus clouds. The grey band is the calculated saturation range. The excess cloud water content is indicated as the grey area, showing large variations. Courtesy Andreas Zahn, IMK, FZ, Karlsruhe, Germany.

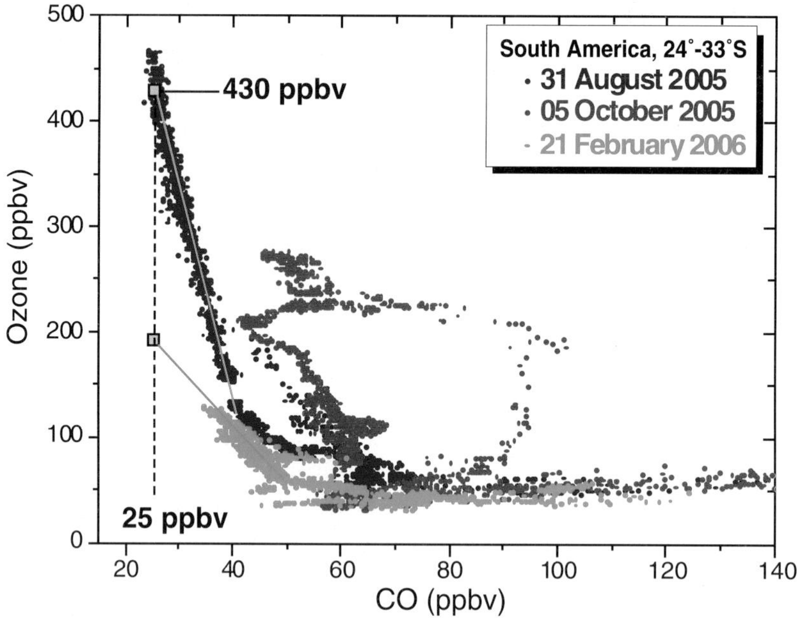

Figure 9. The correlation between ozone and carbon monoxide, showing stratospheric air, tropospheric air and mixtures. Courtesy Andreas Zahn, IMK, FZ, Karlsruhe, Germany.

the slower response of this system. Clearly the mirror has to be heated and cooled as fast as possible to follow fast changes, but there are inherent time constants.

We mentioned that ozone and carbon monoxide are important trace gases, and their mutual relationship is and has been analysed throughout the years many, many times. We also mentioned that the long range aircraft used do intersect the tropopause and many different types of air masses. A nice illustration then is figure 9 correlating ozone with carbon monoxide. In the stratosphere (the lowermost part) we see the well know negative correlation caused by the increase of ozone with altitude (ozone layer) and a concurrent decline in carbon monoxide, due to its destruction in the stratosphere in absence of sources, except the oxidation of mainly methane that altogether leads to a level of here only 25 ppbv. The high linearity means that mixing prevails. Thus what we see is the result of mixing stratospheric with tropospheric air. When chemistry dominates we see some curvature, a more complex behaviour. Furthermore we see for the October flight high carbon monoxide with low ozone. This typically is tropospheric air. The small positive slope points to ozone production in polluted air masses. Carbon monoxide is frequently a good indicator of pollution, including biomass burning, e.g. wildfires, deforestation by fire etc. Finally the October data also show air masses with high variable mixtures, and together with other data, the origin of such variations can be understood.

Finally we show an example of particle detection for a flight from Santiago de Chile to São Paulo, June 2005 in figure 10. Particle number concentrations are around 600 cm^{-3} for particles larger than 12 nanometer. However, upon descending through clouds, this number fluctuates and the number of ultra fine particles, between 4 and 12 nanometer increases and fluctuates strongly. These ultra fine particles have short lifetimes, and their presence is indicative of active processes. The question here is why just below the cloud, the ultrafine particles are present.

6. CONCLUSION AND OUTLOOK

Methods for global scale observations by civil aircraft (in present practice passenger aircraft) are in full development and establishment. The 3 existing projects appear to have excellent scientific scope for

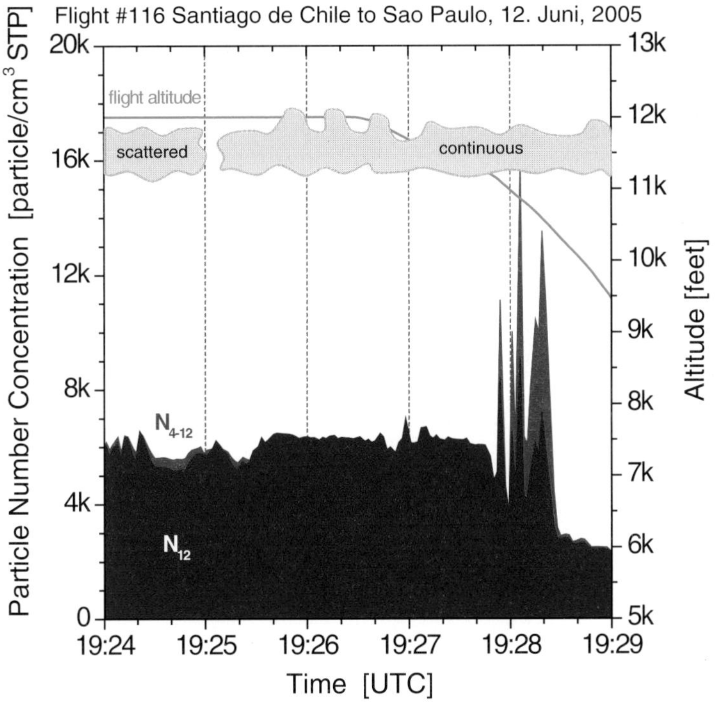

Figure 10. The detection of ultrafine particles (N_{4-12}) after descending through a cloud on approach of the airport in São Paulo. The cloud information is based on recordings by the video camera. Courtesy Markus Hermann, IfT, Leipzig, Germany.

contributing to our understanding of the chemistry of the troposphere and UTLS. The aircraft supply unique information that cannot be obtained otherwise, and here we point out the importance of the UTLS, the role of convection in uplifting surface air with all its complexities and local differences and the large scale transport and chemical modification processes. In a wider sense the projects provide just that type of data that are needed for helping to advance the development of global atmospheric chemistry models in the future. This is the monitoring component of JAL, MOZAIC and CARIBIC which entails a longer term effort. Atmospheric chemistry research no doubt is changing. How in the end all information will flow into global (and regional) models is at present not clear. Aircraft campaign based data give tremendous detail but cannot be executed sufficiently frequently to follow systematically annual and decadal changes in the atmosphere. The science of monitoring is still in development, and beyond doubt will involve an increasing component of engineering (which has to be funded somehow) on which the advance of atmospheric science will depend. Because the basic concept of the civil aircraft deployment is sound (and costs are acceptable), and the need for systematic coherent detailed global scale information increases, it is wise to develop projects like CARIBIC.

Acknowledgement

The author can hardly do justice in thanking all his colleagues (including young scientists) sufficiently for making CARIBIC so successful.

References

[1] http://www.ecmwf.int/products/catalogue/
[2] http://www.cmdl.noaa.gov/
[3] http://www.wmo.ch/web/arep/gaw/gaw_home.html
[4] http://www.jal-foundation.or.jp
[5] http://mozaic.aero.obs-mip.fr/web/
[6] http://www.fz-juelich.de/icg/icg-ii/iagos
[7] http://www.eumetnet.eu.org/
[8] http://www.caribic-atmospheric.com

J. Phys. IV France **139** (2006) 337–360
© EDP Sciences, Les Ulis
DOI: 10.1051/jp4:2006139022

Rayleigh temperature lidar applications: Tools and methods

P. Keckhut[1]

[1] *Service d'Aéronomie, Institut Pierre-Simon Laplace, BP. 3, 91371 Verrières-le-Buisson, France*

Abstract. After 27 years of continuous operations of the Rayleigh temperature lidar, many tools have been developed. The lidar is well adapted to study the dynamics at several time scales. It includes gravity waves, atmospheric tides, planetary waves and student stratospheric warmings, atmospheric reentry, interannual changes and satellite validations. Methods have been explained as well as the limitations and uncertainties of these analyses.

1. INTRODUCTION

Backscattering lidar signals are primarily due to molecules and particles. Above 30 km, atmospheric composition is purely molecular and atmospheric density and the temperature profile can be deduced [1, 2]. The main uncertainty in lidar measurements is directly related to the photon noise. This uncertainty in the number of photons detected is related to the statistic of rare events and is described by the Poisson Law. In this case the standard error is expressed as the square of the number of photon received.

The signal S(z) is a combination of the backscattered laser photons N(z) and the parasite signal from the sky background B_{PM} and detector noise B_{SB}.

$$S(z) = N(z) + B_{PM} + B_{SB} \tag{1}$$

The uncertainty of the density measurement is then given by the following equation:

$$\frac{\Delta n(z)}{n(z)} = \frac{\Delta S(z)}{S(z)} = \frac{\sqrt{(N(z) + B_{PM} + B_{SB})}}{N(z)} \tag{2}$$

At middle altitude (<60-70 km), where the Rayleigh backscattering signal is predominant, and noise comes exclusively from the photon counting, the uncertainty is only related to the number of photons:

$$\frac{\Delta n(z)}{n(z)} = \frac{\Delta T(z)}{T(z)} = \frac{1}{\sqrt{N(z)}} = \sqrt{\frac{C_2}{n(z)\Delta z n_t}(z - z_0)^2} \tag{3}$$

The principal improvement of the photon noise consists in increasing the number of photons received. This can be achieved by the instrumental design or this can also be obtained in increasing vertical and temporal resolution in:

- Increasing the vertical resolution in averaging several individual layers.
- Increasing the number of shots in integrating several successive shots.

So depending to the spatio-temporal scales of the processes under investigation, the vertical resolution and the temporal integration need to be adapted respectively from 10 m to 3 km and from few minutes to nightly means (12 hours). Also temperature lidar is free of any calibration and so is a perfect instrument for long term monitoring. So Rayleigh lidars allow observing short-scale structures such as gravity waves, tides, as well as longer time scales such as daily changes, climatology and long-term monitoring.

2. GRAVITY WAVES

The gravity waves, also known as a buoyancy waves, result from the stability of air in a stable stratified atmosphere. They are formed in the troposphere above mountains (lee waves), convective clouds and when the flow is not in geostrophic equilibrium as for instance in fronts. After being generated in the troposphere, waves propagate into the stratosphere, growing in size as they move upward (because of decreasing air density in the stratosphere). Waves can propagate vertically inducing temperature fluctuations but can also dissipate. They can propagate upwards into the middle atmosphere until their phase speed is equal to the wind speed or they reach critical amplitudes with respect to the static or dynamical stability of the atmosphere. Gravity waves break and transfer their momentum flux to the mean flow. The breaking of gravity waves explains the main features of the mesospheric circulation and the thermal structure. They explain the paradoxical very cold temperature observed at the summer polar mesopause. An inversion of the vertical temperature gradient from negative to positive in the mesosphere observed frequently during several days, are also thought to be due to gravity waves [3]. The amplitude of the temperature inversion may reach values as high as 40 K. There are a lot of scientific questions associated with gravity waves and their exact role in climate. A lot of investigations were performed to found some universal form of the waves spectrum [4]. Also during some events, some characteristics of some specific modes were derived as well as their vertical propagation [5]. Because of the scales, some parameterizations were developed to take into account gravity waves into climate models. Also some statistical approaches were performed to estimate their climate impact, and geographical differences associated to mean flow, convection activity or orography.

Gravity waves are difficult to observed because of their scale and lidar is one of the best instrument to derived fluctuations associated to these waves. Main tools consist in Fourier analyses, Wavelet decomposition or energy estimates.

2.1 Fourier Transform analyses

The fluctuation of the air mass associated with gravity waves for adiabatic displacements is related to temperature fluctuations by the following formula:

$$\xi = \frac{g}{N^2}\frac{T'}{T_0} \tag{4}$$

With g as the gravity, N the Brunt-Väisälä frequency associated with the static stability, T_0 the mean temperature and T' the temperature anomaly. The available potential energy per unit mass, E_p is given by

$$E_p = \frac{1}{2}\left(N^2\overline{\xi^2}\right) = \frac{1}{2}\left(N^2\overline{\left(\frac{g}{N^2}\frac{T'}{T_0}\right)^2}\right) \tag{5}$$

With N^2 equal to:

$$N^2 = \frac{1}{T_0}\left(\frac{\partial T}{\partial z} + \frac{g}{C_p}\right) \tag{6}$$

Where the vertical temperature gradient can be estimated over the mean background temperature profile.

The first difficulty concerns the estimate of the anomalies associated with gravity waves. Anomalies can be derived as a function of time or vertically. In both cases, temperature fluctuations can be due to other processes and it is always difficult to estimate a background that does not include gravity waves with long periods or wavelengths and include fluctuations from other sources that can be considered as the mean atmospheric condition regarding gravity waves propagation. Usually a low pass filtering is performed with a third order polynomial fit or a smother function. In this case, vertical domains are equal or smaller than 15 km. For larger vertical domains, the procedure for trend extraction is less accurate.

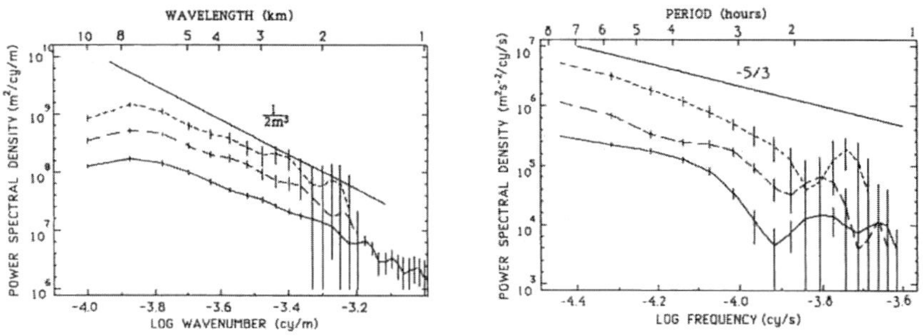

Figure 1. Power spectral density, versus vertical wavenumber (a) and versus apparent frequency (b), of the relative density fluctuations scaled by g/N2, in the stratosphere (30-45 km, strait line), lower mesosphere (45-60 km, dashed line), and middle mesosphere (55-70 km, dotted line) deduced with lidar data obtained at Bordeaux (France) in January 1989 and processed by Wilson et al. [4]. Strait lines of N2/2m3 and of –5/3 slopes, are plotted respectively for comparisons.

The 1D spectra are obtained by applying a frequency Fourier Transform over both the successive vertical or temporal anomalies. The 2D spectra are obtained by applying a frequency Fourier Transform over the successive vertical wavenumber spectra. The 2D spectrum thus describes, for each wavenumber, the phase evolution of the sinusoidal functions, Fourier Transforms of the vertical profiles of the density perturbations. The altitude or/and temporal range Δ used to derive spectrum give the frequency resolution that is close to $1/\Delta$. However the rectangular window gives an impulse response as a cardinal sinus that provides large wings and that spread some energy in the spectrum. For this reason other windows are used to reduce this effect without enlarge the frequency resolution: Hanning, Bartlett, Tukey, Parzen, ...

One of the issues in these spectrum analyses is the background estimate that is mainly due to statistical counting noise. If we consider the noise of the lidar signal as random it correspond to a white noise and its Fourier Transform is a random variable X with a probability density P Gaussian centered on 0 with a standard deviation E.

$$P(X)\,dX = \frac{1}{E\sqrt{2\pi}} esp\left(-\frac{X^2}{2E^2}\right) dX \tag{7}$$

Then the mean noise is related to E as:

$$B_m = \int w \exp\left(-\frac{w}{2E^2}\right)\frac{dw}{2E^2} = 2E \tag{8}$$

The noise can then be removed easily from the spectrum if we consider density rather than temperature. For this reason gravity waves estimates are mainly based on density rather than temperature signals. Noise can be estimated rather by theory based on the signal level itself (2) or in investigating the ranking of individual spectrum points; the probability to found energy larger than E_k can be express as:

$$Q(E > E_k) = \exp\left(-\frac{E_k}{B_m}\right) \tag{9}$$

Spectrum has been estimated using the method described above for 5 long nights. These analyses allow deriving a mean shape of the gravity wave spectrum.

2.2 Wavelet analysis

Because the gravity waves spectrum is composed by different modes that change with time and altitude domain, the running window analysis introduced by Gabor in 1946 [6] is well adapted to study non-stationary signal as such as gravity waves. The position of the window gives an additional dimension, instead of the standard Fourier transform:

$$S(m) = \int_{-\infty}^{+\infty} s(z)\, e^{-2\pi jmz}\, dz \tag{10}$$

the following formula depends on both to altitude and wavenumber:

$$C_{a,b} = \int_{-\infty}^{+\infty} s(z) h_{a,b}(z)\, dz \tag{11}$$

The both parameters a and b are related to the wavenumber and spatial or vertical position. One of the problems with the rectangle windows is that as far as the wavenumber increase the window content more cycles. To avoid this problem Morlet [7] introduces the wavelet decomposition. The wavelet contents always the same numbers of cycles in a Gaussian envelope:

$$\psi_{a,b}(z) = \frac{1}{\sqrt{a}} \psi\left(\frac{z-b}{a}\right) \tag{12}$$

b is the translation parameter (permitting to locate in altitude or time), $1/a$ is the dilatation parameter related to the wave number. The wavelet transform is thus defined by the convolution product:

$$C_{a,b}(z) = \frac{1}{\sqrt{a}} \int_{-\infty}^{+\infty} \psi\left(\frac{z-b}{a}\right) S(z)\, dz \tag{13}$$

The basic wavelet has the following form

$$\phi_\lambda(z) = \exp\left(-\frac{z^2}{2\tau^2}\right) \exp\left(2\pi i \frac{z}{\lambda}\right) \tag{14}$$

Where τ is the parameter that is used to optimize

- The altitude (or time) localization
- The wavelength of the dominant modes
- The boundary effect.

The Fourier Transform also called filter can be express as a function of the wave number m:

$$\hat{\phi}_\lambda(m) = \tau\sqrt{2\pi} \exp\left(-2\pi^2\tau^2 \left(m - \frac{1}{\lambda}\right)^2\right) \tag{15}$$

This is a Gaussian that can be write in the following form:

$$\hat{\phi}_\lambda(m) = \frac{1}{\sigma\sqrt{2\pi}} \exp\left(-\frac{\left(m - \frac{1}{\lambda}\right)}{2\sigma^2}\right) \tag{16}$$

with the $\sigma = 1/2\pi^*\tau$

If σ is large (τ small), the spectral resolution is coarse but the vertical (or time) identification is good. In opposite with small σ, one can get a fine spectral resolution but a bad vertical localization and this case is closer to standard Fourier Transform (as seen in 14 for large τ). Some tests have been performed to study the wavelet ability to detect simultaneously 3 modes with wavelength equal to 2, 5, and 9 km. The best compromise of vertical identification, spectral resolution and boundary effect mainly if the vertical domain is limited to 30-60 km, are obtained for wavelet with around 6 oscillations. It corresponds to τ close to 1.

2.3 The variance method

An alternative method to compute the gravity wave potential energy is based on the variance of the backscattered Lidar signal. Unlike a number of authors [4, 5, 8, 9, 10, 11, 12] that investigate the gravity wave energy, in using spectral methods and integrate it on a given spectral domain, the variance method does not required to calculated any spectrum and is relatively strait forward to apply and has the advantage to be performed directly on the Lidar raw signal (photon counts). While it does not give the spectral distribution, it is more sensitive and free of data processing errors and so better adapted for climate study [13].

The method consists of computing the signal perturbations for short time and vertical intervals using the Lidar raw data. The variance of the atmosphere is then estimated from the difference between the observed variance and the instrumental noise variance. The instrumental noise variance is estimated by assuming that the noise follows a Poisson law of statistics of the signal. We consider the signal of an incoherent backscatter Lidar operated in photon count mode. This signal can be summed up in small time-vertical intervals ($\Delta t, \Delta z$). We assume that the vertical profile of the signal may be separated in a sum of a smooth mean profile plus short scale perturbations, with amplitude much smaller than the mean profile. The relative perturbation is described as:

$$S'\left(z_i, t_j\right) = \frac{S(z_i, t_j) - {}^1\!/_2\left(S(z_{i-1}, t_j) + S(z_{i+1}, t_j)\right)}{{}^1\!/_2\left[S(z_i, t_j) + {}^1\!/_2\left(S(z_{i-1}, t_j) + S(z_{i+1}, t_j)\right)\right]} \tag{17}$$

where $S(z_i,t_j)$ and $S'(z_i,t_j)$ are the signal and the perturbation at altitude z_i and time t_j in raw counts. Perturbations are due either to instrumental noise or the atmosphere.

We now consider a larger time-vertical interval (ΔT, ΔZ) obtained by grouping N_t time intervals by N_z altitude elementary intervals such as $\Delta T = \Delta t.N_t$ and $\Delta Z = \Delta z.N_z$. The observed variance of the signal in the large interval is defined as:

$$V_{obs} = \frac{1}{N_z N_t} \sum_{N_z} \sum_{N_t} \left[S(z_i, t_j)\right]^2 \tag{18}$$

The observed variance is the sum of the instrumental noise and the atmospheric variances:

$$V_{atm} = V_{obs} - V_{inst} \tag{19}$$

In photon counting mode the signal obeys the Poisson law and, except for saturation effects due to strong Lidar return signal, the instrumental noise variance is estimated as:

$$V_{inst} = \frac{1}{N_z N_t} \sum_{N_z} \sum_{N_t} \frac{S(z_i, t_j) + \frac{1}{4}\left(S(z_{i-1}, t_j) + S(z_{i+1}, t_j)\right)}{\left[{}^1\!/_2 S(z_i, t_j) + {}^1\!/_4\left(S(z_{i-1}, t_j) + S(z_{i+1}, t_j)\right)\right]^2} \tag{20}$$

The estimation of the atmospheric variance is obtained from the difference between the observed variance and the instrumental noise variance. If the signal $S(z_i,t_j)$ is varying weakly around its mean value $\overline{S}$, (20) can be simplified as:

$$V_{inst} \cong \frac{3}{2\overline{S}} \tag{21}$$

When applying the method of variance, it is important to determine if the estimated atmospheric variance is significantly above 0. This is equivalent in determining if V_{obs} is significantly different from V_{inst}. In order to do that, we will first require to estimate the standard deviation of V_{obs} in case where the perturbations are only due to instrumental noise. As soon as the average Lidar signal $\overline{S}$ is large enough (more than about 100 photon counts per elementary interval), we can apply the central limit theorem and the perturbation $S'(z_i,t_j)$ converges to a Gaussian distribution with mean value 0 and standard deviation $\sqrt{\frac{3}{2S}}$.

The gravity wave energy is computed using the above method by averaging the atmospheric variance V_i in the altitude range 30-40 km using a layer of thickness $\Delta z = 2.4$ km. The potential energy density, which is a measure of the gravity wave activity for the night, is given by:

$$E_P = \overline{V}_i \times \frac{g^2}{2N^2} \tag{22}$$

where $\overline{V_i}$ is the mean atmospheric variance of the Lidar signal for the night, g is the acceleration due to gravity and N is the Brunt-Vaïsälä frequency [14].

3. ATMOSPHERIC TIDES

Atmospheric tides are global-scale oscillations that are primarily induced by the diurnal variation of solar radiation absorption, mainly by water vapor in the troposphere and ozone in the stratosphere. In the middle atmosphere, tide theories predict the dominance of the 24-hour (diurnal) and 12-hour (semidiurnal) westward migrating modes. Atmospheric tides have been studied extensively through theoretical models [15-17]. Recent progress have occurred concerning the development of more realistic thermal excitation and numerical simulation models [18] though the observation of tides still reveals complex features that are not well reproduced by those models. The source of discrepancy is not yet resolved and may originate either from observational deficiencies (aliasing effects, temporal coverage) or from the spatiotemporal variability of the tides (spatial variability: zonal inhomogeneity, latitudinal influence; time variability: seasonal, inter-annual). More investigations based on both observation and modeling are thus needed to improve the consistency between numerical predictions and observational evidences.

Middle atmospheric tides are believed to induce temperature anomalies lower than 5-10 K. Despite our ability to simulate atmospheric tides, some important issues associated with tidal effects have appeared to be a source of error in the analysis of temperature measurements in the middle atmosphere. Specific areas where the tidal effects are of importance include: satellite adjustment procedures [19] and inter-instrumental comparisons of physical quantities such as temperature, measured systematically at different local solar times [20, 21];

- assimilation of observational data; specific issues that can affect the assimilation of tidal information are the miss-treatment of asynoptic data and/or the use of measurements that are inadequate to measure atmospheric tides [22];
- long-term trend estimates from time series with varying time of measurements [19]
- studies of planetary waves performed on time series built with a non-regular sampling: the non-uniform distribution of the data can generate spurious spectral signals on the spectral signature of processes with periods less than 10 days.

However, until recently, few observations of middle atmospheric tides were available. Observational data amenable to middle atmosphere tidal analyses were divided into wind radar measurements, primarily covering heights between 70 and 100 km, and rocket measurements between 30 and 60 km. Unfortunately, temperature measurements from rockets have a poor temporal coverage – as measurements are limited to a few launches a day in most cases [23] and are subject to errors induced by radiation from exposed components of the rocketsonde sensor [24]. Until recently, satellite measurements have been inadequate to observe the middle atmospheric tides. The measurements were actually coming from sun-synchronous satellites, which limited the measurement at any latitude to two fixed local solar times, insufficient to separate completely tidal phases and amplitudes [25]. However, Dudhia et al. [26] have lately used the temperature measurements obtained from ISAMS (Improved

Stratospheric And Mesospheric Sounder) experiment to measure the tides between 15 and 80 km altitude. In fact, this experiment has allowed the sampling of the atmosphere throughout the diurnal cycle over the course of a month. Tidal characteristics have been determined assuming that the tidal structures do not change over the period of observation. Yet, the constraining integration over periods of a month imposed by the satellite does not allow either carrying out observations with a high temporal resolution or taking into account the local changes.

In contrast, several studies based on Lidar temperature measurements [27-31] ha.e shown that the Rayleigh Lidar instrument can be a useful tool for observing tides between 30 and 80 km. Lidars can actually provide long and quasi-continuous nighttimes measurements, with a high resolution in both time and space [27]. However the Lidar measurements are limited to nighttimes and clear sky conditions. Consequently, the tides need to be studied over several nights using appropriate methods. Two analysis methods able to estimate the mean nighttime evolutions of the temperature for the period of observations have been reported.

Two analysis methods are usually employed to extract tidal information from Lidar observations. Both methods estimate the mean nighttimes evolution of Lidar temperature deviations from the nightly mean value. The first method has been proposed and described by Leblanc et al. [30, 31]. In this method, half-hour raw data for all the nights are averaged to obtain mean raw data profiles at given local solar time. Temperature retrieval is then performed. This process reduces noise from geophysical and instrumental origins, which is expected to occur randomly from one night to another, and therefore, improves the signal-to-noise ratio and the quality of the data to be used. Since expected tidal vertical wavelengths are about 20 km or more, short vertical wavelengths (less than 6 km) are filtered out using a low-pass Butterworth filter with a 6-km cut-off wavelength.

When the different nights of measurements extend over different periods it is difficult to estimate the mean and bias can be introduce. An alternative method has been developed by Morel et al. [32]. The mean temperature for a given night is calculated by using a specific algorithm that takes into account the bias due to the different time coverage from night to night and/or the lack of measurements during some hours for each night.

The mean temperature for each night of measurement is calculated

$$\overline{T_n} = \frac{1}{H(n)} \sum_{H=Hd(n)}^{Hf(n)} T_{h}, n \qquad (23)$$

temperature anomalies $\Delta T^{i}_{h,n}$ are calculated for each night

$$\Delta T^{I}_{h,n} = T_{h,n} - \overline{T}_n \qquad (24)$$

Then within a loop the mean $\overline{\Delta T_h}^{i}$ is calculated by averaging the $\Delta T^{i}_{h,n}$ values over the analysis period

$$\overline{\Delta T_h}^{i} = \frac{1}{N(h)} \sum_{n=1}^{N(h)} \Delta T^{i}_{h,n} \qquad (25)$$

Since total error level on temperature is less than 1 K at the lower limit ($\sim$30 km) and climbs rapidly to 30 or 40 K at the top of the profile, values are averaged taking into account those error values. This process reduces the importance of the terms with larger errors. Then the bias of the mean temperature for a given night is calculated, i.e., the difference between the initial value $\overline{T_n}$ and the value found when taking into account the mean temperature deviations $\overline{\Delta T_h}^{i}$ from the nightly mean due to the tidal effects

$$E^{i}_{n} = \frac{1}{H(n)} \sum_{h=Hd(n)}^{Hf(n)} \overline{\Delta T_h}^{i} \qquad (26)$$

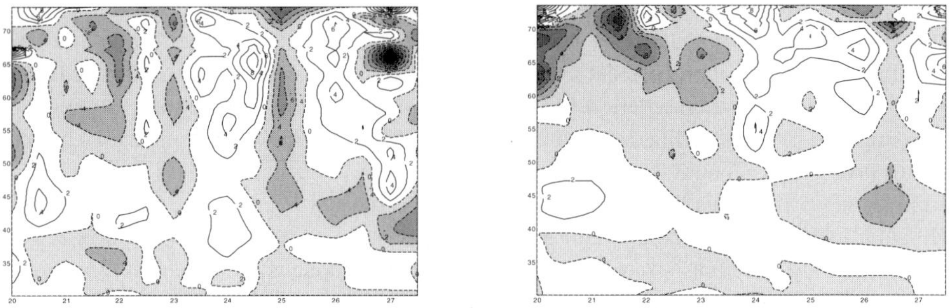

Figure 2. Half-hourly mean Lidar temperature differences from their nighttime average at La Réunion Island (20.8°S) during 11 nights taken from November 07-30, 1995 estimated using method 1 (a) and 2 (b).

The new $\Delta T_{h,n}$ values are then calculated

$$\Delta T^{i+1}_{h,n} = \Delta T^{i}_{h,n} + E^{i}_{n} \tag{27}$$

and these operations are carried out while:

$$\left| E^{i}_{n} - E^{i+1}_{n} \right| < \varepsilon \text{ for each night n} \tag{28}$$

ε is the steady criterion taken as 10^{-2}.

The averaged temperature differences have been derived using the both methods (figure 2) by Morel et al. [32]. Similarities are found between the two methods, in particular large temperature variations, showing warm and cold temperature bands with a pattern of downward phase propagation. This behavior appears to be related to the presence of tidal components. The averaged temperature difference values calculated with method 1 undergo some significant and rapid transitions that affect the tidal signature. One possible explanation for such a discrepancy is that method 1 does not allow correcting the mean temperature for a given night. Since the mean temperature calculated for a given night is sensitive to the time span of the data, mean temperature values are different for long and short nights, so that, for each half-hour slice, the value of the half-hour temperature deviation from the nightly mean differs from night to night. This comparison indicates that better results are obtained with method 2 than with method 1, when nights with different measurement windows are used. It also confirms that method 1 might be applied rather to time series with quasi-identical measurement windows from night to night [30, 31].

4. DAILY FLUCTUATION: PLANETARY WAVES AND STRATOSPHERIC WARMINGS

The main variability of the middle atmosphere is caused by the continuous propagation of Rossby planetary waves, which are generated in the troposphere by meridian motions due to the meridian gradient of the Coriolis parameter. Rossby waves can propagate only westward relative to the mean flow and have a relatively slow phase speed [32]. They are therefore blocked by easterly stratospheric winds in summer. In winter, planetary waves can propagate through the westerly stratospheric flow and, due to the exponential decrease of the atmospheric density with height, there relative amplitude increases until it reaches a critical amplitude leading to the wave breaking and a non linear interaction with the zonal flow at the origin of sudden warmings of the polar stratosphere. One of the main problems to investigate daily fluctuations comes from the fact that lidar measurements depend of the weather. Then continuous measurements cannot be obtained and lidar series consist of irregular temperature profiles (figure 3). The difficulty of interpolated missing data come from the fact that the use of splines can induce artificial variance while linear interpolations induce broken changes.

An algorithm allows to interpolate missing temperature data T_I. at the time t_i have been developed. It consists of a linear combination of the value given by the linear interpolation T_L and the value given

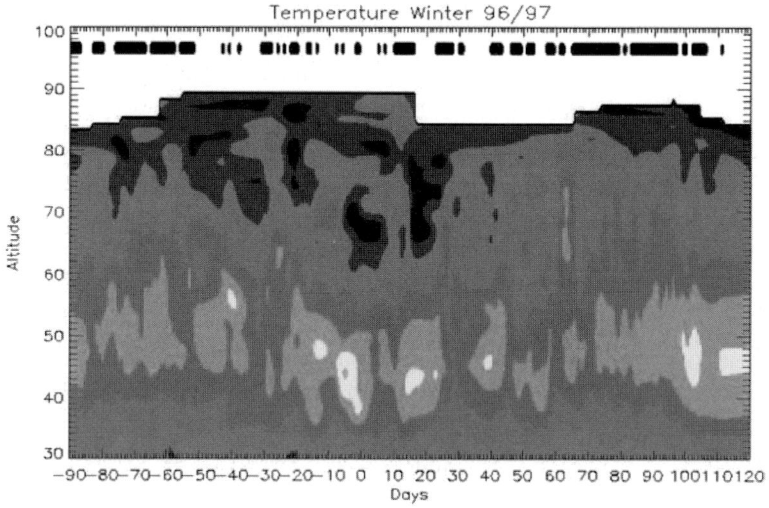

Figure 3. Evolution of the temperature at OHP during winter 1996-1997. The days with lidar measurements are indicated with vertical bars at the top of the figure.

by the cubic spline interpolation T_S. This combination depends of the time between two successive measurements.

$$T_I = \frac{k_S T_S + k_L T_L}{k_S + k_L} \tag{29}$$

Weighting functions k_S and k_L have been chosen as parabolic functions of the ratio $\Delta t/\Delta t_C$ with

$$\Delta t = \frac{t_i - t_p}{t_n - t_p} \tag{30}$$

t_p and t_n being the time for the previous and next measurements around the time for required interpolation t_i. .Δt_C being an adjusted parameter corresponding to the critical gap. The weighting functions k_S and k_L are analytically determined in forcing several constraints. The derivative at the time t_p and t_n equal to the one given by the cubic spline. For $\Delta t = 0$ and $\Delta t = 1$ then $k_S = 1$ and $k_L = 0$.

When the interval between two measurements is equal to the critical gap ($\Delta t_C = t_n - t_p$) at the middle of the interval ($\Delta t = 0.5$), we want the interpolation be half the linear and half the spline interpolation ($k_S = k_L = 1$).

These constrains permit to calculate the both weighting functions k_S and k_L

$$k_s = 1 \text{ and } k_L = 4\frac{\Delta t}{\Delta t_c^2}(1 - \Delta t)\left(t_n - t_p\right)^2 \tag{31}$$

Then the analytic form of the interpolation can be rewrite as follow

$$T_I = \alpha T_S + (1 - \alpha) T_L \tag{32}$$

with

$$\alpha = \frac{1}{1 + 4\frac{\Delta t}{\Delta t_c^2}(1 - \Delta t)\left(t_n - t_p\right)^2} \tag{33}$$

The cubic spline used here is a smoothing spline that is well adapted to approximate noisy data. Each data is associated with a weight corresponding to the uncertainty estimates (mainly statistical noise). The spline algorithm come from the IMSL mathematical library and is described in de Boor [33].

5. INTER-ANNUAL CHANGES AND TRENDS

Uncertainty and bias associated with trend estimates are due to the limited length of the time series, the presence of instrumental noise or the existence of sources of natural variability. Decades of data are often necessary to detect environmental changes whereas important time series such as those available from satellites cover at the most the last 25 years. The use of a multi-regression approach is widely accepted for trend analyses [34, 35]. Most results in assessing changes have been obtained in this way. The selected method and the choice of variables to be included in the multi-regression analyses, varies: it is clearly related to the altitude range considered and the variables. For example, while the tropospheric temperature is more likely to be influenced by El Niño Southern Oscillation (ENSO), and less by the solar flux, it will be the opposite for upper stratospheric/mesospheric studies. The short-term variability masks potential trends and its influence should decrease when the time series is long enough. The multi-regression approach is a general method that will be applied to future analyses as the identification of other sources of forcings are investigated. For example, recent research suggests that dynamics may have a greater effect on past ozone changes than previously thought. In this case, new variables such as tropopause height, Arctic Oscillation (AO) index, Eliasen Palm (EP) flux... may be added to the statistical trend models to account for this potential contribution. Rigorous assessments of the additional uncertainties due to their possible interactions in the regression analysis have been investigated following methods described here. Investigations of statistical biases depending on the noise level, and the length of data series have been carried out by Weatherhead *et al.* [36, 37] and Frederick [38]. They show that the number of years of data necessary to detect a specific trend depends firstly on the magnitude of the variance of the data noise, and secondly on the autocorrelation coefficient of the noise. They find that, in order to detect a 5% change in the slope of the trend, between 10 to 20 years of data are necessary, depending on the latitude and the variability of the data. They also address the problem of a discontinuity in the data series due to an instrumental change and they concluded that about 50% additional years of data are necessary to detect a trend. Reinsel *et al.* [39] investigated a turnaround in the downward trend of ozone in their trend model. Additional uncertainties are introduced by the interferences of different source terms in the multi-regression analysis. Many uncertainties still exist in identifying the mechanisms related to natural variability; a linear relationship with the different variables is used as a first approximation, even though it is not proven that the relationship is linear. Furthermore, in the middle atmosphere, proxies are used to represent the geophysical parameters, such as the equatorial wind for the Quasi Biennial Oscillation (independently of the site considered), the UV flux or 10.7 cm flux for solar activity, SST for ENSO and NAO or the optical thickness for stratospheric aerosols. These functions, however, may lead to miss-interpretations in the multi-regression analysis if the variability is based on non-linear effects. In supplement to assuming a linear relationship, we assume that this linear factor is constant with time for the whole length of the dataset even though for the past decades, a nonlinear trend in the temperature of the lower stratosphere has been found in the analysis of radiosonde data. The breaking point between the 2 different slopes of the trend occurred around 1980, and is clearly due to the amplification of the cooling due to ozone depletion. The first cooling trend before the early 80s is only due to CO_2 : after that, cooling is influenced by both increase of CO_2 and the decrease of ozone. Now the recovery of ozone leads us expect a change in the trend both for ozone and temperature, independently of the CO_2 increase. Therefore, a search for nonlinear trends is needed.

Using an approach with artificial signals Kerzenmacher *et al.* [40] have shown that even when the length of the data series increases, the biases decrease only slowly and depend on the relative phase between the considered forcings. For example, if the solar term is not considered, it leads to an oscillating pattern in the resulting biases. Even without noise, a 40% bias can be found after several decades. In all the simulated examples used, the increase in noise reduces the precision of the data analysis. When looking for a trend, the presence of a volcanic signal induces a bias whatever the length of the data series, whereas taking the QBO component into account is not very important and this term can be omitted if the data series is longer than 20 years. There is a marked improvement in the analysis of all the experiments if an adequate model is used. For weak noise, the signal agrees well with the model; only about 1 decade of data is needed to reduce the bias under 10%. If the noise ratio is above 2, the data series should be at least 3-4 decades long.

The tests confirm the importance of using a regression model based on several functions, rather than a basic slope for quantifications of trends for data series of limited length. The use of too many components may lead to spurious results, even if a good agreement between the data series and such a complex model is obtained. However, several limits can be discussed, for example, the number of terms and the choice of proxies. The approach using artificial signals permits biases resulting from the interference of an additional forcing term to be deduced. By use of standard forcings, it is shown that the methodological error could be neglected after about 10 years of data, even for the aerosol and solar cases which were expected to show large errors in the analyzed data. However, in this simulation it is assumed that the atmosphere responds to both these proxies. There is some evidence that dynamical feedback may induce some non-linearity in the response. For explorative work, the choice of the functions does not matter and simple (linear) models are adequate even though, for quantitative estimates, this choice may affect the results adversely. For this reason, it is suggested that the dataset for the 1-2 years following a volcanic eruption be suppressed because of the aerosol load in the stratosphere. These simulations also show that the introduction of a changing trend increases the bias in the estimates of that trend when five additional years of flat signal are added. For higher noise ratios, the non-linear term added in the trend analysis becomes less significant. If the data series is longer than five years, a nonlinear trend model has to be used for the analysis because the non-linear term in the analysis becomes statistically significant. The simulations that include a change in the slope of the trend indicate that a bias increases after 5 years of change if the preceding data extends over a sufficiently long period, and if the noise ratio is moderate. These results can be used to evaluate the time required to detect a change in the trend term for various types of data (e.g. the recovery of ozone). In order to test the methodology, Kerzenmacher *et al.* [40] performed the analysis on artificial signals that have been generated using known combinations of certain proxies and noise. Then the resulting bias in a particular forcing term in the analysis, when the other terms are excluded from the model fit, is calculated, by taking the difference between the signal derived by the regression model respectively with one and two proxies. With such an approach the contribution of noise is eliminated and a bias can be estimated. Therefore the "methodological bias" estimated is only due to the analysis and is expressed as a function of certain parameters such as the length of the data series or the noise.

The analysis of artificial signals composed by the combinations of various components have been conducted in order to estimate the influence on trend detection and solar responses of missing terms in the multi-regression fit. Methodological biases purely based on interferences between the forcing terms are estimated. Furthermore, effects of a changing trend on the linear analysis have been taken into account. The data length required to detect a nonlinear trend will obviously need to cover a longer period. For such a detection, data sets require both to cover continuously a long period of time extended over at least a decade and to provide quasi-homogeneous series including no spurious changes associated with non purely climatologic changes [41]. Successive instrumental evolution made on ground-based systems have improved data quality, however it may has induced artificial long term changes when considering the continuity over the total length of the data set.

In the upper mesosphere, one of the uncertainties comes from the initialization of the pressure profile. The error is estimated at 15% at the mesopause level. The calculation of the uncertainties [1] shows that this error becomes rapidly negligible (due to the exponential decrease of the atmospheric pressure) as compared to the statistic noise, which increases with the altitude. Lidar systems that operate during more than a decade are usually improved and the altitude of initialization increases. So a systematic trend may exist. Currently around 60-70 km, the error due to normalization may largely be considered as negligible. Normalization was made at 80 km during the first years of measurements. Then the errors at 60 and 70 km were smaller with a factor of 30 to 3, respectively, than the initializing error due to the uncertainty of the model in this height range. When climatological data obtained from 1984 to 1989 by lidar to the CIRA 86 model, differences of ±10 K around 80 km. This means that during the first years of operation, systematic variations of 3 and 0,3 K could have been induced, respectively, at 60 and 70 km, that is, for a linear increase in range, an apparent trend smaller than 0,3 to 0,003 Kyr^{-1}.

In the stratosphere the two main sources of error come from the presence of aerosols, and the overlapping between laser beam and telescope field of view. Two lidars located both in south of France respectively located at Biscarosse (Centre d'Essais des Landes: 44°N, 1°W) and at St. Michel de l'Observatoire (Observatoire de Haute Provence: 44°N, 6°E) provide simultaneously some routine measurements from March 1986 to February 1994. The both stations are located at the same latitude from 550 km apart in longitude which is a short distance compared to the wavelength of the expecting decadal structures. A statistical comparison of 169 quasi-simultaneous profiles from 1986 to 1990 has revealed a mean difference smaller than 2K in the mesosphere and 1K in the stratosphere [21]. Instrumental failures on one of the lidars are reported during some specific periods [2, 42] that suggest that instrumental bias may have changed with time caused by successive improvements performed on these instruments. The lidar data is computed in integrating the measurements over the full time of operation during the night. The both profiles are not obtained during the similar portion of the night due to different local weather and some deviations between the measurements of the both instruments can be expected from both systematic changes (tides) or more random atmospheric variability such as the one induce by the wave propagation. For some seasons, a large variability of the daily differences can be noted when compare measurements at both sites that reduces the possibilities to detect the differences associated to instrumental changes. The selection of the data is done for the period from April to September reducing considerably the variability induced by planetary waves which can propagated mostly during winter and due to gravity waves and induced-mesospheric inversions above 60 km which present a strong seasonal component. When compare, profiles from both sites presenting an integration period having an overlapping better than 50 % of the total operating time, a minimum of variability of 2K is observed for data acquired between April and September. In opposite, when data obtained quasi-simultaneously between October and Mars are compared a larger variability can be noted in the upper stratosphere and in the upper mesosphere due to the dynamical perturbations. Data integrated over very different periods exhibit a larger variability mainly around 45-50 km and in the mesosphere where tidal effects are expected to be large. According to the successive major instrumental improvements [2] a continuous decrease with time of the standard deviation is observed. The major disturbance is suspected to be due to miss-alignment between the laser beam and the field of the collector telescope through parallax or defocusing effects. This effect is probably caused by OHP system, which is based on a bistatic configuration while the Lidar at Biscarosse is a monostatic system. In January 1991 a major technical improvement concerning the alignment has been implemented at OHP that permits to reduce problems of miss-alignment and to check their amplitudes. The comparison of the both data sets before and after this change reveals very different mean variability. In one hand in the mesosphere, the root mean square of the difference is reduced due to the combined effects of the regular laser power increase and of the decrease of sky background noise. In the other hand, in the lower part on can note a reduction of the variability of the difference that can be attributed to the reduction of miss-alignment events. The comparison of quasi-simultaneous summer data from April to September (19 events) reveals a mean difference (<1K), which appears to be really significant around 50 km according to the standard error. As no instrumental effect is expected in this region, this mean difference is supposed to be due to atmospheric causes such as the impact of orographic gravity wave energy or non-migrating tidal modes.

Individual lidar profiles consist in a mean profile integrated during the night over several hours. This period depends of the cloud cover, the protocol measurements and sometime the availability of the operator. The time of measurements and the integration period fluctuated quite a bit. Due to tidal motions, these changes of the measurement time may have induced some residual temperature changes. In the stratosphere during summer temperature tides consist mostly of a diurnal cycle with maximum amplitude around the stratopause of ± 2 K with a maximum at 18:00 solar local time [43]. Temperature anomalies have been estimated according to these tidal characteristics and the period of the lidar sounding at OHP. This effect has induced some inter-annual changes smaller than 1K with a main structure between 1986 and 1991. The residual drift is smaller than $+0.02$ K per year, and smaller than the expected trends. These residual artificial trends are negligible compared to long-term trends derived with the data sets [44, 45].

6. ATMOSPHERIC REENTRY

Atmospheric entry is the process by which vehicles that are outside the atmosphere can penetrate that atmosphere and reach the surface in good conditions. For human made vehicles, the deceleration through atmospheric friction and drag (aerobraking) is preferred than landing solely with some active powered braking firing in the opposite direction of the path because it would require a large amount of fuel that have to be lifted into orbit first. Vehicles that undergo this process include spacecrafts, suborbital home-based vehicles, and ballistic missiles and nuclear weapons. All have their requirements according to the reentry phase. For a valuable spacecraft, decelerations in the atmosphere and landing must be gentle enough, and the inside of the spacecraft must be kept at a safe temperature. Ballistic missiles are only guided during the powered phase of flight and the laws of the ballistic mechanic govern their courses. The challenge consists to reduce the aerothermic heating to prevent the full vaporization of the warhead, to deliver a warhead to a predetermined target and to show the smallest atmospheric signature to avoid the early detection of the nuclear warhead to the companion heads. The entry phase in Earth's atmosphere correspond to an altitude range from 120 km to the ground, however the critical phase corresponds to the peak heat flux that occur in the mesosphere around 60-80 km. This phase is characterized by large variations of environmental conditions.

The consequences of atmospheric friction and drag lead to incredibly hot (several 1000 K), leading to dissociated and ionized air in the shock layer. The ablative heat shield functions by lifting the hot shock layer gas away from the heat shield's outer wall through blowing. This thermal protection system sublimates through the process of pyrolysis. The head trails a stream of vaporized metal making it very visible to radar. This effect, allows early detections of the nuclear weapons overly susceptible to anti-ballistic missile systems. For this reason, several heads are included in the payload that exhibits very different shapes. The aerodynamic perturbation acting on a spacecraft during its orbital or atmospheric descent can be described by the drag and lift component of the aerodynamical force. Due to the energy dissipating character of the drag deceleration, natural orbital motion below altitude of 120 km cannot be sustained in the Earth atmosphere. The value of the drag and lift coefficient are depending on the spacecraft shape and dimensions, and on the flow conditions which can be characterized by the dimensionless Mach number, Reynolds number and Knudsen number. The forces acting on the vehicle are the aerodynamic forces, the gravity force and the thrust force. It depends on the control surface deflections, aerodynamic angles, angular rates, Mach number, Knudsen number, air density, and air temperature. There are several basic shapes used in designing entry vehicles: Sphere or spherical section, biconic, or delta wing.

The knowledge of the temperature profile from 120 km to the ground and manly in the mesosphere is required. The occurrence of the mesospheric inversions [3] that is believe to be due to gravity waves braking induces in this region a large variability that is crucial for the trajectory of the vehicles. Gravity waves and tides themselves with their amplitudes growing with heights due to the density reduction, lead to large fluctuations (Figure 4).

The challenge in measuring temperature profiles according to reentry operation comes from the fact that it is virtually impossible to measure temperature along the concerning body path. The investigations on reentry phase require temperature information and uncertainties. To extrapolate temperatures along

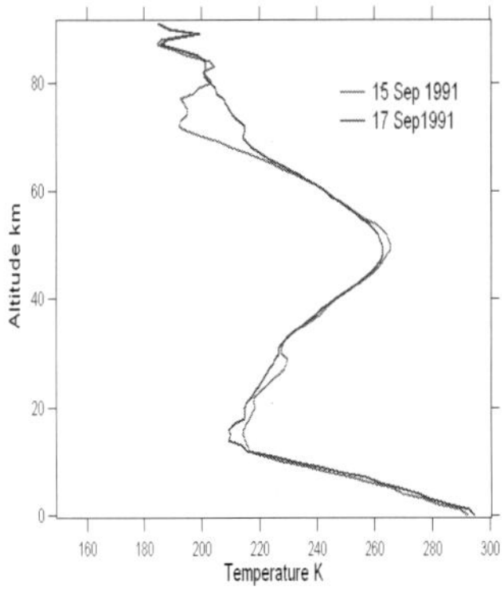

Figure 4. Examples of two successive measurements that illustrate the presence of a mesospheric inversion.

the path from vertical continuous measurements, gravity waves and tides need to be estimated. Such an estimates are possible from models or measurements itself.

As it is virtually impossible to cover altitudes above 100-110 km with lidar, the measurements can be coupled with an atmospheric model above 80-100 km and with radiosonding below 30 km. The coupling can be performed in weighting both the model profile and the observations. Weight is given by the inverse of the square of the uncertainty variance. The temperature lidar uncertainty is given by statistical noise and initialization, radiosonde uncertainties are given by international intercomparison campaigns while the uncertainty associated with the model can be given by the observed variability (section 6.2).

$$m_z = \frac{\sum_i m_{i,z} w_{i,z}}{\sum_i w_{i,z}}, wi = \frac{1}{\sigma_i^2} \tag{34}$$

On the synthetic profile from ground to 120 km, uncertainties due to short-term fluctuations can be added. They can be estimated from lidar continuous observations as described previously (section 2, 3) or can be estimated according to time and space differences.

7. ATMOSPHERIC MODELS

7.1 Mean state

Among many other purposes, atmospheric numerical models and satellite retrievals require as far as possible accurate description of the mean states of the studied medium. This is also crucial to better understand the atmosphere as mean states is the mixture of many atmospheric processes. Climatological data sets for the middle atmosphere are also necessary for constraining the behavior of numerical models. These models provide a useful format for organizing and making widely available the results of satellite missions providing large amount of data. These models facilitate data comparisons and theoretical calculations requiring background atmosphere as well as providing convenient engineering solutions.

Several generations of climatologic models on the stratosphere and mesosphere have been computed with an increased degree of complexity motivated by either or both the scientific needs and the availability of new data sets. First studies of the structure of the middle atmosphere above 30 km are based on in situ rocket measurements. The standard atmosphere of 1962 assumes the division of the atmosphere into 11 layers and linear variation of temperature with altitude. Assuming the model of spheric Earth the relations between pressure, density and altitude can be extrapolated. This model lack accuracy but it allows fast computations. Temperatures as a function of altitude starting from 288.15 K, at ground level (h = 0), are than given by the following linear functions:

- DT/dh = −6, 5°/km for 0 < h < 11 km
- 0°/km for 11 < h < 20 km
- +1°/km for 20 < h < 32 km
- +2.8°/km for 32 < h < 47 km
- 0°/km for 47 < h < 51 km
- -2.8°/km for 51 < h < 71 km and
- -2.0°/km for 71 < h < 85 km.

The earliest comprehensive climatologies for the middle atmosphere were the 1964 and 1972 COSPAR reference atmospheres (CIRA), which were based largely on data interpolation of single balloon and rocket stations. Barnett and Corney [46-48] have settled up the COSPAR International Reference Atmosphere (CIRA 1986) from PMR data over the 1975-1978 period. It gives one of the best available representations of the latitudinal, longitudinal and seasonal variations of the temperature, at least up to 65 km. An updated version of CIRA in 1986 included early satellite observations of the stratosphere and mesosphere and has served as a community standard since the MSIS model.

Comparisons between lidars and CIRA show discrepancies of ±4 K in the mesosphere [49]. Conclusions are difficult to draw because differences can be expected due to long-term changes if comparisons do not cover the same periods and also interferences with tides may exist.

Non-static homosphere models such as MSISE-90 allow for variations of temperature profiles and resulting densities with diurnal, longitudinal-latitudinal, seasonal-latitudinal, and solar/geomagnetic activity effects. MSIS [50] model is based on similar data than CIRA but differs by generating analytical functions instead of extrapolating the data.

More recently, satellite observations from the Upper Atmosphere Research Satellite (UARS), launched in 1991 and continuing to operate in 2002, have provided additional climatological data sets for the middle atmosphere. A new climatologic model URAP (UARS Reference Atmosphere Project) was proposed (*http://www.sparc.sunysb.edu/*). In the frame of the SPARC (Stratospheric Processes and their Role in Climate) systematic comparisons have been performed [51].

7.2 Atmospheric variability

However, with few exceptions, there is no information about standard deviations from the mean values reported on existing climatologies. This is a serious deficiency in nearly all reference and standard atmospheric models for any parameters. This lacuna prohibits quantitative assessment of uncertainties when climatologic models are compared themselves or with numerical models or observations. Atmospheric variability is also crucial for the design of space vehicles to estimate the possible and extreme heating effects during phase of atmosphere re-entry. Satellite experiments alone are able to give a global coverage of the temperature field but have a relatively low horizontal and vertical resolution. While atmospheric variability can be derived from those data, the low resolution and/or the tidal effects represent some serious limitations for such estimates.

GRAM (Global Reference Atmospheric Model) an empirical model of NASA Marshall Space Flight Center, is accurate but because of its complexity and detail computations with this model are relatively slow. The "NASA/MSFC Global Reference Atmospheric Model-1995 Version" uses a specially developed set of data based on Middle Atmosphere Program (MAP) data. Above 90 km the Marshall Engineering Thermosphere model (MET) is used. Fairing techniques assure smooth transitions among the models and sets of data in the overlap height ranges. GRAM incorporated a new variable-scale perturbation model that provides both large-scale (wave) and small-scale (Stochastic) deviations from mean values for thermodynamic variables and horizontal and vertical wind components.

Also, the technique of Rayleigh lidar has been shown capable of obtaining temperature profiles with a good vertical resolution from 30 to 90 km and to follow the temporal evolution of this parameter at all time scales from fractions of an hour to years. On an intermediate time scale, the daily variability have been already quantify from lidar sites showing a strong seasonal and vertical pattern. The variability of the middle atmosphere on a daily basis is mainly associated with stratospheric warming and mesospheric inversion events but also with permanent inertio-gravity and Rossby wave propagation. The more recent investigations of the temperature variability with lidar [52] include some data from different locations reporting an important latitudinal gradient with decreasing variability from the mid-latitude to the lower latitude.

In the Northern Hemisphere several Rayleigh lidar have been operated continuously on a multi-year basis. A climatologic model of the variability has been computed using five data sets well distributed from tropic to polar vortex border. The model is composed by some analytic functions to permit horizontal extrapolation and an easy implementation in any tools that may need a variability estimates according to latitudes and seasons. Some simple functions describing seasonal and latitudinal changes, have been found. In the second section, lidar data sets used here have been described.

The investigation of the changes of the variability along the year clearly exhibits a seasonal component. The component of the variability can be represent with cosine functions, which required only 2 parameters: amplitude and phase of the maximum. The variability estimates are fitted with a constant value, an annual and semi-annual waves using a least square fitting method. The values are quite small for the tropical site at Hawaii, however significant components have been obtained from 2 to 15 Kelvin at mid-latitude and for high latitude sites. Semi-annual component is generally smaller than the annual one but sometime can also have a clear contribution. Times of the maximum are generally in phase with the annual cycle. For each level five parameters are calculated and used to represent the seasonal changes of the variability at each individual station.

Table 1. Coefficients used for the model of the temperature variability. These parameters need to be used with formula given above. Coefficients are reported in assuming latitudes in degrees. Phases $\varphi(\Phi)$ are reported.

altitude [km]	mean component A_0			annual component A_1			semi-annual component A_2	
	α	β	χ	δ	ε	γ	η	κ
30	−0.0003	0.06	1.5	0.0004	0.028	−0.28	−0.012	0.47
35	−0.0004	0.10	1.1	0.0002	0.093	−1.19	0.039	−0.78
40	−0.0011	0.18	−0.1	−0.0014	0.258	−4.26	0.040	−0.38
45	−0.0007	0.13	0.9	−0.0005	0.148	−2.19	0.033	0.01
50	0.00013	0.04	2.4	−0.0006	0.135	−2.33	0.017	0.38
55	0.0009	0.0	3.1	−0.0007	0.103	−1.36	0.033	−0.12
60	−0.0025	0.29	−1.6	−0.0001	0.094	−1.75	0.033	−0.16
65	−0.0044	0.43	−2.7	−0.0002	0.067	−0.97	0.022	0.09
70	−0.0061	0.53	−2.5	−0.0009	0.157	−2.25	0.024	0.11
75	−0.0058	0.50	0.0	0.0011	−0.037	1.17	0.030	−0.39
80	−0.0018	0.14	7.2	0.0021	−0.108	1.94	−0.028	2.60

The temperature variability $\sigma(\Phi)$ can be given by the combination of 3 simple functions: 2 cosines for the previous function that need 5 parameters: A_1,φ_1 A_2,φ_2.

$$\sigma(\Phi) = A_0(\Phi) + A_1(\Phi).\mathrm{Cos}[Wt + \phi_1(\Phi)] + A_2(\Phi).\mathrm{Cos}[wt\phi_2(\Phi)] \quad (35)$$

with Φ, being the latitude and w the annual frequency.

Each of these five parameters can be represent by some polynomes with degrees equal to a constant for phases (φ_1,φ_2) month of the maximum) to 3 for the means (A_1, A_2). The evolution of annual amplitude is a parabolic function (A_1) while the semi-annual is represented by a linear function.

$$A_0(\Phi) = \alpha\Phi^2 + \beta\Phi + \chi \quad (36)$$

$$A_1(\Phi) = \delta\Phi^2 + \varepsilon\Phi + \gamma \quad (37)$$

$$A_2(\Phi) = \eta\Phi + \kappa \quad (38)$$

$$\varphi_1(\Phi) = 1 \quad (39)$$

$$\varphi_2(\Phi) = 0.5 \quad (40)$$

8. SATELLITE VALIDATION

8.1 General considerations

What we called validation is a set of comparisons between similar measurements from a new instrument to be validated and a reference instrument already validated or at least well having well known limitations. The objectives of these inter-comparisons consist to better identify systematic bias and quantified the precision and compare these two quantities with the ones estimated. The temperature lidar can be used as a reference atmosphere as it provides autocalibrated data. The errors are mainly located in the lower part of the profile around 30 km due to aerosols, linearity of the counting and miss-alignment of the emission and reception. The upper part of the profiles, around 80 km, is limited by the background noise and the initialization. The noise is mainly due to the statistical photocounting and then permits to retrieved temperature with an accuracy of 1K in the height range of 30-80 km.

The strategy and terminology of validations are review recently by von Clarmann [53]. Let x_h* be a vertical temperature profile obtained at the time t, sampled on a discrete vertical grid h considered as the true atmospheric state. Let x_h be an independent measurements of the temperature. The accuracy a_h of the measurement is the square root of the mean of the squared difference of the N associated couples of the reference and the measurement to be validated.

$$a_h = \sqrt{\sum_N \left(x_h - x_h^*\right)^2} \quad (41)$$

The bias b_h at the altitude h is the systematic difference that is given by the mean difference of the N associated couples of the reference and the measurement to be validated.

$$b_h = \sum_N \left(x_h - x_h^*\right) \quad (42)$$

The precision p is characterized by the reproducibility of the measurements and is often associated with the random error or noise.

$$p_h = \sqrt{\sum_N \left(x_h - \sum_N \frac{x_h}{N}\right)^2} \quad (43)$$

Satellite experiments usually provided estimates of the total error covariance matrix σ_t, the systematic error covariance matrix σ_s and the random covariance matrix σ_r. The diagonal elements of these matrices are the related variances [54, 55].

Part of the validation consists in verifying that a, b and p respectively match the σs, σt and σr, and when it is not the case to quantify a better estimate of σ_s, σt and σ_r. This can happen due to some underestimates of some error sources or more often due to some miss functioning associated with unexpected problems in space. These problems can occur during the take off, during the deployment in orbit or due to slowing degradation of the instrument characteristics.

8.2 The NDSC Network

While the measurements from space offer the best opportunity to provide a coherent measured field because the same instrument is used, some bias are geographically dependent and mainly according to latitudes because the pointing of the instrument is different or the geometry of the field of view. For this reason, it is valuable to use several ground based instruments place at different strategy place. The best well-known effort of network building was performed by the Network of the Detection of Stratospheric Changes (NDSC) [56] now named NDACC (Network for the Detection of Atmospheric Composition Changes). While this network is dedicated to the atmospheric composition, temperature was defined as a key parameter because most of the chemical equilibrium rates depend strongly of the temperature. It was suggested to measure temperature with Rayleigh lidar above 30 km and metrological radiosonde below.

To provide the optimum latitude coverage within the obvious constraints of quality, funds and resources, it was proposed that the network would consist of 5–7 primary stations fully equipped with all the NDSC-defined instruments: polar, mid-latitude, and tropical stations in both hemispheres, plus an equatorial station. Depending on specific site characteristics such as geography, orography or meteorology, a composite station may be formed with individual or a limited group of instruments at different sites within a given latitudinal or regional zone. Some complementary sites and mobile instruments are also associated with this network to expand the geographic coverage.

For satellite validation, horizontal homogeneity of the correlative measurements is required in order to be able to quantify possible latitudinal bias of the instrument in space. The NDSC instruments are all research prototypes and have some differences in their design and final capabilities. Systematic differences that are of particular relevance for data quality must be detected, critically analyzed, and ruled out. The Steering Committee of the NDSC has therefore proposed to use, when and wherever possible, a mobile system alternately visiting the different stations of the network. Since the official inception of the network in 1991, several campaigns have been organized employing the NASA/Goddard Space Flight Center

(GSFC) mobile lidar. However, comparisons with the mobile lidar are not the only way to determine the quality of the data and other techniques have been investigated simultaneously in some campaigns, or more routinely at some sites. Comparisons with co-located radiosondes, or microwave radiometers are often performed, and also with close-by satellite measurements. There is no absolute reference instrument and the individual capabilities, and the network homogeneity, cannot be assured or quantified without undertaking all the NDSC validation exercises. Instruments launched into space are continually improving and their capabilities are close to what is obtainable from the ground in terms of sensitivity. Validation using ground-based instruments is therefore pushed to the limit. The use of assimilation allows, in theory, freedom from most of the possible discrepancies induced by geophysical causes and nonperfect spatio-temporal coincidences. However, some weighting of the data input to the model needs to be decided. The use of the reference data requires provision of standard values for bias and noise.

Many intercomparisons of temperature profiles have been performed under the framework of the NDSC-NDACC [57]. Since the instruments were designed independently, it offers a good opportunity to evaluate the overall capability of such instruments for atmospheric monitoring. While the use of a mobile instrument traveling from one station to another is the preferred method for comparisons, satellites can also provide a good spatial transfer for a quasi-simultaneous comparison of several lidars. Temperatures in the upper stratosphere and mesosphere, suffer from tidal effects that cause systematic changes and can induce bias in the data comparisons when they are not performed simultaneously. Also for temperature, summertime is more appropriate for comparisons as the variability has a strong seasonal component at mid and high latitudes. Variability is smaller in the tropics and does not exhibit a

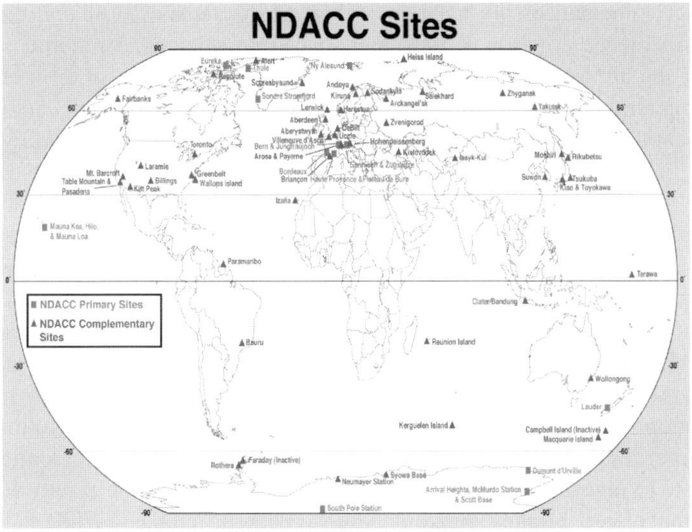

Figure 5. Location of the NDSC-NDACC stations.

strong seasonal cycle. During winter, at high latitude the permanent daylight restricts the use of lidars. Few systems have been designed for daylight operations. In the lower stratosphere the presence of aerosols required the development of more sophisticated methods to provide non-biased data. However, these techniques have been developed more recently and the quality of the data provided in this altitude range has been less investigated. The comparisons conducted with the mobile lidar, as well as algorithm inter-comparisons, allow improvements to individual systems by sharing the knowledge of the different groups involved in the NDSC. While the standard deviations of the individual temperature differences were close to the estimated noise, biases of 1 K are reported. Larger biases are reported below 30–35 km, due to stratospheric aerosols and around the top of the profile 75–85 km probably due to initialization and noise extraction. The use of satellite data is difficult for assessing temperature lidar measurements because tidal effects bias comparisons. To avoid such effects, solar occultation experiments would be preferred. However the number of coincidences would then be drastically reduced.

8.3 Satellite validations

In addition to these general statement about validation, there are several factors that complicate a straightforward validation: Different smoothing or girding effects between the reference and the measurement to be validated, impact of apriori information, and the spatio-temporal coincidences.

The geometry of the reference measurements and the measurements from space to be validated is often quite different. The vertical smoothing is described by the kernel functions, it is then easy to degrade lidar data into a smooth temperature profile having the same vertical resolution. Usually it improves the lidar accuracy as the vertical accuracy of the lidar is directly associated with the number of photon averaged. If the measurements include a priori information, both profile have to used the same a priori profile or it has to be estimated and included into the covariance matrix [58]. The horizontal and temporal smoothing is more difficult to adjust as lidar need long integration time to improve their accuracy. For temperature 15 minutes to hours are required and are then larger than the time required by the satellite. In opposite, depending of the technique, satellite measurements are associated with bad horizontal resolution compare to the local nature of the lidar. Usually, it is considered that the horizontal inhomogeneity and time smoothing induce by lidar integration are booth mainly due to gravity waves and compensated each other to give a similar mean temperature.

The non-perfect coincidences are a larger source of discrepancy. Usually, only profiles, which meet a certain co-incidence criterion in time and space, are selected. Variability of the temperature as well as most of the atmospheric variables, is composed by a functional term and a random term. The residual random term could be characterized by its covariance matrix and can be considered as an additional instrumental error. However, the functional term should be corrected by some appropriate parameterizations. For temperature, several process ca induce systematic differences: Planetary waves, gravity waves and tides. During winter, planetary waves can induce strong horizontal gradients and sudden changes during stratospheric warming. However, these temperature changes can be estimated horizontally with satellite measurements and with time with two daily successive lidar measurements. Gravity waves, induce reversible fluctuations with small scales and most of the time the fluctuations can be considered as random processes and smooth out by either horizontal resolution or time integration for lidar. However gravity waves also interact with the mean flow and can induce some irreversible effect on the circulation and the thermal structure of the atmosphere. Both interactions and gravity wave sources present some geographical pattern. In these cases and if space coincidences follow systematically a similar spatial differences then gravity wave effects can induce bias. Different configurations regarding wave activity need to be investigated to test the potential contribution of the gravity waves. Tides are one of the most important effects because they induce systematic differences. While the general theory for the non-migrating tides is well known the variability of tidal characteristics is large due to the variability of the mean atmospheric state. In addition, migrating tides are probably more important than expected.

Lidar data have been used to validate temperature measurements from the four UARS temperature sounders [59-61]. The Upper Atmospheric Research Satellite (UARS), launched on September 12, 1991, which has a non-sun-synchronous orbit, covers, for a single location, a full 24-hour cycle in about 1 month. The comparisons include time-of-day adjustments based on a local tidal model derived from MLS data. The purpose of such comparisons is to evaluate the improvements given when tidal effects are considered and to estimate the quality of the temperature measurements available through UARS experiments [43]. Although both techniques are differently influenced by tidal changes, very good agreements are obtained for diurnal and semidiurnal temperature variations, calculated from lidar measurements in southern France and from MLS data. A large variability of the tidal characteristics is reported by both data sets. Seasonal changes are detected and are the main cause of variability for the diurnal component. An analytical model, including a seasonal sinusoidal function, is computed in fitting the 3 years of tidal climatology of MLS. The diurnal component presents substantial agreement with the published results of a two-dimensional numerical model. Some discrepancies exist at 34 km and simulated amplitudes in summer are smaller than those observed here by 20% around 45-50 km. However, the differences around the stratopause are not incompatible with the observed interannual variability. In winter, larger variability is reported that can be related to large-scale dynamic changes, but more work is needed to establish clear relationships with background parameters. Also, non-migrating tides could have a significant contribution to the observed variability. Further analyses are needed to separate migrating tides, non-migrating tides, and aliasing effects due to planetary waves, using both zonal and local harmonic analyses. The semidiurnal component is in total disagreement with theory that predicts small amplitude. However, the good agreement between MLS and lidar suggests that this observed component might be real. An artificial harmonic superimposed on the real signature of the semidiurnal modes is suspected. Correlative analyses of gravity waves and tidal changes should be investigated in the stratosphere in the future to better explain such discrepancy between models and observations. From April to September, the temporal and horizontal variability of the stratospheric temperature above mid-latitude is smaller than in winter, providing a better environment for data comparisons. Biases (if any) are expected to appear with better confidence. However, from April to September, SD larger than 2 K are still observed. Time-of-day adjustments applied on the data reduced mean differences and the SD (0.5-1 K) between 45 and 55 km. However, below 45 km, data comparisons have not been improved by tidal adjustments. Significant residual mean differences, of 1.1-3.2 K around 35-43 km, remain with a similar sign for the four UARS experiment-lidar comparisons. However, no significant mean differences have been observed when ISAMS was compared with simultaneous lidar integrated over 1 hour. This result suggests that around 40 km, tidal effects are not properly reproduced in our model, and even with a simple time-of-day adjustment, comparisons between non-simultaneous data continue to be biased as compared with simultaneous comparisons. Sunrise/sunset differences for HALOE-lidar comparison and comparisons between our model and the numerical simulations also suggest that our tidal model might be not adequate in the lower part, probably due to the large tidal variability and the presence of non-migrating tides. Lidar comparisons reveal that a simple model such as the one developed here may take account for not more than one third of the tidal variability. However, such improvements seem to be insufficient to neglect tidal contributions on data comparisons.

8.4 Long-term validations

One of the issues with satellite is to estimate global trends. The difficulties is associated with the fact that the successive satellites provide different techniques and even with a succession of similar satellite they present their own true calibration curve that need to be estimated from the ground. OHP lidar data and NCEP stratospheric temperature analyses provide long and continuous data bases for the middle and upper stratosphere that are highly valuable for long-term studies. However each data set has limitations. Comparisons between lidar data from 1979 to 1993 and NCEP data interpolated from the global analyses to the lidar location reveal significant mean temperature differences [19]. Insight into the origin of the differences, offers an opportunity to improve the overall quality of temperature monitoring in the stratosphere. Instrumental effects can explain some of the differences in the lidar system. In the stratosphere most of the limitations in lidar temperatures appear below 35-40 km, due to events of lidar misalignment (as large as 10 K) or to the effects on lidar data of volcanic aerosols (as large as 15 K). Changing biases between lidar and NCEP temperatures above 5 hPa coincide with replacement of satellites used in the NCEP analyses. However, some bias differences in upper stratospheric temperatures remain even after NCEP adjustments are made, based on rocketsonde comparisons. While these biases have been already suspected, they had never been explained. The comparison with lidar suggests that the remaining bias (2-4 K) is caused by tidal influences, heretofore not accounted for by the NCEP adjustment procedure. Lidar profiles have been filtered in their lower part for misalignment and aerosol contamination. Long term changes have been compared and a factor of two in trend differences have been reported. No significant trends (at 95% confidence) have been detected except with lidar around the stratopause and with NCEP analyses at 5 and 10 hPa. According to instrumental limitations of both data sets temperature trend may vary from 1 to 3 K with altitude (10-0.4 hPa). The only global coherent source of temperature, in the middle atmosphere, on a long-term basis (decades), is provided by the TIROS/NOAA operational vertical sounders (TOVS). A series of TOVS instruments (which includes the Microwave Sounder Unit-MSU and the Stratospheric Sounder Unit-SSU) has been put into orbit onboard a succession of operational satellites since late 1978. These instruments do not yield identical radiance measurements for a variety of reasons, and derived temperatures may change substantially when a new instrument is introduced. The situation was even more complex with the inclusion of a new and improved version of the spectrometer (Advanced Sounder Unit-AMSU) in 1998. Temperatures are derived from these data, with different techniques and tools. Finger *et al.* [42] have compared the operationally derived temperatures with collocated rocketsondes and lidar observations and find systematic biases of the order of 3–6 K in the upper stratosphere. These biases furthermore change with the introduction of new operational satellites, and Finger *et al.* [42] provide a set of recommended corrections to the temperature data, which have been used by NOAA. In spite of the application of the recommended adjustments, time series of temperature anomalies from the NOAA analyses still exhibit significant discontinuity near the times of satellite transitions due to tidal interferences [19]. Other uncertainties associated with the satellite data, in general, include the effects due to longitudinal drifts that cause i/ spurious trends as the diurnal cycle is sampled at earlier or later times for a single satellite, and ii/ changing solar shadowing effects on the instrument, in turn causing heating or cooling of the radiometer. While estimates for the correction and attempts to remove the biases are made, this factor does introduce a potential uncertainty in the trend determination. Lidar data and NCEP data have been compared for the OHP site from 1979 to 1993. Some of the differences can be explained by either an instrumental bias in the OHP lidar system (as the presence of volcanic aerosols injected in the stratosphere or misalignment effects) or tidal effects in the upper stratospheric NCEP data. Biases of 2 to 5 K, morning versus afternoon orbit effect have been suggested through NCEP-lidar comparisons. Tidal changes appear to be the possible source of such discrepancy.

More recently NOAA has produce new temperatures series following the "Nash" methodology [59]. The preliminary investigations of SSU-AMSU temperature series show a change in the cooling rate at the time of the AMSU inclusion. The "Nash" data set consists of brightness temperatures from observed (25, 26, and 27) and derived (47X, 36X, 35X, 26X, and 15X) channels of the SSU) and High-Resolution Infrared Sounder (HIRS) 2 instruments on these same satellites. The weighting function for the SSU channels are typically 10–15 km thick. One complication with satellite data is the discontinuities in the time series owing to the measurements being made by different satellites monitoring the stratosphere since 1979. Adjustments have been made in the Nash channel data to compensate for radiometric differences, tidal differences between spacecraft, long-term drift in the local time of measurements, and spectroscopic drift in channels 26 and 27. NOAA has continued to produce temperature series applying the "Nash" methodology. However, more recently, the second generation of instrument AMSU has been used to insure the sounding continuity. AMSU is an improved version of the SSU and provides more radiance channels exhibiting thinner weighting functions. To insure the continuity, NOAA has constructed synthetic SSU Channels with the AMSU channels. The vertical resolution of the lidar data has been reduced to fit the SSU-AMSU series. The weighting functions presented previously have been used. However, as most stations cannot provide temperature below 30 km the weighting functions have been forced to 0 below 30 km and normalized. The anomalies have been considered and have been calculated in removing annual and semi-annual sinusoidal functions.

The temperature series show cycles that can be associated to seasonal changes, to the Quasi Biennale Oscillation (QBO) and to the 11-year solar cycle. Trends are also clear at some levels. However, SSU reveal a plateau after 1998 when AMSU where introduced and careful examination are required. The comparisons with SSU and the OHP lidars (Figure 6) reveal a similar temporal behavior, showing a cooling up to a null-trend around 1998. For the channels covering lower altitudes, a similar behavior

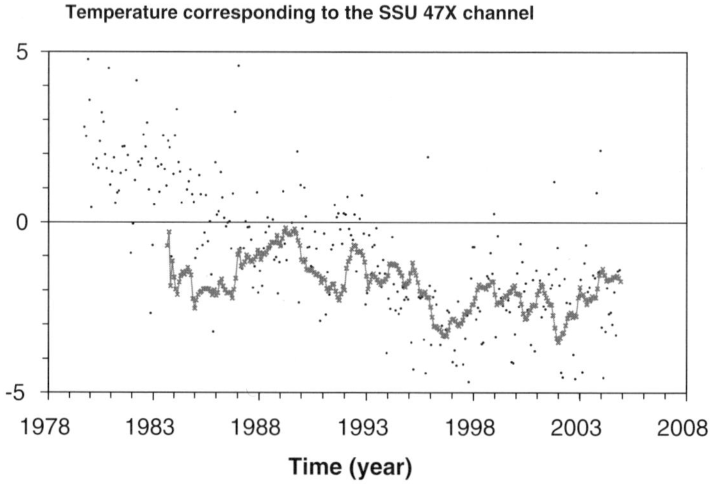

Figure 6. Zonal monthly mean temperature derived from SSU (Dots) compared to corresponding monthly mean of weighted OHP lidar (line).

can be noted with smaller amplitudes. However over the full period for OHP no significant residual trend is reported. Residual trends during a smaller period 1994 to 2004 are equal to –0.7±0,9 Kelvin respectively for Observatory of Haute Provence. At lower altitude (channel 26) the OHP temperature lidar data are warmer up to 1987 probably due to the effect of miss-alignment already reported. Residual trend over the full period is equal to –0,9Kelvin/decade over the 1994-2004 period. For the two other intermediate channels (36X and 27), OHP temperature series show residual trends close to the limit of confidences respectively of $1,6 \pm 1,2$ Kelvin/decade.

References

[1] Hauchecorne A. and M.-L. Chanin, *Geophys. Res. Lett.* **7** (1980) 565-568.
[2] Keckhut P., A. Hauchecorne, and M.L. Chanin, *J. Atmos. Ocean. Technol.* **10**, (1993) 850-867.
[3] Hauchecorne A., M.L. Chanin, and R. Wilson, *Geophys. Res. Lett.* **14** (1987) 933-936.
[4] Wilson R., M.L. Chanin, and A. Hauchecorne, *Geophys. Res.*, **96**, (1991) 5169-5183.
[5] Wilson, R., M.L. Chanin, M.L. and A. Hauchecorne, *J. Geophys. Res.*, 96, (1991) 5153-5167.
[6] Gabor D., *J. Inst. Electr. Eng.* **93**, (1946).
[7] Morlet J., Acoustic Signal/Images Processing and Recognition, (C. H. Chen, Ed., NATO ASI Series, 1, Springer-Verlag, 1983) pp. 233–261.
[8] Shibata, T., S. Ichimori, T. Narikiyo, and M. Maeda, *J. Meteorol. Soc. Jpn.* **66** (1998) 1001-1005.
[9] Marsh, A. K. P., N. J. Mitchell, and L. Thomas, *Planet. Space Sci.* **39** (1991) 1541-1548.
[10] Mitchell, N. J., L. Thomas, and A. K. P. Marsh, *Ann.Geophysicae* **9** (1991) 588-596.
[11] Whiteway, J. A., and A. I. Carswell, *J. Atmos. Sci.* **51** (1994) 3122-3136.
[12] Whiteway, J. A., and A. I. Carswell, *J. Geophys. Res.* **100** (1995) 14113-14124.
[13] Hauchecorne, A., N. Gonzalez, C. Souprayen, et al., *J. Atmos. Terr. Phys.* 56 (1994) 1765-1778.
[14] LeBlond P. H. and L. A. Mysak, Waves in the Ocean (Elsevier Scientific Publishing Company, Amsterdam, 1980).
[15] Forbes, J. M., *J. Geophys. Res.* **87** (1982) 5222-5240.
[16] Vial, F., *J. Geophys. Res.* **91** (1986) 8955-8969.
[17] Hagan, M. E., J. M. Forbes, and F. Vial, *Geophys. Res. Lett.* **22** (1995) 893-896.
[18] Forbes, J. M., *J. Atmos. Terr. Phys.* **46** (1984) 1049-1067.
[19] Keckhut, P., J.D. Wild, M.E. Gelman, et al., *J. Geophys. Res.,* **106** (2001) 7937-7944.
[20] Wild, J.D., M.E. Gelman, A.J. Miller, M.L. et al., *J. Geophys. Res.* **100** (1995) 11105-11111.
[21] Keckhut, P., A. Hauchecorne, and M.L. Chanin, *J. Geophys. Res.* **100** (1995) 18887-18897.
[22] Swinbank, R., R. Orris, and D. L. Wu, *J. Geophys. Res.* **104** (1999) 16,929-16,941.
[23] Forbes, J. M., Middle atmosphere tides (*MAP HANDB. 18*, Univ. Of Ill., Urbana, 1985) pp. 50-56.
[24] Hoxit, L. E., and R. M. Henry, *J. Atmos. Sci.* **30** (1973) 922-933.
[25] Hitchman, M. H., and C. B. Leovy, *J. Atmos. Sci.* **42** (1985) 557-561.
[26] Dudhia, A., S.E. Smith, A.R. Wood, and F.W. Taylor, *Geophys. Res. Lett.* **20**(1993) 1251-1254.
[27] Gille, S.T., A. Hauchecorne, and M.L. Chanin, *J. Geophys. Res.* **96** (1991) 7579-7587.
[28] Dao, P.D., R. Farley, X. Tao, and C.S. Gardner, *Geophys. Res. Lett.* **22** (1995) 2825-2828.
[29] Meriwether, J. W., X. Gao, V. B. Wickwar, T., Wilkerson, K. Beissner, S. Collins, and M. E. Hagan, *Geophys. Res. Lett.* **25** (1998) 1479-1482.
[30] Leblanc, T., I.S. McDermid, and D.A. Ortland, *J. Geophys. Res.* 104 (1999) 11,917-11,929.
[31] Leblanc, T., I.S. McDermid, and D.A. Ortland, *J. Geophys. Res.* 104 (1999) 11,931-11,938.
[32] Morel, B., H. Bencherif, P. Keckhut, S. Baldy, and A. Hauchecorne, *J. Atmos. Sol. Terr. Phys.* **64** (2002) 1979-1988.
[33] Charney J.G., and P.G. Drazin, *J. Geophys. Res.* **66** (1961) 83-109.
[34] de Boor, C., A Practical Guide to Spline (Springer-Verlag, New-York, 1978) pp. 235-243.
[35] Ramaswamy, V., M.L. Chanin, J. Angell, et al., *Rev. Geophys.* **39** (2001) 71-122.
[36] Beig, G., Keckhut, P., Lowe, et al., Rev. Geophys., 41 (2003) 1015.

[37] Weatherhead, E.C., G.C. Reinsel, G. C. Tiao, et al., *J. Geophys. Res.*, **103** (1998) 17149–17161.
[38] Weatherhead, E.C., G.C. Reinsel, G. C., Tiao, et al., *J. Geophys. Res.* **105** (2000) 22201-22210.
[39] Frederick, J. E., Measurement requirements for the detection of ozone trends, (Ozone orrelative measurements workshop, NASA Conf. Publ. 2362:B1-B19, 1984).
[40] Reinsel G.C., E.C. Weatherhead, G.C. Tiao et al., *J. Geophys. Res.* **107**(2003).
[41] Kerzenmacher T.E., P. Keckhut, A. Hauchecorne, and M.L. Chanin, J. Environ. Monit. **8** (2006) 682-690.
[42] Karl T.R., R. Q. Quayle and P. Y. Groisman, J. Clim., 6 (1993) 1481–1494.
[43] Finger, F.G., M.E. Gelman, J.D. Wild, et al., *Bull. Amer. Meteor. Soc.* **74** (1993) 789-799.
[44] Keckhut, P., M.E. Gelman, J.D. Wild, et al., *J. Geophys. Res.* **101** (1996) 10299-10310.
[45] Hauchecorne, A., M.L. Chanin, and P. Keckhut, *J. Geophys. Res.* **96** (1991) 15297-15309.
[46] Keckhut P., A. Hauchecorne, and M.L. Chanin, *J. Geophys. Res.* 100 (1995) 18.887-18.897.
[47] Barnett, J.J. and M. Corney, Temperature data from satellites (Middle Atmosphere Program, Handbook for MAP," Vol. 16, ed. K. Labitzke, J. J. Barnett and B. Edwards, 1985) pp. 2-11.
[48] Barnett, J.J. and M. Corney, Middle atmosphere reference model from satellite data ("Middle Atmosphere Program, Handbook for MAP," Vol. 16, ed. K. Labitzke, J.J. Barnett and B. Edwards, 1985) pp. 47-85.
[49] Fleming, E.L., et al., Zonal mean temperature, pressure, zonal wind, and geopotential height as functions of latitude, COSPAR International Reference Atmosphere 1986, Part II: Middle Atmosphere Models, *Adv. Space Res.*, (10), **12**, 11-59, 1990.
[50] Leblanc, T., I.S. Mcdermid, P. Keckhut, et al., *J. Geophys. Res.* **103** (1998) 17191-17204
[51] Hedin, A.E., *J. Geophys. Res.* **88** (1983) 10,170-88.
[52] Randel, W., Udelhofen, P., Fleming, et al., *J. Clim.* **17** (2004), 986-1003.
[53] Keckhut P., A. Hauchecorne, S. Henot, O. Coesnon, S. McDermid, T. Leblanc, von Cossart, and Von Zahn, *Recent Res. Devel. Geophysics* 4 (2002) 7736-0760.
[54] Von Clarmann, T., *Atmos. Chem. Phys. Discuss.*, **6** (2006) 4973-4994.
[55] Rodgers, C.D., *J. Geophys. Res.,* **95** (1990) 5587-5595.
[56] Rodgers, C.D., Inverse methods for atmospheric sounding: Theory and practice, vol. 2 of series on atmospheric (Oceanic and Planetary Physics, edited by: Taylor, F.W., World Scientific, 2000).
[57] Kurylo M.J. and S. Solomon, Network for the Detection of Stratospheric Change, (NASA Report, Code EEU,1990).
[58] Keckhut P., S. McDermid, D. Swart, et al., *J. Environ. Monit.* **6** (2004) 721-733.
[59] Rodger, C.D., and B.J. Connor, *J. Geophys. Res.* **108** (2003) 4116.
[60] Fishbein, E.F., Cofield, R.E., Froidevaux, L., et al., *J. Geophys. Res.* **101** (1996) 9983-10016.
[61] Gille, J.C., P.L. Bailey, S.T. Massie, et al., *J. Geophys. Res.* 101 (1996) 9583-9601.
[62] Hervig, M.E., J.M. Russell III, L.L. Gordley, et al., *J. Geophys. Res.* **101** (1996) 10277-10285.
[63] Nash, J. and G.F. Forrester, *Adv. Space Res.* **6** (1986) 37-44.

J. Phys. IV France **139** (2006) 361–372
© EDP Sciences, Les Ulis
DOI: 10.1051/jp4:2006139023

DIAL lidar for ozone measurements

A. Pazmiño[1]

[1] *Service d'Aéronomie, Université Pierre et Marie Curie, 4 place Jussieu, 75252 Paris Cedex 05, France*
e-mail: andrea.pazmino@aero.jussieu.fr

Abstract. This chapter presents a general overview of ground-based lidar system for ozone measurements in the troposphere as well as in the stratosphere. The DIAL technique, generally used by the lidar teams for ozone measurements is explained. Afterwards, the different elements of the DIAL system are assessed in this work. Finally, a brief description of the ozone DIAL systems for different altitudes ranges operated at the Haute Provence Observatory (OHP) in Southern France as well as some scientist results are showed.

1. INTRODUCTION

The ozone is a minor constituent present in the atmosphere. Its vertical structure shows a maximum peak of ozone concentration in the lower-middle stratosphere between 20 and 30 km. This atmospheric region is called the ozone layer, which has two main consequences for life on Earth. First, the ozone layer acts as a filter of harmful solar ultraviolet radiation reaching the Earth's surface [1]. Second, the ozone is responsible for the positive temperature gradient in the stratosphere by a warming effect due to the photo-dissociation and recombination cycle of ozone [2].

Besides of stratospheric ozone, which represents about 90% of total column ozone in a specific place, about 10% is distributed in the troposphere. Tropospheric ozone acts as a pollutant affecting directly living organisms. It also plays an important role in the climatic change as a greenhouse gas [3].

The ozone evolution in the stratosphere as well as in the troposphere has been a major topic for the scientific community to highlight the effects of anthropogenic activity, particularly after the Montreal and Kyoto Protocols signatures (1987 and 1997 respectively). Various instruments are used for monitoring ozone. The lidar instrument has been used for ozone measurements since the end of the seventies [e.g., 4] as for others atmospheric constituents and parameters [e.g., 5]. The international lidar network NDACC (Network for the Detection of Atmospheric Constituents Changes)[1], allows worldwide long-term series of vertical ozone measurements. The ground-based lidar measurements are characterized by their reliability, accuracy and high vertical and temporal resolutions. These characteristics make the lidar system a good monitoring instrument and a good reference for the validations of different satellite products.

2. PRINCIPLE OF LIDAR

The LIDAR (acronym of LIght Detection And Ranging) is a remote sensing instrument. It operates according to the same principle as the RADAR (RAdio Detection And Ranging) but functioning at optical wavelengths. The lidar technique is frequently used to study the chemical or physical parameters in the atmosphere. It is based on the interaction of optical radiation with the atmosphere. Since the lidar provides the necessary radiation for the measurement, it is considered as an active instrument in contrast to the passive instruments which use natural radiation sources like the sun or the moon. The diversity of lidar systems allows the observation of a variety of atmospheric parameters such as aerosols,

[1] Former called NDSC (Network for the Detection of Stratospheric Changes), http://www.ndacc.org/.

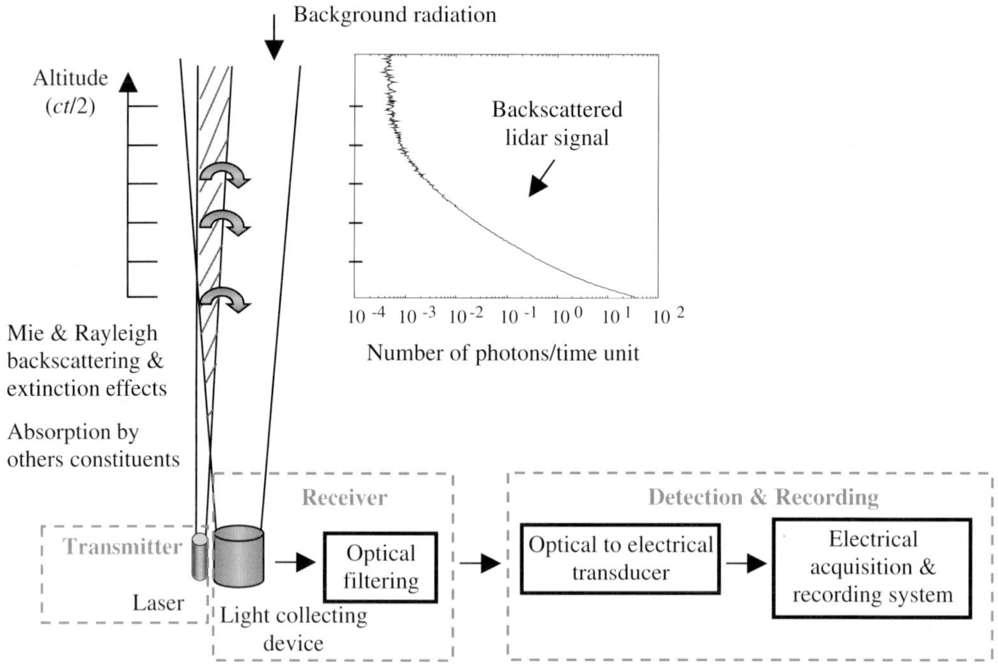

Figure 1. Schematic view of a generic lidar system in a monostatic configuration and biaxial arrangement.

temperature, wind speed and direction, and the concentration of atmospheric constituents (e.g., water vapor, ozone).

The lidar instruments provide range-resolved measurements by emitting pulsed laser radiation into the atmosphere at specific wavelengths. A schematic view of the principle of a generic lidar system is shown in figure 1. As each light pulse travels through the atmosphere, it interacts with atmospheric particles and molecules. A small part of the radiation is scattered back to the lidar receiver system. The backscattered light is collected generally by a telescope and transmitted to the detector which transforms the irradiance into electrical current. The range is then determined by the round-trip time t of the scattered light pulses ($z = ct/2$, with c the light speed).

Figure 1 shows the principal elements of a generic lidar instrument: 1) a transmitter, generally a pulsed laser at specific wavelength (assumption of monochromatic emission), 2) a receiver, generally a telescope and spectrometer, 3) a device for signal detection such as a photomultiplier and 4) an electronic unit for signal acquisition, signal processing and on-line recording. Two basic configurations characterize the lidar ground-based instruments: bistatic, where the transmitter and the receiver are far away and the monostatic, with the transmitter and receiver at the same location. In addition, the later configuration can be divided in two different arrangements: coaxial, where the axis of the laser beam is coincident with the axis of the receiver, and biaxial, where telescope's field of view (FOV) fully overlaps the laser beam beyond some altitude range. This altitude depends on the FOV, the diameters of the telescope and laser beam, the relative distance between the optical axes and laser divergence.

The simplified lidar equation for a monochromatic laser pulse emission and only considering elastic interactions (no wavelength shifting) [e.g., 6] can be written as:

$$N(\lambda, z) = K \cdot \frac{\Delta z}{z^2} \cdot \beta(\lambda, z) \cdot e^{-2\tau(\lambda,z)} \qquad (1)$$

where $N(\lambda,z)$ is the received number of photons at wavelength λ from backscattering cell at a range z and of thickness Δz; K is the instrumental constant; β the total volume backscatter coefficient (Rayleigh + Mie) in the layer Δz at the wavelength λ and altitude z; and τ the optical depth that is given by the following equation:

$$\tau(\lambda, z) = \underbrace{\int_0^z \left(\sigma_{O_3}(\lambda, z') \cdot n_{O_3}(z')\right) \cdot dz'}_{\tau_{O_3}(\lambda, z)} \underbrace{\int_0^z \left(\alpha_{mol}(\lambda, z') + \alpha_{aer}(\lambda, z') + \sum_e \sigma_e(\lambda, z') \cdot n_e(z')\right) \cdot dz'}_{\tau_{no\_O_3}(\lambda, z)} \quad (2)$$

where $\tau_{O_3}(\lambda, z)$ is the integrated optical thickness due to absorption by ozone at a wavelength λ, that is determined in integrating the product of the ozone cross section coefficient σ_{O_3} and the ozone molecular density n_{O_3} up to altitude z. The optical depth due to others parameters $\tau_{no\_O_3}(\lambda, z)$ takes into account: α_{mol} and α_{aer} corresponding to the molecular and aerosol extinction coefficients, respectively; and the absorption by others gases that interfere with ozone measurements due to their spectral and concentration characteristics $\sum_e \sigma_e(\lambda, z') \cdot n_e(z')$.

$\tau_{O_3}(\lambda, z)$ can be retrieved from the lidar signal according to the following equation:

$$\tau_{O_3}(\lambda, z) = \frac{1}{2} \cdot \left[- \ln(N(\lambda, z) \cdot z^2) + \ln(K \cdot \Delta z) + \ln(\beta(\lambda, z) - 2 \cdot \tau_{no\_O_3}(\lambda, z))\right] \quad (3)$$

The ozone number density is retrieved from the derivation of the different terms of equation (3) and then dividing by the ozone cross-section. In order to determine $n_{O_3}(\lambda, z)$, it is necessary to use simultaneous measurements of different parameters as atmospheric backscattering and extinction or standards atmospheric models.

3. OZONE MEASUREMENT BY THE DIAL TECHNIQUE

3.1 Principle of the DIAL lidar system

The atmospheric parameter or constituent to be measured by a lidar determines the characteristics of the different elements of the system, depending on the targeted light-atmosphere interaction, e.g., Rayleigh and Mie scattering, Raman inelastic scattering or light absorption by atmospheric species. In the case of minor constituents like ozone, the differential absorption technique initially proposed by Schotland in 1974 [7], is used. It consists of the emission into the atmosphere of two wavelengths differently absorbed by ozone. The intensities of the two wavelengths scattered back to the system are compared to determine the optical attenuation by ozone. This technique introduces the idea of "differential" that allows to limit the influence of atmospheric variables others than ozone using a close spectral pair of wavelengths. Besides, the differential absorption cross-section of ozone has to be important in order to detect the ozone amounts. Then, the DIAL (acronym of DIfferential Absorption Lidar) technique requires that one of the emitted wavelengths been strongly absorbed by ozone (λ_{on}) and the other weakly absorbed (λ_{off}). The last one is considered as the reference wavelength.

The ozone retrieval from the lidar signals after correction of the background radiation is given in equation 4:

$$n_{O_3}(z) = n_{O_3}^{meas}(z) + \delta n_{O_3}(z) \quad (4)$$

where $n_{O_3}^{meas}(z)$ represents the measured ozone and $\delta n_{O_3}(z)$ represents the contribution of atmospheric parameters others than ozone to the lidar signal at the two wavelengths, called the complementary term.

The first term of equation 4 can be expressed as follows:

$$n_{o_3}^{meas}(z) = \frac{1}{2 \cdot \Delta\sigma_{o_3}} \cdot \frac{d}{dz} \ln \frac{\left(N(\lambda_{off}, z) - N_b(\lambda_{off}, z)\right)}{\left(N(\lambda_{on}, z) - N_b(\lambda_{on}, z)\right)} \tag{5}$$

Equation 5 represents the measured term corresponding to the spatial derivative of the logarithm of the ratio of the lidar signals already corrected from background radiation signal ($N_b(\lambda,z)$). For any parameter X depending on the wavelength in lidar equation, ΔX represents the difference $X(\lambda_{on}) - X(\lambda_{off})$. The DIAL technique is self-calibrated since the ozone retrieval does not depend on the instrumental constant K (c.f. equation 5). The complementary term $\delta n_{o_3}(z)$ can be written as:

$$\delta n_{o_3}(z) = \frac{1}{\Delta\sigma_{o_3}} \cdot \left[\frac{1}{2} \cdot \frac{d}{dz} \ln \frac{\beta(\lambda_{on}, z)}{\beta(\lambda_{off}, z)} - \Delta\alpha_{mol}(z) - \Delta\alpha_{aer}(z) - \sum_e \Delta\sigma_e(z) \cdot n_e(z)\right] \tag{6}$$

Two types of errors appear in the ozone measurement by the DIAL technique: a statistical error which is related to the randomness of the detection process and a systematic error, related to the uncertainties of the different parameters included in the complementary term (equation 6). Many instrumental considerations are taken into account in order to minimize these errors.

3.2 Description of a DIAL system

The DIAL system can be divided into three different sub-systems as shown in figure 1 for a simple lidar set-up: the transmitter (in the DIAL case with at least two emissions), a receiver with one or more optical receptors, and the detection and acquisition of two or more different wavelength signals. The ozone DIAL instruments generally use the monostatic configuration in its both types of arrangements (coaxial and biaxial) according to the altitude range to be measured (c.f. section 3.2.2).

3.2.1 Transmitter sub-system

Most of ozone measurements by lidar in the troposphere and particularly in the stratosphere are performed in the ultraviolet (UV) spectral region where ozone absorption is most efficient (some examples are shown in section 4). Nevertheless some DIAL systems for ozone measurements in the lowermost troposphere work in the ozone infrared (IR) absorption band 9-11 μm [e.g. 8], especially useful when aerosol influence is significant (less affected by differential aerosol scattering). The low values of ozone cross section in the IR avoid measurement at higher altitudes. Henceforth, only UV DIAL systems will be explained.

The value of the wavelength pair has to be adapted to the ozone altitude range to fulfill all the optimization criteria to reduce the statistical and systematical errors according to λ_{on}, λ_{off} and $\Delta\lambda$ (λ_{off}-λ_{on}). Minimizing the statistical error requires high $\Delta\sigma_{o_3}$ (high $\Delta\lambda$ value) while minimizing the systematic error needs to reduce the influences of other parameters (low $\Delta\lambda$ value). A compromise has thus to be reached. Mégie and Menzies [6] have shown that the DIAL technique presents the best sensitivity when the ozone optical depth at the absorbed wavelength λ_{on} is close to unity over the specific measurement range, assuming that the absorption at the reference wavelength λ_{off} and the background signal at both wavelengths are negligible and the influences of others parameters at λ_{on} equal those at λ_{off}. As the ozone optical depth depends on the ozone profile (left panel of figure 2) and the ozone cross section (right panel of figure 2), the optimal λ_{on} vary as a function of altitude. The tropospheric measurements require strong UV absorption to detected weak ozone amounts. The optimal ozone absorbed wavelength is found in the 260-300 nm spectral range. In the case of stratospheric ozone measurements, the laser emissions have to be weakly attenuated when traveling in the troposphere and the optical absorbed wavelength is found between 300 and 310 nm.

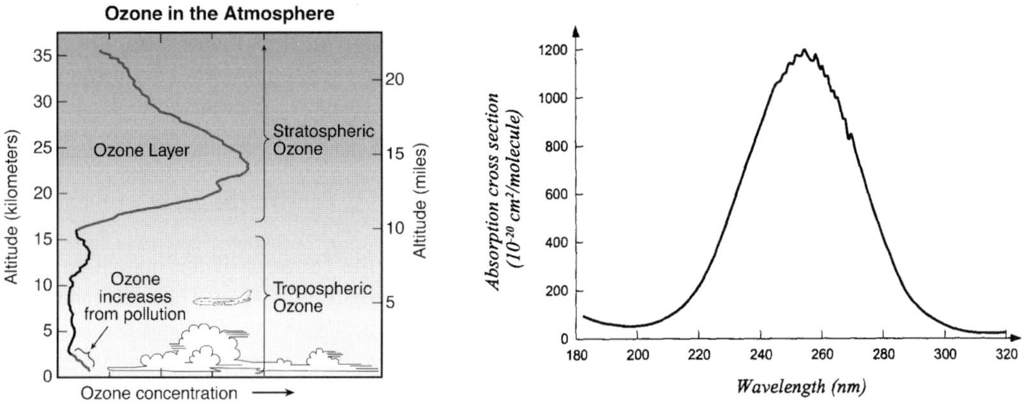

Figure 2. Left panel: atmospheric ozone profile [from 9]. Right panel: ozone absorption cross section in the spectral region 180-320 nm [from 10].

The statistical error can also be minimized by a high repetition rate and a high energy laser pulse, particularly for stratospheric ozone measurements where photons originating from high altitude layers have to be detected.

At present, most of DIAL systems use the powerful excimer and/or reliable Nd:YAG lasers that emit in the optimal spectral ranges. Although fully tunable laser (dye-laser or tripled Ti:Sa solid state laser) are not mandatory for tropospheric ozone measurement, they offer more flexibility in Mie wavelength selection process and can be used for other pollutant measurement (e.g., SO_2). Tropospheric ozone measurements can be performed with KrF excimer laser emitting at 248.5 nm [11] or with frequency-quadrupled Nd:YAG laser, followed in both cases by a stimulated Raman shifting generally using hydrogen (H_2), deuterium (D_2) or a combination of both gases. An example of ozone measurements using a Nd:YAG is shown in section 4.

In the case of stratospheric ozone measurements, the XeCl excimer laser is generally used for the generation of the absorbed wavelength radiation at 308 nm. The key feature of this kind of lasers is the high energy and repetition rate that enable the detection of photons originating from the upper stratosphere. The main difference between the stratospheric lidars sources is in the generation of the reference wavelength. The different wavelengths used are:

1) The 1st Stokes radiation of the Raman shift of the XeCl wavelength in H_2 (353 nm) [12] or in D_2 (339 nm). The advantage of this technique is the use of only one laser but its disadvantage is the loss of power energy after Raman shifting (about 50%);
2) The emission wavelength of the XeF excimer laser at 351 nm [13]. The advantage is the powerful energy but it suffers of maintenance problem and cost;
3) The third harmonic of a Nd:YAG laser at 355 nm. The advantage of this type of lasers is the reliability and operating cost. It provides a lower energy than the excimer laser but enough for stratospheric ozone measurements. An example of DIAL system using this laser source is shown in section 4.

Optionally, beam expanders (afocal systems) can be used to reduce the divergence of transmitted laser beams.

3.2.2 Receiver sub-system

The functions of the receiver sub-system are: 1) to collect the photons backscattered from the atmosphere and 2) to separate the wavelength radiations. The different optical elements of the receiver sub-system have to be optimized for the chosen spectral range.

The first optical element is generally a telescope. Its mirror collects a fraction of the backscattered light and focuses it to a small spot. A large surface area of the mirror contributes in minimizing the statistical error by increasing the probability of collection of backscattering photons. In the case of tropospheric ozone measurements, the diameter can vary from 10 cm up to 1 m depending of the required altitude range. In order to probe the whole stratosphere, a larger mirror surface is necessary to detect the weaker backscatter light from high altitudes associated with the sharp decrease of the atmospheric number density with altitude. Such a large telescope mirror is quite expensive; the use of a set of smaller mirrors can be preferred instead.

The light is transmitted from the telescope's focal plane to the entrance of the optical analyzing device (the second optical element of the receiver) by an optical fiber for instance. Then, the radiation is separated into λ_{on}- and λ_{off}-signals and directed to the detection sub-systems. This spectral separation can be carried out with holographic gratings or beam-splitter combined with narrow-band interference filters. The advantage of the former ones is their ability to separate close wavelengths. It is very useful in the case of multi wavelength detection. The grating allows a great attenuation of the straylight with a great transmittance (40 - 50%). In the second case, the use of beam-splitter is cheaper and easier to carry out, but it provides low transmittance (20 - 35%) and may produce crosstalk when close wavelengths have to be detected, especially for daytime measurements.

The high dynamic range of the lidar signals from tropospheric as well as from stratospheric measurements can be a problem, depending on the required accuracy, when one wants to handle the λ_{on}- and λ_{off}-signals in the whole considered altitude range with only one detector for each signal. This difficulty can be solved by determining 2 channels: one for lower altitudes and the other one for the higher altitudes. Two methods can be applied. The first one is the use of two or more distant telescopes adapted for the different altitudes ranges to probe. The coaxial arrangement is used for measurements in the first atmospheric layers and the biaxial arrangement mostly for the higher altitudes. The second way to solve this problem is based on the optical separation of λ_{on}- and λ_{off}-signals in a low and high energy channels.

The strong backscattering radiation from the lowermost atmospheric layers can produce signal-induced noise (SIN) in the detector (photomultipliers). This effect is highlighted at high altitudes where the signal to noise ratio is quite low. In order to limit the SIN, a high speed mechanical chopper synchronized with the lasers triggering and the acquisition sub-system can be inserted at the entrance of the optical analyzing device to stop the incoming light for some micro-seconds (1 μs $\equiv$ 150 m) and protect all the detectors of the DIAL system.

3.2.3 Detection and Acquisition sub-system

The goal of this sub-system is to convert the receiving light from the optical analyzing device into an electrical signal that will be recorded by an electronic device.

The first element of this sub-system is the detector. The typical detector of the lidar signals in the UV spectral range is the Photomultiplier Tube (PMT). The PMT is a high sensitive detector for low-intensity applications (high internal gain and low noise). The PMT provides current pulses that are produced by photons impacting its photocathode. Indeed, the output signal of a PMT consists of the addition of 3 signals: one corresponding to photons arriving to the cathode; and the others two signals associated to the thermal emission of electrons inside the PMT (dark current or dark counts) and the SIN.

In section 3.2.2 the use of a mechanical chopper in lidar systems as a protection of PMTs from high-intensity light sources coming from the first few kilometers has been discussed. Another type of protection is the electronic gating that can be applied to each PMT. The gating consists in the application of high inverse voltage to the PMT's dynode chain in order to avoid the electrons acceleration. Electronic gating is easy to apply to a selected PMT and presents the advantage of reliability because it no requires moving parts as a mechanical chopper. However, gating can cause difference in the PMT gain.

The output signal of PMT is amplified and converted from current signal to a voltage signal according to the input levels required in the acquisition-recording device. The signal can be electronically recorded in two different ways: the analog detection to probe the lower atmospheric layers, and photo-counting (single-photon detection) for the higher altitudes levels.

Analog detection uses fast analog transient digitizers to convert the output current from the PMT into a digital form for the acquisition device. This device performs averages of a few laser shots depending on the required temporal resolution and accuracy. Hardware averagers as well as software averaging can be utilized. Then, the acquisition device stores the signal on a hard disk for example, for a posteriori evaluations. The acquisition device has to be synchronized with the lasers trigger for ranging. Acquisition vertical resolution better than 30 m can be obtained with this technique.

Photo-counting technique utilizes the PMT ability to detect single photon events. It requires a fast comparator (discriminator) to remove a substantial number of dark counts from the signal count. Then a multi-channel counter adds the pulses arriving from the discriminator to its active memory according to the lasers trigger. In this way, the photons number density as a function of height is performed. The signals are then stored in a hard disk. In this technique, many profiles are averaged to increase the signal to noise ratio. The loss of counts due to the pulse pile-up effect [14] can happen, which has to be corrected by the signal processing algorithm. The acquisition vertical resolution is typically of few hundreds of meters. For this technique as well as for analog detection, on-line processing and visualization are generally implemented.

4. THE OHP DIAL INSTRUMENTS

The OHP lidar station[2] (43.9°N, 5.71°E, 670 m asl) was one of the pioneer stations of the NDACC international network for ozone, stratospheric aerosols and temperature measurements by lidar. Regular measurements of ozone profiles are performed at OHP by lidar systems since 1990 for the free troposphere and since 1986 for the whole stratosphere. Since the mid-1990s, more than 120 profiles per year are measured by these systems.

4.1 Stratospheric ozone DIAL lidar system

Figure 3 shows a schematic overview of the present UV DIAL instrument for stratospheric ozone measurements at OHP. The lidar system includes a Lambda Physik XeCl excimer laser emitting at the absorbed wavelength (308 nm) and a Continuum Nd:YAG laser at 1.06 μm which third harmonic is used as the reference wavelength (355 nm). Both lasers operate at 50 Hz in an alternate way in order to avoid possible cross-talk in the optical analyzing device. In addition, two beam expanders are used.

The backscattering radiation of the emitted laser pulses is collected by four Newtonian telescopes of 0.5 m diameter mirrors. An optical fiber transmitting UV radiation is placed in the focus of each telescope. The multiple-fiber collector concept is applied, where the optical fibers are put together at the optical analyzing device entrance. As the transmission of emitted laser pulses is placed in the centre of the 4 telescopes set, the system configuration is quasi-coaxial, with an equivalent mirror surface of 1 m diameter.

The optical analyzing device includes a mechanical chopper rotating at 24000 rpm to obstruct the first 5 km; and a spectrometer. Lens for optical adaptation of the output of the optical fiber to the spectrometer entrance are also used. The spectrometer consists of a collimated mirror for the incoming light, a holographic grating (3600 grooves per mm and efficiency of 52%) for wavelength separation and additional mirrors for light transmission inside the spectrometer.

---

[2] Since October 2005 the lidar station at OHP is called Gérard Mégie station.

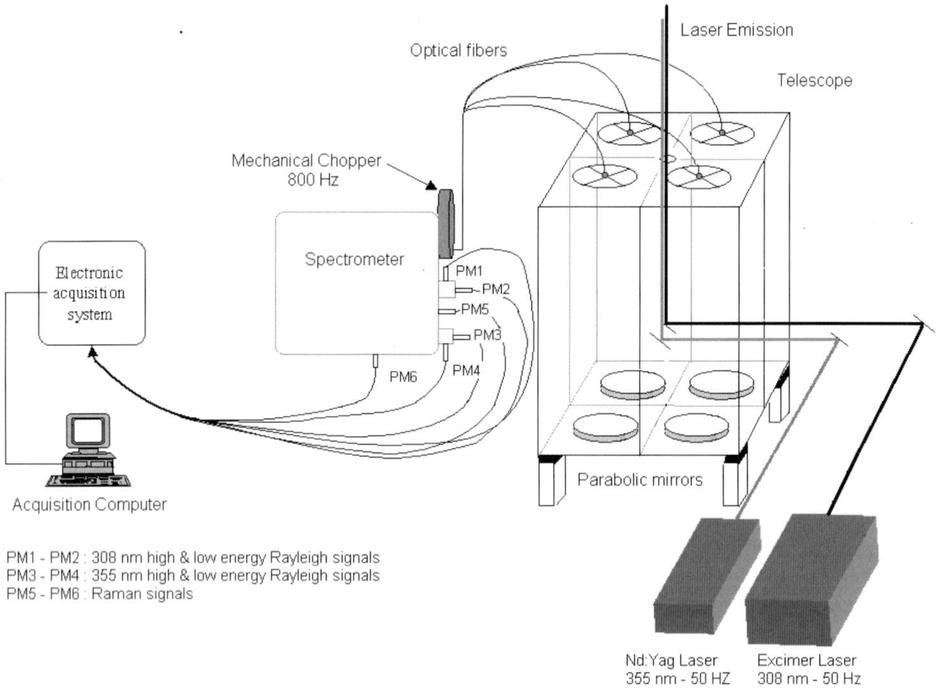

Figure 3. Set-up of the OHP lidar system for stratospheric ozone measurements [from 15].

The spectrometer separates the backscattering radiation in 4 detected wavelengths: the emitted laser ones 308 and 355 nm (Rayleigh signals) and the corresponding 1st Stokes wavelengths in the nitrogen vibrational Raman spectrum (331.8 and 386.7 nm, respectively). The Rayleigh signals are separated in a high (90%) and low (10%) energy channels to handle the dynamic of lidar signals in the whole altitude range. The 4 Rayleigh signals are routinely used. In case of high stratospheric aerosols loading (after volcanic aerosols emissions), the error due to the backscattered term could reach a value of 100% in lidar equation [15]. Then, the Raman wavelengths that suppress the aerosol contribution to the backscatter component are employed [16]. The shortcoming of the Raman signals is that they are much weaker than Rayleigh ones ($\sigma_{o_3\_Ray} \equiv 3 \cdot \sigma_{o_3\_Ram}$) producing a loss of accuracy. Nevertheless the low density of Raman signals is ordinary used to check the linearity of Rayleigh channels in the lower altitude range during signal processing.

The 6 optical signals are detected by Hamamatsu PMTs suitable for photo-counting. The PMTs of the high energy Rayleigh signals are electrically gated up to 20 km. Thereafter, the 6 electrical signals are amplified by a 150 MHz bandwidth amplifier. The 6 independent channels are directed to the acquisition device that was designed by the Service d'Aéronomie (SA) laboratory as well as the amplifiers. The acquisition system includes for each channel: a discriminator and 2 high speed counters (300 MHz) working in parallel in order to avoid dead time between two memories bins. The lidar is controlled by a PC program developed at SA that enables on-line visualization and recording. The master clock is provided by the chopper that is taken into account by the program to trigger the lasers. The trigger of the counters is set by the emitted light pulse detected by a photodiode for the corresponding laser. For example, the emitted pulse of the excimer laser triggers 3 channels: 308 nm low and high energy and the corresponding 331.8 nm Raman channel. For more details of this lidar instrument and its signal processing algorithm, see Godin-Beekmann *et al.* [15].

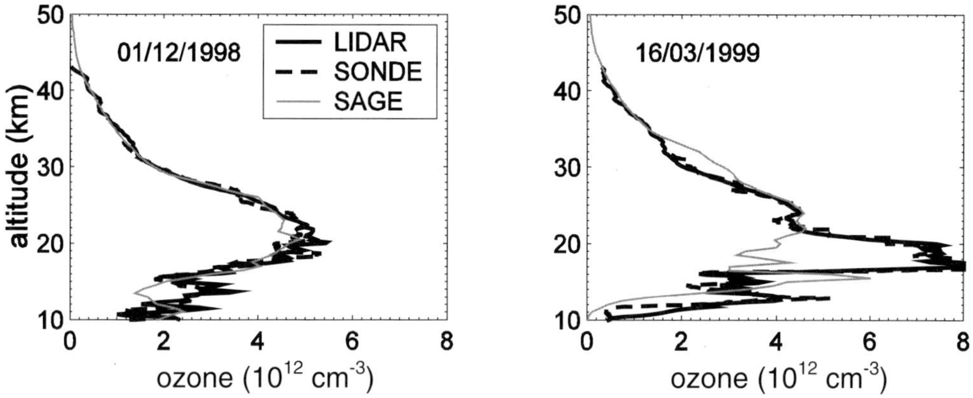

Figure 4. Ozone profile comparisons between the DIAL lidar system and ozonesondes at OHP and the SAGE II satellite instrument [from 15].

This DIAL lidar system performs measurements of the vertical ozone distribution from 10 to 50 km with a typical temporal resolution of 5 hours. In the signal processing algorithm, the vertical acquisition resolution of 150 m has to be degraded as a function of height resulting in a total accuracy of 5% below 20 km and 15-30% above 45 km. The final vertical resolution ranges from 0.5 km at 15 km to 5 km at 45 km. The DIAL lidar system permits the detection of small vertical structures of ozone in the lower stratosphere as shown in figure 4 for December 1, 1998 and March 16, 1999. In this figure, quasi-coincident ozone measurements obtained by the DIAL system and by ozonesonde are compared. The performance of the DIAL system is highlighted by the good agreement with the ozonesonde measurements characterized by a high vertical resolution. In figure 4, the closest in time and in space SAGE II satellite measurements are also displayed. The agreement with SAGE II observations is quite good in the middle and high stratosphere except for March 16. This day, polar air masses with large ozone amounts have passed over OHP. The satellite measurement was 10° far away OHP location missing the polar ozone tongue that has been detected by the others two instruments.

4.2 Lidar for ozone measurements in the free troposphere

An UV DIAL lidar system was developed at OHP for ozone monitoring in the free troposphere. The present tropospheric lidar instrument has the same concept as ALTO (Airborne Lidar for Tropospheric Ozone) system [17] that was installed at OHP station in summer 2004 for ozone campaigns. The block diagram of the OHP lidar instrument is shown in figure 5.

The transmitter sub-system consists in a Continuum Nd:YAG laser emitting at 1.06 μm. Its fourth harmonic (266 nm) is used to stimulate Raman shifting in a single high pressure cell filled with D_2. Then, 2 simultaneous emitted wavelengths are sent in the atmosphere: the 1st and 2nd Raman Stokes lines of 266 nm (289 and 316 nm, respectively). The system configuration is bi-axial with a distance between the beam laser and the telescope axes of 50 cm.

A Cassegrainian telescope with a 80 cm principal mirror is used. The telescope collects the backscattered radiation and focus the light in a UV optical fiber. The radiation coming from the optical fiber is separated in the 2 emitted wavelengths by a spectrometer. The spectrometer consists in a holographic grating, field lens to adapt the output beam to the grating entrance. It also allows background sky-light filtering (bandwidth less than 1 nm) especially at the larger wavelength which is not in the solar-blind spectral region. In order to handle the dynamic range of the lidar signal, each wavelength radiation is separated in low (10%) and high (90%) energy that are then detected by Hamamatsu PMTs.

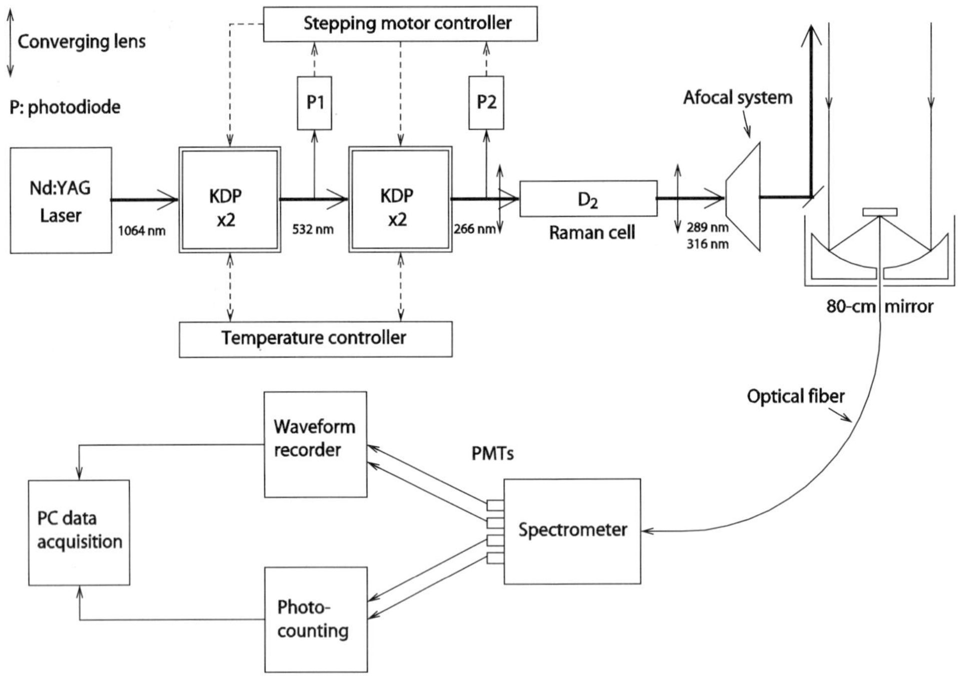

Figure 5. Set-up of OHP lidar system for ozone measurements in the free troposphere [adapted from 17].

The 2 PMTs corresponding to the low-energy signals are adapted for analog detection. Then, the output signals are amplified according to the input range of the waveform recorder (10 MHz, 12 bits analog-to-digital conversion units) developed at SA laboratory. For the high-energy signals, the photo-counting technique is used. After the detection by PMTs suitable for photo-counting, the 2 lidar signals are amplified by homemade 500 MHz bandwidth amplifiers. Then, each signal is discriminated and recording by a similar acquisition sub-system utilized for stratospheric ozone measurement previously described. The laser's trigger at 20 Hz is controlled by a program installed in a PC and the photo-counting and waveform devices triggers are set by the laser light pulse detected by photodiodes. This program allows on-line signal processing, visualization and recording (further details in Ancellet and Ravetta [17]).

The OHP lidar system provides measurements of vertical ozone distribution in the free troposphere from ~3 up to ~14 km during nighttime and up to ~9 km during daytime. The vertical acquisition resolution is variable and set by the operator. According to the sampling periods of the waveform recorder (100 ns) and photo-counting units (250 ns), the maximum vertical acquisition resolution are 15 m and 37.5 m, respectively. Since the signal to noise ratio decrease with height, the final vertical resolution is 300 m at 6 km increasing to 900 m at 11 km [18]. The temporal resolution is variable as a function of the desired ozone accuracy (1 hour for seasonal variation and year-to-year trend, 2-5 min for process studies as shown in figure 6).

Figure 6 shows the evolution of the ozone vertical distribution measured by a lidar system at OHP (left panel) during a European campaign in March 1995 to study the tropopause fold. Each ozone profile correspond to 2 min averaged laser shots. The ozone evolution is compared with potential vorticity (PV) evolution obtained by a mesoscale model version 5 MM5 (right panel). The ozone and PV fields show similar characteristics. The intrusion of ozone-rich air from the stratosphere to the troposphere

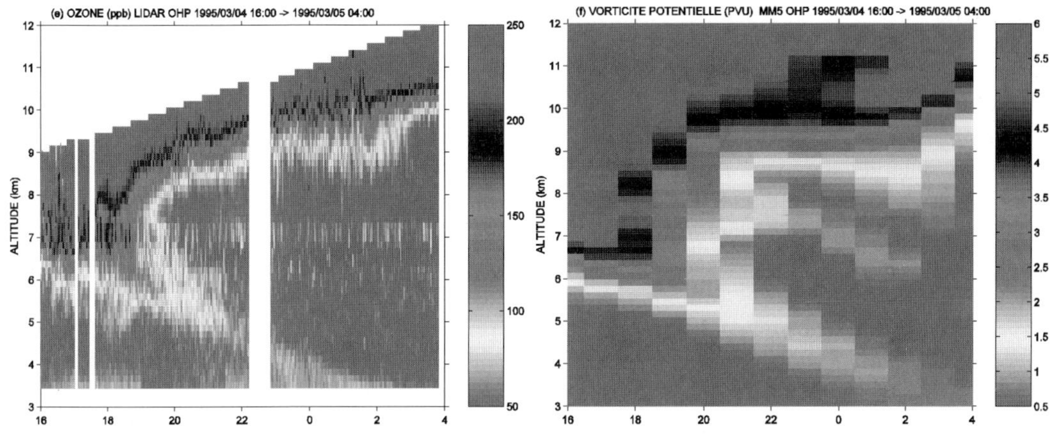

Figure 6. Left panel: ozone mixing ratio averaged over 2 min (ppb) measured by lidar system at OHP. Right panel: potential vorticity (pvu $= 10^{-6}$ m^2 s^{-1} K kg^{-1}) simulated by the MM5 model [from 18].

is highlighted in figure 6 between 16 to 20 UTC. This effect is also seen in the PV field showing the tropopause fold in the same region and with comparable extension and shape.

5. CONCLUSION

The ozone vertical distribution can be retrieved with good accuracy and good temporal and vertical resolution from lidar measurements. The DIAL technique is suited for measurements of minor atmospheric constituents as ozone since it minimizes the influences of other atmospheric parameters. In addition, the DIAL method does no require instrumental constants assessments. DIAL lidar instruments compared to ozonesondes as well as satellite instruments provide measurements of the temporal ozone evolution in time ranges of several hours.

The altitude range of the measurements depends on the different elements of the DIAL system (e.g., emitted wavelengths and power of the lasers, telescope characteristics, type of detection). Tropospheric ozone measurements require strong UV absorption in order to detect small ozone amounts as shown in figure 6 which displays ozone measurements by the tropospheric lidar system at OHP. In the case of stratospheric ozone measurements, powerful lasers are necessary to probe the whole stratosphere. In addition, the laser emitted wavelengths do not have to be strongly absorbed by ozone in order to reach the upper stratosphere.

The DIAL system for ozone measurements is a powerful instrument well suited to address different topics related to climatology and trend analyses, air pollution monitoring, transport studies based on ozone as a quasi-passive tracer (e.g., stratosphere/troposphere mass exchanges, vortex occurrences) among others. At present, many lidar groups tend to measure simultaneously ozone and the other parameters that appear in the lidar equation (equation 6) in order to improve the ozone profile retrieval.

Acknowledgments

I wish to thank S. Godin-Beekmann and G. Ancellet for their comments as well as J. Lefrère and R. Wilson.

References

[1] Hartley, W.N., *Chem. News*, **42** (1880) 268-274.

[2] Andrews, D.G., Holton, J.R. and Leovy, C.B., *Middle Atmosphere Dynamics.* International Geophysical Series, **40**, San Diego, Calif. Academic Press, (1987).

[3] IPCC (Intergovernmental Panel on Climate Change), *IPCC Third Assessment Report: Climate Change 2001*, Cambridge, (2001).

[4] Mégie, G., Allain, J.Y., Chanin, M.L. and Blamont, J.E., *Nature*, **270** (1977), 329-331.

[5] Measures, R.M., in *Laser Remote Sensing*, edited by John Wiley & Sons, Inc. (1984).

[6] Mégie, G. and Menzies, R.T., *Appl. Opt.*, **19** (1980) 1173-1183.

[7] Schotland, R.M., *J. Appl. Meteorol.*, **13** (1974), 71-77.

[8] Asai, K., Itabe, T and Igarashi, T., *Appl. Phys. Lett.*, **35** (1979) 60-62.

[9] WMO (World Meteorological Organization), *Scientific Assessment of Ozone Depletion: 2002, World Meteorological Organization Global Ozone Research and Monitoring Project, Report N° 47*, Geneva, 2003.

[10] Inn, E. C. and Tanaka Y., *J. Opt. Soc. Amer.*, **43** (1953) 870-873.

[11] Kempfer, U., Carnuth, W., Lotz, R. and Trickl, T., *Rev. Sci. Instrum.*, **65** (1994) 3145-3164.

[12] McDermid, I.S., Godin, S. and Lindquist, L.O., *Appl. Opt.*, **29** (1990) 3603-3612.

[13] McGee, T.J., Gross, M.R., Singh, U.N., Butler, J.J. and Kimvilakani, P.E., *Opt. Eng.*, **34** (1995) 1421-1430.

[14] Donovan, D.P., Whiteway, J.A. and Carswell, A.I., *Appl. Opt.*, **32** (1993) 6742-6753.

[15] Godin-Beekmann, S., Porteneuve, J. and Garnier, A., *J. Environ. Monit.*, **5** (2003) 57-67.

[16] McGee, T.J., Gross, M., Ferrare, R., Heaps, W. and Singh, U., *Geophys. Res. Lett.*, **20** (1993) 955-958.

[17] Ancellet, G. and Ravetta, F., *Appl. Opt.*, **37** (1998) 5509-5521.

[18] Ravetta, F., Ancellet, G., Kowol-Santen, J., Wilson, R. and Nedeljkovic, D., *Mon. Weather Rev.*, **127** (1999) 2641-2653.

J. Phys. IV France **139** (2006) 373–390
© EDP Sciences, Les Ulis
DOI: 10.1051/jp4:2006139024

Astronomical observations with OHP telescopes

J. Patris[1] and A. Sarkissian[2]

[1] *Université Paul Cézanne, FST Centre Montperrin, avenue du Pigonnet, 13090 Aix-en-Provence, France*
[2] *Service d'Aéronomie, Institut Pierre-Simon Laplace, 91373 Verrières-le-Buisson, France*

Abstract. ERCA students have the opportunity to use one of the main European-based astronomical observing facilities: the "Observatoire de Haute Provence"but before a general presentation of astronomy and astrophysics is necessary. Because astronomy is an old science, starting from history is determinant. Observation tools, i.e. telescopes but also data analysis tools, are presented here as they are presented during on-site courses. This paper can be used by students before courses, for a better preparation during observations with telescopes, but also after courses, for better completion of their formation.

1. INTRODUCTION

Looking at the stars is fascinating, and almost every human being has once taken pleasure in observing these shiny little dots of the terrestrial night, picturing the wide empty spaces, dreaming about other forms of life or remembering all he's been told about galaxies and black holes.

Astronomical observations continue to be the pleasure of many. Meanwhile astronomy is now a very important science, which has progressed spectacularly since the first attempts to understand the sky. In this paper we aim to explain the basis of modern astronomical observations, and particularly to enable ERCA students to enjoy thoroughly their visit to the "Observatoire de Haute Provence". We want them to be able to make use of the different telescopes themselves, and to understand what's going on in the astronomical observatory. Thus, we'll present a brief overview of the history of astronomical ground based observation and a *very* brief portrait of our universe, as we now understand it. Then we'll explain the basic physics of telescopes, present OHP instruments and modern techniques of observation.

2. OBSERVING: A LONG (HI)STORY

Astronomy is one of the oldest sciences: we have evidence of astronomical observations before the beginning of history. By "astronomical observations", we mean that the prehistorical man not only looked at the sky above him, but studied it, compared the observations of several nights during the year, discovered periodicities in the sky's movements, and probably linked them to the seasonal periodicities. He was able to build according to his deductions, as at the site of Stonehenge in UK (3100 BCE). For an interesting hypothesis on Stonehenge, see the historical paper from Hawkins on archeoastronomy [4] - although we now think the astronomical achievement of Stonehendge was not so important as this author thought.

The history of astronomy presented in this section is short, and is centered on earth based observations. For a detailed history of observations, the reader may refer to references [18] (in French) or [5].

2.1 The naked eye period

1800 BCE : Our first useful astronomical data were recorded almost 4,000 years ago by the Babylonian priests [15]. The Mesopotamian clay tablets show different sets of continuous observations of positions of stars and planets, dispersed along the 2000 years of this very longed-lived civilization. Tablets dated

Figure 1. Early printed version of Ptolemaic system (Christian Aristotelian Cosmos. From Peter Apian, Cosmographia, 1524).

from the Seleucid period, around 300 BCE, even show computations of the position of the full moon using linear interpolation from measured positions (tablet from the British Museum # 32651 [11]).

150 BCE : After the Mesopotamian period, astronomy was taken up by ancient Greek philosophers. This is the epoch of the first known “scientists”. Althought ancient Greek thinkers are more known for their theoretical and philosophical studies, some of them dedicated themselves to observations. Among those, we shall mention Aristarchus (310 BCE - 230 BCE) who measured the size of the moon, supposed the earth was in circular motion around the sun, and tried to compute the distance to the sun; Hipparchus (190 BCE - 120 BCE) who computed a catalogue of stars with very accurate positions; and Ptolemy (85 - 165) who computed new positions and published in *the almagest* the whole of astronomical knowledge at that time. The figure 1 show Ptolemy’s world: all is centered on the Earth, “center of the massive bodies”, the Moon, the Sun and the planets turning on a complicated system of epicycles (computed to reproduce better the observations). The final celestial sphere, called “sphere of fixes” carries all the stars and the Zodiacal system and turns around the Earth in 24 hours.

800 AD: After the fall of Greek civilization (and the symbolic fire which destroyed the library of Alexandria), scientific activities revived under the Abassid caliphs of Baghdad, and especially with the foundation of the “House of Wisdom” (Bayt al-Hikma) by the caliph Al-Ma’mun around 820. Al-Ma’mun also conducted, on the plains of Mesopotamia, two astronomical operations intended to determine the value of a terrestrial degree. Among the most important activities of the House of Wisdom were the tasks of translating Greek manuscripts into Arabic, and gathering knowledge from different epochs and countries. The introduction of mathematical tools from India (sine, “Arabic” numbers, . . .) and the improvements of computing and observing devices (triquetum, astrolab) made the Arabic times a very rich epoch in the history of astronomy. Arabic scientists developed systematic observations of the sky, catalogues of stars and position of planets. A lot of stars were named during this period (Albireo, Altair, Mizar, etc.). The first mention of the fuzzy spot which today is identified as the Andromeda galaxy, the only visible galaxy in northern skies, was made by Arabic scientist Al-Sufi in 964. The Arabic period lasted more or less till 1350, when Ibn Al-Satir made a synthesis of all the previous models of the universe which he published in a clear and understandable version. The Arabic improvements of Ptolemy's model and the very precise observations that had been made during this epoch were so that they didn't match anymore : a new set of theory was felt to be needed.

1530: The Copernician revolution marks the begining of European scientific activity. Arabic observations and theoretical tools had reached the point where, pushed by the winds of many changes in Europe, they could lead to a radically different conception of our world, where the earth - and thus we humans - were no more in the center of the universe. Althought the Copernician theory was not immediatly accepted, more and more people recognized it as a "good mathematical tool" to compute the position of the stars. Many philosophers got interested in astronomy, among whom were Tycho Brahe (1546 - 1601), the last great observer with naked eyes in his observatory on the island of Hveen. Tycho Brahe performed a long series of continuous observations of stars, planets and comets; he discovered that comets were astronomical phenomena (and not meteorological ones, as was previously thought) and his works enabled his student, Johannes Kepler (1571 - 1630) to propose his famous theory of elliptic orbits.

2.2 Astronomy with telescopes

1610 Galileo's refractor: Galileo Galilei (1564-1642) was already a recognized scientist in 1609, working on physics, from temperature to magnets. In 1609 he decided to observe the sky with a recently invented instrument, made of two lenses placed in an afocal configuration. With his telescope, Galileo made several very important observations: the strange shape of planet Saturn, sun spots, the changes in Mars's size, the phases of planet Venus, four satellites turning around Jupiter. The first of this observation he wasn't able to explain: the surprinsing and changing shape of Saturn through the telescope would only correctly be interpreted by Huygens (1629-1695) half a century later.

The other observations, however, were of the utmost importance for the history of astronomy. They all gave an observational evidence to the Copernician theory. The phases of Venus, as the change of size of Mars and Venus, are easily explained only if we suppose these planets to be in orbit around the Sun. The satellites of Jupiter show that the antic paradigm of the Earth as "the center of the massive bodies" was false. If Jupiter could have bodies turning around it, then the much bigger Sun could be placed at the center of the turning planets.

Galileo was prosecuted by the church, and his condamnation in 1633 marked the end of his researches in astronomy. The reason why he was a particular target for the ecclesiastic authorities was partly because he turned the "mathematical", theoritical model of Copernic into a practical truth. While prudent Copernic wrote in Latin, Galileo chose to write in Italian, in a very pedagogical and easy-to-understand writing. He made the heliocentric model a topic of enthusiastic discussion among the cultivated of his time.

Thus, the all-important result of the first observation with a telescope was to change drastically the image the occidental man had of the universe, and of his own position into his world.

1700 Newton's reflector: The prosecutions of the church couldn't stop the evolution of astronomy. When Newton got interested in astronomy, he took the heliocentrism and Kepler's elliptic orbits as his basis. He was a great physisist as well as a great mathematician, and constructed a new type of telescope, called a refractor, in which the biggest lens (the objective lens) was changed to a concave parabolic mirror. His observations confirmed those of Galileo as those of Tycho Brahe. Supported by precise data, he elaborated his theory of Universal Gravity, and gave a very simple and elegant mathematical model

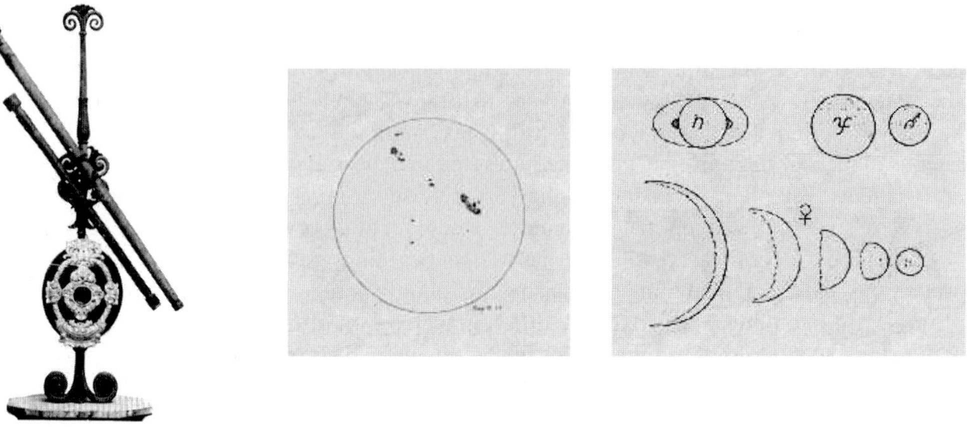

Figure 2. Galileo's telescope (image courtesy of IMSS in Florence, Italy) and a few sketches from Galileo's *Il Saggiatore* showing the sunspots, and his observation of Saturn, Jupiter, Mars, and the phases of Venus.

to explain all previous observations. The elliptic orbits of the planets, the movements of the satellites, as well as Galileo's studies on the free fall on Earth were explained by a single theory.

In addition to his famous law, Newton created a tradition of astronomical observations in England.

1800 Herschel and the begining of astrophysic: the advances in astronomy in the XVIIIth century concerned mainly mathematical researches in the resolution of the N-bodies problem posed by Newton's theory. This aspect of astronomy called Celestial Mechanics continued progressing all along of the XIXth and the XXth century, with scientists as Lagrange (1736-1813), Laplace (1749-1827), Le Verrier (1811-1877), Poincaré (1854-1912), Kolmogorov (1903-1987).

On the other hand, the end of the XVIIIth century sees the beginning of a new aspect of astronomy, which explores the universe behind the local solar system, and makes use of the results of physics to apply them to extraterrestrial phenomena. This new branch of astronomy is called astrophysics.

One of the first actor of this new science is William Herschel (1738-1822), a german musician came to London when he was 19 years old. Herschel got intersted to astronomy via musics and mathematics. He constructed his own telescope, and discovered a new planet, Uranus, in 1781. This important discovery (it was the first planet discovered since antiquity) enabled him to turn to astronomy full-time. He then worked, with his sister Caroline Herschel (1750-1848) - the first woman officially recognized for a scientific position - on comets, binary systems, nebulae. He also discovered the infrared radiations.

The interaction between physics and astronomy continued to grow. The sky became a source of questions and problems for the physicist to solve, along with the mathematicians and also other scientists, chemists, biologists. The telescope is a trigger for technological advances. Astronomical observations were one of the main scientific use of the photography in the XIXth century. Later, observations also supported the developpement of radio communication, electronics, semiconductors, and more recently computer science and signal processing.

2.3 OHP: An observatory of the XXth century

This section is a presentation of the Observatoire de Haute Provence and its astronomical instruments. Most of the information in this chapter is from the OHP website [12]: thanks to Sergio Ilovasky, webmaster, and Michel Bauer, actuel director of the observatory.

The Observatory is owned by the Centre National de la Recherche Scientifique (CNRS) and is funded by the CNRS and the Institut National des Sciences de l'Univers (INSU). OHP is part of the Marseille-Provence Astronomical Observatory (OAMP) federation.

The decision of building the Observatoire de Haute-Provence (OHP) followed the creation of CNRS by the "front populaire" governement in 1936. It was intended by astronomer Jean Perrin to be a national facility for French observers. The location was chosen as being one of the best skies on France, by the dryness of the mediterranean climate and the frequency of dry northern wind Mistral, far enough from any big city. The construction began in 1937, but was slowed by World War II.

The first telescope installed was the 80 cm, folowed by the arrival of the 120 cm telescope, which had been buit in Paris in 1877 and was moved to Haute Provence in 1943. The first observations then took place in Observatoire de Haute Provence, and the first research paper dates from 1944. Meanwhile, during World War II, the OHP director hid from the Nazis a young jewish student named Evry Schatzman, who would later become one of the greatest radio astronomists.

The facilities of the OHP were made available for foreign visiting astronomers in 1949. The OHP's biggest instrument, the 1.93 optical telescope, was built in 1958. It was equipped with Elodie Spectrometer from end of 1993 to July 2006 and with Sophie since August 2006 Elodie, and has adaptative optics. It was the instrument which enabled Major and Queloz [9] to discover the first planet outside our solar system in 1995; a historical discovery of the importance of Herschel's discovery of Uranus which made the OHP a world famous place, and saw it enter the history of astronomy.

3. ORDERS OF MAGNITUDES IN PRESENT DAY UNIVERSE

This paper is not a Theoritical course on astronomy: we'll just make a quick portrait of our universe. We aim to help the unprofessional eye to picture the distances and understand the global structure of our living place.

3.1 The large scale universe

The universe is defined as the space-time into which we are plunged, characterised by the consistancy of the physical laws. Till now, we have never found a place were these laws are different. Black holes or the first moments of our universe are challenging frontiers to our understanding, but we don't think they represent a place with a different set of physical laws - rather, they are extreme conditions of our physics. Thus, the universe is all we know to be existing.

The science which studies the large scale universe is called cosmology. Modern cosmology tells us that a "flat" (of euclidian geometry), infinite (since there is no evidence of any boundary), homogeneous 3-D space in accelerated expansion is probably a good approximation to the large scale universe. Expansion means that, at a given instant, any two points of the universe are getting appart at a speed in proportion to their distance. This picture of our universe is one of the main results of the XXth century astronomy.

Of this infinite universe, only a part is visible. To an observer on Earth, the observable universe is a globe centered on us and having a radius of more or less 14 billions light-years. The reason why we can't observe farther than this distance is a combination of two physical phenomena: the light velocity is finite, and our universe has an history. The universe is not constant in time as it is homogeneous in space: if we go back in the past, we find a denser universe, getting denser and denser to a theoretical "starting instant" called the Big Bang. This name is rather improper if we consider that the singularity is only in time, and not in space - so that though there was a special moment, there was no special point from which the universe began: the universe is and was infinite in space.

The finite speed of photon ($c = 299,792,458$ m/s) causes a delay of information: the light from the moon reaches us 1 second after it was sent, the light from the sun 8 minutes after it was emitted, it takes 4.22 years to photons from Proxima Centauri to reach earth and light from the Andromeda galaxy travel for about two million years before reaching us. Thus, the farther an astronomical object is from us, the longer its light has to travel before reaching us, and the younger we see it. As an example, the faintest galaxies seen by the Hubble Space Telescope are situated so that it took almost 13 billion years to the

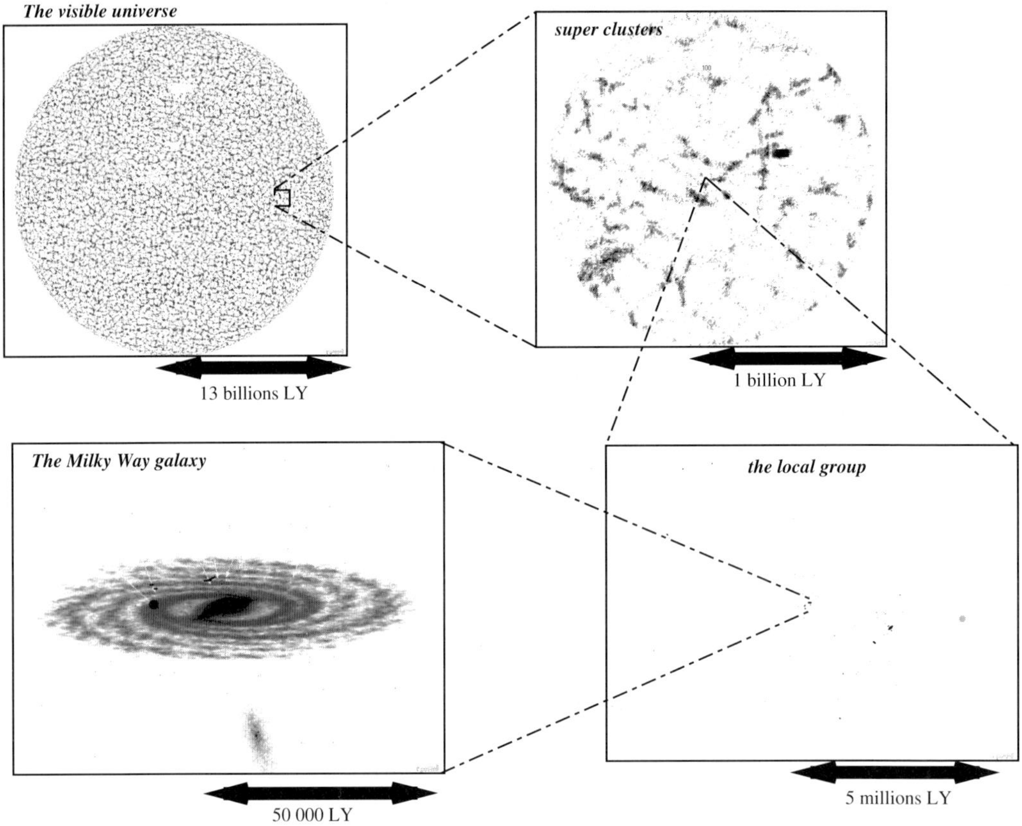

Figure 3. Some orders of magnitude in large scale structure: from the visible universe to our galaxy (Maps from Richard Powell [16]).

light to reach us. These galaxies are seen as they were at the begining of the universe's history - they are the first galaxies ever formed. If we try to see farther, we will reach the time photons - our main source of information - were formed: we can't observe farther than that. Our present day's "horizon" is thus roughly given by the age of the universe, expressed in light years.

3.2 Main structures

In this observable universe, the main visible structures are galaxies, in clusters or isolated. Much smaller than the length scale of the observable bubble, galaxies are seen as luminous concentration of matter. The distance between neighbour galaxies ranges from zero (interacting galaxies) to less than a million light year (l.y.) in cluster cores or 100 million l.y. between two clusters.

Our Milky Way is an example of a large spiral galaxy of a diameter of 100.000 l.y. situated, with its twin neighbour Andromeda, in the "local group", a group of about 50 galaxies spanning 5 million l.y. The Milky Way is a set of stars (about 90% of its total mass) and gaz (10% in mass) orbiting around a dense core.

One of the 10^{11} stars of the Milky Way is our Sun. A very unexceptionnal sphere of hydrogen in fusion, our star is situated in the Orion arm of the Milky Way, about 20.000 l.y. from its core. The closest star to the sun is Proxima Centauri, a mere 4.2 l.y. away. Our sun is surrounded by a planetary system, including planetoids (comets, asteroids) and eight large planets. The extreme boundaries of the solar

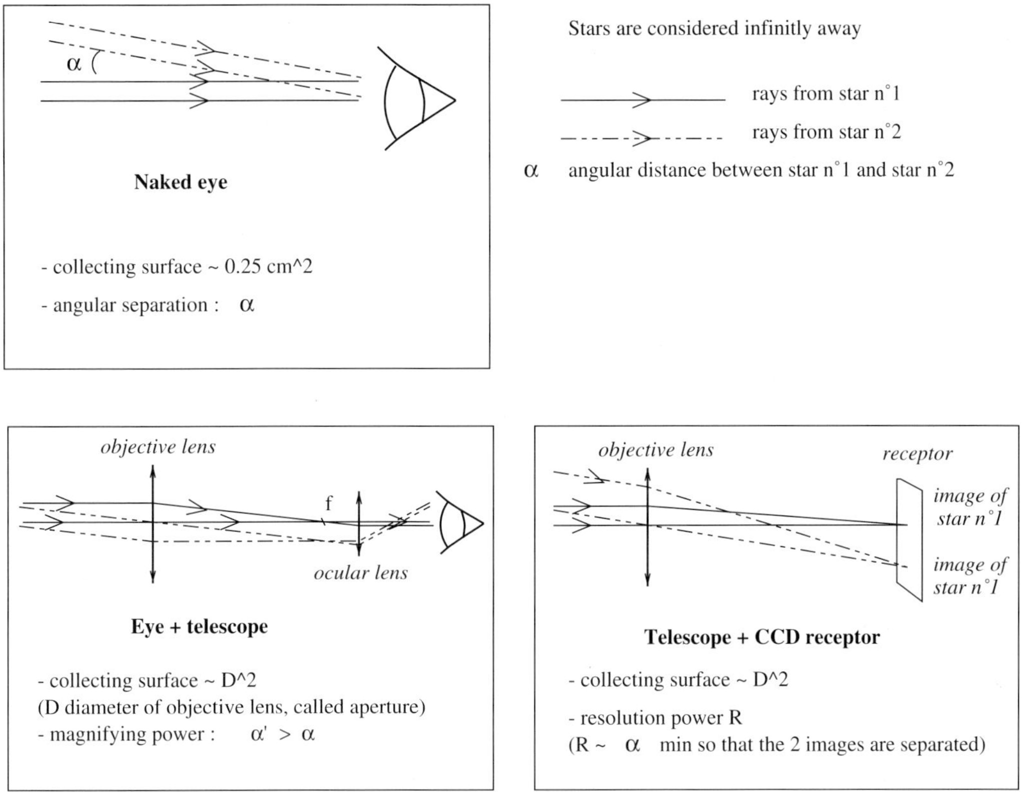

Figure 4. Physical principle and caracteristics of a refracting telescope.

system extend as far as 2 light years (100.000 AU, i.e. 100.000 times the distance from the Earth to the Sun), though the orbit of Neptune is only 30 AU aroud the Sun.

4. OBSERVATION'S TOOLS

Today's observing instruments, from the Earth, normally include a telescope, set under a dome to protect it, a receptor device (adapted to the informations needed) and a computer software to reduce and analyse the data.

The modern telescope is not very different from the instrument invented by Newton. It is built on the same principle, the main difference is the size of the primary mirror. On the contrary, the receptor have greatly evolved from the eye to the photographics plates in the XIXth century, and to the CCD captors of the modern times.

4.1 Telescopes optics

When the photon's receptor was the eye, the goal of the telescope was twofold: to enlarge astronomical objects (so the phases of Venus could be seen) and to collect more photons, showing what was too faint (Jupiter's satellites). Now that we have other types of receptors - mainly electronic - the real purpose of the telescope is to collect photons. Figure 4 shows the optics of observations with naked eyes, with Galileo's telescope, and with a telescope equipped with a receptor.

In a reflector telescope, the objective lens is changed into a concave paraboloic mirror. The optical principle is the same, but the great advantage of this system is that the light crosses no glass before

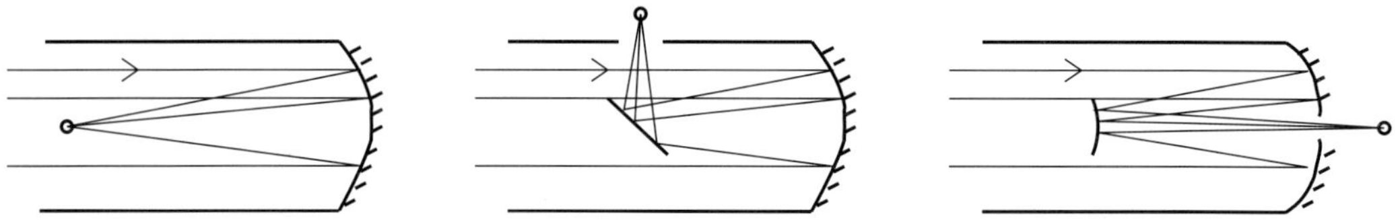

Figure 5. The most classical focuses of a reflector telescope.

reaching the detector: this telescope is much more luminous, and easier to construct. The largest lenses used in astronomy cannot exceed a diameter of 1 meter or so, whereas telescopes' mirrors constructed nowadays have a diameter of more than 8 meters (the four mirrors of the Very Large Telescopes are 8.2 m wide [21]).

However, the reflector telescope has the disavantadge that the focus point is right in the middle of the optic tube. The several solutions used to displace the focus point are illustrated on figure 5.

4.2 Coordinates systems and telescopes mountings

Be it with a telescope or with naked eye, to observe a particular object in the sky it is necessary to have a reference frame. As no depth is visible in the sky, atronomical objects are seen on the surface of a sphere, of an arbitrarily large radius, centered on the Earth, called the celestial sphere. Points in the celestial sphere are given by two angular coordinates - very like a point at the surface of the Earth is given by its longitude and its latitude. But, because the Earth is moving with respect to the stars, a coordinate of time is also necessary. Thus in any system of coordinate in the sky, two data of position and one of time are needed.

The equatorial system is the most common of the different systems existing. The axis of reference of this system is the rotation axis of the Earth. Thus, the equatorial system is a very natural one, and very well adapted to the problem.

The two angular coordinates are:

- α, the right ascension (RA) is measured eastward along the celestial equator, in hours, minutes, and seconds of sidereal time, starting from 0h at the vernal equinox.

- δ, the declination (dec) is counted in degrees, arcminutes and arcseconds from the equator. It is positive towards the North (from 0 to 90) and negative towards the South.

The references of the equatorial system are fixed to a very good approximation: the axis of the poles, as the vernal point, are mainly affected by the precession of the equinoxes, a recurring, spinning-top like movement of a period of roughly 26 000 years. Thus, on a few hundred years these movements are negligible and the equatorial coordinates are fixed for each star. This is why this system is usually used in catalogues of stars.

However, for the observation we need to know which star is were at a precise moment of the night. The time used in astronomy is called sidereal time (ST) and is roughly proportionnal to universal time by a factor of $(1 + \frac{1}{365})$. ST can be computed using tables or a computer software. It is usually given in any observatory.

We construct then another system of angular position from the equatorial system and the sidereal time:

- δ is unchanged (not affected by the rotation of the Earth)

- α is replaced by the Hour Angle (HA): $H = ST - \alpha$, counted in hours, minutes and seconds.

ST is defined so that $H = 0$ for a star passing the local meridian line. This system makes it very easy to find a star, and to define when the star is at meridien for better condition of astronomical observation.

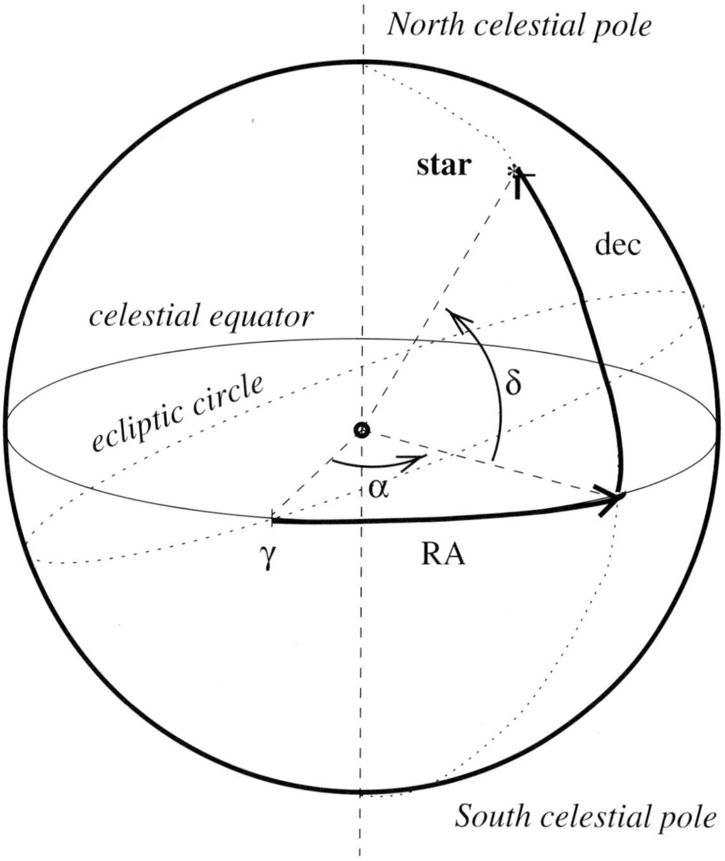

Figure 6. Definition of the equatorial coordinates.

One has to locate the local meridian line (in the northern hemisphere, the circle joining the Polar star to the geographical south on the horizon). If $H > 0$, the star is on the western side of the line.

The mounting of a telescope is the structure that enables the telescope to explore the sky. It is made of two perpendicular axis. The traditional mounting until the Eighties is called "equatorial mounting": one of the axis of the telescope is parallel to the axis of the poles, and the other is included in the equatorial plane. All the telescopes at the OHP are equipped with this kind of mounting.

It's main advantage is that it takes only one rotation (with constant speed of 360° each 24 hour) to compensate the rotation of the Earth, and so to follow the stars on the sky. However, this mounting takes a lot of place, and occasion great mechanical constraints on the structure.

The mounting used in very recent telescopes (VLT, ...) is an alt-azimuth mounting. One axis is vertical and the other in the horizontal plane. It is much less bulky and mechanically more stable. The movement to follow the stars is much more complicated, involving the 2 axis in a combination of variable speeds. This is made possible by today's computer-controlled engines.

4.3 OHP telescopes

The 1.93 m telescope

This telescope, built by Grubb-Parsons and operating since 1958, has Newton, Cassegrain (f/15) and Coudé foci. Only the Cassegrain focus is currently used. Available instruments are : a long-slit Cassegrain spectrograph (Carelec) with all-reflecting optics and a echelle spectrograph (Elodie) in the

Figure 7. Picture of the clocks in the 80 cm telescope's dome at OHP (photo: Torma Csaba, ERCA participant 2006).

first floor which is fed by a fiber-optic bundle from the Cassegrain focus. All instruments are computer-controlled and use thin, back-illuminated CCD chips. On-line image display and pre-processing are available for Carelec using MIDAS software from ESO. Elodie has its own pipeline reduction software from Geneva Observatory. The telescope attitude is digitized, the coordinates are remotely displayed and a detailed pointing model exists, allowing accurate setting.

1.52 m telescope

Built by REOSC, and operating since 1967, this telescope has only a Coudé focus. It is almost a twin of the 1.52 m ESO telescope at La Silla (which in addition has a Cassegrain focus). Currently available is a high-resolution spectrograph (Aurélie). It features a 3" entrance aperture, an efficient Bowen-Walvaren image slicer, has resolutions from 34000 to 120000 and utilizes a thin back-illuminated CCD (2048 × 1024). The spectrograph is computer-controlled and pre-processing with MIDAS is available.

1.20 m telescope

This telescope, the first installed at St. Michel, operates since 1943. It only has a Newton focus (f/6) with two available ports. It is now used with a CCD camera system for direct imaging and photometry using UBVRI filters. A thinned back-illuminated AR-coated SITe 1024 × 1024 CCD chip was installed in January 1996. A remotely controlled auto-guider system allows long exposures to be made. Image display and reduction facilities using MIDAS are available.

0.80 m telescope

This is the oldest telescope in use at OHP. It was first used at Forcalquier for site testing in 1932 and was later moved to St. Michel in 1945. The first foreign visiting astronomers to St. Michel used it in the

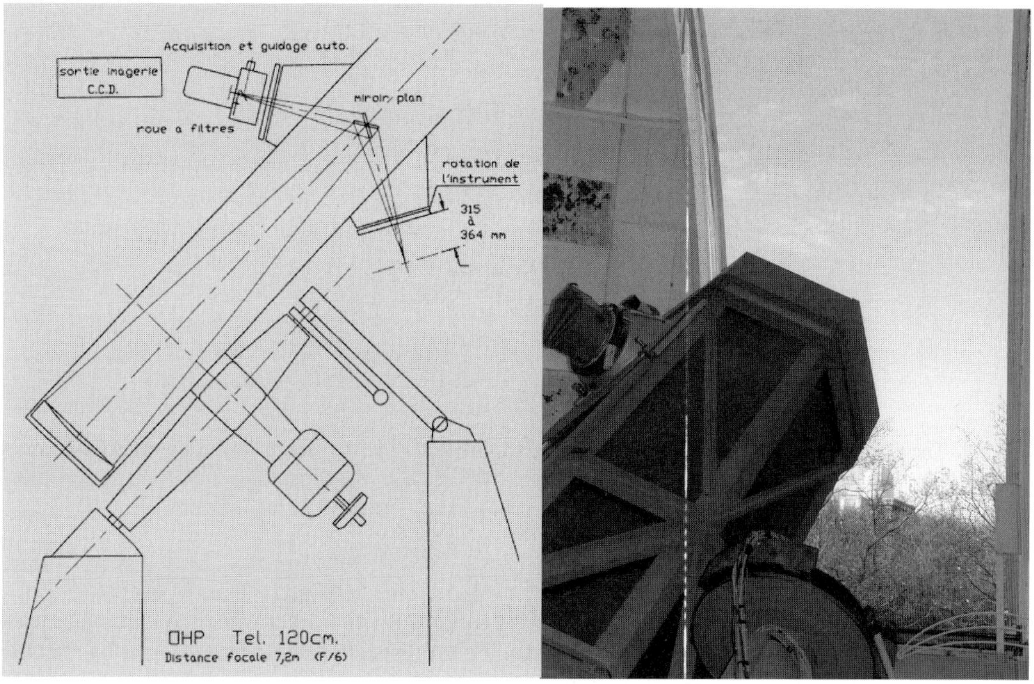

Figure 8. The 120 cm telescope: sketch (OHP scientific team [12]) and picture at dusk (photo: Torma Csaba, ERCA participant 2006).

summer of 1949 (Mr. and Mrs. Burbidge, to be exact). It features Newtonian and Cassegrain (f/15) foci. A camera is mounted on the Cassegrain focus which can be turned off for direct observing through an occular. It now accommodates visitor or student equipment for special needs. Its all-manual pointing is very pedagogical in teaching how to find an object in the sky and understanding systems of coordinates.

Other, specialized telescopes are also located at OHP : a 60/90 cm Schmidt telescope for wide-field direct imaging using direct plates and/or film, a 1-m telescope, operated by Geneva Observatory, with a Coravel radial-velocity spectrometer, and a 50-cm telescope operated by the CNES French Space Agency for satellite surveillance.

5. MAIN TYPES OF DATA

Two main techniques are used with the OHP telescopes: photometry and spectroscopy. They are both in optical light (wavelength roughly between 300 and 900 nanometers - see the CCD efficiency curve figure 9) and both use the CCD (Charge coupled device) as the final receptor.

5.1 Photometry

Photometry is the "simplest" way of using a telescope, that is, the closest to observing with the naked eye. It produces images in two dimensions. It is possible to obtain a colored image by using a set of filters: the same object is observed in three wavebands, and the images are afterward recomposed in one unique colored picture. The standard filters are "cousins" filters, with large wavebands (see figure 9), however it is also possible to use interferential filters with a very thin waveband (about 5 nm) to select one specific emission ($H\alpha$ filters centered at 656 nm, etc.).

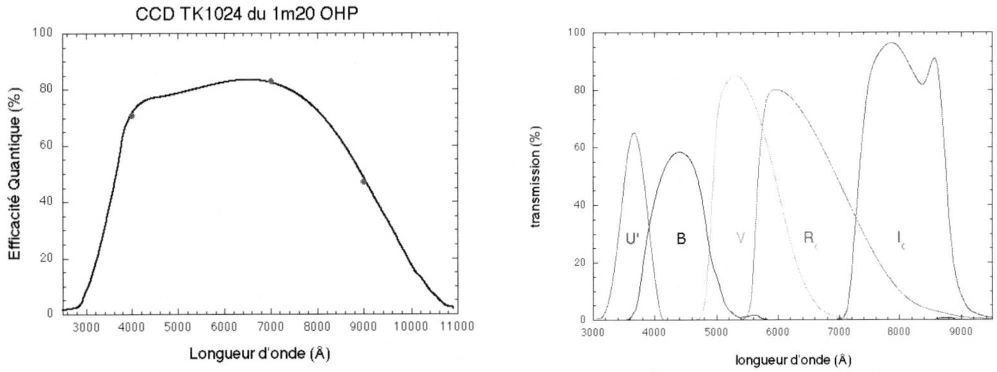

Figure 9. The efficiency curve of the 1.20 m telescope's CCD (left) and the wavebands of the filters used on this same telescope (right) . Documents S. Ilovaisky [12].

5.1.1 Technical devices

The bonette (instrument attached to the telescope) is equipped with a filter wheel containing at least 6 filters. The visiting astronomer usually asks in advance for a specific set of filters (or brings his own) and installs them in the wheel during the day before the observing time. During the night, the computer controlled wheel is turned to place the chosen filter for each exposure.

The length of the exposure is then chosen, and the CCD receives an image of the sky.

5.1.2 Reducing the data

The images can't be used directly: they have to be "reduced".

The different steps of the reductions are the folowing:

- cosmic rays removal: the high energy particles that came to the CCD without going through the telescope are usually detected by their sharpness (they are not convolved by the instrument's response) and removed

- dark substraction: an "dark" exposure is taken with closed shutters to account for the CCD background noise

- flat field convolution: the CCD response may not be exactly flat: some pixels or some zones may have a slightly different sensitivity factor. To remose this effect, a few exposures are taken of the evening sky (because it is not too bright, but has no star visible yet) or of the defocalised dome. The mean response is then used to correct the data.

- flux calibration: this is usually done by measuring the light of a well known star, and calibrate the response curve accordingly.

This is a very brief sketch of what is really done in professional observations: a lot of more complicated techniques are used todays, to account for the atmosphere effects (absorption and distortion), the optical response of the instrument, the CCD defects, etc.

5.1.3 Example

The study of Dolez et al. on stellar seismology is an exemple of what can be achieved with an original use of photometry. The aim of the study is to measure accurately the light variability of a very particular type of star, called a subdwarf, to determine its mode of oscillations. In addition, the observed star

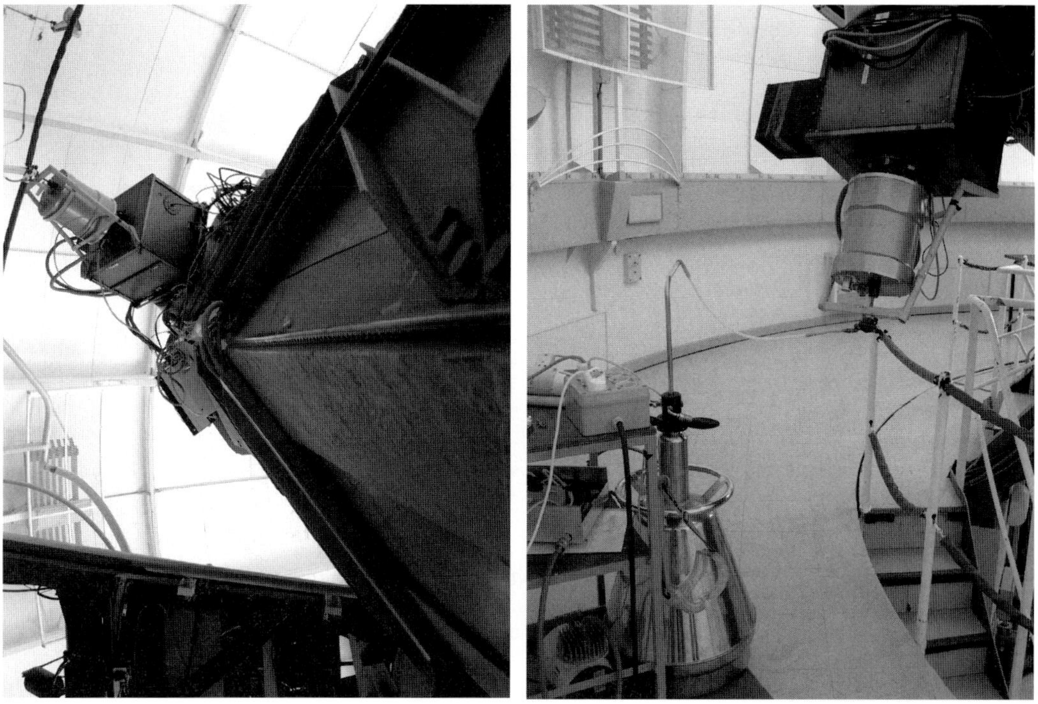

Figure 10. Newton focus of 1.20 m telescope (left) and close-up with the CCD camera and its cooling system (photos: Torma Csaba, ERCA participant 2006).

(PG 1336) is part of a binary system, so that the influence between the companion star and the modes of vibrations can be explored.

The mode of observing a variation of brightness has to adapt to constraints: no filter is necessary, but a high frequency of exposures, a continuous observation over several days, and an accurate luminosity calibration. A net of several observatories (the Whole Earth Telescope) observed the star, allowing for no interruption in the data. This net involves more than 20 countries around the Earth, with telescopes from 0.6 to 2.6 meters wide, amongst which the 193 at the Observatoire de Haute Provence.

The instrument used at OHP 193 telescope was a Chevreton fast photometer. The flux calibration was made throught observing two reference stars at the same time with PG 1336 .

The results are exposed in the paper [7], but also on OHP's webpage [13]. One of the light curves obtained is shown in figure 11.

Two main types of variability can be traced in the light curve: eclipses from a companion star causes the main features, but a periodic variation is also visible. The understanding of these diminutive variable stars could be important for distance calibrations and population synthesis.

5.2 Spectroscopy

The spectrum of an astrophysical object is its brightness versus wavelength. It is a very rich technique, symptomatic of the passage from celestial mechanics to astrophysics: a wealth of information is obtained on point-like stars, at all scales of the analysis.

With a low-resolution spectrum, the general shape gives the blackbody temperature of the surface of the star. A higher resolution shows emission or absorption lines, informing on its chemical composition or "metallicity", on more complicated parameters on its atmosphere. The stars are then classified upon their spectrum.

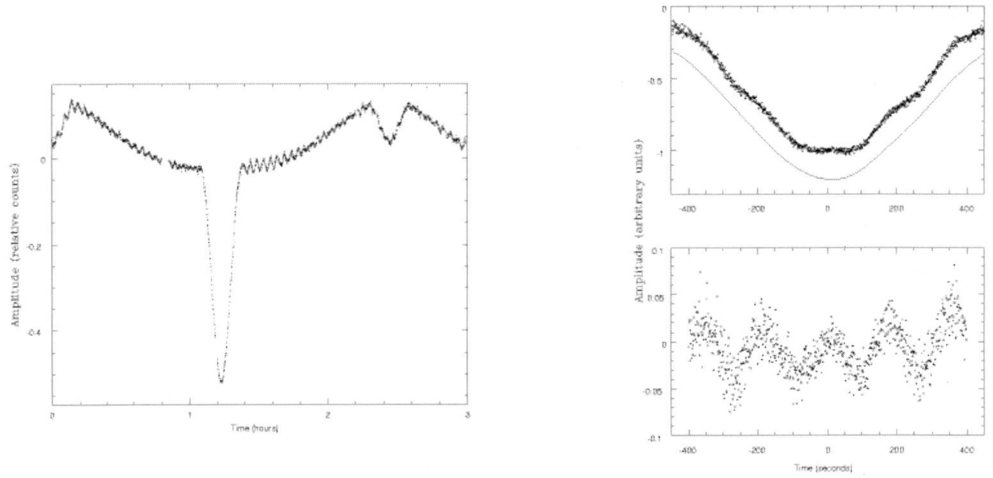

Figure 11. Light curve of PG 1336 as observed with OHP 193 telescope. On the left is the general curve, showing two main peaks due to eclipses and an underlying vibration. On the right is a zoom on the peak: variation in this part of the curve when the star is partially eclipsed show different oscillations (see [13]).

A high resolution spectrum can also serve to measure the star's radial velocity by Doppler effect (see below part 5.2.3).

For other types of astronomical objects spectroscopy is also a very important source of information. With spectroscopy we measure the growth of a supernova remnant, the rotation of a protoplanetary system, we find which star illuminate a reflection nebula.

In extragalactic astrophysics, it is all-important to determine the redshift (and thus the distance) in large scale universe. It also enables astrophysicists to compute the star formation rate of the galaxies, to trace the presence of a central black hole, and their amount of dust.

5.2.1 Technical devices

The usual way is to place a slit and a grating in the focal point of the instrument. This technique leads to a two dimensional spectrum having only one spatial dimension and one spectral dimension (figure 12). The slit can be placed, with a computer controlled device, over one or several lined up objects. There can be several independently positionable slits (multislit spectroscopy).

Several alternatives exist to this slit system. With optic fibers in lieu of the entrance slit, it is possible to keep the two spatial dimensions, resulting in a 3-D final image. One of the Very Large Telescope (ESO latest observatory on mount Paranal, Chile) instrument VIMOS spectrograph, designed and built at OHP and Marseille observatory, is equipped with this technique [19].

5.2.2 Reducing the data

The reduction procedure of a spectrum is similar to a 2 dimensional image, with some additions.

First, a wavelength calibration is needed. This is usually done with a reference lamp. The lamp (usually Ne-Ar), which produces emission lines at known wavelenghts, is observed several times each night through the optical system. A wavelength value is then given to each pixel.

For a long-slit spectrograph, the 1-D spectrum has to be extracted, usually by summing up all the flux for each wavelength, as shown in figure 12.

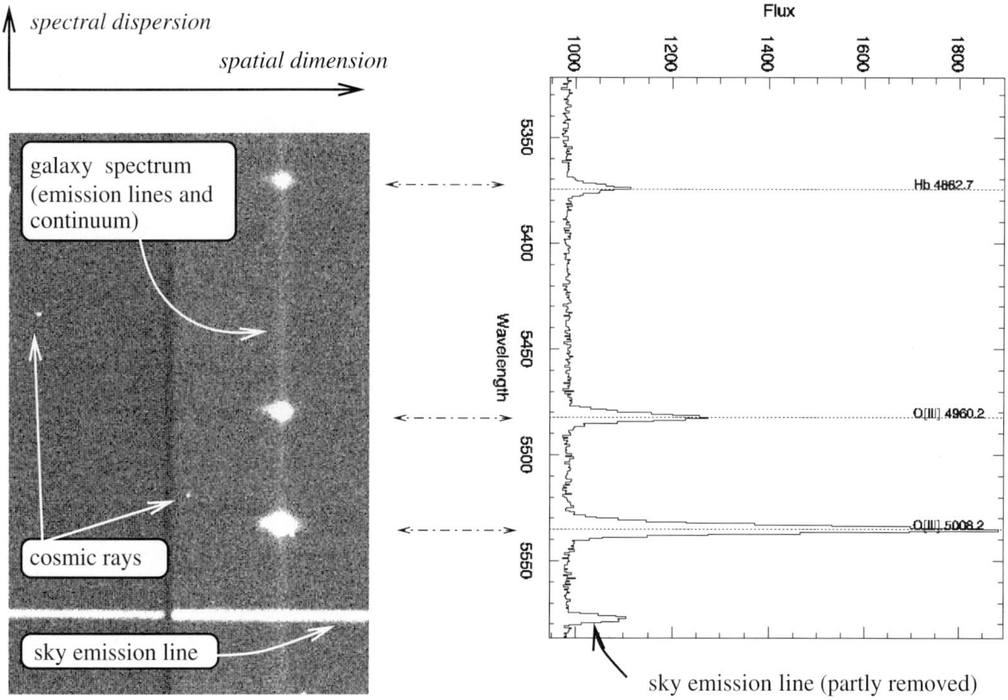

Figure 12. Spectrum of a galaxy. Left: 2-D raw image. Right: 1-D spectrum of the galaxy, showing emission lines characteristical of a star forming object. The position of the line gives the galaxy's redshift, or distance to the observer (data from [14]).

The sky emission is removed (as far as possible), and the features of the object are identified. In the example of figure 12, three emission lines are visible. They have been identified as [OIII] doublet and Hβ Balmer line. One can see that the emission lines are not at their due wavelength: this Doppler effect is caused by the recession velocity of the galaxy. Interpreted with the Expansion theory, this gives the redshift of this galaxy:

$$z = \frac{\lambda_{\text{obs}} - \lambda_{\text{real}}}{\lambda_{\text{real}}} = 0.105$$

If we apply the Hubble law with a Hubble constant of 70 km/s/Mpc we can compute the distance of this galaxy at about 450 Mpc, e.g. about 500 times the distance of Andromeda galaxy.

5.2.3 Example

A very famous example of the use of spectrometry with the 193 cm telescope of the OHP is the discovery of the first exoplanet, 51 peg b in 1995 [9].

The radial velocity of several stars was measured and analysed until one presented a periodical variation of radial velocity compatible with the influence of a large planet around the star.

The accuracy needed to make this discovery was very important, and it was achieved throught the design of Elodie spectrograph, with an accuracy of ± 15 m/s. More information on this historical discovery can be found on the exoplanet website [17].

To this day, 204 exoplanets have been reported with tolerable certainty in 174 planetary systems, including 21 multiple planet systems. In addition to the first discovery, 21 exoplanets have been found at OHP since 1995 (in september 2006).

For an exciting paper about extrasolar habitable worlds, see J. Lunine work in the ERCA 6 book [8].

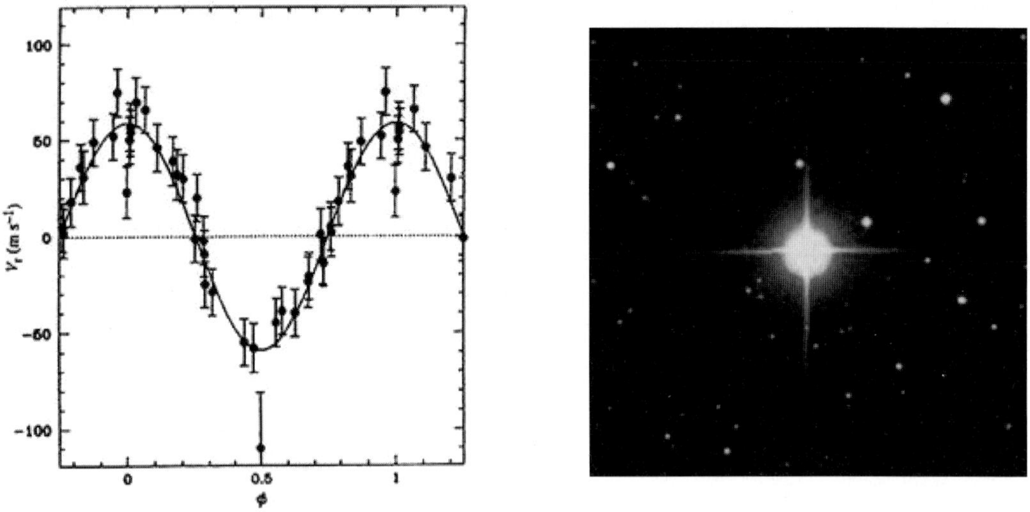

Figure 13. Radial velocity of 51 Peg (left) and field image of the star (data source [12]).

5.3 Astronomy on-line

Today, most of observatories provides on-line access to their observations. Data are usually open for astronomical community after two years of scientific exploitation by the scientists proposing the observations, enough time to make analysis and publications. The next step is, of course, to drive telescopes at observatories from your office: a dream for financial services, a nightmare for astronomers always willing to see -to touch !- what they are measuring or observing. But the reality is here: on request, can you imagine that you could check in a few seconds a new idea, a new concept, without waiting for months writing proposals for dedicated observations, replying to calls to finance your observations (trip and telescope fees) and this, for all existing telescopes, ground-based or space, all spectral ranges and all skies. Astronomy on-line is also now well developed at OHP with the Elodie Archive [**?**]. This section presents OHP activities related to databases and particularly OHP involvement in the Astronomical Virtual Observatory and applications.

5.3.1 The Elodie Archive

The Elodie Archive allows scientists a full access to high resolution spectra of astronomical objects recorded from 1993 to the summer of 2006 with 1.93 cm telescope and Elodie echelle spectrometer. The ELODIE archive contains 34286 spectra, among which 17894 are accessible to the astronomical community. One can access raw and processed data, in HTML and text formats using standard browser or more sophisticated query languages, allowing consultation of Elodie Archive within your "home" developed application. This is very important considering that modern astrophysics uses widely comparison potential of data bases: you can compare spectra obtained with different instruments, different conditions, at different spectral ranges. The Elodie Archive data base is used firstly by astronomers to store and to retrieve their data after observation, but also to built catalogues of reference sources, to make synthetic spectra of galaxies from formation schemes, and as presented below, in the frame of the Virtual Observatory, a new concept in astronomy, now reachable.

5.3.2 Virtual Observatory

The International Virtual Observatory Alliance (IVOA) [6] was formed in June 2002 to enable the international utilisation of astronomical archives as an integrated and interoperating virtual observatory. International coordination and collaboration is therefore necessary for the development and deployment of the tools, systems and organisational structures necessary to reach the objectives. France, like more than 15 other countries or Institutions engaged in the VO, participates actively to standard definitions and tools development. The Centre de Donnes de Starsbourg, CDS, is famous for astronomical tools development and now drives in France VO activities. The most used tools from CDS are 1) Simbad for basic data, cross-identifications and bibliography database for astronomical objects outside the solar system; 2) Vizier, a catalogue access database; and 3) Aladin, an interactive software sky atlas allowing the user to visualize digitized images of any part of the sky and to superimpose entries from astronomical catalogs. Most astronomer use these tools, even if sometimes they don't realise they do. These tools can be launched interactively, locally if you install it on your machine, or called via the net from your machine with your own program.

5.3.3 Example of Workflow with OHP DATA

The workflow presented here is often presented to students as illustration of 51 Peg b exoplanet discovery by Mayor and Queloz at OHP in 1995 using the radial velocity method. For description of the method, the reader can look at OHP site for the 5th NEON school in 2006 presented by Bouchy [**?**]. The workflow makes fast spectral shift calculation on spectra recorded with Elodie, with enough accuracy to retrieve planetary effect in case of 51 Peg. More than 200 spectra of 51 Peg have been obtained with Elodie since 1995. The workflow downloads these spectra on request, make spectral shift computations on several spectral lines (instead of nearly 200000 in the original method), for all spectra, checks properties of the parent star and the exoplanet on Exoplanet Encyclopedia [17] and in Simbad, checks reference spectral line wavelength at BASECOOL and save results in a VO Table for future exploitation and interpretation. The workflow can also be used to analyse other types of series of spectrometric observations, depending of your request. A demo is in preparation at OHP. An exercise for students developed for teaching purpose and based on this workflow is proposed and widely used by the teaching community.

6. CONCLUSION

Astronomical observations at Observatoire de Haute Provence are various and cover all fields of astronomy and astrophysics, including planetology. The telescopes and tools presented here are commonly used at OHP, but also in most observatories. For young scientists, not always familiar with astronomy, such a general approach with selected state-of-the-art research aims to open the students interest, and this is the main purpose of our course at OHP and of the paper.

Thanks to Sergio Ilovaisky, Alain Moulet, Jean-Claude Brunel, Aimé Paul and all the OHP staff for their help at the telescopes.

References

[1] A. Baranne et al. 1995 **A&A Suppl. Ser.** 119, 373 *Elodie: a spectrograph for accurate radial velocity measurements*
[2] 5th NEON School at OHP, http:/www.obs-hp.fr/www/ecole-ete/Exoplanets-FBouchyNEON 2006.pdf
[3] The Elodie Archive, http:/atlas.obs-hp.fr/elodie/
[4] Gerald S. Hawkins, 1963 **Nature** 200, 306-308 *Stonehenge decoded*

[5] M. Hoskin, 1999 **cambridge university press** *The Cambridge concise history of astronomy*
[6] International Virtual Observatory Alliance, IVOA, http:/www.ivoa.net/
[7] D. Kilkenny et al. 2003 **MNRAS** 345, pp 834-846 *A Whole Earth Telescope campaign on the pulsating subdwarf B binary system PG 1336-018 (NY Vir)*
[8] Lunine, J.I. 2004 J.Phys. IV France **121** (2004) 259-268
[9] M. Major & D. Queloz, 1995 **Nature** 378, 355 *A Jupiter-Mass Companion to a Solar-Type Star*
[10] J. Moultaka, S.A. Ilovaisky, P. Prugniel, and C. Soubiran, 2004, **PASP** 116, 693-698, The ELODIE archive
[11] O. Neugebauer, London 1955 *Astronomical cuneiform texts*
[12] Observatoire de Haute Provence: http://www.obs-hp.fr/
[13] What is observed at Observatoire de Haute Provence? http://www.obs-hp.fr/www/observations/pages/dolez-1.html
[14] J. Patris et al. 2003 **A&A** 412, 349 *Spectroscopic follow-up of FIRBACK-South bright galaxies*
[15] A. Pichot, 1991 **folio essais** *La naissance de la science*
[16] *Atlas of the universe* website by Richard Powell http://www.atlasoftheuniverse.com/
[17] J. Schneider *The extrasolar planet encyclopaedia*, http://exoplanet.eu/
[18] Jean-Pierre Verdet, 1990 **Points** *Une histoire de l'astronomie*
[19] VIMOS spectrograph web page http://www.oamp.fr/virmos/vimos.htm
[20] Virtual Observatory France , VO-France, http:/www.france-vo.org/
[21] Very Large Telescope, European Southern Observatory, http://www.eso.org/

J. Phys. IV France **139** (2006) 391–401
© EDP Sciences, Les Ulis
DOI: 10.1051/jp4:2006139025

Sustainable health in a globalised world*

P. Martens[1]

[1] *International Centre for Integrated Assessment and Sustainable Development (ICIS), Maastricht University, PO Box 616, 6200 MD Maastricht, The Netherlands*
e-mail: p.martens@icis.unimaas.nl

Abstract. It is clear that in making the concept of sustainable development concrete, one has to take into account a number of practical elements and obstacles. There is little doubt that integrated approaches are required to support sustainable development. Therefore, a new research paradigm is needed that is better able to reflect the complexity and the multidimensional character of sustainable development. The new paradigm, referred to as sustainability science, must be able to encompass different magnitudes of scales (of time, space, and function), multiple balances (dynamics), multiple actors (interests) and multiple failures (systemic faults). To illustrate the above, we described potential health transitions in a globalised world.

1. WHAT IS SUSTAINABLE DEVELOPMENT?

The essence of sustainable development is simply this: to provide for the fundamental needs of mankind without doing violence to the natural system of life on earth. This idea arose in the early eighties of the last century and came out of a scientific look at the relationship between nature and society. The concept of sustainable development reflected the struggle of the world population for peace, freedom, better living conditions and a healthy environment (Council 1999). During the latter half of the 20th century, these four goals recurred regularly as world-wide, basic ideals.

With the end of the Second World War in 1945, it was widely believed that the first goal of peace had actually been achieved. But then came the arms race and, although a kind of global peace was maintained, the Cold War led to a range of conflicts fought out at the local level. When one looks today at many parts of the world – the Middle East, Middle Africa, for example – it is all too evident that peace is still a long way off.

Under the banner of freedom, people fought for the extension of human rights and for national independence. Today, the poorest two thirds of the world population sees 'development' as the most important goal, by means of which they hope to achieve the same material well-being as the wealthy one third.

But this ideal, upon which so much emphasis has been laid recently, has to reckon with the earth itself. This reckoning began with concern over the exhaustion of our natural resources and only later did it dawn on us that a disturbance of the complex systems upon which our lives depend can have enormous consequences.

The last twenty five years have been characterized by an attempt to link together the four ideals cited above – peace, freedom, improved living conditions and a healthy environment (Council 1999), an ambition which stems from the realization that striving for one of these ideals often means that the others must necessarily also be striven for. This struggle for 'sustainable development' is one of the great challenges for today's society.

*Based on Martens, P. (2005). Duurzaamheid: wetenschap of fictie? Maastricht, Oratie Universiteit Maastricht, Open Universiteit Nederland, Hogeschool Zuyd.
Martens, P. (2006). "Sustainability: science or fiction?" Sustainability: Science, Practice and Policy **2**(1): 1-5.

Sustainable development is a complex idea that can neither be unequivocally described nor simply applied. There are scores of different definitions, but we shall restrict ourselves to the most frequently quoted, that of the Brundtland Committee (1987) (WCED 1987):

"Sustainable development is development which meets the needs of the present without compromising the ability of future generations to meet their own needs."

If we look at the lowest common denominator of the different definitions and interpretations of sustainable development, we note four common characteristics (Rotmans, Grosskurth et al. 2001). The first indicates that sustainable development is an intergenerational phenomenon: It is a process of transference from one to another generation. So, if we wish to say anything meaningful about sustainable development, we have to take into account a time-span of at least two generations. The time period appropriate to sustainable development is thus around 25 to 50 years.

The second common characteristic is the level of scale. Sustainable development is a process played out on several levels, ranging from the global to the regional and the local. What may be seen as sustainable at the national level, however, is not necessarily sustainable at an international level. This is due to shunting mechanisms, as a result of which negative consequences for a particular country or region are moved on to other countries or regions.

The third common characteristic is that of multiple domains. Sustainable development consists of at least three: the economic, the ecological and the socio-cultural domains. Although sustainable development can be defined in terms of each of these domains alone, the significance of the concept lies precisely in the interrelation between them.

The aim of sustainable social development is to influence the development of people and societies in such a way that justice, living conditions and health play an important role. In sustainable ecological development the growth of natural systems is the main focus of concern and the maintenance of our natural resources is of primary importance.

What is at issue here are three different aspects of sustainable development which in theory need not conflict but which in practice often conflict. The underlying principles are also essentially different: with sustainable economic development the concept of efficiency has a primary role, whereas with sustainable social development the same may be said of the concept of justice and with sustainable ecological development it is the concepts of resilience or capacity for recovery that are basic.

The fourth common characteristic concerns the multiple interpretation of sustainable development. Each definition demands a projection of current and future social needs and how these can be provided for. But no such estimation can be really objective and, furthermore, any such estimation is inevitably surrounded by uncertainties. As a consequence, the idea of sustainable development can be interpreted and applied from a variety of perspectives.

As will be apparent from the above, a concept like sustainable development is difficult to pin down. Because it is by its nature complex, normative, subjective and ambiguous, it has been criticized both from a social and from a scientific point of view. One way of escaping from the 'sustainability dilemma' is to begin from the opposite position: that of non-sustainable development. Non-sustainable or unsustainable development is only too visible in a number of intractable problems entrenched in our social systems and which cannot be solved through current policies. These intractable problems are characterized by the involvement of multiple interests as well as their great complexity, lack of structure, structural uncertainty and apparent uncontrollability.

Such problems can be recognized in many national and global economic sectors. One sees them in agriculture, for example, with its many facets of unsustainability becoming manifest in the form of protein-related diseases such as BSE (mad cow disease), and in foot-and-mouth disease. The water sector has to deal with such symptoms as flooding, droughts and problems related to water quality, while the energy sector produces energy in a one-sided manner and – as a direct result – affects the environment. One sees the same symptoms in traffic and transport systems, where atmospheric pollution and traffic queues can be seen as symptoms of unsustainability; and as far as our health is concerned, the spread of SARS, the global increase in the incidence of malaria, malnutrition and its counterpart – the increase in obesity – are all far from sustainable.

These unsustainable developments reflect systemic faults embedded in our society. In contrast to market faults, systemic faults derive from deep-seated lacks or imbalances in society. They cannot be corrected through the 'market' and form a serious impediment to the optimal functioning of our social system. Systemic faults operate at various levels and can be of an economic, social or institutional nature. If such intractable problems are a sign of an unsustainable development, they can only be solved through fundamental changes in our society. Only thus can non-sustainable conditions be transformed and put on a more sustainable basis.

2. INTEGRATED ANALYSIS OF SUSTAINABILITY

This new paradigm has far-reaching consequences for the methods and techniques that need to be developed before an integrated analysis of sustainability can be carried out. These new methods and techniques can also be characterized as follows:

- from supply- to demand-driven
- from technocratic to participant
- from objective to subjective
- from predictive to exploratory
- from certain to uncertain.

In short, the character of our instruments of integrated analysis is changing. Whereas previous generations of these instruments were considered as 'truth machines', the current and future generations will be seen more as heuristic instruments, as aids in the acquisition of better insight into complex problems of sustainability. At each stage in the research of sustainability science, new methods and techniques will need to be used, extended or invented. The methodologies that are used and developed in the integrated assessment community are highly suitable for this purpose.

Roughly, there are a number of different kinds of methods for the integrated assessment of sustainability: analytic methods, participative methods and more managerial methods. Analytic methods mainly look at the nature of sustainable development, employing among other approaches the theory of complexity. In participative research approaches, non-scientists such as policy-makers, representatives from the business world, social organizations and citizens also play an active role. The more managerial methods are used to investigate the policy aspects and the controllability of sustainable transitions.

An example of an analytic instrument for the assessment of sustainability is the integrated assessment model which allows one to describe and explain changes between periods of dynamic balance. This model consists of a system-dynamic representation of the driving forces, system changes, consequences, feed-backs, potential lock-ins and lock-outs of a particular development in a specific area. Another analytic instrument is the scenario that describes sustainable and unsustainable developments, including unexpected events, changes and lines of fracture.

Participatory methods differ according to the aim of the study and its participants. Thus negotiation processes are mimicked in so-called policy exercises, whether or not these are supported by simulations. In the method of mutual learning, the analysis is enriched by the integration of the knowledge possessed by participants from diverse areas of expertise.

An example of a new kind of policy instrument is provided by transition management (Rotmans 2003). Transition management is a visionary, evolutionary learning process that is progressively constructed by the undertaking following steps:

(i) develop a long-term vision of sustainable development and a common agenda (macro-scale)
(ii) formulate and execute a local experiment in renewal that could perhaps contribute to the transition to sustainability (micro-scale)

(iii) evaluate and learn from these experiments
(iv) put together the vision and the strategy for sustainability, based on what has been learned (this boils down to a cyclical search and learn process that one might call evolutionary steering: a new kind of planning with understanding, based on learning by doing and doing through learning).

But now that the first steps towards an integrated sustainability science have been taken, there is a prospect of making some major leaps forward. In the next section we explore some of the above in more detail, while looking at past a future health transitions.

3. THE HEALTH TRANSITION**

The shifts that have taken place in the patterns and causes of death in many countries, and that were described in the precious sections, can be described and explained within a conceptual framework known as the health transition. Previously, the health transition has been covered by two separate terms: 'demographic transition', describing the change from high fertility and mortality rates in less developed societies to low fertility and low mortality rates in 'modern' societies, and 'epidemiological transition', which was introduced to describe the changes in mortality and morbidity patterns (from infectious to chronic diseases) as societies' demographic, economic and social structures changes (Omran 1983). The health transition is a more appropriate term, as it covers the full range of social, economic and ecological changes driving the epidemiological and demographic transition (Caldwell 1978; Beaglehole and Bonita 1997). It comprises several stages characterised by categories in which fertility levels and causes of death are grouped (Omran 1998). Despite its limitations (Mackenbach 1994), the health transition is a useful tool for understanding current health trends and exploring future developments. Below, we use this concept to explore the general trends related to the main forces determining human health.

The age of pestilence and famine

This first stage of the health transition stage (the age of pestilence and famine) is characterised by the kind of mortality that has prevailed throughout most of human history. Epidemic, famines and wars cause huge numbers of deaths. The provision of basic ecological resources, i.e. food and fresh water, is inadequate. The lack of economic means to provide a sufficient infrastructure for adequate health services, schooling and sewage systems leads to low levels of literacy, and high levels of mortality and fertility. Infectious diseases are dominant, causing high mortality rates, especially among children. In this stage, women of childbearing age also face considerable risks due to the complications associated with pregnancy and childbirth (Omran 1983). The combination of a wide range of health risks with the lack of health-care facilities results in very low levels of life expectancy: in Indonesia in 1950, for example, this was as low as age 40. Population growth, improvements in health, and advances in socio-economic development are all limited by the local carrying capacity of the environment. Some developing countries are still in this stage.

The age of receding pandemics

This stage (the age of receding pandemics) began in the mid-19th century in many of what are now developed countries. It involved a reduction in the prevalence of infectious diseases, and a fall in mortality rates. As a consequence, life expectancy at birth climbed rapidly from about 35 to 50. Due to increased economic growth, the first phase of this stage is marked by the improved use of natural

Based on Martens, P. (2002). "Health transitions in a globalising world: towards more disease or sustained health?" Futures **37(7): 635-648.
Martens, P. and M. M. T. E. Huynen (2003). "A future without health? The health dimension in global scenario studies." Bulletin of the World Health Organisation **81**(12): 896-901.

resources; typically, this leads to a sharp fall in deaths from water-borne and food-borne infectious diseases, and from diseases related to malnutrition (e.g. typhoid fever, diarrhoea and measles).

In tandem with these improvements in population health, social factors become increasingly important. Improved social circumstances (e.g. improved hygiene) and community health services causes a further decline in infectious diseases, giving an extra boost to economic development. Finally, the introduction of modern healthcare and health technologies, e.g. immunisation programmes and the introduction of antibiotics enable the control and elimination of group of infectious diseases such as acute bronchitis, influenza and syphilis.

As fertility rates are high, a population grows rapidly at this stage of the health transition. Without moving to the next stage, the carrying capacity of the local ecosystem may be exceeded (McMichael 1993). As population and ecological pressures increase, food and water become scarcer, and the lack of ecological and social resources may cause economic development to stagnate. If there is a surplus of available resources, the transition may be accelerated, but if they are lacking, the transition may slow, or even stagnate in this phase.

The age of chronic diseases

In the third stage (the age of chronic diseases), the elimination of infectious diseases makes way for chronic diseases among the elderly. While improved healthcare means that these are less lethal than infectious diseases, they nonetheless cause relatively high levels of morbidity. Increasingly, health patterns depend on social and cultural behaviour, such as patterns of food consumption and drinking behaviour (McMichael and Powles 1999). Due to low levels of mortality and fertility, there is little population growth. When the health transition is at an advanced stage, life expectancy may exceed 80 years. However, the prevalence of one or more diseases means that such a long life also includes, on average, a relatively long period of morbidity.

This stage occurs at different rates in different nations: in both developed and developing countries, mortality rates are driven by socially determined factors; in developed nations they are also driven by medical technology. In such situations, the aim is not only to reduce mortality rates via the optimised deployment of ecological and social capital, but also to use advance medical technologies for the reduction of morbidity. It becomes necessary to ensure sufficient social and health-care investment for all age groups. At the same time, there is increased demand for healthcare related to the diseases of older people.

Currently, most developed countries are in the third stage of the health transition: fertility rates are low, and the causes of diseases and deaths have shifted from infectious to chronic diseases. All developed countries in Europe, North America and Asia are seen as having arrived in the latter stage of the health transition in the 1970s, although there were large differences with regard to timing, particularly in the onset the decline in fertility (Hilderink 2000). In these countries, declining fertility rates and increased life expectancy have led to the ageing, or so-called ‘greying’, of the population.

The health situation in developing countries varies greatly from one country to another. In most, there is still very low life expectancy; this is due largely to malnutrition and the lack of safe drinking water, which are compounded by poor healthcare facilities. In other countries, however, particularly in Asia and Latin America, chronic diseases have now become more important than infectious diseases (Murray and Lopez 1996). The same large variation is reflected in the demographic situation. In countries such as China and Thailand fertility rates are very low; in others they are very high. Due to sub-national differences of an economic, social or ecological nature, there may also be large differences within a single country.

It is widely believed that, with increasing economic growth, developing countries will follow the same pattern of health transition as Europe and North America. However, although poverty is always a major force in the health transition, mortality levels in all but the very poor populations are determined more by social and ecological resources than by income. Many countries, especially the poorest, will not 'trade' infectious diseases for chronic diseases; instead, they may even suffer a 'double burden' of disease.

In the next section we will explore how future health pathways may look like.

4. THE NEXT STAGE IN THE HEALTH TRANSITION

So what lies ahead? In order to explore possible future global health transitions we have made use of a set of scenarios. Scenarios do not predict the future, but rather paint pictures of possible futures and explore the various outcomes that might result if certain basic assumptions are changed. The scenarios used explore the global and regional dynamics that may result from changes at a political, economic, demographic, technological and social level, and are based on the recent efforts of the IPCC (IPCC 2000) (see Box). The distinction between classes of scenario was broadly structured by defining them ex ante along two dimensions. The first dimension relates to the extent both of economic convergence and of social and cultural interactions across regions; the second has to do with the balance between economic objectives and environmental and equity objectives. This process therefore led to the creation of four scenario 'families' or 'clusters', each containing a number of specific scenarios.

Box: The IPCC SRES scenarios (IPCC 2000).

The first group of scenarios [A1] is characterised by fast economic growth, low population growth and the accelerated introduction of new, cleaner and more effective technologies. Under this scenario, social concerns and the quality of the environment are subsidiary to the principal objective: the development of economic prosperity. Underlying themes combine economic and cultural convergence, and the development of economic capacity with a reduction in the difference between rich and poor, whereby regional differences in per capita income decrease in relative (but not necessarily absolute) terms.

The second group of scenarios [A2] also envisages a future in which economic prosperity is the principal goal, but this prosperity is then expressed in a more heterogeneous world. Underlying themes include the reinforcement of regional identity with an emphasis on family values and local traditions, and strong population growth. Technological changes take place more slowly and in a more fragmented fashion than in the other scenarios. This is a world with greater diversity and more differences across regions.

In the third group [B1], striving for economic prosperity is subordinate to the search for solutions to environmental and social problems (including problems of inequity). While the pursuit of global solutions results in a world characterised by increased globalisation and fast-changing economic structures, this is accompanied by the rapid introduction of clean technology and a shift away from materialism. There is a clear transformation towards a more service and information-based economy.

The fourth group [B2] sketches a world that advances local and regional solutions to social, economic and ecological problems. This is a heterogeneous world in which technological development is slower and more varied, and in which considerable emphasis is placed on initiatives and innovation from local communities. Due to higher than average levels of education and a considerable degree of organisation within communities, the pressure on natural systems is greatly reduced.

We describe developments in the health status of populations according to three potential future 'ages': the age of emerging infectious diseases, the age of medical technology and the age of sustained health. Of course, these stages are imaginary (although some features are already recognisable in some countries) and are not sharply delineated - there is always a continuum. There is also always the possibility that economic, political, social, or environmental crises will cause the process of transition to stagnate, or to go into reverse. Of course, each country follows its own route to the 'ages' in question, described in the sections below.

The age of emerging infectious diseases

In this stage (see also (Olshansky, Carnes et al. 1998)), the emergence of new infectious diseases or the re-emergence of 'old' ones will have a significant impact on health. A number of factors will influence this development: travel and trade, microbiological resistance, human behaviour, breakdowns in health systems, and increased pressure on the environment (Barrett, Kuzawa et al. 1998; Louria 2000). Social, political and economic factors that cause the movement of people will increase contact between people and microbes; environmental changes caused by human activity (for example, dam and road building, deforestation, irrigation, and, at the global level, climate change) will all contribute to the further spread of disease. The overuse of antibiotics and insecticides, combined with inadequate or deteriorating public health infrastructures will hamper or delay responses to increasing disease threats.

As a result, infectious diseases will increase drastically, and life expectancy will fall (as is currently the case in many developing countries due to the AIDS pandemic). Ill health will lead to lower levels of economic activity, and poor countries will be caught in a downward spiral of environmental degradation, depressed incomes and bad health. Control of infectious diseases will be hampered by political and financial obstacles, and by an inability to use existing technologies.

The age of medical technology

To a large extent, increased health risks caused by changes in life-style and environmental changes will be offset by increased economic growth and technology improvements in the age of medical technology. To some extent, this might be comparable with other views on a fourth' stage, e.g. the 'hybristic' stage (Rogers and Hackenberg 1987), and the 'age of delayed degenerative diseases' (Olshansky and Ault 1986).

If there is no long-term, sustainable economic development, increased environmental pressure and social imbalance may propel poor societies into the age of emerging infectious diseases. On the other hand, if environmental and social resources are balanced with economic growth, sustained health may be achieved.

The age of sustained health

In the age of sustained health, investments in social services will lead to a sharp reduction in life-style related diseases, and most environmentally related infectious diseases are will be eradicated. Health policies will be designed to improve the health status of a population in such a way that the health of future generations is not compromised by, for example, the depletion of resources needed by future generations. Although there is only a minimal chance that infections will emerge, improved worldwide surveillance and monitoring systems will mean that any outbreak is properly dealt with. Despite the ageing of the world population, health systems will be well adjusted to an older population. Furthermore, disparities in health between rich and poor countries will eventually disappear.

Let us now look at how different development paths, as described by the four scenarios, can be linked to the 'ages' of human health described above.

A1 scenario group

In this group, a central role will be played by economic growth and technological developments. Populations will grow at a moderate rate and start to decline in the years around 2050, and the average age of the world's population will increase. In many societies, accelerated economic growth and technological advances will enhance health and life expectancy; in developing countries fertility rates will continue to fall. At least in the short-to-medium term, material advances, allied with social modernisation and various healthcare and public health programmes, will lead to gains in overall public health.

Although the high level of average income per capita will contribute to the improvement in the overall health of the majority, increased economic growth may lead to problems of 'social exclusion'.

In many places, income growth will put pressure on global resources and thus lead to environmental degradation. From a health perspective, we might see a divergence between the developed and parts of the developing world. In the developed world, increasing wealth and improvements in healthcare and technology will offset most of the emerging risks; the richest populations may experience particularly pronounced health improvements. The poorest countries, on the other hand, might not be able to advance to the stage of medical technology. As population growth and the depletion of resources increase environmental pressures, there may be a resurgence of old diseases and an increase in new infections.

A2 scenario group

In this group, health in the developed countries will to a large extent be left to individual choice and less of an issue for public policy. The greatest economic growth will take place in the developed regions, and technological advances will benefit only rich countries, as there will be less diffusion of knowledge and economic capital. Developed countries will make increasing investments in education and better welfare. However, globally, the gains in health brought about by increasing economic growth and technology will be partly offset by erosion of social and environmental capital, the global division of labour, the exacerbation of the rich-poor gap between and within countries, and the accelerating spread of consumerism. While some developed countries will be able to counteract part of the threat of emerging infectious diseases by increasing investment in public health and medical care (age of medical technology), the proportion of infectious diseases that contribute to the total burden of disease will increase. The situation will be fragile, and in some countries the burden of infectious disease may rise considerably, with the potential of falling back into the age of emerging infectious diseases. The developed world will experience 'a double burden of disease': an increase in chronic diseases brought about by a longer life expectancy, complicated by a resurgence of infectious diseases. Many developing countries suffer from this double burden now.

In developing countries, levels of health and welfare spending will either remain the same or decline. In poor countries, current barriers to the control of such major diseases as malaria are likely to remain, and the importance of adequate water and food supplies will increase. Fertility rates will decline relatively slowly. This combination of limited economic resources, high population growth, and increasing pressures on the local and global environments will increase the prevalence of infectious diseases, leading to the age of emerging infectious diseases.

B1 scenario group

A central element in the B1 scenario is a high level of environmental and social consciousness, combined with a global approach to sustainable development. In the developed world, life expectancy will be increased by an improved social structure (and the concomitant benefits to individual lifestyles), and by investments focused on decreasing pressure on ecological systems via the sustained management of resources. An extensive welfare net will prevent poverty-based social exclusion. Although the population will age, healthcare systems will be properly adjusted to an older population. Under this scenario, developed countries may well complete the transition towards the age of sustained health.

However, thanks to transfers of knowledge and technology, and a sharp decline in their national debt, the developing world will make it through the age of receding pandemics. Although some countries will arrive at the age of chronic diseases (i.e. the stage at which the developed world finds itself today), technology and knowledge transfer will enable most of them to skip this stage and approach the age of sustained health.

B2 scenario group

In the B2 scenario, there will be increased concern for environmental and social sustainability. Governments will therefore find policy-based solutions to environmental and health problems. Most governments will increase public spending, including that on public health. Environmentally aware citizens will exercise a growing influence on national and local policy. There will be shift to local decision-making, with a high priority being given to human welfare, equality and environmental protection. Education and welfare programs will be widely pursued, thus reducing mortality, and, to a lesser extent, fertility. However, the rate of implementation will vary across regions and countries. Increased expenditure on 'health' and 'environment' will be implemented first in richer countries and it will take time for poorer countries to follow. However, the health transition will be slower than under A1. For developing countries, the situation may become more robust than under A2; developed countries may experience an increase in chronic diseases, moving slowly through the age of chronic diseases to the age of medical technology.

In conclusion, it is important to emphasis the following: future developments will not be the same for all countries, and developing countries are unlikely to follow the same transition path as the developed world. Although improvements in health may take place worldwide, differences in health status between the developing and developed world will to some extent remain, regardless of the future development path.

5. TOWARDS A STRATEGY FOR SUSTAINABLE DEVELOPMENT

The processes of globalisation in today's world - that include socio-economic change, demographic change and global environmental change - oblige us to broaden our conception of the determinants of population health. In recent decades, health conditions have improved and life expectancy has increased in almost all countries. Within many populations or communities, however, the prospects for health are being adversely affected by the diminution of social capital – i.e. the widening gaps between rich and poor and the weakening of public health systems. Furthermore, the biosphere's capacity to sustain healthy human life is beginning to be impaired by the loss of natural capital: this is manifested as climate change, downturns in food-producing systems, the depletion of freshwater supplies, and the loss of biodiversity (McMichael and Beaglehole 2000).

We must therefore be increasingly alert to the impact on health of larger-scale socio-economic processes and systemic environmental disturbances. Our vulnerability to disease could easily be increased by changes in environmental conditions or a breakdown in public health services. New diseases such as AIDS highlight the sharp divisions in our society. As global populations age, the burden imposed by chronic disease will inevitably increase. Poverty, poor nutrition, debt crises and environmental deterioration all demand attention – as does population growth, whose many interactions with poverty increase pressures on the environment. Increasingly, the factors that affect health are transcending national borders and, as they do so, the health needs of diverse countries are beginning to converge. At the same time, national health systems are increasingly being influenced by global processes. In a world where nations and economies are increasingly interdependent, ill health in any population affects all people, rich and poor alike.

Guiding the health transition towards the age of sustained health will require a development policy that includes social, environmental and economic sectors. Managing the health transition effectively will require a micro-approach, taking into account the social, cultural and behavioural determinants of health (beliefs and practices that account not just for illness and poor health, but also for good health). But this micro-approach will only ensure sustained better health in combination with a macro-approach. On the macro-level, the strong (and growing) evidence of the links between poverty reduction, education, lower fertility rates and better health makes a compelling case for programmes to reduce poverty and slow down population growth. A fall in mortality rates will be brought about not only by rapid macro-economic development, but also by policies designed to satisfy the basic needs of the majority of the population. Rapid progress through the health transition urgently requires substantial investments in education and a restructuring of health systems so that they provide better access to poor people. Simultaneously, international action needs to be undertaken to ensure that impacts of global environmental changes on health will be minimal (e.g. reductions of the emission of greenhouse-gases to reduce the impacts on health caused by the anticipated climate change the coming decades).

However, the major determinants of ill health are beyond the direct control of health services (Woodward, Hales et al. 2000). Therefore, new integrated programmes are needed that combine poverty reduction, protection of the global environment, and economic and social policies more effectively. The responses of international health agencies will also have to be embedded within this broader context, a process that will in turn have consequences for health policies worldwide. Since 'sustained good health is a key indicator of how well we are managing our resources, our social relations, and the ecology of our way of life' (McMichael 1997), public health considerations must inform decisions and actions in all policy sectors.

References

Barrett, R. and C. W. Kuzawa, et al. (1998). "Emerging and re-emerging infectious diseases: the third epidemiologic transition." *Annual Review of Anthropology* **27**: 247-271.

Beaglehole, R. and R. Bonita (1997). *Public health at the crossroads.* Cambridge, Cambridge University Press.

Caldwell, J. C. (1978). "Part I: Epidemiologic transition and health transition." *World Health Statistics Quarterly* **51**: 120-136.

Council, N. R. (1999). *Our common journey: a transition toward sustainability.* Washington, D.C., National Academy Press.

Hilderink, H. (2000). *World population in transition: an integrated regional modelling framework.* Amsterdam, Thela Thesis.

IPCC (2000). *Emissions scenarios.* Cambridge, Cambridge University Press.

Louria, D. B. (2000). "Emerging and re-emerging infections: the social determinants." Futures **32**(6): 581-594.

Mackenbach, J. P. (1994). "The epidemiological transition theory." *Journal of Epidemiology and Community Health* **48**: 329-331.

Martens, P. (2002). "Health transitions in a globalising world: towards more disease or sustained health?" *Futures* **37**(7): 635-648.

Martens, P. (2005). Duurzaamheid: wetenschap of fictie? Maastricht, Oratie Universiteit Maastricht, Open Universiteit Nederland, Hogeschool Zuyd.

Martens, P. (2006). "Sustainability: science or fiction?" *Sustainability: Science, Practice and Policy* **2**(1): 1-5.

Martens, P. and M. M. T. E. Huynen (2003). "A future without health? The health dimension in global scenario studies." *Bulletin of the World Health Organisation* **81**(12): 896-901.

McMichael, A. J. (1993). *Planetary overload: global environmental change and the health of the human species.* Cambridge, Cambridge University Press.

McMichael, A. J. (1997). "Integrated assessment of potential health impacts of global environmental change: prospects and limitations." *Environmental Modelling and Assessment* **2**(3): 129-137.

McMichael, A. J. and R. Beaglehole (2000). "The changing global context of public health." *Lancet* **356**: 495-499.

McMichael, A. J. and J. W. Powles (1999). "Human numbers, environment, sustainability, and health." *British Medical Journal* **319**: 977-980.

Murray, C. J. L. and D. L. Lopez, Eds. (1996). *The global burden of disease: Volume 1.* Geneva, World Health Organization/Harvard School of Public Health/Worldbank.

Olshansky, S. J. and A. B. Ault (1986). "The fourth stage of the epidemiological transition: the age of delayed degenerative diseases." *Milbank Memorial Fund Quarterly* **64**(3): 355-391.

Olshansky, S. J. and B. A. Carnes, et al. (1998). "Emerging infectious diseases: the fifth stage of the epidemiological transition?" *World Health Statistics Quarterly* **51**: 207-217.

Omran, A. R. (1983). "The epidemiological transition, a preliminary update." *Journal of Tropical Pediatrics* **29**: 305-316.

Omran, A. R. (1998). "The epidemiological transition theory revisited thirty years later." *World health statistics quarterly* **51**: 99-199.

Rogers, G. R. and R. Hackenberg (1987). "Extending epidemiological transition theory: a new stage." *Social Biology* **34**(3-4): 234-243.

Rotmans, J. (2003). *Transitiemanagement: sleutel voor een duurzame samenleving*. Assen, Koninlijke Van Gorcum.

Rotmans, J. and J. Grosskurth, et al. (2001). Duurzame ontwikkeling; van concept naar uitvoering. Maastricht, International Centre for Integrative Studies.

WCED (1987). *Our Common Future*. Oxford,UK, Oxford University Press.

Woodward, A. and S. Hales, et al. (2000). "Protecting human health in a changing world: the role of social and economic development." *Bulletin of the World Health Organization* **78**(9): 1148-1155.

J. Phys. IV France **139** (2006) 403–411
© EDP Sciences, Les Ulis
DOI: 10.1051/jp4:2006139026

Reconciling adaptation and mitigation to climate change in agriculture*

J.E. Olesen[1]

[1] *Danish Institute of Agricultural Sciences, Research Centre Foulum, 8830 Tjele, Denmark*

Abstract. An effective adaptation to the changing climate at farm, sector and policy level is a prerequisite for reducing negative impacts and for obtaining possible benefits. These adaptations include land use and land management, as well as changes in inputs of water, nutrients and pesticides. Some of the most wide ranging adaptations involve changes in water management and water conservation, which involves issues such as changing irrigation, adoption of drought tolerant crops and water saving cropping methods (e.g. mulching and minimum tillage). Many of these adaptation options have substantial effects on greenhouse gas emissions from agriculture. However, so far few studies have attempted to link the issue of adaptation and mitigation in agriculture. This is primarily because the issues have so far been dealt with by different research communities and within different policy contexts. As both issues are becoming increasingly relevant from a policy perspective, these issues will have to be reconciled. Dealing with these issues requires a highly interdisciplinary approach.

1. INTRODUCTION

Climate change is expected to affect agriculture very differently in different parts of the world [1]. The resulting effects depend on current climatic and soil conditions, the direction of change and the availability of resources and infrastructure to cope with change. There is a large variation across the European continent in climatic conditions, soils, land use, infrastructure, political and economic conditions [2]. These differences are expected also to greatly influence the responsiveness to climatic change [3]. Intensive farming systems in Western Europe generally have a low sensitivity to climate change [4]. On the other hand some of the low input farming systems currently located in marginal areas may be most severely affected by climate change [5].

An effective adaptation to the changing climate at farm, sector and policy level is a prerequisite for reducing negative impacts and for obtaining possible benefits [3, 6]. These adaptations include land use and land management, as well as changes in inputs of water, nutrients and pesticides. Some of the most wide ranging adaptations involve changes in water management and water conservation, which involves issues such as changing irrigation, adoption of drought tolerant crops and water saving cropping methods (e.g. mulching and minimum tillage).

Agricultural activities are among the major contributors to total greenhouse gas emissions (GHG) (up to 9% for EU in 2000 [7]). In particular agriculture have relatively large emission of nitrous oxide (N_2O) and methane (CH_4), which are greenhouse gases that contribute to the global warming. The global warming potentials of N_2O and CH_4 are 310 and 21 times higher than that of carbon dioxide (CO_2) per kg. Methane accounts for 20% of the anthropogenic contributions to global warming, and N_2O accounts for 6% [8]. Thus, mitigation strategies will also be required within the agricultural sector to comply with the reduction targets of the Kyoto Protocol and beyond.

Agriculture has a range of options to reduce greenhouse gas emissions, either directly by reducing energy use and emissions of methane and nitrous oxide or by substitution of fossil energy use and carbon sequestration in soils. Methane emissions can be reduced through changes in animal feeding strategies and through changes in manure handling, e.g. production of biogas from animal slurry [9]. Nitrous oxide

*This paper is partly based on results obtained in the PRUDENCE and GREENGRASS projects and COST action 627.

emissions may be reduced through changes in manure handling, more efficient nitrogen use and changes in crop and soil management [10]. A number of agricultural management options including conservation tillage practices, crop residue management, cover crops and altered crop rotations have been suggested as measures for carbon sequestration in soils [11, 12]. Advantage should be taken of the fact that some of the measures simultaneously may reduce the net emission of several greenhouse gases. However, climate change may affect the emission of greenhouse gases from agriculture [13], particularly through increasing soil organic matter turnover and thereby enhancing CO_2 emissions [14].

Many of the options available for adapting agricultural activities will influence the emissions of greenhouse gases either by enhancing or reducing the fluxes. However, it should be kept in mind that agricultural activities affect several greenhouse gases simultaneously, and it is the net effect on the global warming potential of all gases that should be considered. There may also be differences between short- and long-term responses to introduction of system and management changes, in particular for measures that involve changes in soil management and input of carbon and nitrogen to the soil [15, 16].

There are few studies in the literature dealing with the linkages between adaptation and mitigation in agriculture, and the present paper therefore attempts to give an overview of possible linkages based on some initial evaluation of possible impacts of adaptation measures on the GHG emission from agriculture with a focus on examples from European studies.

2. LAND USE CHANGES

The evaluation of climate change is usually based on simulations with global climate models (GCM) for the IPCC emissions scenarios (SRES scenarios), which describe very different socio-economic futures [8]. The SRES scenarios are grouped into four different categories (A1: world markets, A2: provincial enterprise, B1: global sustainability, B2: local stewardship). The grouping relies upon two orthogonal axes, representing social values (ranging from consumerist to conservationist) and level of governance (ranging from local to global), respectively.

The SRES scenarios for socio-economic development have been adapted to European conditions [17-19], and their main characteristics are outlined in Table 1. Assumptions about future European land use and the environmental impact of human activities depend greatly on the development and adoption of new technologies. For the SRES scenarios it has been estimated that increases in crop productivity relative to 2000 could range between 25 and 163% depending on the time slice (2020 to 2080) and scenario [20]. These increases were smallest for the B2 and highest for the A1FI scenario.

Changes in crop productivity along with changes in demand for agricultural products will affect the total agricultural land area and its use for growing various crops [18, 21, 22]. This is one of the primary adaptations to climate change, which will occur as a result primarily of market forces, but to some extent influenced by policies.

2.1 Projected changes in European land use

Temporally and spatially explicit future scenarios of European land use have been developed for the four core SRES scenarios [22, 23]. These scenarios are based on supply/demand models of market forces, rural development and environmental policies based on qualitative descriptions in the scenarios and the characteristics of the European landscapes. The results show large declines in agricultural land uses resulting from the assumptions about future crop yield with respect to changes in demand for agricultural commodities [21]. Expansion of urban area is similar between the scenarios, whereas forest areas increase in all scenarios (Table 2). The scenarios showed decreases in European cropland for 2080 that ranged from 28% to 47% [21]. The reduction in European grassland for 2080 ranged from 6 to 58%. This decline in agricultural area will make land resources available for other uses such as biofuel production and nature reserves. Over the shorter term (up to 2030) changes in agricultural land use may be small [24].

Table 1. Adaptation of SRES scenarios to Europe [17, 18, 21, 22].

Scenario	Characteristics for Europe
A1. World market	Emphasis on pursuing economic growth and free trade European economic inequalities eradicated and rising income levels Stable political and social climate with good health care and education EU enlargement to include new member states EU is a single market, functionally integrated with other markets
A2. Provincial enterprise	Society is dictated by short-term consumerist values Policy decisions are taken at national and sub-national levels Europe adopts protectionist economic and trade policies Declining equity between European countries EU competences remain as they are today and enlargement is restricted
B1. Global sustainability	Emphasis on international solutions to global environmental problems Enlargement of the EU and development towards a federal structure EU takes over responsibility to solve environmental problems International institutions will adopt social programmes
B2. Local sustainability	Focus on solving environmental problems locally (green technologies) In EU the principle of subsidiarity shifts governance to the local level The enlargement and the deepening of EU is abandoned Decisions are often taken at subnational levels Europe is more heterogeneous, including larger differences in regional incomes

Table 2. Projected changes in land use compared with the baseline for the EU15 countries plus Norway and Switzerland and consequences for the carbon balance for the time period 2080 relative to 1990 for four IPCC SRES scenarios using the HadCM3 GCM [23]. The carbon balance is calculated as the cumulative flux from 1990 to 2080, and positive values denote fluxes to land (i.e. uptake).

	Baseline	B1	B2	A1FI	A2
Land use and change (%)					
Cropland	23.0	−7.0	−6.4	−10.7	−10.4
Grassland	17.2	−1.1	−6.7	−8.7	−10.0
Forest	31.0	3.5	5.6	0.8	0.7
Urban	1.5	0.05	0.06	0.09	0.08
Bioenergy		3.4	7.4	8.7	8.7
Protected		6.1	6.1	6.1	6.1
Surplus (other in baseline)	27.3	1.1	0.0	9.8	10.9
Carbon balance (Pg)					
Cumulative total		2.2	2.4	1.8	3.0
Soil organic carbon (SOC)		−0.1	−0.9	−4.1	−4.4
SOC, cropland		−4.3	−4.3	−5.9	−5.6
SOC, grassland		1.5	−1.2	−2.2	−2.7
SOC, forest		2.8	3.6	1.0	1.1

The climate change scenarios could lead to increases in greenhouse gas emissions from agriculture. Increasing temperatures will speed decomposition where soil moisture allows, so direct climate impacts on cropland and grassland soils will tend to decrease soil organic carbon (SOC) stocks for Europe as a whole [14]. This effect is greatly reduced by increasing C inputs to the soil because of enhanced NPP, resulting from a combination of climate change and increased atmospheric CO_2 concentration. However, decomposition becomes faster in regions where temperature increases greatly and soil moisture remains high enough to allow decomposition (e.g. North and East Europe), but does not become faster where the soil becomes too dry, despite higher temperatures (Southern France, Spain, and Italy) [25]. Overall this will lead to soil carbon losses from croplands, mainly due to reductions in the area devoted to cropland, whereas there is less change in the total SOC stocks, in particular for the B1 and B2 scenarios (Table 2).

2.2 Bioenergy crops

Bioenergy crops and biofuels are often seen as a clean alternative to the use of fossil fuels, i.e. resulting in substantially less CO_2 emissions. In its most simplistic sense the energy content of one litre of petrol may be compared with that of e.g. one litre of bio-ethanol. However, such a comparison does not account for the energy needed to produce either a litre of petrol or a litre of bio-ethanol. Also there will often be substantial additional emissions of other greenhouse gases, in particular nitrous oxide, associated with the production of biofuels, which need to be factored in. The fossil energy used in the production of biofuels may exceed the energy output in the biofuel [26], although some technologies do offer net energy benefits [27]. This situation can be changed considerably when emissions of other greenhouse gases (e.g. nitrous oxide) and effects on soil carbon storage is included [28]. Different production systems of bioenergy crops may therefore result in substantial differences in greenhouse emission, with production systems involving perennial crops generally having lower emissions [29].

There are, however, bioenergy production systems, which may contribute positively to the net energy production, in particular, when other side effects are considered. Thus many biofuel crops have additional uses, e.g. from by-products from their production, and indeed some of the studies reporting negative net energy from biofuels have ignored the byproducts [27].

Energy crops currently contribute relatively little to the total energy produced from biomass each year, but the production is projected to increase over the next few decades. A realistic achievable global potential for energy crops by 2025 may offset 100-2070 Mt CO_2-eq. yr^{-1} [30].

The substitution of food production by energy production through the widespread cultivation of bioenergy crops can become an important consequence of climate change as well as an option for mitigating climate change [31]. Due to the possible large reductions in cropland area, land available for bioenergy crops may expand considerably, especially in the A1FI and A2 scenarios (Table 2). Several temperate and Mediterranean crop species are suitable for various types of biofuels, including oilseed crops, starch crops, cereals and solid biofuel crops. All climate change scenarios show a northward expansion of these species with northern Europe becoming more favourable for most species. However, the choice of energy crops in southern Europe may be severely reduced in future, both due to increased temperatures and reduced rainfall.

3. LAND MANAGEMENT CHANGES

Land management changes include mostly agronomic strategies or adapting to climate change, involving both short-term adjustments and long-term adaptations [32].

The short-term adjustments include efforts to optimise production without major system changes. They are autonomous in the sense that no other sectors (e.g. policy, research, etc.) are needed in their development and implementation. Examples of short-term adjustments are changes in varieties, sowing dates and fertiliser and pesticide use [33-36]. In particular, in Southern Europe short-term adaptations may include changes in crop species (e.g. replacing winter with spring wheat) [36], changes in cultivars and sowing dates (e.g. for winter crops, sowing the same cultivar earlier, or choosing cultivars with longer crop cycle; for summer irrigated crops, earlier sowing for preventing yield reductions or reducing water demand) [38].

Long-term adaptations include breeding of crop varieties, new land management techniques to conserve water or increase water use efficiencies, including irrigation and changes in tillage systems, crop rotations and residue management.

3.1 Changes in production intensity

Changes in productivity of crops will lead to changes in the optimal rates of fertiliser and pesticide inputs needed to support the crops. In those regions where crop productivity declines this should lead to less use of inputs, whereas in regions of increased productivity more inputs will be needed.

Changes in intensity of cropping systems affect both the use inputs and thus the energy needed for providing these inputs. It not only affect the direct CO_2 emissions associated with the energy consumption, but also the indirect emissions from nitrous oxide emissions and changes in soil carbon storage [39-40]. Even at farm scale linear relationships have been observed between production intensity and greenhouse gas emissions per land area [41]. However, it is likely that high increases in nitrogen fertiliser inputs will lead to non-linear increases in nitrous oxide emissions, giving proportionately higher GHG emissions at high input levels [42].

3.2 Crop rotations and cover crops

Changes in crop rotations involve changes not only in the choice of crops and the sequence of crops grown, but also the choice of break crops (including green manure crops and set-aside) and cover crops. Set-aside and green manure crops often involve grasses or leguminous crops that are grown for one or more years with the purpose of increasing soil fertility [43]. Cover crops are crops, which are grown within the same year as the yielding crops, but with the purpose of protecting the soil against erosion or leaching of nutrients. Both types of crops will influence the input of carbon and nitrogen to the soil, and therefore the soil carbon storage [16] and emissions of nitrous oxides [44].

Environmental impacts of agriculture under a changing climate are increasingly becoming recognised. In particular the increased temperature and rainfall changes on soil organic matter [14, 25] and effects on nitrate leaching [45] have been recognised as key issues. To mitigate negative trends in these respects changes in crop rotations may be adopted, in particular increased use of cover crops may be effective in maintaining both SOC levels and reducing nitrate leaching [46]. Simulations have also indicated that increases in GHG emissions with global warming can be effectively mitigated by including more spring cereals and catch crops in the rotation [46] (Table 3).

3.3 Tillage and crop residue management

Several water-conserving practices are commonly used to combat drought. These may also be used for reducing climate change impacts [32]. Such practices include conservation tillage, which is the practice of leaving some or all the previous season's crop residues on the soil surface in combination with non-inversion tillage [47]. This may protect the soil from wind and water erosion and retain moisture by reducing evaporation and increasing the infiltration of rainfall into the soil.

These practices also have major impacts on GHG emissions with increased soil carbon storage [16], reduced fuel use [48] and in some cases increases in nitrous oxide emissions [15]. In the long-term increases in rate of soil carbon sequestration will decline, as will the changes in nitrous oxide emissions [15].

Table 3. Simulated effects on increasing temperature by 4 °C and CO_2 concentration by 50% on soil carbon storage and N_2O emissions for three cereal-based crop rotations in Denmark [46].

	Baseline	+4 °C	+50% CO_2
Change in SOC (kg C ha^{-1} yr^{-1})			
Winter cereals, no cover crop	−43	−76	−2
Spring cereals, no cover crop	−38	−55	−1
Spring cereals, cover crop	277	284	337
N_2O emissions (kg N_2O-N ha^{-1} yr^{-1})			
Winter cereals, no cover crop	3.4	3.6	3.2
Spring cereals, no cover crop	3.3	3.3	3.2
Spring cereals, cover crop	3.5	3.7	3.4

3.4 Irrigation

Irrigation is commonly proposed adaptation option for coping with increased summer droughts. Irrigation management can be used to improve considerably the utilisation of applied water through proper timing of the amount of water distributed. When irrigation is applied this will increase crop productivity and usually also the amount of crop residues returned to the soil, which will increase carbon sequestration. However, the energy use associated with irrigation can often be a major component of the GHG balance of irrigated systems [40].

4. CHANGES IN FARMING SYSTEMS

Climate change will not only affect the crop production, but also livestock production systems through effects on the direct components of those systems, but also on their interactions, e.g. the effect of change in crop quality on animal feed and effects on warming on nutrient turnover in manure storage systems.

4.1 Livestock management

Climate and CO_2 effects influence livestock systems through both availability and price of feed and through direct effects on animal health, growth, and reproduction. Effects of climate change on grasslands will have direct effects on livestock living on these pastures. Results from a simulation study suggest that the impact on milk production for grass-based systems in Scotland would vary depending on the locality. Conversely, for herds grazing on grass-clover swards milk output may increase regardless of site, when the concentration of CO_2 is enhanced [49]. Changes in feed quality are likely to affect the methane emissions from enteric fermentation, in particular for ruminant animals [50].

Livestock production may be negatively affected in the warm months of the currently warm regions of Europe, as has been found for parts of the USA [51]. Warming during the cold period for cooler regions may on the other hand be beneficial due to reduced feed requirements, increased survival, and lower energy costs. Impacts will probably be minor for intensive livestock systems (e.g. confined dairy, poultry and pig systems) because climate is controlled to some degree. Climate change may, however, affect requirements for insulation and air-conditioning and thus increase or decrease housing expenses in different regions. This is also likely to affect the GHG emissions associated with energy requirements.

4.2 Manure management

Climate change will affect the turnover and losses of nutrients from animal manure, both in houses, storages and in the field. An example of this is the increase in ammonia volatilisation with increasing temperature [52].

Manure management systems are a major source of nitrous oxide (N_2O) and methane (CH_4) [53-54]. Methane is emitted during storage of both liquid and solid manure and results from anaerobic fermentation processes. Temperature has a strong influence on methane production with a suggested exponential increase in production with increasing temperature [55-56]. Methanogenesis is inhibited by large concentrations of ammonia [57]. The methane production also depends on the methane production potential of the manures, which in turn depends on animal type, feed ration and addition of organic materials to the manure. The amount and quality of organic matter or volatile solids (VS) in the manure is therefore of key importance. Cattle are efficient in using the energy in feed due to fermentation in the rumen, consequently ultimate methane production is approximately 75% higher for pig slurry VS than for cattle slurry VS [56]. Methane can be oxidised by a specialised group of bacteria. This process is known to occur in both surface crust of manure slurry tanks [58] and in composting manure heaps [59].

N_2O may be produced both during the nitrification of ammonium to nitrate and during the denitrification of nitrate to dinitrogen (N_2). Low oxygen concentrations and low pH are two environmental factors, which is favourable for N_2O production as a bi-product from nitrification. N_2O is a free intermediate during the denitrification process, and net emissions often occur in transition zones from anaerobic to aerobic zones, where the N_2O may escape to the atmosphere before being fully transformed to N_2. Such transition zones may occur both in time and space, e.g. within the heterogeneity of decomposing manure [56, 59]. In general, however, nitrous oxide emissions are low during manure storage, except for some solid manure handling systems.

There are still large uncertainties related to the size of the GHG emissions from the various elements of the manure flows. However, a number of options for reducing emissions of CH_4 and N_2O can be identified, including both technical and management changes. The most effective mitigation options appear to be those measures that target several gases simultaneously. An example is the use of anaerobic digestion of manure which, provided proper operating conditions, can reduce CH_4 and N_2O emissions during storage while also reducing N_2O emissions after field application and substituting fossil fuel [60]. It is likely that such options will also be effective under climate change.

5. PERSPECTIVES ON LINKING ADAPTATION AND MITIGATION

Few studies have so far attempted to link the issue of adaptation and mitigation in agriculture. This is primarily because the issues have so far been dealt with by different research communities and within different policy contexts. However, as both issues are becoming increasingly relevant from a policy perspective, these issues will have to be reconciled. Dealing with these issues requires an interdisciplinary approach, where the research on adaptation in agriculture will need to deal not only with the effects on changes in productivity and economic viability, but also on the related environmental impacts of climate change and adaptation measures. On the other hand the research community dealing with mitigation measures will need to be concerned not only with the efficiencies of mitigation measures, but also with the extent to which these measures and technologies are compatible with changes in climatic conditions and resulting changes in farming systems.

Acknowledgements

Inspiration for this paper was obtained through the PRUDENCE (EVK-CT-2001-00132) and GREENGRASS (EVK-CT-2001-00105) EU-projects and well as COST action 627 on Carbon storage in European grasslands.

References

[1] Parry M.L., Rosenzweig C., Iglesias A., Livermore M. and Fischer G., Global Environ. Change 14 (2004), 53-67.
[2] Bouma J., Varallyay G. and Batjes N.H., Agric. Ecosyst. Environ. 67 (1998), 103-119.
[3] Olesen J.E. and Bindi M., Eur. J. Agron. 16 (2002), 239-262.
[4] Chloupek O., Hrstkova P. and Schweigert P., Field Crops Res. 85 (2004), 167-190.
[5] Reilly J. and Schimmelpfennig D., Clim. Change 43 (1999), 745-788.
[6] Salinger M.J., Stigter C.J. and Das H.P., Clim. Change 43 (1999), 745-788.
[7] EEA, *Environmental signals 2002. Benchmarking the millenium.* Environmental assessment report No. 9. European Environmental Agency, Copenhagen (2002).
[8] Houghton J.T., Ding Y., Griggs D.J., Noguer M., van der Linden P.J., Dai X., Maskell K. and Johnson C.A., *Climate change 2001: The scientific basis. Contribution of working group I to the third assessment report of the Intergovernmental Panel on Climate Change.* Cambridge University Press, Cambridge, UK (2001).
[9] Monteny G.-J., Bannink A. and Chadwick D., Agric. Ecosyst. Environ. 112 (2006), 163-170.
[10] Rosenzweig, C. and Hillel, D., Soil Sci. 165 (2000), 47-56.

[11] Smith P., Powlson D.S., Smith J.U., Falloon P.D. and Coleman K., Global Change Biol. 6 (2000), 525-539.
[12] Freibauer A., Rounsevell M.D.A., Smith P. and Verhagen J., Geoderma 122 (2004), 1-23.
[13] Mosier A.R., Biol. Fertil. Soils 27 (1998), 221-229.
[14] Davidson E.A. and Janssens, I.A., Nature 440 (2006), 165-173.
[15] Six J., Ogle S.M., Breidt F.J., Conant R.T., Mosier A.R. and Paustian K., Global Change Biol. 10 (2004), 155-160.
[16] Ogle S.M., Breidt F.J. and Paustian K., Biogeochem. 72 (2005), 87-121.
[17] Parry, M.L. (ed.), *Assessment of potential effects and adaptations for climate change in Europe: The Europe ACACIA project*. Jackson Environment Institute, University of East Anglia, Norwich, United Kingdom (2000).
[18] Abildtrup J., Audsley E., Fekete-Farkas M., Giupponi C., Gylling M., Rosato P. and Rounsevell M.D.A., Environ. Sci. Pol., in press (2006).
[19] Holman I.P., Rounsevell M.D.A., Shackley S., Harrison P.A., Nicholls R.J., Berry P.M., and Audsley E., 2005. Clim. Change 70 (2005), 9-41.
[20] Ewert F., Rounsevell M.D.A., Reginster I., Metzger M.J. and Leemans R., Agric. Ecosyst. Environ. 107 (2005), 101-116.
[21] Rounsevell M.D.A., Ewert F., Reginster I., Leemans R. and Carter T.R., Agriculture, Ecosystems and Environment 107 (2005), 117-135.
[22] Rounsevell M.D.A., Reginster I., Araújo M.B., Carter T.R., Dendoncker N., Ewert F., House J.I., Kankaanpää S., Leemans R., Metzger M.J., Schmit C., Smith P. and Tuck G. Agric. Ecosyst. Environ.114 (2006), 57-68.
[23] Schröter D., Cramer W., Leemans R., Prentice I.C., Araújo M.B., Arnell N.W., Bondeau A., Bugmann H., Carter T.R., Gracia C.A., de la Vega-Leinert A.C., Erhard M., Ewert F., Glendining M., House J.I., Kankaapää S., Klein R.J.T., Lavorel S., Lindner M., Metzger M.J., Meyer J., Mitchell T.D., Reginster I., Rounsevell M., Sabaté S., Sitch S., Smith B., Smith J., Smith P., Sykes M.T., Thonicke K., Thuiller W., Tuck G., Zaehle S. and Zierl, B., Science 310 (2005), 1333-1337.
[24] van Meijl H., van Rheenen T., Tabeau A. and Eickhout, B., Agric. Ecosyst. Environ. 114 (2006), 21-38.
[25] Smith J., Smith P., Wattenbach M., Zaehle S., Hiederer R., Jones R.J.A., Montanarella L. and Rounsevell M.D.A. Reginster I. and Ewert F., Global Change Biol. 11 (2005), 1-12.
[26] Pimentel D. and Patzek T.W., Natural Res. Res. 14 (2005), 65-76.
[27] Farrell, A.E., Plevin, R.J., Turner, B.T., Jones, A.D., O'Hare, M. and Kammen, D.M., 2006. Ethanol can contribute to energy and environmental goals. Science 311, 506-508.
[28] Marland, G. and Schlamadinger, B., Energy 20 (1995), 1131-1140.
[29] Olesen J.E., Andersen J.M., Jacobsen B.H., Hvelplund T., Jørgensen U., Schou J.S., Graversen J., Dalgaard T. and Fenhann J, *Kvantificering af tre tiltag til reduktion af landbrugets udledning af drivhusgasser*. DJF-rapport Markbrug 48 (2001).
[30] Sims, R.E.H., Hastings A., Schlamadinger B., Taylor G. and Smith P., Global Change Biol. 12 (2006), 1-23.
[31] Tuck G., Glendining M.J., Smith P., House J.I. and Wattenbach M., Biomass Bioenergy, in press (2006).
[32] Easterling W.E., Agric. Forest Meteorol. 80 (1996), 1-53.
[33] Ghaffari A., Cook H.F. and Lee H.C., Clim. Change 55 (2002), 509-533.
[34] Alexandrov V., Eitzinger J., Cajic V. and Oberforster M., Global Change Biol. 8 (2002), 372-389.
[35] Tubiello F.N., Donatelli M., Rosenzweig C. and Stockle C.O., Eur. J. Agron. 13 (2000), 179-189.
[36] Chen C.C. and McCarl B.A., Clim. Change 50 (2001), 475-487.
[37] Mínguez M.I., Ruiz-Ramos M., Díaz-Ambrona C.H., Quemada M. and Sau F., Clim. Change, in press (2006).

[38] Olesen J.E., Carter T.R., Diaz-Ambrona C.H., Fronzek S., Heidmann T., Hickler T., Holt T., Minguez M.I., Morales P., Palutikov J., Quemada M., Ruiz-Ramos M., Rubæk G., Sau F., Smith B. and Sykes M., Climatic Change, in press (2006).
[39] King J.A., Bradley R.I., Harrison R. and Carter A.D., Soil Use Manage. 20, 394-402.
[40] Mosier A.R., Halvorsen A.D., Peterson G.A., Robertson G.P. and Sherrod L., Nutr. Cycl. Agroecosyst. 72 (2005), 67-76.
[41] Olesen J.E., Schelde K., Weiske A., Weisbjerg M.R., Asman W.A.H. and Djurhuus J., Agric. Ecosyst. Environ. 112 (2006), 207-220.
[42] Bouwman A.F., Boumans L.J.M. and Batjes N.H., Global Biochem. Cycles (2002) doi:10.1029/2001GB001812.
[43] Olesen J.E., Askegaard M. and Rasmussen I.A., Acta Agric. Scand. B Soil Plant Sci. 50 (2000), 13-21.
[44] Freibauer A. and Kaltschmitt M., Biogeochem. 63 (2003), 93-115.
[45] Olesen J.E., Carter T.R., Diaz-Ambrona C.H., Fronzek S., Heidmann T., Hickler T., Holt T., Minguez M.I., Morales P., Palutikov J., Quemada M., Ruiz-Ramos M., Rubæk G., Sau F. Smith, B. and Sykes, M., Clim. Change, in press (2006).
[46] Olesen J.E., Rubæk G., Heidmann T., Hansen S. and Børgesen C.D., Nutr. Cycl. Agroecosyst. 70 (2004), 147-160.
[47] Holland J.M., Agric. Ecosyst. Environ. 103 (2004), 1-25.
[48] Robertson G.P., Paul E.A. and Harwood R.R., Science 289 (2000), 1922-1925.
[49] Topp C.F.E. and Doyle C.J., Agric. Syst. 52 (1996), 243-270.
[50] Benchaar C., Pomar C. and Chiquette J., Can. J. Anim. Sci. 81 (2001), 563-574.
[51] Klinedienst P.L., Wilhite D.A., Hahn G.L. and Hubbard K.G., Clim. Change 23 (1993), 21-36.
[52] Sommer S.G., Géneremont S., Cellier P., Hutchings N.J., Olesen J.E. and Morvan, T., Eur. J. Agron. 19 (2003), 465-486.
[53] Lelieveld J., Crutzen P. and Dentener F.J., Tellus 50B (1998), 128-150.
[54] Kroeze C., Mosier A. and Bouwman L., Glob. Biochem. Cycl. 13 (1999), 1-8.
[55] Husted S., J. Environ. Qual. 23 (1994), 585-592.
[56] Sommer S.G., Petersen S.O. and Møller H.B., Nutr. Cycl. Agroecosyst. 69 (2004), 143-154.
[57] Angelidaki I., Ellegaard L. and Ahring B.K., Biotech. Bioeng. 42 (1993), 159-166.
[58] Sommer S.G., Petersen S.O. and Søgaard H.T., J. Environ. Qual. 29 (2000), 744-751.
[59] Sommer S.G. and Møller H.B., J. Agric. Sci., Camb. 134 (2000), 327-335.
[60] Weiske A., Vabitsch A., Olesen J.E., Schelde K., Michel J., Friedrich R. and Kaltschmitt M., Agric. Ecosyst. Environ. 112 (2006), 221-232.

J. Phys. IV France **139** (2006) 413–423
© EDP Sciences, Les Ulis
DOI: 10.1051/jp4:2006139027

Communicating air pollution science to the public and politicians

P. Brimblecombe[1] and E. Schuepbach[2]

[1] *School of Environmental Sciences, University of East Anglia, Norwich NR4 7TJ, UK*
[2] *cabo$_3$/Physical Geography, University of Berne, Switzerland*

Abstract. Air pollution of the 21st century is a problem that involves a large number of chemical species and complex reactions between them. Both the public and politicians finds the science difficult to understand, and so, often mistrust the presentation of data and the scientific principles behind air quality. Yet, there are a range of important issues associated with air pollution that concern lay people and policy makers and hence, they have to be presented in a clear and simple way so that informed judgements can be made. Traditionally, the media was the main way to disseminate scientific discovery, but novel methods for engaging scientists in the transfer of scientific know-how to politicians and the general public have emerged in recent years. Scientists receive relatively little training in the area of communication, and often find engaging in more public debates difficult. These including *V.I.P. meetings, Public Open Forum, Café Scientifique* and various games and role plays. Such outreach events expose us to new challenges, and the skills required to communicate to non-scientists become an increasingly important part of being a scientist.

1. INTRODUCTION

Engagement with stakeholders and politicians is a characteristic of 21st century science. Both see scientists increasingly as an important part of their world – biologists are increasingly called upon to talk about issues relating to genetically modified crops, geologists about the finite nature of our mineral resource, meteorologist and atmospheric scientists about global warning and air pollution. Atmospheric scientists can find themselves interacting with a widening group of specialists: gerontologists about vulnerable groups, sociologists about patterns of behaviour, and philosophers and theologians about questions related to ethics.

Scientists are often uncomfortable with a socio-political role. This is particularly emphasised by the fact that they are called to comment on issues as experts, but the discussion frequently moves away from their areas of expertise – no more so that when they are required to express political or moral judgements and thus making their opinions a personal view rather than from the own research. This transition can occur so easily it is hard to for the scientist or the audience to be aware that the discourse has shifted.

The problem is exacerbated by other issues. On one hand scientists, like most professionals, receive relatively little training in communication. They fear it will incur disapprobation from their colleagues or that their employers will take offence at their views. There are high profile cases of this, such as the debate over Jim Hansen's well attended keynote address to the to American Geophysical Union in December 2005 (see: http://en.wikipedia.org/wiki/James_Hansen). Most of us receive much less notice, so reasonable care over what we say can lead to satisfactory interviews and in many cases colleagues will be pleased that the issue has been laid out clearly.

However, one always has to be aware that media has objectives that are quite different from that of a scientists – this often means selling the human interest or laying out a particular "angle" that satisfies a broad readership. There is often a need for snappy titles which for many scientists distort reality.

Here we review a number of ways in which material is brought into the public and political arenas. We also introduce some of the topics that are often debated. Air pollution is a complex issue which goes beyond the scientific complexities, so this brief review tries to explore the need to integrate air pollution issues with society and the world of policy making. Although many textbooks treat air pollution as though it was simply a problem of chemistry or meteorology, it is also a product of social and political perceptions.

2. COMMUNICATING AIR POLLUTION DATA TO THE PUBLIC

Transparency in factual information has been a key aspect of more open, democratic government in recent decades. Air pollution monitoring networks have grown in parallel across the same period and often the data gathered has been made accessible to the public. In Europe this provision grew in the early 1990s. Initially the data was provided by a telephone link, *teletext* on television and through a range of other less instantaneous sources, which have led to flyers and brochures being made available in libraries and even supermarkets. Direct mailing has often been used to inform about some issues (e.g. reporting smoke from diesel vehicles), while wider use of the internet and email has meant personal computers have become an increasingly important.

2.1 Flavour of communication

The flexibility and immediacy of internet and telecommunication provision cannot guarantee public acceptability. Despite the excellent provisions of the award winning UK Air Quality archive (http://www.airquality.co.uk/archive/index.php), some have complained that it is just a mass of data. The expectations of the general public are extremely high and even what may seem to scientists an excellent information source can be perceived as inadequate in an age of internet gamers!

Many people do not know how to find air pollution information and where they do they frequently complain that the information is unintelligible. Some have even seen it as an attempt of government to blind the public with science. The reality for scientists is that the data collected is at now at a high level of sophistication and monitoring networks may sample many components at a site as frequently as once every 15 minutes. If hydrocarbon data is collected this will provide concentration measurements of many organic compounds which are unfamiliar to the general public. Even quite simple inorganic pollutants such as carbon monoxide or sulphur dioxide are confusing chemical names to many.

This requires that data be simplified. Air pollution is now often expressed as though it is a single entity and we can hear it in statements like: "the air pollution is high today". The reduction of air pollution data to simple expressions about air quality is important for communication. It is nevertheless fraught with difficulties. It may well be argued that simplify data restricts public access to the underlying measurements and obscures the true state of air quality. This is a particular criticism of the widespread adoption in indices and bands to describe air pollution [1].

2.2 Indices and banding systems

In the UK the Department of Environment, Food and Rural Affairs says: "In order to make air quality information more meaningful, a set of criteria are used to classify air pollution levels into bands, with a description associated with each band". These are *low, moderate, high* and *very high* levels of air pollution. The divisions between these bands are defined in terms of the air pollution standard, an information level and an alert threshold. The terms were modified from an earlier system criticised for using value-laden words such as *good* or *poor*. However the UK has now adopted an additional scale that rates air pollution with an integer in the 1-10 range (as in France). This is reminiscent of with the familiar pollen indices or sunburn scales. The US Environmental Protection Agency (EPA) had developed a Pollutant Standards Index (PSI), which rates air pollution on a more finely graded scale where 100 represents the limit value or air quality standard. Indices of this form are now found in many countries [1].

The step function nature of bands creates problems along with issues of averaging time and synergisms between pollutants. Some of the difficulties can be overcome by giving the numerical values of pollutant concentration along side the bands [1]. Despite a number of revisions the UK's system remains misunderstood and has been seen as concealing the real picture by using vague descriptive terms, although the counter view that the numbers are baffling is also a source of complaint. Much has to be done in terms of reporting air pollution data and address the severe difficulties of averaging times and spatial scales. Nevertheless it is probable that careful scientific analysis is not necessarily sufficient to meet with public acceptance.

The public understanding and trust in air pollution information and indices may be very different to the notions of validity or accuracy that are important for a scientist. Although public interest groups and the government were keen that air pollution information is made available to the public, surprisingly little attention is paid to what information is desired or how it will be used. Bickerstaff [2] has emphasised that many monitoring networks may be ill equipped to provide the appropriate information to the public. People's understandings tend to be localised within the immediate physical, social and cultural landscape, while networks provide data in a national or regional sense that is defined by administrative geography. Trust is more likely to derive from personal experiences with the public administration or the political system rather than quality control on the numerical data.

Air quality data would probably need to be more tuned to personal or health interests of individuals at the locations they will occupy if it was to meet public expectations. This would often mean it would have to be forward looking and give pollution for the next day rather than the previous. This demand creates many problems, and although air pollution forecasts are increasingly made at a high temporal resolution (hourly) they had difficulty coping with the heterogeneity of urban air or with background conditions. They could easily fail to pick up short term excursions on a very local level and where this was done it still might not relate the air pollution data to individual experience or feelings. There is also a desire to use the data for very practical purposes, such as deciding which route would expose children to the lowest pollution concentrations on the way to school.

In summary, air pollution data provided to the public needs to have an immediacy and relevance to a localised situation. It has to be presented in simple language and readily translated into actions.

3. COMMUNICATING SCIENCE

Public understanding of science is now widely promoted by scientific organisations, NGOs, and environmental agencies. Research grants often carry a requirement for dissemination that goes beyond the publication of results in scientific journals. Many funding bodies, such as Natural Environment Research Council in the UK require scientists to communicate their findings to a wide audience, but they are also aware that scientist still need training in this area and provide support (e.g. NERC http://www.nerc.ac.uk/insight/support/ideas/index.asp). The large ACCENT project (the European Network in Atmospheric Composition Change: www.accent-network.org) has two major tasks devoted to outreach.

The success of these efforts to promote wider dissemination is limited by the willingness of scientists to engage with outreach activities. These can often seem burdensome, especially when there seem limited rewards or recognition for a commitment to public presentation and the liaison with stakeholders. Even when broad dissemination is a requirement of an award, some scientists feel awkward in presenting scientific results as training material, or to participate in outreach activities. In addition, funding bodies do not seem to measure performance in this area and have few sanctions over failure to enhance the public understanding of science.

On the outside there is cynicism about the desire to enhance the public understanding of science. Campaigning groups can see such initiatives as little more than a form of indoctrination which attempts to gain broader acceptance of already strong links between science-industry and government. They see this as asserting a reductionist and technocratic vision of the world.

Despite these tensions it seems wrong for scientists to disengage themselves from debates about air pollution or indeed environmental issues in general. Fortunately scientists usually realise it is important to keep both politicians and the public informed of broader developments in our understanding of air pollution. However, such issues often go beyond the science and require comment on lifestyle choice and difficult issues relating to personal freedom. We can be seen as overly "green" when criticising the popularity of SUV's, second homes or long-haul holidays.

3.1 Health impacts

The primary focus of our attention on air pollution remains the health effects with asthma probably most prominent in the public mind. Although the winter smogs that characterised London from the 1880's to the 1950's have faded from popular view, there are more recent events, e.g., the new winter nitrogen dioxide smogs, ozone episodes in Europe during the summer heat waves (in 2003, 2006), and the forest fires of South East Asia.

Air pollutants usually affect health in quite subtle ways so it can be difficult to give mechanisms that are clear enough to explain simply. An exception is carbon monoxide which binds with haemoglobin in the blood, such that it prevents the uptake of oxygen. Contemporary pollutants such as nitrogen dioxide cannot really be explained so easily.

Ozone is recognised as a gas that can cause inflammatory responses in the lung, effects which might be described loosely term "lung leakiness". Other products of photochemical smog such as aldehydes and peroxyacetylnitrate have irritant properties.

In recent years fine particles have increasingly been seen as a key variable on the impact of pollutants on urban health. The work of Dockery *et al.* [3] showed a clear relation between daily mortality in an urban population and the presence of fine particles, typically categorised by their diameter in microns as PM_{10} and $PM_{2.5}$ or even smaller. Particle loads in urban atmospheres have been exacerbated by the increasing use of diesel vehicles and the production of secondary particles in photochemical smog.

There is a relatively low public awareness of this shift to seeing fine particles as dominant in controlling the health effects of air pollution. It may be that the mechanism has remained illusive, although there are elements of the picture that still seem easy to explain. Size has a clear role to play but the delivery of particles less than 3 microns deep into the lung is something that can be readily appreciated. There are some useful suggestions about the physiological mechanisms through which fine particulate matter such as the stress placed on the lung by action of macrophages in attempting to remove these fine particles from the alveoli.

More generally there is a public understanding that the diesel engine is problematic, but this is probably derives from a link to the carcinogenicity of the organic materials found in diesel particles. The exact role of chemical composition in inducing short term health effects is still not very clear. In general the public understanding of the differences between short-term effects and the longer term carcinogenic effects is limited.

3.2 Acid rain, dust and forest fires

Although health has typically been the key driver of interest in air pollution, damage to forests and lakes was important during the acid rain debate of the 1980's. This topic was widely taught in schools, but there is much less public interest as part of a decreasing media focus on acidic rain. The acid rain debate promoted a wider interest in the state of the urban fabric and air pollution damage to historic buildings. Although concern in such non-health issues is not obvious its latent presence can be revealed in opinion surveys where there is a ready association between the state of historic buildings and air pollution [4]. It should be said though that this concern can readily shift to a focus on to health affects. For example, when visitors are questioned about dust in libraries or historic interiors they will often express fears of asthma in these locations [5].

In Europe and North America sulfur deposition is mostly in decline, although the decreases aren't matched by equivalent improvements in the amount of acid brought down in rain. Lower sulfur emissions have not always been accompanied by lower emissions of the nitrogen oxides, which give rise to nitric acid in precipitation. Furthermore, calcium was once more abundant and hence available to neutralise some of the acidity. In recent decades, the amount of alkaline particulate material has declined, perhaps because there is less dust from unsealed roads and less grit from industry and power generation.

Acid rain is now more of a problem along the Asia-Pacific rim where vast quantities of coal are burnt as a part of industrial expansion. Not all that has been learnt in temperate regions is easily applicable in the tropical context. Research and regulation needs to face a different acid rain problem here. Entirely new ecosystem will be confronted by acidic deposition, although we have to recognise some of the novel factors in these regions have been present for many centuries. In places such as Japan and Korea, alkaline dust offers the potential to buffer acids in rainfall. Forest fires can produce acids, but also liberate large amounts of alkaline material that disperses along with the acids. However, such neutralisation processes are not always well understood. The greater extension of acid rain into the tropics, where soils are often deeply weathered, makes available new routes for mobilizing toxic metals within ecosystems.

In the last years of the 20th century it became clear that haze was widespread across continents. Most recently there has been much comment on the Asian brown-haze. This seems a product of emissions from forest fires, wind blown soil dusts, cooking smoke and industrial emissions. The concern about this does not arise simply from its broad scale, but it also reminds us of the belief that particles from any source, natural or anthropogenic that may impose health risks. This particulate material adds to or reacts with pollutants already present in urban air [6].

3.3 Local and global

In the last fifty years air pollution problems have become more global, and even local pollution issues have begun to have global similarities. There has been a convergence of the types of pollution experienced and the regulatory responses to these issues [7]. In terms of public understanding there is a wide appreciation of an enhanced greenhouse effect, acid rain and the ozone hole. It has to be said though, that these issues (e.g., ozone pollution in the lowest part of the atmosphere and the ozone hole) are often confused. Some people have become quite fearful about issues such as global dimming which suggest global warming resulting from the greenhouse effect may be much underestimated. A particular problem in discussing issues such as the ozone hole or the greenhouse effect, which occur over long timescales, is the desire for many to relate these to weather events on a day to day basis.

It is important that contemporary society is able to appreciate that pollution extends far beyond urban areas and we have to consider problems such as trans-boundary movement of air pollutants and windblown dusts along with issues of the emission of greenhouse gases and ozone depleting substances. Nevertheless these are difficult to keep on social and political agendas, because issues are often interpreted at local scale and local perceptions often define the changes people want. It can be difficult to maintain interest in larger spatial scales, and over long time periods.

3.4 Importance of secondary pollution

The changes in pollutants with time and their role in affecting human health, have had an impact on the way we develop air pollution policy. Perhaps the most important transformation in air pollution of the 20th century was the shift from primary pollutants to secondary pollutants within urban air. To some extent this may have arisen from success in ameliorating direct pollutant emissions, but more generally it is the outcome of a switch from solid fuels burnt in stationary sources to liquid fuels burnt in mobile sources.

The most abundant secondary pollutant is ozone, which has no significant primary source, so its presence in photochemical smog of the 20th C has tended to decouple air pollution from the obvious sources of air pollutants. This is of great importance to the social and political perceptions of air pollutants. There is no longer an immediate link between the pollutant and its source. In the past environmental degradation could easily be related to the obvious and visible smoke from a factory chimney. Stopping the pollution was simply a matter of emission control: stopping the pollutants coming from the chimney would stop the problem. Photochemical atmospheres usually have to be modelled and we cannot intuit the consequences of policy decisions. A knowledge of atmospheric chemistry becomes important in framing policy, despite the difficulty in explaining the detail.

The importance of secondary pollutants remains absent from popular understanding of air pollution. Pollutants are still seen as arising directly from polluting vehicles or industrial plants. The limited understanding probably derives from the fact that the chemical transformations in the production of ozone in smog, involve an interaction between ozone and the nitrogen oxides and further complications that come about through the role of peroxy radicals in oxidizing nitric oxide to nitrogen dioxide.

3.5 Monitoring networks

The late 20th century saw the development of many excellent air pollution monitoring networks. The availability of automated equipment allowed these to be reliable and accurate, giving good records of the ambient concentrations of pollutants. The location of monitoring stations has always been difficult. Regulation has usually required them to be placed in large cities (typically with populations of a quarter of a million), rather than in rural areas. In cities the monitoring record might reasonably represent the exposure of large populations. Some favour urban background sites, while others prefer their presence at sites with the greatest pollution. The establishment and maintenance of a network must also meet financial requirements. As a consequence of local public and administrative pressures that influence the location of monitors, the Swiss NABEL network has been able to locate monitors at urban, rural and remote sites.

3.6 Exposure to pollutants indoor and out

Even more difficult questions with monitoring emerge when we consider that the main purpose is often said to reflect the health impacts on an exposed human population. Typically the monitors do not reflect air pollution exposure of individuals. Surveys indicate that people spend large amounts of time indoors. Even in attractive climates such as California, residents spend, on average, 87% of their time indoors, 7% in enclosed transit and 6% outdoors. The dominating influence of indoor environments is especially true of vulnerable groups such as the old, young and unwell. In addition to this some gender differences remain whereby women are more likely to spend more time in domestic settings and often polluted parts of the house, such as the kitchen. Other indoor problems such as sick building syndrome raise sociological issues as they can affect disenfranchised users of interior air spaces most severely.

The 20th century brought a number of important developments that enhanced indoor air pollution. Energy conservation in cold countries imposes requirements for lower amounts of heating and has led to decreased air exchange rates. Thus indoor concentrations are expected to be higher under very cold conditions when the exchange rates are low and the heating requirement is highest. The use of novel materials in the construction and furnishing of buildings has introduced a range of compounds into the indoor environment. These have typically been organic compounds from simple carbonyls such as formaldehyde through to more complex compounds that originate from glues and polymers. Carpets have been an especially problematic material.

Governments have often found indoor air a tricky problem to deal with; partially because it provokes the issue of personal freedom and the potential of numerous transactions in a regulatory sense. However, beyond this it frequently crosses departmental boundaries, which makes administration especially complex. It so often demands attention from a combination of departments of environment, housing, health, worker safety and even cultural heritage. In Europe the new programme from the European Commission that will direct air pollution concerns, CAFÉ (Clean Air For Europe) will unfortunately only look at indoor issues where they relate to advected outdoor pollutants.

3.7 Modes of communication with the public

As with the data provision through TV or the internet, there are a wide range of media available to assist in the public understanding of science. Nevertheless scientists often have limited access to these. This is particularly true for television which is probably the prime vehicle for communication with a broad audience. Additionally they are frequently troubled by the reduction to sound bites imposed by time hungry television when trying to communicate subtle ideas. Longer television prog ammes are more useful here, but even with these the "angle" adopted by the programme makers can often be irksome. The BBC's program about global dimming while useful in raising the issue of the role of particles in climate change, gave a fairly apocalyptic vision of the future. Channel Four's secret history: *The Great Fog* took a conspiratorial view of the London Smog of 1952.

It is important to recognise the significance of elements in a debate that trigger public interest (see pollution from Battersea power station as an example [8]). Public imagination is often captured by icons that may be images such as the colourful renditions of ozone depletion over the Antarctic or powerful combinations of words such as *ozone hole*, *acid rain* and *greenhouse effect* (see also [9]).

Although reaching a smaller audience, science columns of newspapers can present more measured accounts and popular science journals such as Scientific American and New Scientist can give even more detail. The last decade has seen the development of the *Public Open Forum* or *Café Scientifique* model, which is particularly interesting. These reach only a small number of people, but are potentially very effective.

The *Café Scientifique* allows for close interaction between the general public and scientists (e.g http://www.sciencecafesheffield.org/venue.htm). These open events are normally held in a public café, town squares or perhaps a more academic or institutional venues. They allow the audience to questions scientists, and to express their thoughts and fears about issues such as air quality. They are especially effective where it is possible to include local politicians or administrators. A *Café Scientifique* may start with an overview presentation given by an acclaimed expert, who introduces the topic to lay people. The audience is then invited to participate in a discussion led by a moderator. In other forms the scientists may share a table with a small group and lead a discussion.

The *Café Scientifique* is increasingly popular and has been adopted by the EU funded ACCENT project and also forms a regular part of the British Council's programmes. They can be particularly challenging for scientists. Cafés and public squares can be noisy and exposed. The events attract people with a very broad range of interests and background knowledge. Questions can be disarmingly fundamental and difficult to answer (e.g. I was explaining how positively and negatively charged entities attract and I was asked "why?"). A very relaxed approach seems best in such an unpredictable setting.

As with the *Café Scientifique*, the *Public Open Forum* is designed for local audiences. They also aim to bring people together to discuss their views. In the context of a Swiss-British young scientists meeting in Switzerland in December 2003, a *Public Open Forum* was held on the summer 2003 heat wave. The forum included scientists (experts on climatology, statistics, air pollution, and health) along with politicians, economists and stakeholders in the field of tourism, hydro-electricity, and wine-making. The *Public Open Forum* made it possible for the audience to ask questions, and to express fears and thoughts about the environmental conditions during the past summer. Scientists from Switzerland and the U.K. tried to gauge the fears and attempt to answer the questions. Early-career scientists played a key role in this event as part of their training in communicating to non-scientists. They mixed with the audience and framed the questions that the lay people wanted to pose to the experts.

Web-based discussion forums can form to involve both science experts and local people. Increasingly many large research projects have publicly oriented websites (see the ACCENT site http://www.accent-network.org/portal/outreach-tasks/public-information-and-policy-support) although many of these seem very good, their impact is difficult to assess. There are a range of other approaches to social research such as Citizen Juries (http://www.soc.surrey.ac.uk/sru/SRU37.html), that attempt to judge issues or game-playing (see Fig. 1) or role playing such as in *Democratos*.

Figure 1. Game playing as a method of communication to non scientists.

4. INTERFACING WITH POLITICIANS

It is equally import that politicians gain information about air pollution. While some of the material they need is the same as that required by the public, they usually wish to tailor it to political objectives. There is a strong desire for definitive answers to complex problems which places particular pressures on scientists.

There is, of course a long-standing preoccupation with health. Broader issues of environment, aesthetics or cultural landscapes as driving forces in decision making have always been more difficult to bring to the top of agendas. Long term or global problems can also present a difficulty for politicians, who need rapid solutions that lead to tangible changes at a local level that embraces the interests of their electorate. It has been said that a politician's vision is only as far as the next election. This has meant that democratic structures have problems coping with long-term issues such as global warming which take place over many decades often far outlasting political careers. Collaboration between administrators on air quality management and climate protection at the national level is recommended.

4.1 Modes of communication

As with the general public there are particular approaches to communicating with policy makers. Newspapers and television provide an important backdrop to their understanding, but they often value face-to-face presentations, lectures or briefing papers. A direct method of communication is the *V.I.P. Meeting*. An example for this could be seen at the Ministerial Conference on Global Change, Environment and Development run by United Nations Environment Protection in Seoul. Here scientists met in parallel with high level policy meetings. They were part of an Eminent Scientist's panel and politicians were encouraged to join in. Such interactions to a better understanding of the problems each profession faces but also increased the potential for future collaborations to address air pollution and climate change issues. However, politicians are busy at such high level meetings and often only a small number are able to find time to get to the science fora that they often regard as peripheral.

Politicians wish to translate science to policy. This requires the information to be clear and unambiguous, so the doubt about results so often felt by scientists makes politicians uncomfortable [10].

4.2 Speed and timing

One problem with in political life is the timing of scientific discovery and research output. Clearly this is not the same as cycles in politics which are defined by the seasonal activities of government and the terms for election. Speed of action over environmental issues is a matter of continual concern. Even local air pollution issues are frequently debated over very long periods of time before there is action.

The UK Clean Air Act 1956 was the result of four years work after the London Smog of 1952 and a long legislative history that went back into the 19th century. The creation of the required clean air zones, after this Act, was often slow even where there was a clear desire to create them. Some cities that wanted smokeless zones did not complete them until the 1990's.

The concern over fine particles arose with such rapidity that it affected developing legislation, which tried to incorporate new research as it developed. This presented many difficulties for legislators even in terminology as within the European Commission particles were variously referred to as: (1) suspended particulate matter (2) fine suspended particulate matter and (3) PM_{10} and $PM_{2.5}$. The correlation between fine particles and health effects was so striking that the significance of this work was not missed by politicians and regulatory agencies. In recent years the number of air pollution deaths among the elderly during summer heat-waves [11] has rapidly risen on political agendas.

4.3 Choice of regulated pollutants

The European Directive 96/62/EC on *Air Quality Monitoring and Management* included: sulfur dioxide, nitrogen dioxide, PM10/2.5, lead, ozone, benzene, carbon monoxide, PAH, cadmium, arsenic, nickel and mercury. Some countries tried to have air pollutants such as fluoride and 1,3-butadiene included, without success. The metals within the directive seem quite old fashioned given the rise in interest in other metals, such as antimony and copper from brake-linings that now deposit at the roadside [12]. They probably represented local concerns rather than those of all Europe. However, it is worth reflecting that some of the pollutants considered important were not included.

4.4 Secondary pollutants and air quality management

Contemporary strategies to control air pollution have to address an increasingly complex situation. Photochemical pollutants mean that regulations have to account for chemical reactions and a weaker link to obvious sources. Simple emission control is not adequate, so approaches increasingly involve *Air Quality Management* (AQM), a principle that has become increasingly important within European Union policy as seen in Directive EC/96/62 on air quality monitoring and management. The importance of chemical reactions make it more difficult for politicians to understand the basis for air pollution control policy, because these have involved arguments about issues such as photochemical ozone creation potential (POCP), aromatization (increases in benzene and toluene) of fuels and the complexity of the equilibrium and kinetics of NO_x reactions.

4.5 Challenging individual freedom

The creation of smokeless zones under the UK Clean Air Act of 1956 required people not to use coal in their homes. Such an action had to confront problems of personal freedom, which worried politicians. This was a challenge in the 1950's just as it is today when the freedom to drive a car equally concerns society.

The creation of a zone of congestion charges in London under Mayor Ken Livingstone (although not primarily an air pollution initiative) had to resist legal challenges and considerable opposition from car owners. Opponents argued that public transport could not cope with the load and additionally, the expanded bus fleet would create more pollution. There have fortunately been some modest gains in air pollution and a less congested city as a result [13].

4.6 Science and the policy maker

The interface between science and policy is particularly uncomfortable. Few politicians would claim that they take no cognisance of scientific advice, although ultimately, they may not understand it or choose to ignore it. The reasons why politicians ignore advice of their analysts stems from their different roles and personalities. Policy makers are expected to be decisive so they like advice that has universal applicability and is efficient. Absolutisms, clear and definitive answers that ignore uncertainty are preferred. Scientists can feel misgivings about offering opinions completely free of doubt.

4.7 Technological vs sociological change

The 21st century has seen the increasing fragmentation of air pollution sources. For example, the transitions of last fifty years moved pollution sources from solid fuel based stationary furnaces to liquid fuelled automobiles. This placed pollution generation in the hands of the general public, most particularly the motorist. It has been difficult to approach the sorts of emission reduction required from a multitude of sources.

Engineers and administrators have often opted for technical fixes, better furnaces or catalytic converters on cars. However, sociological fixes may well be more effective. It might be easier to solve air pollution problems in cities if their citizens did not wish to drive their cars. Traditionally it has been argued such re-education is perhaps effective, but it is too slow for the required pace of change, and very difficult to achieve.

In the end we may need to be more imaginative. Some industries manage to alter public behaviour very rapidly. Fashions in particular change quickly. Enormous campaigns create enthusiasm for new cars, but rarely aim at creating a parallel enthusiasm for public transport, in cities where it is poorly used. Indoor heating accounts for about a third of the UK carbon emission, but between 1970 and 2000 the average temperature in centrally heated homes increased from 14.5 °C to 18.1 °C. People now wear fewer clothes indoors, yet if the fashion industry persuaded us to wear warmer clothes indoors a 1 °C decrease in indoor temperatures could give a 2% reduction in UK carbon emissions.

5. CONCLUSIONS

Anthropogenic activities emit numerous components into the atmosphere leading to air pollution that is of concern to everyone, especially because of health effects. While atmospheric scientists are trained to understand the underlying chemistry, physics, or meteorology, they receive less training about communicating scientific results and the associated uncertainties to non-scientists (politicians, the public). Yet, lay people need more information to understand the complex issue of air pollution, and to develop behavioural patterns that help improve air quality. The science of air pollution is, however, so sophisticated that it becomes hard for the public to interpret. Also, scientists operate in a world where complexity requires them to reveal both the difficulties and disagreements in the understanding of the polluted atmosphere. This can be confusing and increase uncertainty in situations where clear political and social decisions are required. Air pollution science hence creates problems for policy makers, as they like to act in situations where there is agreement and clarity. These difficulties come at a time when science is not accepted as the sole source of information on environmental issues. New forms of knowledge transfer are available to communicate air pollution science to non-scientists. *Public Open Forum, Café Scientifique* and *V.I.P. meetings* create environments to encourage direct and open communication to politicians and the general public. Common to all these forms is the desire for sincere engagement between all involved. In this role, scientists need to be able to give simple accounts of air quality that help non-scientists develop informed appreciation of air pollution problems.

Acknowledgements

Funding from ACCENT (the European Network of Excellence in Atmospheric Composition Change), EC Contract GOCE-CT-2004-505337, and the Swiss Secretariat for Science Education and Research (SER), Contract No. 03.0430-2 is gratefully acknowledged.

References

[1] D. Shooter, and P. Brimblecombe, *The International Journal of Environment and Pollution* (accepted).
[2] K. Bickerstaff, *Environment International* **30**, 827 (2004).
[3] D.W. Dockery, C.A. Pope III, X. Xu, J.D. Spengler, J.H. Ware, M.E. Fay, B.G. Ferris Jr., and F.E. Speizer, *New England Journal of Medicine* **329**, 1753 (1993).
[4] C.M. Grossi, and P. Brimblecombe in Air Pollution and Cultural Heritage, edited by C. Saiz-Jimenez (Balkema, Leiden 2004).
[5] K. Lithgow, and P. Brimblecombe, *National Trust Views* **39**, 47 (2003).
[6] X. Hou, G. Zhuang, Y. Sun, and Z. An, *Atmospheric Environment* **40**, 3251 (2006).
[7] P. Brimblecombe, *Globalizations* **2**, 429 (2005).
[8] C. Bowler, and P. Brimblecombe, *Atmospheric Environment* **25B**, 143 (1991).
[9] S.A. Nicholson-Cole, Computers, *Environment and Urban Systems*, **29**, 255 (2005).
[10] P. Brimblecombe, and M. Cashmore, *Clean Air Society of Australia and New Zealand, Quarterly Journal* **36**, 31 (2002).
[11] P.H. Fischer, B. Brunekreef, and E. Lebret, *Atmospheric Environment* **38**, 1083 (2004).
[12] D. Hjortenkrans, B. Bergback, and A. Haggerud, *Environmental Monitoring and Assessment*, **117**, 85 (2006).
[13] S.D. Beevers, and D.C. Carslaw, *Atmospheric Environment*, **39**, 1 (2005).

Index

ERCA
Contents of the 1st volume
"Topics in Atmospheric and Interstellar Chemistry"

ERCA
Contents of the 2nd volume
"Physics and Chemistry of the Atmospheres of the Earth and Other Objects of the Solar System"

ERCA
Contents of the 3rd volume
"From Urban Air Pollution to Extra-Solar Planets"

ERCA
Contents of the 4th volume
"From Weather Forecasting to Exploring the Solar System"

ERCA
Contents of the 5th volume
"From the Impacts of Human Activities on Our Climate and Environment to the Mysteries of Titan"

ERCA
Contents of the 6th volume
"From Indoor Air Pollution to the Search for Earth-Like Planets in the Cosmos"

List of Colloquia

1991

1	Third Atomic Data Workshop	(C1-1991)	Price: 41 €
2	Proceedings of the Eighth European Conference on Chemical Vapour Deposition (*EUROCVD 8*)	(C2-1991)	Price: 99 €
3	3rd International Conference on Mechanical and Physical Behaviour of Materials under Dynamic Loading (*DYMAT 91*)	(C3-1991)	Price: 99 €
4	European Symposium on Martensitic Transformation and Shape Memory Properties	(C4-1991)	Price: 60 €
5	Colloque Weyl VII International Conference on Metals in Solution	(C5-1991)	Price: 53 €
6	2nd International Workshop on Beam Injection Assessment of Defects in Semiconductors	(C6-1991)	Price: 41 €
7	2nd International Laser M2P Conference	(C7-1991)	Price: 73 €

1992

8	Second French Conference on Acoustics	(C1-1992)	Price: 129 €
9	European Workshop on Glasses and Gels	(C2-1992)	Price: 44 €
10	35e Colloque de métallurgie : Matériaux magnétiques, progrès et perspectives	(C3-1992)	Price: 60 €

1993

11	Workshop on Complex Liquid Systems	(C1-1993)	Price: 44 €
12	International Workshop on Electronic Crystals (*ECRYS-93*)	(C2-1993)	Price: 48 €
13	The 9th European Conference on Chemical Vapour Deposition (*EUROCVD 9*)	(C3-1993)	Price: 74 €
14	The 4th International Workshop on Positron and Positronium Chemistry (*PPC 4*)	(C4-1993)	Price: 48 €
15	Optics of Excitons in Confined Systems	(C5-1993)	Price: 51 €
16	International Symposium on (e, 2e) Collisions, Double Photoionization and Related Processes	(C6-1993)	Price: 35 €
17	The 3rd European Conference on Advanced Materials and Processes (*EUROMAT 93*)	(C7-1993)	Price: 269 €
18	IX International Conference on Small Angle Scattering	(C8-1993)	Price: 59 €
19	3rd International Symposium on High Temperature Corrosion and Protection of Materials	(C9-1993)	Price: 213 €

1994

20	Journées d'électrochimie; récents développements en électrochimie fondamentale et appliquée	**(C1-1994)**	**Price: 38 €**
21	European Workshop on Piezoelectric Materials: Crystal Growth, Properties and prospects	**(C2-1994)**	**Price: 38 €**
22	36e Colloque de métallurgie de l'INSTN: Changement de phases et microstructures	**(C3-1994)**	**Price: 45 €**
23	3rd International Laser M2P Conference	**(C4-1994)**	**Price: 99 €**
24	3rd French Conference on Acoustics	**(C5-1994)**	**Price: 179 €**
25	Proceedings of the First European Workshop on Low Temperature Electronics (*WOLTE 1*)	**(C6-1994)**	**Price: 45 €**
26	8th International Topical Meeting on Photoacoustic and Photothermal Phenomena	**(C7-1994)**	**Price: 91 €**
27	International Conference on Mechanical and Physical Behaviour of Materials under Dynamic Loading (*EURODYMAT 94*)	**(C8-1994)**	**Price: 99 €**
28	European Symposium on Frontiers in Science and Technology with Synchrotron Radiation	**(C9-1994)**	**Price: 53 €**

1995

29	Europhysics Industrial Workshop EIW-12	**(C1-1995)**	**Price: 38 €**
30	IIIrd European Symposium on Martensitic Transformations (*ESOMAT'94*)	**(C2-1995)**	**Price: 68 €**
31	37e Colloque de métallurgie de l'INSTN : Microstructures et recristallisation	**(C3-1995)**	**Price: 45 €**
32	Approches microscopique et macroscopique des détonations	**(C4-1995)**	**Price: 76 €**
33	Tenth European Conference on Chemical Vapour Deposition (*EUROCVD 10*)	**(C5-1995)**	**Price: 228 €**
34	Nonlinear Phenomena in Microphysics of Collisionless Plasmas. Application to Space and Laboratory Plasmas	**(C6-1995)**	**Price: 30 €**
35	Second International Conference on Ultra High Purity Base Metals (*UHPM – 95*)	**(C7-1995)**	**Price: 53 €**
36	International Conference on Martensitic Transformations (*ICOMAT 95*)	**(C8-1995)**	**Price: 151 €**

1996

37	International Seminar on Mechanics and Mechanisms of Solid-Solid Phase Transformations (*MECAMAT'95*)	**(C1-1996)**	**Price: 60 €**
38	38e Colloque de métallurgie de l'INSTN : Les intermétalliques superalliages aux quasicristaux	**(C2-1996)**	**Price: 45 €**
39	Second European Workshop on Low Temperature Electronics (*WOLTE 2*)	**(C3-1996)**	**Price: 53 €**
40	Rayons X et matière : 100 ans déjà . . . (*RX'96*)	**(C4-1996)**	**Price: 121 €**
41	43rd International Field Emission Symposium (*IFES'96*)	**(C5-1996)**	**Price: 64 €**

42	Local Approach to Fracture (*EUROMECH-MECAMAT'96*)	**(C6-1996)**	**Price: 68 €**
43	39ᵉ Colloque de métallurgie de l'INSTN : Multicouches métalliques	**(C7-1996)**	**Price: 38 €**
44	Eleventh International Conference on Internal Friction and Ultrasonic Attenuation in Solids (*ICIFUAS 11*)	**(C8-1996)**	**Price: 144 €**

1997

45	7th International Conference on Ferrites (*ICF7*)	**(C1-1997)**	**Price: 144 €**
46	9th International Conference on X-Ray Absorption Fine Structure (*XAFS IX*)	**(C2-1997)**	**Price: 146 €**
47	International Conference on Mechanical and Physical Behaviour of Materials under Dynamic Loading (*EURODYMAT'97*)	**(C3-1997)**	**Price: 121 €**
48	XXIIIrd International Conference on Phenomena in Ionized Gases (*ICPIG*)	**(C4-1997)**	**Price: 89 €**
49	IVth European Symposium on Martensitic Transformations (*ESOMAT'97*)	**(C5-1997)**	**Price: 128 €**
50	Surfaces et interfaces des matériaux avancés (*SIMA*)	**(C6-1997)**	**Price: 59 €**

1998

51	Proceedings of the First Conference on the Active Superconducting Composites (*CCAS1*)	**(Pr1-1998)**	**Price: 19 €**
52	Soft Magnetic Materials 13 (*SMM13*)	**(Pr2-1998)**	**Price: 144 €**
53	3rd European Workshop on Low Temperature Electronics (*WOLTE 3*)	**(Pr3-1998)**	**Price: 85 €**
54	40ᵉ Colloque de métallurgie de l'INSTN : Comportement mécanique et effets d'échelle	**(Pr4-1998)**	**Price: 74 €**
55	Rayons X et matière (*RX'97*)	**(Pr5-1998)**	**Price: 137 €**
56	International Conference on Disorder and Chaos	**(Pr6-1998)**	**Price: 68 €**
57	3rd International Workshop: Microwave Discharges – Fundamentals and Applications	**(Pr7-1998)**	**Price: 100 €**
58	2nd European Mechanics of Materials Conference (*EUROMECH-MECAMAT'97*)	**(Pr8-1998)**	**Price: 114 €**
59	2nd European Meeting on Integrated Ferroelectrics (*EMIF 2*)	**(Pr9-1998)**	**Price: 73 €**

1999

60	Diffusion des neutrons aux petits angles (*DNPA*)	**(Pr1-1999)**	**Price: 45 €**
61	6ᵉ Journées nationales de microélectronique et optoélectronique III-V (*JNMO'97*)	**(Pr2-1999)**	**Price: 42 €**
62	9th SolarPACES International Symposium on Solar Thermal Concentrating Technologies	**(Pr3-1999)**	**Price: 149 €**
63	41ᵉ Colloque de métallurgie de l'INSTN : Ségrégation interfaciale dans les solides	**(Pr4-1999)**	**Price: 60 €**

64	4e Colloque sur les sources cohérentes et incohérentes UV, VUV, et X	**(Pr5-1999)**	**Price: 44 €**
65	International Conference on Coincidence Spectroscopy	**(Pr6-1999)**	**Price: 60 €**
66	Innovative Options in the Field of Nuclear Fission Energy	**(Pr7-1999)**	**Price: 49 €**
67	Proceedings of the Twelfth European Conference on Chemical Vapour Deposition	**(Pr8-1999)**	**Price: 274 €**
68	3rd European Mechanics of Materials Conference on Mechanics and Multi-Physics Processes in Solids: Experiments, Modelling, Applications (*EUROMECH-MECAMAT'98*)	**(Pr9-1999)**	**Price: 99 €**
69	International Workshop on Electronic Crystals (*ECRYS-99*)	**(Pr10-1999)**	**Price: 94 €**

2000

70	Diffusion quasiélastique des neutrons	**(Pr1-2000)**	**Price: 45 €**
71	Twelfth European Conference on Chemical Vapour Deposition	**(Pr2-2000)**	**Price: 275 €**
72	Organic Superconductivity "20th Anniversary"	**(Pr3-2000)**	**Price: 42 €**
73	42e Colloque de métallurgie de l'INSTN : Matériaux pour les machines thermiques	**(Pr4-2000)**	**Price: 79 €**
74	The 1999 International Conference on Strongly Coupled Coulomb Systems	**(Pr5-2000)**	**Price: 128 €**
75	Microstructural Design for Improved Mechanical Behaviour of Advanced Materials	**(Pr6-2000)**	**Price: 89 €**
76	International Workshop on Dynamics in Confinement	**(Pr7-2000)**	**Price: 100 €**
77	Sixième colloque sur les lasers et l'optique quantique	**(Pr8-2000)**	**Price: 70 €**
78	6th International Conference on Mechanical and Physical Behaviour of Materials under Dynamic Loading (*EURODYMAT'2000*)	**(Pr9-2000)**	**Price: 198 €**
79	Rayons X et matière	**(Pr10-2000)**	**Price: 137 €**
80	Fourth International Workshop on Ram Accelerators	**(Pr11-2000)**	**Price: 68 €**

2001

81	43e Colloque de métallurgie de l'INSTN : Matériaux pour le nucléaire : les enjeux des prochaines décennies	**(Pr1-2001)**	**Price: 72 €**
82	7th International Conference on X-Ray Lasers (X-Ray Lasers 2000)	**(Pr2-2001)**	**Price: 150 €**
83	Thirteenth European Conference on Chemical Vapor Deposition	**(Pr3-2001)**	**Price: 190 €**
84	4th European Mechanics of Materials Conference on Processes, Microstructures and Mechanical Properties (*EUROMECH-MECAMAT'2000*)	**(Pr4-2001)**	**Price: 95 €**
85	5th European Mechanics of Materials Conference on Scale Transitions from Atomistics to Continuum Plasticity (*EUROMECH-MECAMAT'2001*)	**(Pr5-2001)**	**Price: 85 €**

86	Sciences de la matière et microgravité	**(Pr6-2001)**	**Price: 79 €**
87	5e Colloque sur les sources cohérentes et incohérentes UV, VUV et X. Applications et développements récents (*UVX 2000*)	**(Pr7-2001)**	**Price: 52 €**
88	Fifth European Symposium on Martensitic Transformations and Shape Memory Alloys (*ESOMAT 2000*)	**(Pr8-2001)**	**Price: 150 €**
89	Neutrons et magnétisme	**(Pr9-2001)**	**Price: 67 €**
90	Journées d'étude des équilibres entre phases (*XXVII JEEP*)	**(Pr10-2001)**	**Price: 71 €**
91	International Conference on Thin Film Deposition of Oxide Multilayers Hybrid Structures (*TFDOM 2*)	**(Pr11-2001)**	**Price: 88.5 €**

2002

92	Voies nouvelles pour l'Analyse des Données en Sciences de l'Univers	**(Pr1-2002)**	**Price: 61.5 €**
93	44e Colloque de métallurgie de l'INSTN : Matériaux pour les énergies propres	**(Pr2-2002)**	**Price: 53 €**
94	5th European Workshop on Low Temperature Electronics (*WOLTE 5*)	**(Pr3-2002)**	**Price: 56 €**
95	Thirteenth European Conference on Chemical Vapour Deposition – *Late News*	**(Pr4-2002)**	**Price: 41 €**
96	Septième colloque sur les lasers et l'optique quantique (*COLOQ 7*)	**(Pr5-2002)**	**Price: 103 €**
97	Rayons X et matière	**(Pr6-2002)**	**Price: 132 €**
98	The Fourth International Symposium on Hazards, Prevention, and Mitigation of Industrial Explosions (*IV ISHPMIE*)	**(Pr7-2002)**	**Price: 119 €**
99	Structural Materials for Hybrid Systems: A Challenge in Metallurgy	**(Pr8-2002)**	**Price: 90 €**
100	International Workshop on Electronic Crystals (*ECRYS-2002*)	**(Pr9-2002)**	**Price: 100 €**
101	From the Impacts of Human Activities on our Climate and Environment to the Mysteries of Titan (*ERCA 5*)	**(Pr10-2002)**	**Price: 84 €**
102	40e anniversaire du laboratoire de mécanique appliquée	**(Pr11-2002)**	**Price: 105 €**

2003

103	Neutrons et matériaux (*JDN 10*)		**Price: 100 €**
104	7th International Conference on X-Ray Microscopy (*X-Ray Microscopy 2002*)		**Price: 160 €**
105	6th European Mechanics of Materials Conference on Non-Linear Mechanics of Anisotropic Materials (*EUROMECH-MECAMAT'2002*)		**Price: 100 €**
106	45e Colloque de métallurgie de l'INSTN : Surfaces, interfaces et rupture		**Price: 53 €**

107	XIIth International Conference on Heavy Metals in the Environment	**Volume I** **Volume II**	**Price: 150 €** **Price: 150 €**
108	6e Colloque sur les sources cohérentes et incohérentes UV, VUV et X. Applications et développements récents (*UVX 2002*)		**Price: 80 €**
109	8th Workshop on Photoacoustics and Photothermics		**Price: 33 €**
110	7th International Conference on Mechanical and Physical Behaviour of Materials under Dynamic Loading (*EURODYMAT'2003*)		**Price: 200 €**
111	Neutrons et systèmes désordonnés (*JDN 11*)		**Price: 95 €**
112	International Conference on Martensitic Transformations (*ICOMAT'02*)	**Part I** **Part II**	**Price: 140 €** **Price: 140 €**

2004

113	Journées d'Étude des Équilibres entre Phases (*XXIX JEEP*)	**Price: 45 €**
114	The Fifth International Symposium on Crystalline Organic Metals, Superconductors and Ferromagnets (*ISCOM 2003*)	**Price: 175 €**
115	7th European Mechanics of Materials Conference on Adaptive Systems and Materials: Constitutive Materials and Hybrid Structures (*EUROMECH-MECAMAT'2003*)	**Price: 105 €**
116	Quantum Gases in low dimensions (*QGLD 2003*)	**Price: 80 €**
117	9th Workshop on Photoacoustics and Photothermics	**Price: 34 €**
118	Rayons X et Matières 2003 (*RX 2003*)	**Price: 120 €**
119	Huitième colloque sur les lasers et l'optique quantique (*COLOQ 8*)	**Price: 92 €**
120	2nd International Conference on Thermal Process Modelling and Computer Simulation (*SF2M*)	**Price: 195 €**
121	From Indoor Air Pollution to the Search for Earth-like Planets in the Cosmos (*ERCA 6*)	**Price: 75 €**
122	Journées d'Étude des Équilibres entre Phases (*XXX JEEP*)	**Price: 70 €**

2005

123	Proceedings of the 9th International Seminar on the Physical Chemistry of Solid State Materials (*REMCES IX*)	**Price: 95 €**
124	Quatrième Colloque Franco-Libanais sur la Science des Matériaux (*CSM4*)	**Price: 95 €**
125	International Conference on Photoacoustic and Photothermal Phenomena (*13thICPPP*)	**Price: 180 €**
126	European Workshop on Piezoelectric Materials (*4thEWPM*)	**Price: 45 €**
127	7ième Colloque sur les Sources cohérentes et incohérentes UV, VUV et X (*UVX 2004*)	**Price: 59 €**

128	Electro-Active Materials and Sustainable Growth (*EMSG 2005*)	**Price: 59 €**
129	34th Winter School on Wave and Quantum Acoustics	**Price: 80 €**
130	Neutrons et Biologie (*JDN 12*)	**Price: 69 €**
131	International Workshop on Electronic Crystals (*ECRYS-2005*)	**Price: 99 €**

2006

132	10th International Conference on the Formation of Semiconductor Interfaces (*ICFSI-10*)	**Price: 99 €**
133	Inertial Fusion Sciences and Applications 2005 (*IFSA-2005*)	**Price: 245 €**
134	8th International Conference on Mechanical and Physical Behaviour of Materials under Dynamic Loading (*EURODYMAT-2006*)	**Price: 259 €**
135	Neuvième colloque sur les lasers et l'optique quantique (*COLOQ 9*)	**Price: 75 €**
136	The International Workshop (*NUCPERF 2006*)	**Price: 75 €**
137	35th Winter School on Wave and Quantum Acoustics	**Price: 82 €**
138	8e Colloque National sur les Sources Cohérentes et Incohérentes UV, VUV, et X (*UVX 2006*)	**Price: 66 €**

© EDP Sciences
Directeur de la publication : Jean-Marc Quilbé

Imprimé en France. JOUVE, 11 boulevard Sébastopol, 75001 PARIS
Dépôt légal : décembre 2006

EDP SCIENCES

An International Scientific Publisher

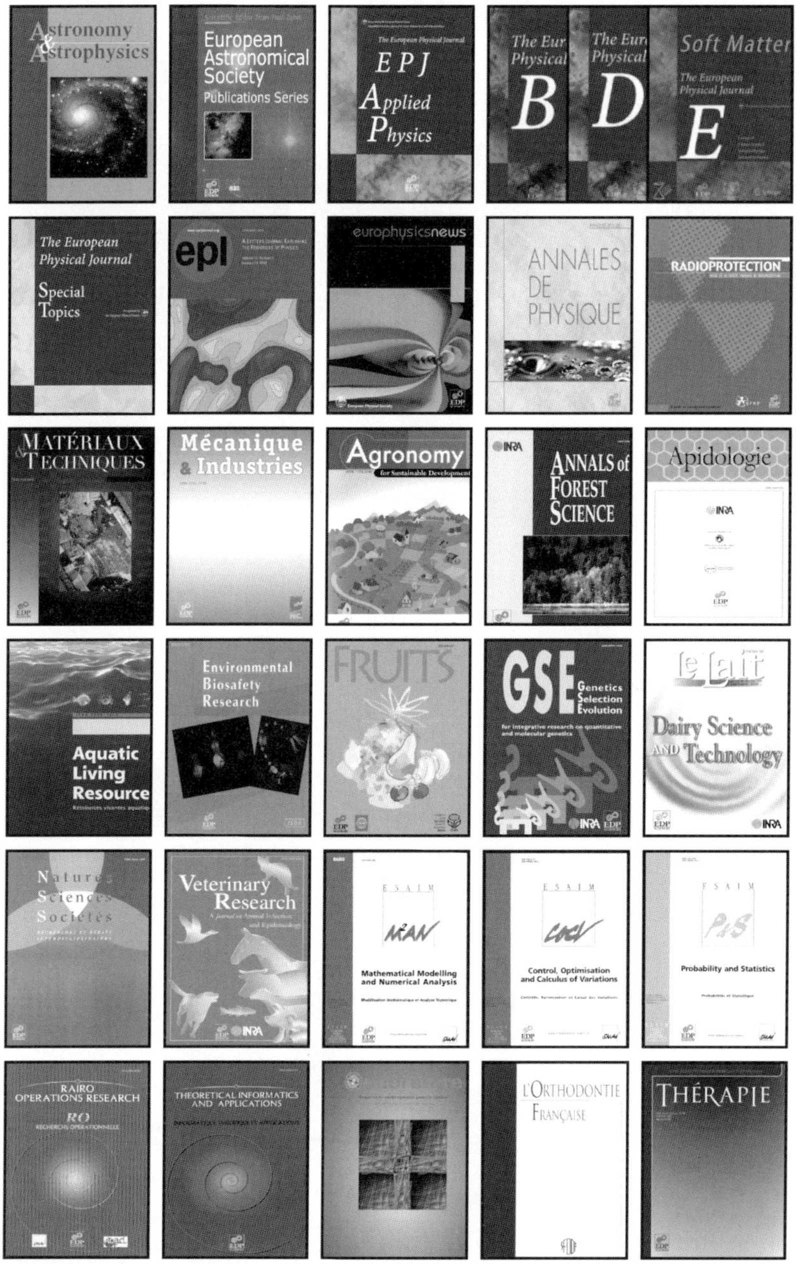

Astronomy & Physics:
- Astronomy & Astrophysics
- EAS Publications Series
- European Physical Journal (The) - Applied Physics
- European Physical Journal (The) - B, D, E
- European Physical Journal (The) - Special Topics
- Europhysics Letters
- Europhysics News
- Annales de Physique
- Radioprotection
- Matériaux & Techniques
- Mécanique & Industries

Life Sciences:
- Agronomy for Sustainable Development *(Formerly Agronomie)*
- Annals of Forest Science
- Apidologie
- Aquatic Living Resources
- Environmental Biosafety Research
- Fruits
- Genetics Selection Evolution
- Le Lait - Dairy Science & Technology
- Natures Sciences Sociétés
- Veterinary Research

Mathematics:
- ESAIM: Mathematical Modelling and Numerical Analysis
- ESAIM: Control, Optimisation and Calculus of Variations
- ESAIM: Probability & Statistics
- RAIRO: Operations Research
- RAIRO: Theoretical Informatics and Applications
- Quadrature

Medical:
- Orthodontie Française (L')
- Thérapie

E-journals:
- J3eA
- ESAIM: Proceedings

crossref member
CROSSREF.ORG
THE CITATION LINKING BACKBONE

J3eA
PROC.

EDP Sciences: 17 avenue du Hoggar • P.A. de Courtabœuf • B.P. 112 • 91944 LES ULIS Cedex A
Phone: +33 (0)1 69 18 75 75 • Fax: +33 (0)1 69 86 07 65 • Email: edp@edpsciences.org
www.edpsciences.org

EDP SCIENCES

Physics at EDP Sciences

International journals with selected research papers

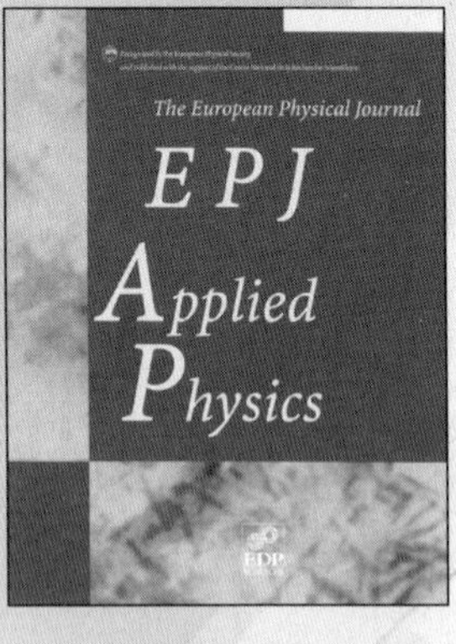

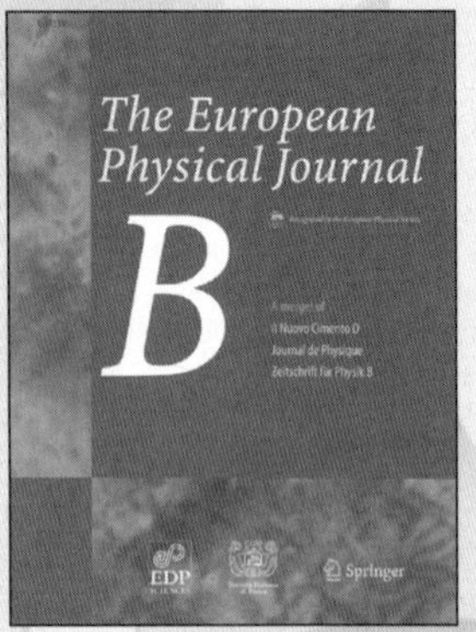

and also...

j3eA

www.edpsciences.org